ADVANCED SHIP DESIGN FOR POLLUTION PREVENTION

PROCEEDINGS OF THE INTERNATIONAL WORKSHOP "ADVANCED SHIP DESIGN FOR POLLUTION PREVENTION", SPLIT, CROATIA, 23–24 NOVEMBER 2009

Advanced Ship Design for Pollution Prevention

Editors

C. Guedes Soares
Instituto Superior Técnico, Technical University of Lisbon, Lisbon, Portugal

Joško Parunov
Faculty of Mechanical Engineering and Naval Architecture, University of Zagreb, Zagreb, Croatia

CRC Press
Taylor & Francis Group
Boca Raton London New York Leiden

CRC Press is an imprint of the
Taylor & Francis Group, an **informa** business
A BALKEMA BOOK

Cover photograph: New generation oil product tanker – photograph provided by journal "Shipbuilding"

Education and Culture
TEMPUS

CRC Press/Balkema is an imprint of the Taylor & Francis Group, an informa business

Typeset by Vikatan Publishing Solutions (P) Ltd., Chennai, India
Printed and bound in Great Britain by Antony Rowe (a CPI Group Company), Chippenham, Wiltshire

Published by: CRC Press/Balkema
P.O. Box 447, 2300 AK Leiden, The Netherlands
e-mail: Pub.NL@taylorandfrancis.com
www.crcpress.com – www.taylorandfrancis.co.uk – www.balkema.nl

ISBN: 978-0-415-58477-7 (Hbk)
ISBN: 978-0-203-84687-2 (eBook)

Advanced Ship Design for Pollution Prevention – Guedes Soares & Parunov (eds)
© 2010 Taylor & Francis Group, London, ISBN 978-0-415-58477-7

Table of contents

Preface ix

Organization xi

Modelling of the environment and environmental loads

Global structural loads induced by extreme waves on ships 3
N. Fonseca & C. Guedes Soares

Safety of ice-strengthened shell structures of ships navigating in the Baltic Sea 17
P. Kujala

Some aspects of quasi static hydro structure interactions in seakeeping 23
F.X. Sireta, Š. Malenica, F. Bigot, Q. Derbanne, V. Hsu & X.B. Chen

Application of an advanced beam theory to ship hydroelastic analysis 31
I. Senjanović, S. Tomašević, N. Vladimir, M. Tomić & Š. Malenica

Methods for estimating parametric rolling 43
A. Turk, J. Prpić-Oršić, S. Ribeiro e Silva & C. Guedes Soares

Effect of water depth and forward speed on ship dynamic behaviour in waves 55
I. Kolacio, C. Guedes Soares & J. Prpić-Oršić

Comparison of various classification societies requirements regarding bow area subject to wave impact loads 65
D. Bajič & J. Prpić-Oršić

Dynamic analysis of structures partially submerged in water 71
D. Sedlar, Ž. Lozina & D. Vučina

Ship structural reliability with respect to ultimate strength

Reliability assessment of intact and damaged ship structures 79
A.P. Teixeira & C. Guedes Soares

Improving the ultimate strength of ship hull 95
K. Žiha, J. Parunov, M. Ćorak & B. Tušek

Extended IACS incremental-iterative method for calculation of hull girder ultimate strength in analysis and design 103
S. Kitarović, J. Andrić & V. Žanić

Performance of the Common Structural Rules design formulations for the ultimate strength of uniaxially loaded plates and stiffened panels 113
M. Ćorak, J. Parunov, A.P. Teixeira & C. Guedes Soares

Bending and torsion of stiffeners with L sections under the plate normal pressure 121
R. Pavazza, F. Vlak & M. Vukasović

Fatigue reliability and rational inspection planning

Risk based maintenance of deteriorated ship structures accounting for historical data 131
Y. Garbatov & C. Guedes Soares

Hull girder fatigue strength of corroding oil tanker 149
J. Parunov, K. Žiha, P. Mage & P. Jurišić

Fatigue crack propagation in stiffened panels under tension loading 155
Ž. Božić, J. Parunov & M. Fadljević

High cycle and low cycle fatigue in ship structures 163
S. Rudan

Design optimization and product reliability—responsibility of designers and producers 173
V. Grubišić & N. Vulić

Maritime component reliability assessment and maintenance using bayesian framework and generic data 181
J. Barle, D. Ban & M. Ladan

Collision and grounding as criteria in design of ship structures

Ship collision as criteria in ship design 191
P. Varsta, S. Ehlers, K. Tabri & A. Klanac

Autonomous guidance and navigation based on the COLREGs rules and regulations of collision avoidance 205
L.P. Perera, J.P. Carvalho & C. Guedes Soares

Oil spill incidents in Portuguese waters 217
J.A.V. Gouveia & C. Guedes Soares

Probabilistic approach to damage stability

Probabilistic approach to damage stability 227
T.A. Santos & C. Guedes Soares

Aircraft carrier stability in damaged condition 243
V. Slapničar & B. Ukas

The effect of L/B and B/T variation on subdivision indexes 253
I. Munić, V. Slapničar & V. Stupalo

Vibration & measurements

Evaluation of propeller-hull interaction in early stage of ship design 261
R. Grubišić

Investigation of propulsion system torsional vibration resonance of a catamaran vessel in service 269
A. Šestan, B. Ljubenkov & N. Vladimir

Active vibration control using PI regulated electro-dynamic actuator—bond graph approach 275
I. Tomac, I. Pehnec & Ž. Lozina

Measurements for reliability of ship production and operation interaction 281
B. Ljubenkov, T. Zaplatić, K. Žiha & F. Bosančić

Ship design

AFRAMAX tanker Baltic Ice class—power determination doubts 291
D. Bezinović

Conceptual changes in the CRS classification and Croatian statutory technical rules 299
N. Vulić & A. Vidjak

Ship geometry modelling using radial basis functions interpolation methods 307
D. Ban, B. Blagojević & J. Barle

Author index 317

Advanced Ship Design for Pollution Prevention – Guedes Soares & Parunov (eds)
© 2010 Taylor & Francis Group, London, ISBN 978-0-415-58477-7

Preface

This book collects the papers presented at the International Workshop "Advanced Ship Design for Pollution Prevention", ASDEPP, which was held in Split 23 & 24 November 2009. The Workshop was the closing event of three and a half years Joint European Project JEP-40037-2005 (ASDEPP) financed by the European Commission within the programme Tempus.

Although several of papers were written by experts from the industry, authors of the most of papers are from consortium members of ASDEPP project, which are:

- Instituto Superior Tecnico (Lisbon, Portugal) (Project Grantholder)
- Helsinky University of Technology (Finland)
- Bureau Veritas (France)
- University of Zagreb, Faculty of Mechanical Engineering and Naval Architecture (Croatia) (Project Coordinator)
- University of Rijeka, Technical Faculty (Croatia)
- University of Split, Faculty of Electrical Engineering, Mechanical Engineering and Naval Architecture (Croatia).

The papers contained in this book reflect the teaching materials for advanced courses developed within ASDEPP project, which are implemented in curricula of MNA (Master of Naval Architecture) studies at University of Zagreb and University of Rijeka. Due to the specialized nature of the courses, they were also used for Phd students. As the book includes other important contributions to the wider objectives of the ASDEPP project, that are advanced ship design methods and environmental protection, the 32 papers are finally categorized into the following themes:

- Modelling of the environment and environmental loads
- Ship structural reliability with respect to ultimate strength
- Fatigue reliability and rational inspection planning
- Collision and grounding as criteria in design of ship structures
- Probabilistic approach to damage stability
- Vibration & measurements
- Ship design.

The papers were accepted after a review process, based on the full text of the papers. Most of the papers are reviewed by the Technical Programme Committee, what is gratefully acknowledged. Thanks are also due to the additional anonymous reviewers. We hope that their constructive comments to the authors contributed to a good level of papers included in the book.

Carlos Guedes Soares
Joško Parunov

Advanced Ship Design for Pollution Prevention – Guedes Soares & Parunov (eds)
© 2010 Taylor & Francis Group, London, ISBN 978-0-415-58477-7

Organization

WORKSHOP CHAIRS

C. Guedes Soares, Instituto Superior Técnico, Lisbon, Portugal
J. Parunov, University of Zagreb, Croatia

TECHNICAL PROGRAMME COMMITTEE

V. Čorić, University of Zagreb, Croatia
N. Fonseca, Instituto Superior Técnico, Lisbon, Portugal
I. Grubišić, University of Zagreb, Croatia
P. Kuyalla, Helsinki University of Technology, Finland
Š. Malenica, Bureau Veritas, France
J. Prpić-Oršić, University of Rijeka, Croatia
R. Pavazza, University of Split, Croatia
I. Senjanović, University of Zagreb, Croatia
A. Teixeira, Instituto Superior Técnico, Lisbon, Portugal
P. Varsta, Helsinki University of Technology, Finland
N. Vulić, Croatian Register of Shipping, Croatia
A. Zamarin, University of Rijeka, Croatia
V. Žanić, University of Zagreb, Croatia
K. Žiha, University of Zagreb, Croatia

LOCAL ORGANIZING COMMITTEE

F. Vlak, University of Split, Croatia, (chair)
V. Slapničar, University of Zagreb, Croatia
D. Ban, University of Split, Croatia
B. Blagojević, University of Split, Croatia
V. Cvitanić, University of Split, Croatia
S. Škoko Gavranović, secretary

The ASDEPP Workshop was organized under auspices of the Croatian Academy of Sciences and Arts.

Modelling of the environment and environmental loads

Advanced Ship Design for Pollution Prevention – Guedes Soares & Parunov (eds)
© 2010 Taylor & Francis Group, London, ISBN 978-0-415-58477-7

Global structural loads induced by extreme waves on ships

N. Fonseca & C. Guedes Soares
Centre for Marine Technology and Engineering (CENTEC), Technical University of Lisbon, Instituto Superior Técnico, Lisboa, Portugal

ABSTRACT: The paper presents a review of the work carried out mainly by the authors on the motions and global structural loads induced on ships by extreme and abnormal waves. The calculation method is based on a nonlinear time domain strip theory where the radiation forces are represented by convolution integrals of impulse response functions, together with infinite frequency added masses and radiation restoring coefficients. The radiation and diffraction forces are assumed linear, while the Froude-Krylov and hydrostatic forces are computed exactly on the instantaneous hull wetted surface. The time domain code is applied to calculate the ship responses to deterministic wave traces that include abnormal waves. These wave traces were measured in the ocean and they include abnormal wave events. Several conclusions were obtained regarding: the dynamics of the ships encountering abnormal waves, the relations between the ship responses and the characteristics of the waves, the effect of ship length and the severity of the global ship responses.

1 INTRODUCTION

The design of ship structures follows several steps and one of them consists on the determination of the design wave induced bending moment. This reference value, weighted by an appropriate safety factor, is added to the still water bending moment and the result is used to verify the structure ultimate strength.

The design wave bending moment represents the maximum expected moment that the hull will be subjected during the operational lifetime. There are different methods to calculate this moment, while the simplest is given by the Classification Societies as an empirical formula. Over the last years efforts have been made to develop methods based on first principles so that they can be incorporated in structural reliability formulations. A review of such methods is presented by Guedes Soares et al. (1996), Guedes Soares (1998) and Parunov and Senjanovic (2003).

The trend to rely on more direct methods of load assessment implies that the non-linear effects associated with large amplitude motions need to be properly accounted for. This has been achieved by methods such as the one of Fonseca and Guedes Soares (1998a, b) and few others of similar nature as reported in Watanabe and Guedes Soares (1999).

This nonlinearity of the response needs to be reflected on the long term distributions and an approach to achieve that has been proposed by Guedes Soares and Schellin (1996).

However many of the methods developed for assessing nonlinear wave induced load effects are on the time domain. This made it necessary to develop methods based on the time domain, i.e. to define wave sequences instead of defining a wave spectrum of storm conditions.

Some of the proposed procedures use, on a first step, linear methods to determine the wave sequences expected to produce the largest bending moments. On a second step, time domain nonlinear seakeeping codes are applied for the determined wave sequences resulting on the design wave bending moment such as the Critical Wave Episodes Method, (Torhaug et al., 1998), the Most Likely Response Method, (Adegeest et al., 1998, and Pastoor, 2003). These wave sequences were derived from storm sea states that would represent relevant design conditions and they were based on linear and second order wave theories.

Current design methodologies do not consider explicitly the wave conditions associated with the encounter of the ship with abnormal, or freak, waves. This is because the probabilistic models describing the waves do not include the abnormal waves that modify the upper tail of those distributions.

Some of the mechanisms that can generate abnormal waves have been identified and reviews of the literature about them can be found in Kharif, and Pelinovsky (2003) and in Guedes Soares et al. (2003) for example. The definition of which waves should be considered as abnormal ones is also not commonly agreed as discussed in those references, but a criterion adopted by some authors is the ratio of the wave height by the significant wave height being larger than 2.0.

Although current design methods do not consider explicitly the abnormal wave conditions, there are some reports from accidents that resulted from the encounter with waves that were much larger than those of the seastate in which they occurred. It is also believed by some authors that such abnormal waves were responsible for the sinking of some ships. For this reason Faulkner and Buckley (1997) suggest that the methods to determine the design loads should be revised to account for the effects of the abnormal waves on the ship structure.

In order to consider the structural loads induced by these conditions, a method was proposed by Fonseca and Guedes Soares (2001) to calculate the structural loads induced by deterministic wave traces of abnormal waves, where the ship responses are calculated by a nonlinear time domain seakeeping code.

Using this approach Guedes Soares et al. (2006) investigated the global structural loads induced on a FPSO platform by a deterministic wave trace that includes an abnormal wave with a height of 26 m, and compared the results with experimental data obtained in a tank with a scaled model. The agreement between numerical simulations and experimental data was good which confirms that the method can represent such extreme conditions.

Guedes Soares et al. (2006) discussed the definition of abnormal waves and their probabilities, presented results for a containership and a FPSO, and compared results with rule minimum requirements and with long term predictions of the vertical bending moments.

Fonseca et al. (2006, 2009) calculated the motions and vertical bending moments induced on a containership and on a FPSO by a set of 20 wave records that include abnormal waves. These wave records have been measured in different places and occasions. The results are compared with rule minimum requirements and estimates of extreme loads by other methods. Fonseca et al. (2008) compared the moments induced by the same set of abnormal waves on the FPSO with the maximum loads induced by design storms with 3 hours duration. It was concluded that the largest moments were induced by groups of waves without abnormal wave characteristics.

Fonseca et al. (2007) applied the same procedure to compare the probability distributions of maximum bending moments induced by the seastates where the abnormal waves were measured, with the moments induced by the abnormal waves themselves.

The present paper starts by presenting a summary of the formulation behind the method to calculate the responses induced by measured wave traces with abnormal waves. Then, some comparisons between experimental data and predictions for a containership in large amplitude waves are presented. The second part presents the most important results and conclusions regarding the global dynamics of the same containership encountering abnormal waves.

2 FORMULATION

This section presents the formulation leading to the nonlinear time domain strip theory. Consider a coordinate system fixed with respect to the mean position of the ship, $X = (x,y,z)$, with z in the vertical upward direction and passing through the centre of gravity of the ship, x along the longitudinal direction of the ship and pointing to the bow, and y perpendicular to the later and in the port direction. The origin is in the plane of the undisturbed free surface.

The rigid body motions as the ship advances in waves consist of three translations and three rotations. It is assumed that the longitudinal (fore-aft) hydrodynamic unsteady forces are negligible (strip theory hypotheses). Furthermore only heave and pitch motions are considered. The pitch angles are considered relatively small; however the relative motions between the ship and the waves may be of large amplitude leading to large variation of the submerged volume with associated non-linear effects.

Neglecting the viscous effects the hydrodynamic problem may be formulated in terms of potential flow theory, thus the fluid velocity vector may be represented by the gradient of a velocity potential. Assuming a slender hull at slow forward speed and small enough amplitudes of the unsteady motions and incident waves, the potential can be expressed in the linearized form:

$$\Phi = -Ux + \Phi^S + \Phi^I + \Phi^D + \sum_{i=1}^{6} \Phi_j^R \tag{1}$$

where the first two terms represent the steady flow due to the presence of the ship advancing through the free surface being U the forward speed of the ship and Φ^S the steady perturbation potential. Φ^I is the incident potential and Φ^D is the diffracted potential related to the perturbation in the incident wave due to the presence of the hull. Φ_j^R is the radiated potential due to an unsteady motion in the j-mode.

Substitution of the potential (1) into the linearized Bernoulli´s equation results in the hydrodynamic pressure. Integration of the oscillatory pressure terms over the wetted surface of the hull results in two groups of hydrodynamic forces, namely the radiation and the exciting forces. The latter are further decomposed in Froude Krylov forces associated with the incident wave field and diffraction forces associated with the wave scattering. Integration of the hydrostatic pressure over the oscillatory wetted hull results on the hydrostatic forces.

It is assumed that the nonlinear effects on the vertical motions and global structural loads are dominated by buoyancy effects, therefore Froude-Krylov and hydrostatic forces are nonlinear, while the radiation and diffraction forces are kept linear.

2.1 *Radiation forces*

The representation of the radiation forces in the time domain is different from the other hydrodynamic forces, since they have a time dependency of the previous history of the fluid motion. In fact the existence of radiated waves implies a complicated time dependence of the fluid motion and hence of the pressure forces. Waves generated by the body at time t will persist long after, as well as the associated pressure forces.

The time domain formulation derived by Cummins (1962) is used to represent the radiation forces in terms of unknown velocity potentials. The basic assumption is the linearity of the radiation forces. The radiation force in the k-direction due to an oscillatory motion in the j-mode is:

$$F_{kj}^{R}(t) = -\ddot{\xi}_j(t)A_{kj}^{\infty} - C_{kj}^{m}\xi(t) - \int_{-\infty}^{t}\left\{K_{kj}^{m}(t-\tau)\dot{\xi}_j(\tau)\right\}d\tau \tag{2}$$

where A_{kj}^{∞}, C_{kj}^{m} and $K_{kj}^{m}(t)$ represent respectively infinite frequency added masses, radiation restoring coefficients and memory functions.

The restoring coefficients in (2) represent a correction to the hydrodynamic steady forces acting on the ship due to the steady flow. The convolution integrals represent the effects of the whole past history of the motion accounting for the memory effects due to the radiated waves. The memory functions and the radiation restoring coefficients are obtained by relating the radiation forces in the time domain and in the frequency domain by means of Fourier analysis:

$$K_{kj}^{m}(t) = \frac{2}{\pi}\int_{0}^{\infty}\left\{B_{kj}(\omega)\cos\omega t\right\}d\omega \tag{3}$$

$$C_{kj}^{m}(t) = \omega^2[A_{kj}^{\infty} - A_{kj}(\omega)] - \omega\int_{0}^{\infty}K_{kj}^{m}(\tau)\sin(\omega\tau)d\tau \tag{4}$$

where $A_{kj}(\omega)$ and $B_{kj}(\omega)$ are the frequency dependent added mass and damping coefficients calculated with a strip theory method (Salvesen et al., 1970).

2.2 *Wave exciting forces*

According to the linear wave theory, the wave potential of a harmonic wave advancing with an arbitrary direction with respect to the coordinate system is:

$$\Phi^{I}(x,y,z,t) = \zeta_a\varphi^{I}(y,z)e^{ikx\cos\beta}e^{i\omega_e t} \tag{5}$$

$$\varphi^{I}(y,z) = \frac{ig}{\omega_0}e^{k_0 z}e^{iky\sin\beta} \tag{6}$$

where ζ_a is the wave amplitude, $k = \omega_0^2/g$ and ω_0 are respectively the wave number and wave frequency, $\varphi^I(y,z)$ is the velocity potential acting on the ship cross sections of an harmonic incoming wave of unit amplitude. The relation between wave frequency and encounter frequency ω_e is $\omega_0 = \omega_0 - kU\cos\beta$, where β represents the ship heading with respect to the incoming waves ($\beta = 0$ for head waves).

The free surface elevation of a harmonic wave represented on the reference system advancing with the ship steady speed (X), is obtained by introducing the wave potential in the dynamic free surface boundary condition:

$$\begin{aligned}\zeta(x,y,t) &= \mathrm{Re}\{-\zeta_a e^{ik(x\cos\beta + y\sin\beta)}e^{i\omega_e t}\} \\ &= -\zeta_a\cos(\omega_e t + kx\cos\beta + ky\sin\beta)\end{aligned} \tag{7}$$

Following St. Denis and Pierson (1953) the wave elevation of an irregular and stationary seastate may be represented by the superposition of an infinite number of harmonic wave components. A wave frequency and random phase angle define each harmonic component. For an irregular and bidimensional seastate the wave elevation is:

$$\zeta(x,y,t) = -\sum_{n=1}^{N}(\zeta_a)_n\cos\begin{bmatrix}(\omega_e)_n t + k_n x\cos\beta \\ + k_n y\sin\beta + \varepsilon_n\end{bmatrix} \tag{8}$$

where $(\zeta_a)_n$, k_n and ε_n are respectively the amplitude, wave number and random phase angle of the harmonic component n, $(\omega_e)_n$ is the encounter frequency of the harmonic component n. The random phase angles are uniformly distributed between 0 and 2π. For deep water, the dispersion relation relates the wave frequency and wave number, $(\omega_0)_n^2 = k_n g$.

The exciting forces due to the incident waves are decomposed into a diffraction part F_k^D and the Froude-Krilov part F_k^K. The diffraction part is kept linear. It results from the solution of the hydrodynamic problem of the ship advancing with constant speed through the incident waves and restrained at her mean position. Since this is a linear problem and the exciting waves are known a priori, it can be solved in the frequency domain and the resulting transfer functions be used to generate a time history of the diffraction heave force and pitch moment. The diffraction forces in harmonic waves for heave and pitch are calculated by a strip method as follows:

$$F_3^D(t) = \mathrm{Re}\left\{ e^{i\omega_e t} \zeta_a \int_L \left(e^{ikx\cos\beta} f_3^D \right) dx \right\} = \zeta_a \left[D_3^a \cos(\omega_e t - \phi_3) \right] \tag{9}$$

$$F_5^D(t) = \mathrm{Re}\left\{ \left(e^{i\omega_e t} \right) \left[-\zeta_a \int_L e^{ikx\cos\beta} \left(x f_3^D + \frac{U}{i\omega} f_3^D \right) dx \right] \right\} = \zeta_a \left[D_5^a \cos(\omega_e t - \phi_5) \right] \tag{10}$$

where integration is over the ship's length, f_3^D is the complex sectional heaving diffraction force for unit amplitude incident waves and D_k^a is the amplitude of the diffraction force in the k-direction for unit amplitude waves and ϕ_k is the respective phase angle.

Since it is assumed that the diffraction forces are linear, in irregular waves they are calculated superposing the diffraction contribution from every harmonic component defining the irregular wave:

$$F_k^D(t) = \sum_{n=1}^{N} (\zeta_a)_n \left(D_k^a \right)_n \cos\left[(\omega_e)_n t - (\theta_k)_n + \varepsilon_n \right] \tag{11}$$

where the subscript n stands for the harmonic component n.

The Froude-Krilov part is related to the incident wave potential, and results from the integration at each time step of the associated pressure over the wetted surface of the hull under the undisturbed wave profile:

$$F_3^I(t) = \int_L \int_{C_x} p^I(t,x,y) n_3 d\varsigma dx \tag{12}$$

$$F_5^I(t) = -\int_L x \int_{Cx} p^I(t,x,y) n_3 d\varsigma dx \tag{13}$$

Integrations are over the ship length L and the cross sections wetted contour C_x, n_3 represents the vertical component of the outward unit normal vector to the cross section and p^I is the incident wave field pressure. In irregular waves the incident wave field pressure results from the superposition of all the harmonic components used to represent that wave:

$$p^I = -\rho g \sum_{n=1}^{N} (\zeta_a)_n e^{k_n z} \cos\left[\begin{matrix} (\omega)_n t + k_n x \cos\beta \\ + k_n y \sin\beta + \varepsilon_n \end{matrix} \right] \tag{14}$$

Within linear theory of harmonic waves the wave pressure is defined up to the mean waterline ($z = 0$), and do not extend upwards. For this reason an approximation is assumed to account for the wave pressure above the mean waterline when the crests pass through the hull. The pressure is zero at the free surface and it is assumed to be hydrostatic between the free surface and $z = 0$.

2.3 *Restoring forces*

The hydrostatic force and moment, F_3^H and F_5^H, are calculated at each time step by integration of the hydrostatic pressure over the wetted hull under the undisturbed wave profile:

$$F_3^H = \rho g \int_L \int_{C_x} z n_3 d\varsigma dx \tag{15}$$

$$F_5^H = -\rho g \int_L x \int_{C_x} (z n_3) d\varsigma dx \tag{16}$$

2.4 *Water on deck forces*

The vertical forces associated with the green water on deck, which occurs when the relative motion is larger than the free board, are calculated using the momentum method (Buchner, 1995). The vertical force per unit length is:

$$f^{gw}(x,t) = \left(\frac{\partial m_{gw}}{\partial t} \right) w + \left(g\cos\xi_5 + \frac{\partial w}{\partial t} \right) m_{gw} \tag{17}$$

which includes one term that accounts for the variation of mass of water on the deck, the hydrostatic component and one term associated with the acceleration of the deck. m_{gw} represents the mass of water on the deck per unit length and w is the vertical velocity of the deck. The mass of water on the deck is proportional to the height of water on the deck, which is given by the difference between the relative motion and the free board of the ship. Finally the heave green water force and pitch green water moment are:

$$F_3^{gw}(t) = -\int_L f^{gw}(x,t) dx \tag{18}$$

$$F_5^{gw}(t) = \int_L x f^{gw}(x,t)\,dx \tag{19}$$

2.5 *Equations of motion and global structural loads*

Equating the hydrodynamic external forces to the mass and gravity forces one obtains the equations of motion. These equations, which combine linear and nonlinear terms, are solved in the time domain by a numerical procedure. For heave and pitch the equations are:

$$\begin{aligned}(M + A_{33}^{\infty})\ddot{\xi}_3(t) &+ \int_{-\infty}^{t} [K_{33}^{m}(t-\tau)\dot{\xi}_3(\tau)]d\tau + C_{33}^{m}\xi_3(t) \\ &+ A_{35}^{\infty}\ddot{\xi}_5(t) + \int_{-\infty}^{t} [K_{35}^{m}(t-\tau)\dot{\xi}_5(\tau)]d\tau + C_{35}^{m}\xi_5(t) \\ &+ F_3^{H}(t) - Mg + F_3^{gw}(t) = F_3^{D}(t) + F_3^{K}(t)\end{aligned} \tag{20}$$

$$\begin{aligned}\left(I_{55} + A_{55}^{\infty}\right)\ddot{\xi}_5(t) &+ \int_{-\infty}^{t} \left[K_{55}^{m}(t-\tau)\dot{\xi}_5(\tau)\right] d\tau \\ &+ C_{55}^{m}\xi_5(t) + A_{53}^{\infty}\ddot{\xi}_3(t) + \int_{-\infty}^{t} \left[K_{53}^{m}(t-\tau)\dot{\xi}_3(\tau)\right] d\tau \\ &+ C_{53}^{m}\xi_3(t) + F_5^{H}(t) + F_5^{gw}(t) = F_5^{D}(t) + F_5^{K}(t)\end{aligned} \tag{21}$$

where ξ_3 and ξ_5 represent respectively the heave and pitch motions and the dots over the symbols represent differentiation with respect to time. M is the ship mass, g is acceleration of gravity and I_{55} represent the ship inertia about the y-axis.

The wave induced global structural dynamic loads at a cross section are given by the difference between the inertia forces and the sum of the hydrodynamic forces acting on the part of the hull forward of that section. The vertical shear force and vertical bending moment are given respectively by:

$$V_3(t) = I_3(t) - R_3(t) - D_3(t) - K_3(t) - H_3(t) - G_3(t) \tag{22}$$

$$M_5(t) = I_5(t) - R_5(t) - D_5(t) - K_5(t) - H_5(t) - G_5(t) \tag{23}$$

I_k represents the vertical inertia force (or moment) associated with the ship mass forward of the cross section under study. As assumed for the calculation of the ship motions, the radiation (R_k) and diffraction (D_k) hydrodynamic contributions for the loads are linear, and the Froude-Krylov (K_k) and hydrostatic (H_k) contributions are non-linear since they are calculated over the "exact" hull wetted surface at each time step. G_k represents the contribution for the structural loads of the green water on deck forces.

The formulation to calculate all hydrodynamic contributions for the loads is consistent with the formulation applied to solve the unsteady motion problem. The convention for the loads is such that the sagging shear force and hogging bending moment are positive.

2.6 *Responses to deterministic wave traces*

The objective of the method presented here is to calculate the ship motions and associated global structural loads induced by extreme waves (abnormal waves) that were in fact measured in the ocean. These real wave traces were measured in one fixed point in space. In order to apply the time domain calculation method described in the previous sections, it is necessary to derive a representation of the incident wave field which is consistent with the time history of the wave elevation defined at a particular point. It is assumed that the frequency content of the wave field can be determined by the Fourier transform of a time signal from a single point measurement of the wave elevation. The dispersion relation propagates the wave field in space consisting of deep water unidirectional waves. Furthermore it is assumed that the kinematics of the waves may be represented by superposition of linear harmonic components. Basically the wave trace is Fourier analysed to obtain a set of harmonic components, the dispersion relation propagates them in space, and then the wave exciting forces to that wave trace are calculated with equations (9) to (14).

The real wave traces are measured in one point fixed in space and they can be used to calculate the dynamic responses of stationary platforms. It can be assumed, for example, that this wave elevation is felt at midship. On the other hand, a ship with forward speed cannot encounter the whole wave trace measured in one point fixed in space. However the ship can be forced to pass through that particular point in space at the time instant when the abnormal wave was measured. The wave field represented on a reference system whose origin coincides with the measurement point is given by equation (8). The same wave field on a reference system advancing with the ship speed U and whose origin coincides with the measurement point for $t = 0$ is:

$$\zeta(x,y,t) = -\sum_{n=1}^{N} (\zeta_a)_n \cos\begin{bmatrix}(\omega_0)_n t + k_n(x+Ut)\cos\beta \\ +k_n y\sin\beta + \varepsilon_n\end{bmatrix} \tag{24}$$

Since the encounter ω_e and the wave ω_0 frequencies can be related as $\omega_e = \omega_0 + kU\cos\beta$, then equation (19) may be represented in terms of encounter frequency components by:

$$\zeta(x,y,t) = -\sum_{n=1}^{N}(\zeta_a)_n \cos\begin{bmatrix}(\omega_e)_n t + k_n x\cos\beta \\ +k_n y\sin\beta + \varepsilon_n\end{bmatrix} \quad (25)$$

If for $t = 0$ the origin of the moving reference system is not coincident with the measurement point, but is located at a distance $d^* = Ut^*$ in the negative x-axis direction, then the wave field is given by:

$$\zeta(x,y,t) = -\sum_{n=1}^{N}(\zeta_a)_n \cos\begin{bmatrix}(\omega_0)_n t + k_n\left(x + Ut - Ut^*\right)\cos\beta \\ +k_n y\sin\beta + \varepsilon_n\end{bmatrix} \quad (26)$$

Observing equations (24) and (26) it is clear that the moving reference system will be at the measurement point encountering the desired event after the period of time t^* has passed.

3 RESPONSES FOR A CONTAINERSHIP

This section presents a summary of the main results and conclusions obtained by the authors on the vertical motions and global loads induced by extreme waves on the S175 containership. This ship is well known because it was used repeated times for experimental and numerical investigations (see for example ITTC, 1978, 1981). Like modern containerships, it has a pronounced flare on the bow and stern, therefore nonlinear effects are associated to the vertical motions.

The displacement of the ship is 24,742 ton., the length between perpendiculars, beam and draught are 175 m, 25.4 m and 9.5 m, the centre of gravity is located 2.43 m aft of midship, the block coefficient is 0.572 and the pitch radius of gyration is 42 m. Figure 1 shows the bodylines.

3.1 *Comparison with experimental data*

A comprehensive experimental program with a model of the S175 was carried out in a seakeeping laboratory to identify the nonlinear effects on the vertical motions and global structural loads in large amplitude waves. Fonseca and Guedes Soares (2004a, 2004b) present the results and conclusions in regular waves and in irregular seastates, while in Fonseca and Guedes Soares (2005) the results from the nonlinear time domain seakeeping code are systematically compared with the experimental data. Figure 2 presents the containership model under testing.

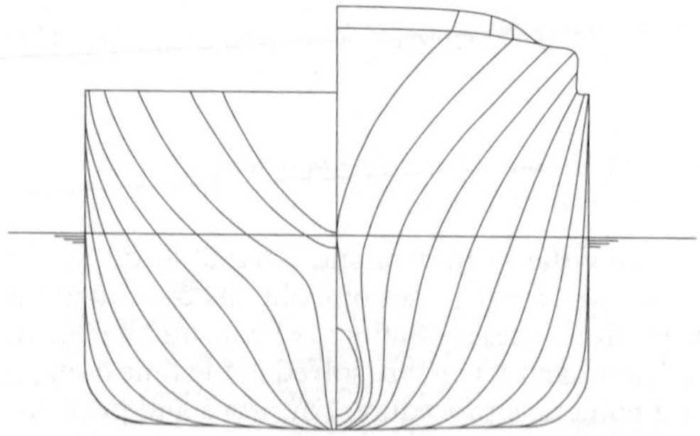

Figure 1. S175 containership hull bodylines.

A few examples of the type of results obtained are shown in figures 3 to 6. Figure 3 presents the heave amplitudes nondimensionalised by the wave amplitude as function of the nondimensional wave frequency. The square and triangle symbols represent experimental data in regular waves with steepness of respectively $2\zeta_a/\lambda = 1/120$ and $2\zeta_a/\lambda = 1/40$ (ζ_a and λ are the wave amplitude and wavelength), where the first group of data corresponds to small amplitude waves and the second to large amplitude waves. The continuous lines stand for the nonlinear numerical results and the dashed lines represent linear results. One observes that the nonlinear numerical results compare very well with the experimental data, including the reduction of the resonance peak with the wave amplitude. On the other hand, the linear results overestimate the experimental resonance peak, especially for large amplitude waves.

Figure 4 presents the sagging peaks (negative), hogging peaks and mean values of the vertical bending moment at midship time records in regular waves. The wavelengths are equal to the ship length between perpendiculars and the results are presented as function of the wave steepness ($k\zeta_a$).

Regarding nonlinear numerical results, large dashed lines correspond to the inviscid model and the continuous lines to the model accounting for viscous effects on the vertical responses (see Fonseca and Guedes Soares 2005). It is clear that the magnitude of the experimental sagging peaks is much larger than of the hogging peaks and the difference increases with the wave steepness. This behaviour is captured only by the nonlinear model, although it tends to overestimate the experimental results.

Figure 5 shows the probability cumulative distributions of the vertical bending moment at midship peaks in irregular waves with a significant wave height of 6.13 m and a peak period of 11.5 s.

One observes a very good correlation between the experimental data and the nonlinear numerical results, while it is clear that the Rayleigh distribu-

Figure 2. Containership model under testing in regular waves.

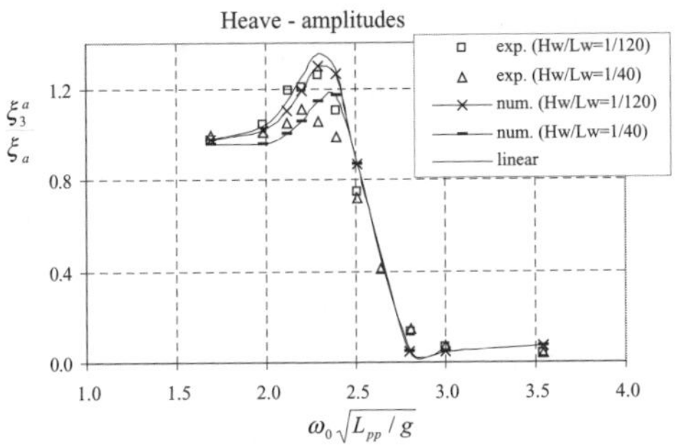

Figure 3. Amplitudes of heave as function of the wave frequency. Comparison of experimental and numerical results for two wave steepness, Fn = 0.25 (Fonseca and Guedes Soares 2005).

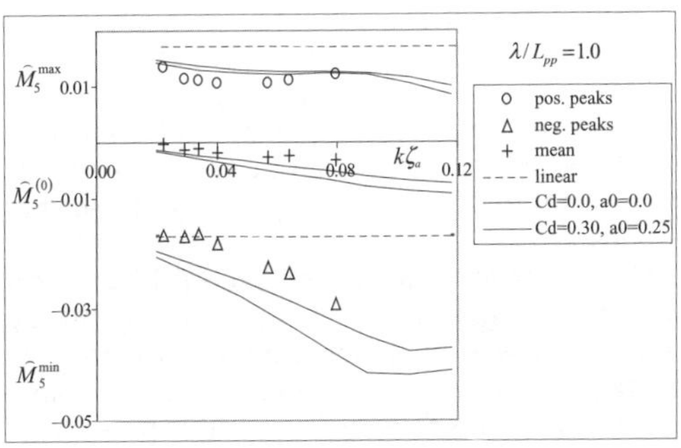

Figure 4. Vertical bending moment at midship as function of the wave steepness. Comparison of numerical and experimental results of the sagging and hogging peaks and mean values, Fn = 0.25 (Fonseca and Guedes Soares 2005).

tion fails to represent the probability distribution of this nonlinear ship response.

The comparative analysis of results leads to the conclusion that the time domain model is able to capture all nonlinear characteristics that were identified in the experimental data. This means that the most important nonlinear contributions for the vertical motions and loads are included in the mathematical model. The numerical results represent the reduction of heave and pitch transfer function amplitudes with the incident wave amplitude, the important second and third order harmonics of the structural loads in regular waves and the asymmetry of the time records of the moment.

Although all nonlinear characteristics of the experimental data are qualitatively represented in the numerical results, there are also discrepancies, which show that the mathematical model does not represent completely the physical process. The discrepancies that can be identified as systematic are the overestimation of the vertical bending moment. It is believed that to improve the results it is necessary to account more completely for the interactions between steady and unsteady flows, account for nonlinearities on the radiation and diffraction forces and introduce a more realistic method to estimate evolution of the height of water on the deck.

3.2 *Responses to abnormal waves*

The objective of the method described in Section 2 is to calculate the global structural loads induced by measured wave traces which include abnormal waves. The responses induced by the New Year Wave (NYW wave) were investigated by Fonseca et al. (2001) and Guedes Soares et al. (2008). The related wave trace was measured during a severe storm occurred in the Central North Sea (Haver and Karunakaran 1998).

In the simulation, the S175 advances in head seas with a reduced speed of 13 knots and it is

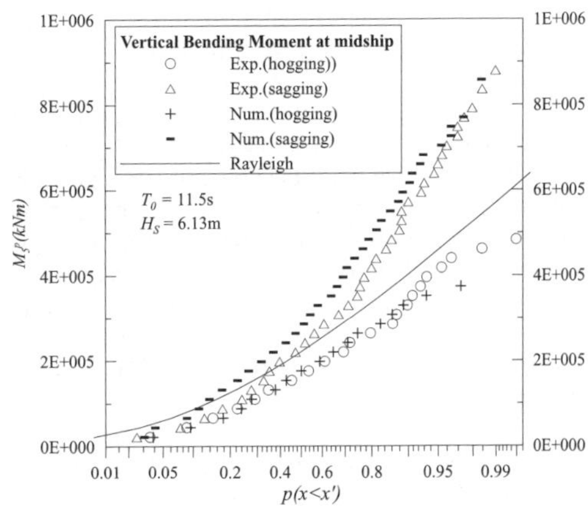

Figure 5. Probability distributions of the experimental and numerical vertical bending moment at midship, Fn = 0.25 (Fonseca and Guedes Soares 2004b).

forced to pass through the point in space where the wave trace was measured when the abnormal wave crest is generated. Figure 6 presents simulations of the wave elevation (ζ) at the centre of gravity (CG), heave (ξ_3), pitch (ξ_5), relative motion at the forward perpendicular (ξ_r), green water on deck vertical force at the forward perpendicular (f^{gw} per meter), vertical shear force (V_3) and vertical bending moment (M_5) at midship.

The wave elevation graph shows an abnormal wave crest of approximately 18 m occurring at $t = 168$ s. The dashed line short sequence represents wave elevation measured at the point fixed in space, while the continuous line represents the simulation at the ship's moving reference system.

To better assess the importance of nonlinear effects on the response simulations, nonlinear results (continuous lines) are plotted together with corresponding linear simulations (dashed lines).

The graphs show that the ship motions are only slightly nonlinear around the abnormal wave. The forces due to water on deck at the forward perpendicular reach a very large value, which is higher than 2000 kN/m. This occurs when the ship immerses the bow deeply into the water after the passing through the large wave. The estimated height of water on deck is close to 15 m at the forward perpendicular.

It is also observed that the vertical bending moment at midship is highly asymmetric, with the largest nonlinear sagging peak showing approximately three times larger magnitudes than the corresponding linear ones. This peak occurs when the ship immerses the bow after encountering the large crest. Finally it is interesting to note that the water on deck vertical forces tend to reduce the sagging moment, since these forces act downward to produce a hogging moment.

Fonseca et al. (2006) generalized the previous study by considering 20 wave traces which include abnormal waves. These wave traces were measured during storms in the Gulf of Mexico, Northern North Sea and Central North Sea. Table 1 presents the significant wave heights (Hs) and peak periods (Tp) of the wave records that include the abnormal waves. AI_u and AI_d represent the abnormality indexes calculated using respectively zero up-crossing and zero down-crossing. The abnormality indexes were defined by the ratio between abnormal wave height and the significant wave height for the record.

The time domain seakeeping code was run for all wave traces. It was concluded that the maximum hogging moments always occurs when the abnormal wave crest passes trough midship, while the maximum sagging peaks occur after the ship passes the abnormal crest and the bow dives into the next wave.

Figure 7 shows the correlation between the maximum hogging peaks (squares) and sagging peaks (triangles) and the abnormal wave height. The maximum moments increase almost linearly with the wave height and one observes that the sagging magnitudes are approximately double the hogging ones. Similar correlation with respect to the wave slope leads to the conclusion that the maximum moments are almost independent from the slope of the abnormal wave.

3.3 *Effect of wavelength*

The maximum bending moments depend on the relation between the ship length and the wavelength; therefore the same abnormal waves will induce different nondimensional bending moments for different ship lengths. In other words, an abnormal wave which is not critical in terms of induced global bending moment for a certain ship length may be critical for a similar ship with a different length. To clarify this aspect, Fonseca et al. (2007) investigated the influence of the length of a containership on the vertical bending moments induced by the 20 abnormal wave records presented before. Five ships similar to the S175 were considered, with lengths between 100 m and 350 m. Similar ships mean that the geometric and mass properties are Froude scaled between each other. Figure 8 shows the profile of the five ships, which gives an impression of the dimension relations.

As an example of comparative results, Figure 9 shows simulations of: the wave elevation corresponding to one of the wave record (1 NA), heave and pitch motions, relative motion at the forward perpendicular, and the nondimensional vertical bending moment at midship. Each graph includes simulations for three ships, namely with lengths of 100 m, 250 m and 350 m. Table 2 includes the abnormal wave crest elevation, the abnormal wave height and the maximum values of the ship responses.

The abnormal wave in this record has a crest with an elevation of approximately 13 m and the wave height is 18.2 m. The simulation represents the wave elevation on the reference system advancing with the ship forward speed, therefore, since the speeds are different, the time histories of the wave elevation are slightly different.

One observes that the heave and pitch motions are much larger for the smaller ship, and they reduce as the ship length increases. This is because the smaller ship, being much smaller than the wavelength (100 m compared to 257 m), is able to follow the wave both in terms of vertical displacement and angular displacement. Looking at the relative motion at the bow graph, by far the largest magnitudes are attained by the middle size ship, namely

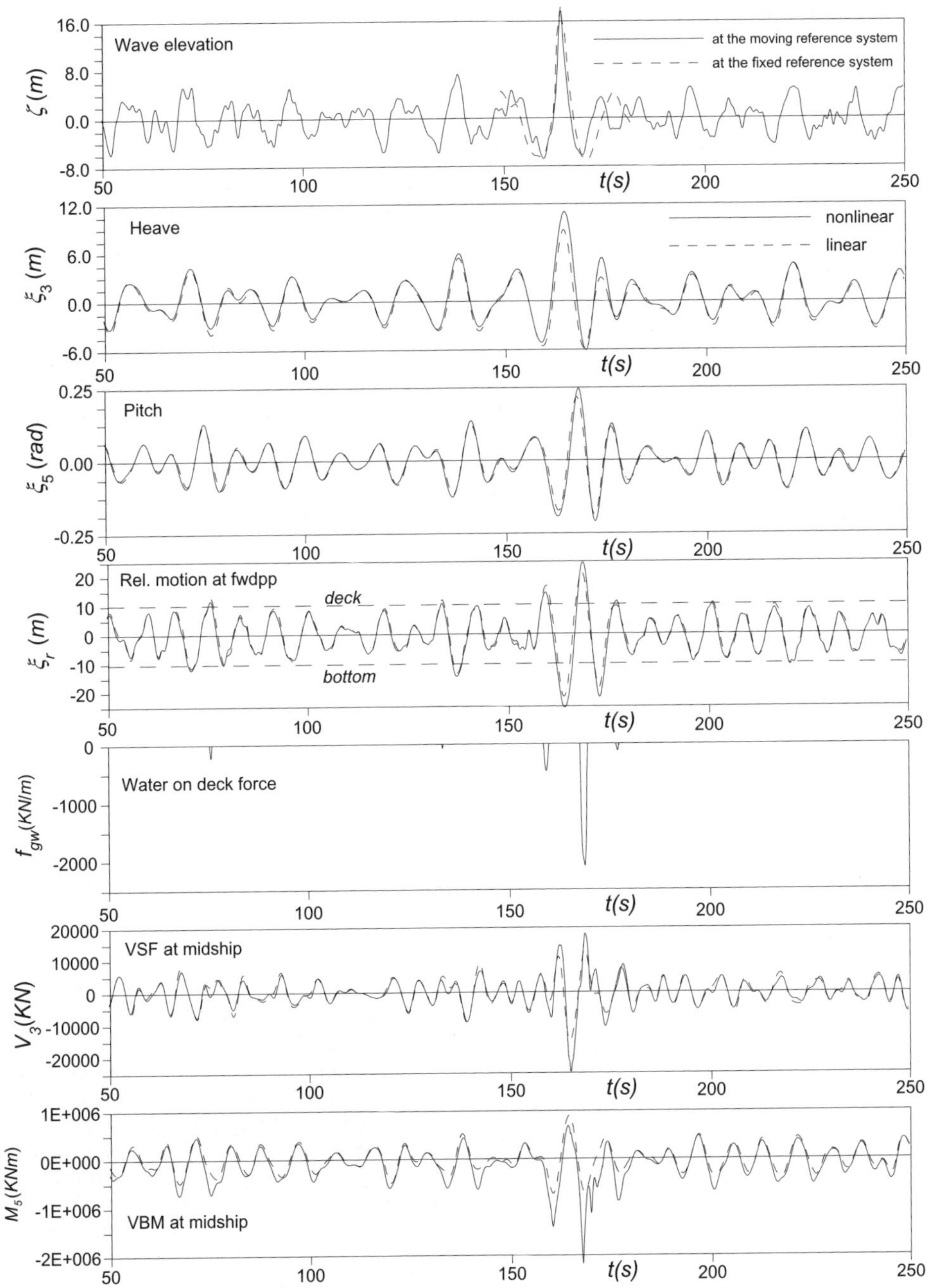

Figure 6. Simulation of vertical responses in head long crested waves. $F_n = 0.16$, significant wave height of 11.4 m and zero upcrossing mean wave period of 11.3 s (Guedes Soares et al. 2008).

the one with a length approximately equal to the abnormal wavelength (250 m compared to 257 m).

Since the magnitude of the vertical bending moment is very much correlated to the magnitude of the relative motion at the bow, the maximum normalized vertical bending moments are obtained for the middle size ship. To compare moments for different ship lengths, they are normalized by $\rho g B L_{pp}^2$ (symbols represent: fluid density, acceleration of gravity, ship beam and length).

Table 1. Characteristics of the abnormal waves.

	H_s	T_p	AI_u	AI_d
01 NA	8.11	13.00	2.29	2.31
02 NA	9.01	11.96	2.04	2.03
03 NA	9.36	12.23	2.02	2.23
04 DR	11.92	13.24	2.23	2.19
05 NA	7.17	12.60	2.26	2.38
06 NA	8.16	12.90	2.08	2.12
07 NA	5.51	12.23	2.13	2.18
08 NA	6.50	13.11	2.21	2.20
09 NA	8.75	12.70	2.03	2.14
11 NA	5.92	9.99	2.13	1.88
12 NA	7.85	12.70	2.12	1.94
14 NA	6.47	12.14	1.78	2.07
15 NA	7.84	13.43	2.27	1.90
16 NA	9.63	12.32	1.93	2.02
17 NA	7.41	12.90	2.19	1.87
19 NA	10.39	13.88	1.75	1.88
20 NA	8.29	13.11	1.98	2.18
24 NA	7.81	12.60	1.96	2.13
37 CA	10.65	13.47	2.32	2.22
41 CA	10.69	13.52	2.10	1.79

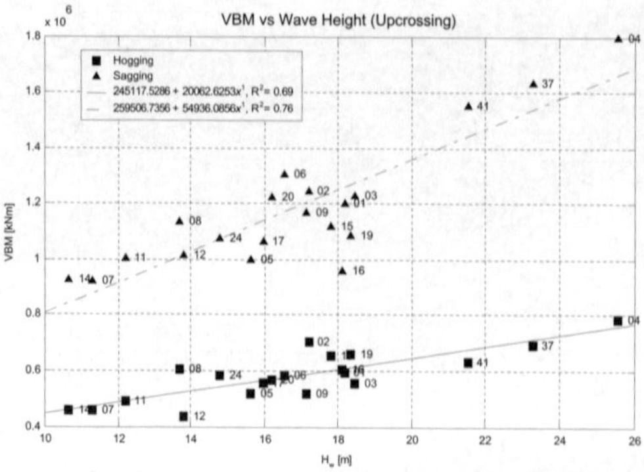

Figure 7. Correlation between the largest sagging and hogging peaks and the height of the abnormal waves (Fonseca et al. 2006).

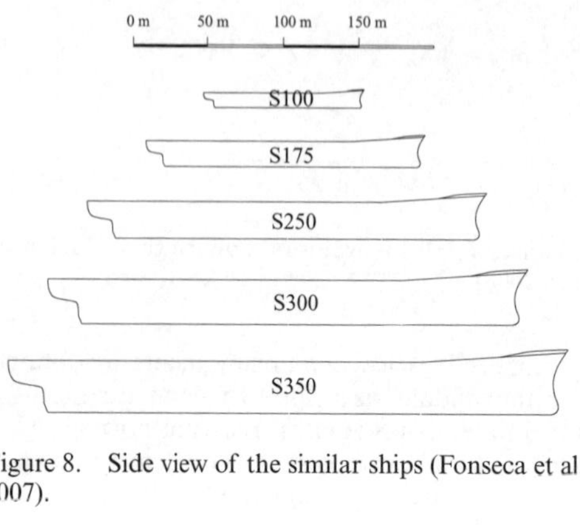

Figure 8. Side view of the similar ships (Fonseca et al., 2007).

Table 2. Maximum values of the abnormal wave and ship responses.

	S100 Lpp = 100 m	S250 Lpp = 250 m	S350 Lpp = 350 m
Wave crest (m)	13.2	13.2	13.2
Wave height (m)	17.3	17.7	17.7
Heave (m)	8.7	5.4	2.9
Pitch (deg)	15.4	5.6	2.2
Relative mot. fwdpp (m)	10.5	15.4	9.2
VBMsag / $\rho g B L_{pp}^2$	–0.012	–0.019	–0.011
VBMhog / $\rho g B L_{pp}^2$	0.005	0.010	0.008

The nonlinear simulations were repeated for all abnormal wave traces and all ships. The maximum moments were identified and the normalized results plotted together as function of the height of the abnormal wave. This is shown in figure 10, where each graph corresponds to one ship length (smaller ship on the left), and the lines are linear regressions of the results.

The graphs show a good correlation between the induced moments and the height of the abnormal waves. It is interesting also to conclude that the rate of increase of the maximum moments with the wave height changes much with the ship length. The linear regression line is less steep for the smaller ship, the steepness increases with the ship length and then it tends to remain constant for lengths above 300 m.

3.4 *Comparison with rule values*

A comparison of interest is with the rules minimum requirements in terms of design wave bending moments. The formulae from the IACS *Common Structural Rules* have used for ships operating without restrictions in the Northern North Sea for a period of 25 years. The calculations have been done for the S175 advancing at 13 knots.

The graph on Figure 11 represents maximum sagging and hogging peaks versus the abnormal wave height for the 20 wave traces analyzed before (solid triangles and squares). Linear and nonlinear long term distribution (LTD) sagging values and the rule minimum requirement values are also represented in the graph as horizontal lines. LTD represent the expected maxima for an exceedance probability of 10^{-8}, which corresponds approximately to a return period of 25 years. Linear and nonlinear LTD means that the seakeeping calculations are either linear or nonlinear (see Guedes Soares and Moan 1991, Guedes Soares 1996).

One observes in Figure 11 that the maximum hogging moments induced by the group of abnormal waves are always bellow the rule value, while a few of the sagging results are larger than the one required by the rules. On the other hand, the nonlinear LTD value is considerably larger than rule requirement. The fact that design bending moments calculated with methods based on first principles are larger than the rule requirements has been identified before. This does not mean that ship structures according to the Classification Society rules are not safe. Past experience has demonstrated that ship structures are safe.

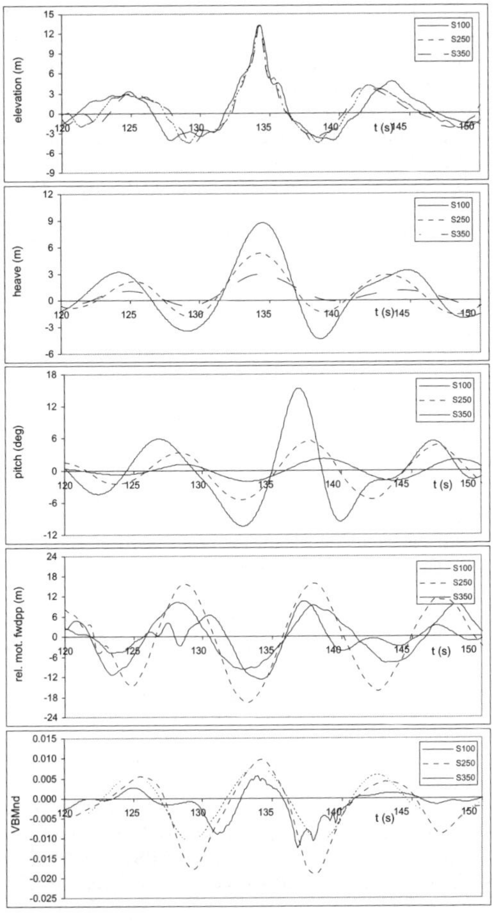

Figure 9. Simulations of abnormal wave elevation and ship responses for 3 ship lengths (Fonseca et al., 2007).

The discrepancy between rule bending moments and calculated design moments is still a topic of research. Possible factors that may explain the discrepancy are, for example, the heavy weather avoidance by ships, the changing of course and speed in the higher seastates, and the fact that the rule design procedure follows a sequence of steps with several safety factors involved which, when combined, might compensate for the underestimated wave bending moment.

Comparing the maximum moments induced by the abnormal waves with the nonlinear LTD values (both set of results are based on the same time domain code), leads to conclude that the maximum expected sagging moment during a period of 25 years in the Northern North Sea is considerably larger than all of the moments induced by the abnormal waves. Assuming that the set of realistic wave traces used is representative of the abnormal waves that the ship could encounter, the former result seems to indicate that there are “non-abnormal” wave conditions that induce larger bending moments than the abnormal waves.

4 CONCLUSIONS

The paper presents a method to calculate the global responses induced by abnormal waves on ships and offshore structures and presents a summary of main results and conclusions regarding the dynamic effects of these waves on moving ships.

Extensive comparisons between results from the time domain seakeeping code and various sets of experimental data from model tests in large amplitude waves lead to the conclusion that the code is able to represent the most important nonlinear effects.

It is concluded that the maximum sagging moments on the containership induced by the abnormal waves are approximately double in magnitude of the hogging moments. The maximum sagging moments occur after the ship passes the abnormal wave crest and the bow dives into the next wave, while the maximum hogging always occur when the abnormal crest is at midship.

The analysis of results in a large set of abnormal waves led to the conclusion that the maximum

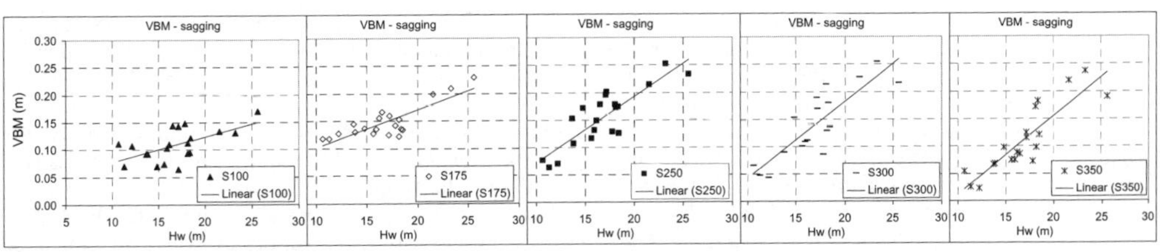

Figure 10. Maximum sagging bending moments induced by all abnormal waves as function of the abnormal wave heights. Different graphs for different ship lengths (Fonseca et al., 2007).

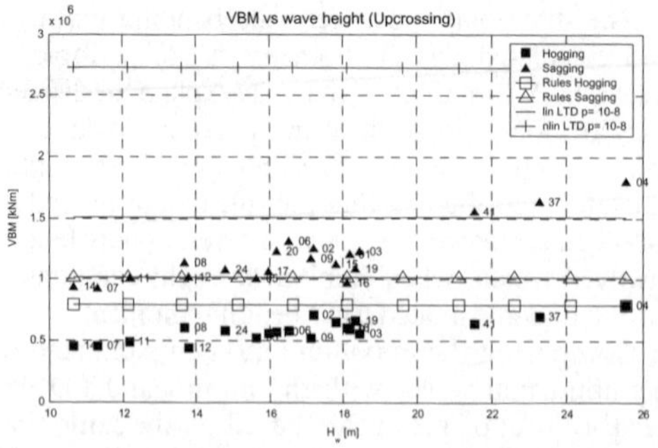

Figure 11. Comparison between the largest sagging and hogging peaks and the rule minimum requirements (Fonseca et al., 2006).

moments depend almost linearly on the abnormal wave height, while there no clear correlation with the steepness and the length of the same abnormal waves.

Calculations were repeated for geometrically similar containerships, however with different lengths. One observes that the nondimensional maximum moments depend much of the ship length.

Finally the maximum moments induced by the abnormal waves were compared to the moments induced by the most severe seastates that the ship would encounter in the Northern North Sea and the later are clearly larger. On the other hand, a few of the moments induced by the abnormal waves are superior to the rules minimum requirement.

ACKNOWLEDGMENTS

This paper presents part of the teaching material prepared and used in the scope of the course "Modelling of Environment and Environmental Loads" within the project "Advanced Ship Design for Pollution Prevention (ASDEPP)" that was financed by the European Commission through the Tempus contract JEP-40037-2005.

REFERENCES

Adeegest, L., Braathen, A. and Vada, T. 1998, "Evaluation of Methods for Estimation of Extreme Non-linear Ship Responses Based on Numerical Simulations and Model Tests", *Proc. 22nd Symp. Naval Hydrodynamics*, Washington D.C. Vol.1, pp. 70–84.

Buchner, B., 1995, "On the Impact of Green Water Loading on Ship and Offshore Unit Design", *Proceedings 6th International Symposium on Practical Design of Ships and Mobile Units (PRADS'95)*, H. Kim, J.W. Lee (Editors), the Society of Naval Architects of Korea, Vol. 1, pp. 430–443.

Cummins, W.E. 1962, "The impulse response function and ship motions", *Schiffstechnik*, **9**, 101–109.

Faulkner, D. and Buckley, W.H. 1997, "Critical Survival Conditions for Ship Design", *International Conference on Design and Operation for Abnormal Conditions*, RINA, paper no. 6, pp. 1–25.

Fonseca, N. and Guedes Soares, C. 1998a, "Time-Domain Analysis of Large-Amplitude Vertical Motions and Wave Loads", *Journal of Ship Research,* Vol. 42, No. 2, pp. 100–113.

Fonseca, N. and Guedes Soares, C. 1998b, "Nonlinear Wave Induced Responses of Ships in Irregular Seas", *Proceedings 17th International Conference on Offshore Mechanics and Arctic Engineering (OMAE'98)*, ASME, New York, paper 98 0446.

Fonseca, N., Guedes Soares, C. and Pascoal, R. 2001, "Prediction of Wave Induced Loads in Ships in Heavy Weather", *Proceedings of the International Conference on Design and Operation for Abnormal Conditions II*, RINA, London, 6–7 November.

Fonseca, N. and Guedes Soares, C. 2004a, "Experimental Investigation of the Nonlinear Effects on the Vertical Ship Motions and Loads of a Containership in Regular Waves", *Journal of Ship Research*, Vol. 48, No. 2, June pp. 118–147.

Fonseca, N. and Guedes Soares, C. 2004b, "Experimental Investigation of the Nonlinear Effects on the Vertical Ship Motions and Loads of a Containership in Irregular Waves", *Journal of Ship Research*, Vol. 48, No. 2, June pp. 148–167.

Fonseca, N. and Guedes Soares, C. 2005, "Comparison between experimental and numerical results of the nonlinear vertical ship motions and loads on a containership in regular waves", *International Shipbuilding Progress*, 52, nº.1, pp. 57–91.

Fonseca, N., Guedes Soares, C. and Pascoal, R. 2006, "Ship Structural Loads Induced by Abnormal Wave Conditions on a Containership", *J. of Mar. Sci. and Technol.*, Vol. 11, pp. 245–259.

Fonseca, N., Guedes Soares, C. and Pascoal, R. 2007, "Effect of ship length on the vertical bending moments induced by abnormal waves", *Advances in Marine Structures*, Guedes Soares & Das (Eds), Taylor & Francis Group, pp. 23–31.

Fonseca, N., Pascoal, R. and Guedes Soares, C. 2008, "Global Structural Loads Induced by Abnormal Waves and Design Storms on a FPSO", *Journal of Offshore Mechanics and Arctic Engineering*, Vol. 130, Iss. 2, pp. 9.

Fonseca, N., Guedes Soares, C. and Pascoal, R. 2009, "Global Loads on a FPSO Induced by a set of Freak Waves", *Journal of Offshore Mechanics and Arctic Engineering*, Vol. 131, pp. 8.

Guedes Soares, C. 1996, "On the Definition of Rule Requirements for Wave Induced Vertical Bending Moments", *Marine Structures*, Vol. 9, nº 3–4, pp. 409–426.

Guedes Soares, C. 1998, Ship Structural Reliability, *Risk and Reliability in Marine Structures,* Guedes Soares, C., Ed. Balkema; pp. 227–244.

Guedes Soares, C., Cherneva, Z. and Antão E.M. 2003, "Characteristics of Abnormal Waves in North Sea Storm Sea States", *Applied Ocean Research,* Vol. 25, No. 6, pp. 337–344.

Guedes Soares, C., Dogliani, M., Ostergaard, C., Parmentier, G. and Pedersen, P.T. 1996, Reliability Based Ship Structural Design. *Transactions SNAME,* Vol. 104, pp. 357–389.

Guedes Soares, C., Fonseca, N., Pascoal, R., Clauss, G., Schmittner, C. and Hennig, J. 2006, "Analysis of wave induced loads on a FPSO due to abnormal waves", *J. Offshore Mech. and Arct. Eng.*, Vol. 128, No. 3, pp. 241–247.

Guedes Soares, C., Fonseca, N. Pascoal, R. 2006, "Characteristics of abnormal waves and their induced global load effects in marine structures", *Transactions SNAME,* Vol. 114, pp. 140–166.

Guedes Soares, C., Fonseca, N. and Pascoal, R. 2008, "Abnormal Wave Induced Load Effects in Ship Structures", *Journal of Ship Research,* Vol. 52, No. 1, pp. 30–44.

Guedes Soares, C. and Moan, T. 1991, "Modern Uncertainty in the Long Term Distribution of Wave Induced Bending Moments for Fatigue Design of Ship Structures", *Marine Structures,* Vol. 4, pp. 294–315.

Guedes Soares, C. and Schellin, T.E. 1996, Long-Term Distribution of Non-Linear Wave Induced Vertical Bending Moments on a Containership, *Marine Structures,* Vol. 9(3–4), pp. 333–352.

Haver, S. and Karunakaran, D. 1998, "Probabilistic Description of Crest Heights of Ocean Waves", *Proceedings 5th International Workshop on Wave Hindcasting and Forecasting*, Melbourne, Florida.

ITTC, 1978, 15th ITTC Seakeeping Committee Report, *Proceeding of the 15th ITTC,* The Hague.

ITTC, 1981, 16th ITTC Seakeeping Committee Report, *Proceeding of the 16th ITTC,* Leningrad.

Kharif, C. and Pelinovsky, E. 2003. Physical mechanisms of the rogue wave phenomenon, *European Journal of Mechanics B/Fluids* 22, 603–634.

Parunov, J. and Senjanovic, I. 2003, "Incorporating Model Uncertainty in Ship Reliability Analysis", *Transactions SNAME,* Vol. 111, pp. 376–408.

Pastoor, W., Helmers, J.B. and Bitner-Gregersen, E. 2003, "Time simulation of ocean going structures in extreme waves", *Proceedings 22th International Conference on Offshore Mechanics and Arctic Engineering (OMAE'03)*, ASME, Paper OMAE03-37490.

St. Denis, M. and Pierson, W.J. 1953, "On the Motions of Ships in Confused Seas", *Transactions SNAME,* Vol. 61, pp. 302–316.

Torhaug, R., Winterstein, S.R. and Braathen, A. 1998, "Non-linear Ship Loads: Stochastic Models for Extreme Response", *Journal of Ship Research,* Vol. 43, No. 1, pp. 46–55.

Watanabe, I. and Guedes Soares, C. 1999, Comparative Study on Time Domain Analysis of Non-Linear Ship Motions and Loads, *Marine Structures,* Vol. 12, No. 3, pp. 153–170.

Advanced Ship Design for Pollution Prevention – Guedes Soares & Parunov (eds)
© 2010 Taylor & Francis Group, London, ISBN 978-0-415-58477-7

Safety of ice-strengthened shell structures of ships navigating in the Baltic Sea

P. Kujala
Helsinki University of Technology, Espoo, Finland

ABSTRACT: The aim of this paper is to discuss the safety level of a ship hull operating in the Baltic ice conditions. First the typical damage cases are discussed, then the limit state equations for the permanent deflection of the plating and for the development of the three plastic hinge mechanism in the frame are presented. These equations are used for the safety index analysis, which is applied for transversely framed plating and transverse frames at the bow part of a typical ice-strengthened vessel. Long term ice load distributions are based on the full scale measurements.

1 INTRODUCTION

Extreme consequences, such as loss of life, environmental damage and ship losses, that can be the result of severe structural damages in ice, are almost non-existent in the Baltic Sea. These problems have been practically overcome e.g. by the development of the winter navigation system, composed of reasonable ice class rules (regarding the strengthening of the ship hull structure against ice loads), traffic restrictions and efficient icebreaker assistance.

The basis of the present FSICR (Finnish-Swedish ice class rules) is in the feedback from damages sustained by high ice class ships in the 60's. At this time the year-round navigation to also the northernmost Finnish ports started and many ships had ice damage. Insurance companies found the situation intolerable and changes were required. The change in the winter navigation system was in the ice rules—these were updated based on calculation of loads causing the damages. The design point was selected on the boundary of the damage—non damage in the plots of the load carrying capacity of the shell structure versus the ship size.

At present the further major updating process of the FSICR should be based on risks, starting from deciding the design point to be used. Once the design point is set, the analysis of response must be carried out. If the design point is based on allowing some plasticity, the development of response equations especially for framing becomes a demanding task. These two tasks (design point and response formulation) must be carried out before any progress towards scantlings can be made.

In this paper few examples of risk analysis are presented having the plastic collapse of shell plating as limit states. The long term loads needed in the analysis are based on the long-term full scale measurements.

2 DEFINITION OF THE DESIGN POINT

The typical damage case in ice is a permanent dent of the plating or frames on the ice-strengthened region or just below or above the ice-strengthened region. Typically the dents below the ice-strengthened region are caused by the ice-breaking process when ships navigate in ballast conditions. Whereas the dents above the ice-strengthened region are caused e.g. by the bow wave which can increase the contact height between ship and ice.

Figure 1 illustrates typical damages observed on ships navigating in the Baltic ice conditions. The ice induced loads can be considered as some kind of accidental loads. Ships navigate most of the time in open water as the winter season is usually fairly short i.e. few months. Therefore the principle of the ice class rules is that small yielding and even plastic deformations are allowed on the shell structures when ships navigate in ice. First yielding is a clear design point and it is also easy to determine by straightforward elastic response calculation of the shell structures applying e.g. linear finite element modeling. After yielding, plastic deformations start to take place and this requires analysis of the nonlinear response of the structure. It is not so clear what is a good design point when some plasticity is allowed. In the following, the risk for permanent deformation of the shell plating is analysed to find out what is the level of safety according to the present FSICR approach.

Figure 1. Typical ice induced damages on the shell structures of an ice-strengthened ships (Kujala, 1991, Hänninen, 2004).

3 DEFINITION OF THE LIMIT STATE FUNCTION FOR ICE DAMAGE ON THE PLATING

As can be seen from Fig. 1, ice induced loads typically cause some permanent deflections on the ice-strengthened plating and frames, the depth of the dents are typically 30–100 mm (Kujala, 1991).

In estimation of the load causing permanent deflections in the side plating, an approach developed by Hayward (2001) is used. The approach is based on extensive finite element calculations to find out a correction f_{DT}, which takes into account the effect of the load height on the permanent deflection. The starting point for the analytical expressions is the formulations developed by Jones (1972) considering yield line theory for uniform pressure on plating. So the Hayward (2001) approach for the required line load, q, has the following form when $w_p/t \leq 1$:

$$q = \frac{p_c h_c}{f_{DT}}\left[1+\frac{w_p{}^2}{3t^2}\left(\frac{\zeta_0+(3-2\zeta_0)^2}{3-\zeta_0}\right)\right] \quad (1)$$

and when $w_p/t > 1$:

$$q = \frac{2p_c h_c w_p}{tf_{DT}}\left[1+\frac{\zeta_0(2-\zeta_0)}{3-\zeta_0}\left(\frac{t^2}{3w_p{}^2}-1\right)\right] \quad (2)$$

where t is the plate thickness, h_c is the load height, w_p is the permanent deflection in the plating, σ_y is the yield strength of the plate material. The threshold pressure p_c causing double Y-shaped yield line is (Jones, 1972):

$$p_c = \frac{48M_p}{s^2\left(\sqrt{3+\left(\frac{s}{l}\right)^2}-\left(\frac{s}{l}\right)\right)^2} \quad (3)$$

where s is frame spacing, l is the frame span, M_p is the plastic moment of the plating:

$$M_p = \sigma_y \frac{t^2}{4} \quad (4)$$

the shape parameter, ζ_0, has the following form:

$$\zeta_0 = \frac{s}{l}\left(\sqrt{3+\frac{s^2}{l^2}}-\frac{s}{l}\right) \quad (5)$$

The correction factor f_{DT} in equations (1) and (2) has the following form:

$$f_{DT} = -0.1330x_T{}^2 + 0.6701x_T \quad (6)$$

where

$$x_T = \frac{h_c}{s}\left(\frac{s}{t}\right)^{0.2} \quad (7)$$

4 DEFINITION OF THE LONG TERM ICE LOAD

The main challenge for reliable risk based analysis is the determination of long term ice loads. The ice loads have a strong statistical natures with a lot of affecting parameters such as: time in ice, variation in ice conditions and variations in the operations principles of the ship in ice.

The long term loads used in the analysis are based on the long term measurements onboard MT Kemira (Kujala, 1989a, Muhonen 1991). It is assumed that the line load follows the Gumbel I distribution. Cumulative distribution function, CDF, of the Gumbel I distribution $G(y_n)$ is (Ang and Tang, 1985):

$$G(y_n) = e^{-e^{-c(y_n-u)}} \tag{8}$$

where c is the inverse measure of dispersion of the measured maxima, u is the characteristic largest value and y_n is the extreme value at the measured 12 hour intervals. The parameters c and u and can be determined once the mean, μ_{yn}, and standard deviation, σ_{yn}, of the measured 12 hour maximum values are known:

$$c = \frac{\pi / \sqrt{6}}{\sigma_{y_n}}$$
$$u = \mu_{y_n} - \frac{\sqrt{6}}{\pi}\gamma\sigma_{y_n} \tag{9}$$

where γ is the so called Euler constant and has the numerical value of 0.577.

The long term extreme value distributions after N events can now be determined from the relationship:

$$G(y_n) = \left(e^{-e^{-c(y_n-u)}}\right)^N \tag{10}$$

Usually the long term loads are presented as a function of the so called return period, $T(y_n)$, which defines the required time in ice [days] to achieve the estimated long term load level and it can be solved from the relationship:

$$T(y_n) = 0.5\frac{1}{1-G(y_n)} \tag{11}$$

when the 12 hour maximum values form the basic measured data base. Figure 2 shows a typical maximum ice load values at the bow frame during one winter. The winter started at January and continued until May. As can be seen from Fig. 2, there is high scatter on the measured 12 hour maximum values as the ice conditions and ship operation events vary a lot during each voyage in ice.

Figure 3 shows typical long ice load distribution, which is based on the long term measurements onboard MT Kemira during winters 1985–1991 (Kujala et al., 2007a). Figure 3 gives the measured load at the bow frame of the ship and the load is divided by the frame spacing of the ship, 0.35 m, to get the line load/m. The Gumbel parameters are in this case u = 97.87 kN/m and c = 0.01256 (Muhonen, 1991). As can be seen the Gumbel fits very well on the measured data up to the return period level of 20–30 days. There is higher scatter with the high load values and longer return periods and this due to the fact the highest load typically takes place in an extreme

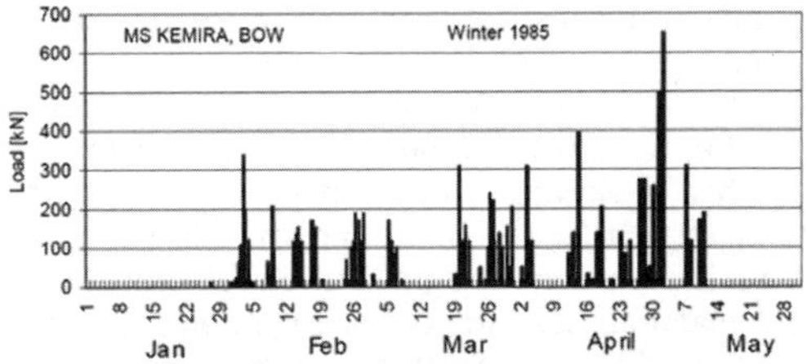

Figure 2. Typical variation of 12 hour maximum load values at the bow frame of MT Kemira when navigating in ice (Kujala, 1994).

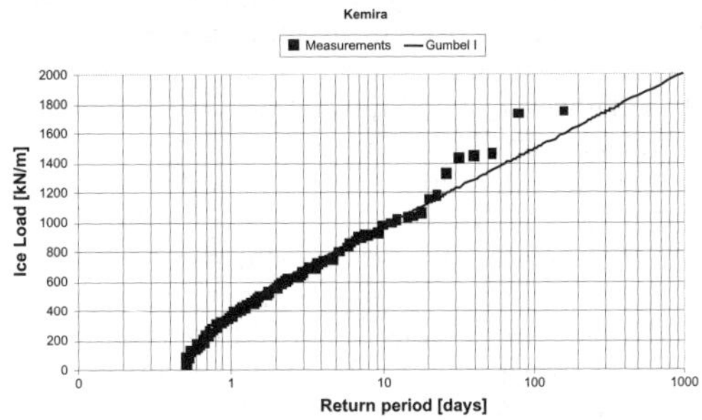

Figure 3. Measured long term ice loads onboard MT Kemira and Gumbel 1 distribution fitted on the data from the winters 1985–1990 (Muhonen, 1991, Kujala et al., 2007a).

ice condition e.g. when we have heavy ridges or moving thick ice. These occur seldom and therefore the distribution fitted on the long term data can underestimate these rare events. It is believed, however, that the fitted distribution estimate well the basic load events during ship's lifetime and also the measured distribution gets closer the fitted distribution if there have been longer measuring period.

5 CALCULATION OF THE SAFETY LEVEL

Based on the so called safety index approach as described in further detail e.g. by Kujala (2008), the safety level of for the permanent deflection of the shell plating of MT Kemira is determined here when MT Kemira operates in ice. The ship life time of 25 years is used to determine the long term loads distributions. The ship is sailing between Kokkola and some European ports. In this period the ship will be about 980 days in ice from which about 200 days on the Gulf of Finland. The basic equations to determine the safety index require that the distributions are normally distributed. In the case of non-normal distributions such as

Gumbel 1, the distributions are presented as the so called equivalent normal distributions at the failure point., see e.g. Ang and Tang (1985), Kujala (1989b), Kujala et al. (2007b) for further details.

When looking at the used equations for the strength part (eq. 1–7), we can see that the equations include load height, structural dimensions and material yield strength as variables. As analysed by Kujala (1989b), the statistical variation for the structural dimensions are small compared to the loads and material yield strength, therefore these are taken here as deterministic variables. The actual height of the ice load during the ice-breaking is naturally varying a lot, unfortunately there are not many measured observations of the load height to determine the statistical characteristics of it. However, it is known that the load height is typically small and the assumption to take the ice load as a line like loading is widely accepted (Kujala et al., 2007b).

Consequently, the main statistical variables in this study are the yield strength and the ice load. The used ice load distribution is shown in Figure 3 for the Bay of Bothnia. The mean value of the yield strength is assumed to be 290 MPa and the standard deviation is 22.40 MPa (Kujala, 1989b). These values are based on the measured data by the steel manufacturer Rautaruukki. The strength is assumed to be normally distributed as the normal distribution gave the best statistical fit on the measured yield strength database. The plate thickness used for the bow of MT Kemira has been 20 mm and the transverse frames at the bow are assumed to be HP 260*10 profiles with a frame spacing of 350 mm. The span of the frames is assumed to be 3.2 m. These are the minimum requirements for 1A Super class ship. The ice load is assumed to be line like with the load height of 0.075 m (Kujala et al., 2007b). With these parameters, the limit state equations (1–2) get the following forms as a function of the relative permanent deflection w_p/t:

a) plating with $w_p/t \leq 1$:

$$0.00472\sigma_y(1+\frac{w_p^{\ 2}}{3t^2}2.5407)-q_{ice}=0 \qquad (12)$$

b) plating $w_p/t > 1$

$$0.00944\sigma_y\frac{w_p}{t}(1+0.11483\left(\frac{t^2}{3w_p^2}-1\right)-q_{ice}=0 \qquad (13)$$

where q_{ice} is the ice-induced line load (see e.g. Figure 3). An example of the obtained distributions for the strength and load is shown in Figure 4 for the plating.

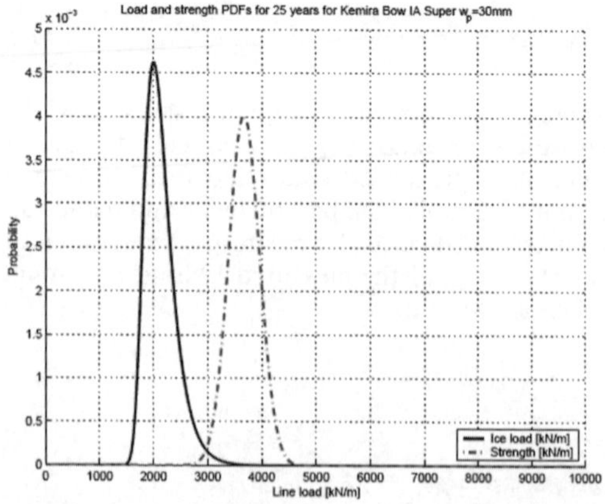

Figure 4. Obtained distributions for the strength and ice induced load.

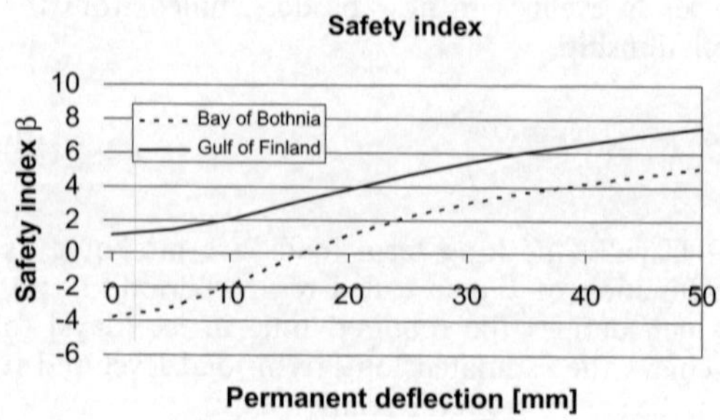

Figure 5. Obtained safety index as a function of the permanent deflection of the plating (Kujala et al., 2007b).

In the case of Figure 4, the permanent deflection of 30 mm is used as the limit state of the plating. Once the permanent deflection is used as a parameter, the obtained safety indices are shown in Figure 5. The safety index is calculated using the whole measured data up to the Bay of Bothnia as one case and only the measured data on the Gulf of Finland as the second case. In the Gulf of Finland case, Gumbel parameters are u = 92.8 kN/m and c = 0.0187 for the bow frame (Muhonen, 1991).

Figure 5 clearly indicates that small dents on the plating take place frequently when ships navigate in ice, especially when navigating in the northern Baltic Sea. If safety index 2 is considered as a good level, this is achieved by using permanent deflection of 25 mm as the limit state. This means w_p/t relation getting the value of 1.25 which seems to be a fairly reasonable level.

6 CONCLUSIONS

Level 2 reliability analysis is conducted to study the probability of ice induced damages on a typical ice-strengthened vessel navigating in ice in the Baltic Sea. Dents on the plating with varying depth are used as limit states for the plating. The measured long term ice loads onboard MT Kemira is used to represent the statistical distribution of ice induced loads.

The results show that small dents take frequently place on the plating when navigating in ice, especially in the northern part of the Baltic Sea. However, the level of safety seems to be adequate as the level 2 of the safety index β is achieved with permanent deflection w_p/t getting the value of 1.25.

The level 2 approach gives a good basis to study the proper level of ice-strengthening for ships on the various sea areas of the Baltic Sea. The statistical approach gives reliable basis to determine proper scantlings for the shell structures of ice-strengthened vessels so that the adequate safety is achieved.

At present the further major updating process of the FSICR should be based on risks, starting from deciding the design point to be used. Once the design point is set, the analysis of response must be carried out. If the design point is based on allowing some plasticity, the development of response equations especially for framing becomes a demanding task. These two tasks (design point and response formulation) must be carried out before any progress towards scantlings can be made.

It is expected that the arctic shipping will increase remarkably in the near future when the oil and gas exploration moves to the arctic areas. The present ice strengthening principles are based on deterministic approach without proper analysis of the environmental effects together with the statistical nature of the ice induced loads and structural response. In future also the ice strengthening should be based on the risks, starting from deciding the design point to be used and continuing to the analysis of response and finally aiming to the proper Goal Based Standard (GBS) type approach.

ACKNOWLEDGMENTS

This paper presents part of the teaching material prepared and used in the scope of the course "Modelling of Environment and Environmental Loads" within the project "Advanced Ship Design for Pollution Prevention (ASDEPP)" that was financed by the European Commission through the Tempus contract JEP-40037-2005.

REFERENCES

Ang, A.H.-S. and Tang W.H., 1985. *Probability concepts in engineering planning and design. Vol. 2—Decision, risk and reliability*. New York, John Wiley and sons. 562 p.

Hayward, R., 2001. *Plastic response of ships shell plating subjected to loads of finite-height*. Master's Thesis, Memorial University of Newfoundland, St. John's, New Foundland, Canada.

Hänninen, S., 2004. *Incidents and Accidents in Winter Navigation in the Baltic Sea*, Winter 2002–2003. Winter Navigation Research Board, Research Report No. 54.

Jones, N., 1972. Review of the plastic behavior of beams and plates. *International Shipbuilding Progress*, 19(218), 313–327.

Kujala P., 1989a. *Results of Long-Term Ice Load Measurements on Board Chemical Tanker Kemira in the Baltic Sea during the winters 1985 to 1988*, Winter Navigation Research Board, No. 47, Espoo, 55 p. + Appendixes.

Kujala, P., 1989b. *Probability based safety of ice-strengthened ship hull in the Baltic sea*. Licenthiate's Thesis, Helsinki University of Technology, Espoo, Finland, 109 p. + app. 14p.

Kujala, P., 1991. *Damage statistics of ice-strengthened ships in the Baltic Sea 1984–1987*. Winter Navigation Research Board. Report. No. 50. 61 p. + app. 5 p.

Kujala, P., 1994. *On the statistics of ice loads on ship hull in the Baltic. Dissertation. Acta Polytechina Scandinavica*, Mechanical Engineering Series No. 116. Helsinki. 98 p.

Kujala, P., Valkonen, J. and Suominen, M., 2007a. Maximum ice induced loads on ships in the Baltic Sea. *PRADS'07. The 10th International Symposium on Practical Design of Ships and Other Floating Structures, Houston, October 1–5.*

Kujala, P., Suominen, M. and Jalonen, R., 2007b. (Edited) *Increasing the safety if ice-bound shipping. EU-funded SAFEICE project, Final scientific report*, Vol. 1. Helsinki University of Technology, Ship Laboratory. Resarch report M-302. Otaniemi. 424 p.

Kujala, P., 2008. Reliability of ice-strengthened shell structures of ships navigating in the Baltic Sea. *Journal of Structural Mechanics*, Vol. 41, No. 2:108–118.

Muhonen, A., 1991. *Ice Load Measurements on Board the MS Kemira, Winter 1990. Number 109 in M-Series*. Helsinki University of Technology, Faculty of Mechanical Engineering, Laboratory of Naval Architecture and Marine Engineering, Espoo, Finland.

Advanced Ship Design for Pollution Prevention – Guedes Soares & Parunov (eds)

Some aspects of quasi static hydro structure interactions in seakeeping

F.X. Sireta, Š. Malenica, F. Bigot, Q. Derbanne, V. Hsu & X.B. Chen
Bureau Veritas Research Department, Neuilly sur Seines, France

ABSTRACT: The paper deals with the methods for evaluation of the quasi static ship structural response under the action of the sea waves. Coupling in between the seakeeping code based on potential flow theory (Hydrostar) and the general 3DFEM structural code (Nastran) is discussed. Several types of application are concerned namely: linear, nonlinear, frequency domain, time domain, partial and complete structural models, internal tanks... The theory is followed by the practical examples of validation on some typical merchant ships.

1 INTRODUCTION

The difficulties related to the balancing of the 3D FEM structural model, in the context of hydro-structure interactions in linear seakeeping are discussed. Different philosophies in modeling the structural and the hydrodynamic parts of the problem usually lead to very different meshes (hydro and structure) which results in an unbalanced structural model and consequently in doubtful results for structural responses. The procedure usually employed consists in using different kinds of interpolation schemes to transfer the total hydrodynamic pressure from the centroids of hydrodynamic panels to the structural finite elements. This approach is both very complex for complicated geometries and also rather inaccurate. The method proposed here is based on two main ideas:

1. Pressure recalculation instead of interpolation
2. Separate transfer of different pressure components (incident, diffraction, radiation & hydrostatic variation).

The first point removes the difficulties related to the interpolation techniques, and allows for a very robust method of pressure transfer. The second point ensures the perfect equilibrium because the body motions are calculated after integration over the structural mesh, which means that the equilibrium is implicitly imposed.

This method is first applied on a complete 3D FEM model for the transfer of linear and nonlinear seakeeping loads. The robustness and flexibility of the method make it easily adaptable not only to the case of partial 3D FEM model but also to the transfer of coupled seakeeping and sloshing loads.

It should be noted that this procedure is not completely straightforward and that several numerical "tricks" need to be introduced. However, once these difficulties are solved, the final numerical code is extremely robust and can be easily coupled with any of the general 3D FEM packages.

2 LINEAR SEAKEEPING

First we briefly recall the basics of the linear and non linear seakeeping problem in frequency and time domain.

2.1 *Linear seakeeping analysis in frequency domain*

The linear seakeeping problem is formulated in frequency domain under the potential flow assumptions. The total velocity potential is decomposed into the incident, diffracted and 6 radiated components:

$$\varphi = \varphi_I + \varphi_D - i\omega \sum_{j=1}^{6} \xi_j \varphi_{Rj} \tag{1}$$

where:

ϕ_I incident potential
ϕ_Δ diffraction potential
$\phi_{P\varphi}$ radiation potential
ξ_φ *j*-th rigid body motions

Within the Bureau Veritas's numerical code HYDROSTAR, the Boundary Integral Equation (BIE) method based on source formulation, is used to solve the Boundary Value Problems (BVP) for different potentials (see Chen (2004)). In the case of zero forward speed, the general form of the BVP is:

$$\left.\begin{array}{lll} \Delta\varphi = 0 & & \text{in the fluid} \\ -\nu\varphi + \dfrac{\partial\varphi}{\partial z} = 0 & & z = 0 \\ \dfrac{\partial\varphi}{\partial z} = V_n & & \text{on } S_b \\ \lim = \left[\sqrt{\nu R}\left(\dfrac{\partial\varphi}{\partial R} - i\nu\varphi\right)\right] = 0 & R \to \infty & \end{array}\right\} \quad (2)$$

where V_n denotes the body boundary condition which depends on the considered potential:

$$\frac{\partial\varphi_D}{\partial n} = -\frac{\partial\varphi_I}{\partial n}, \quad \frac{\partial\varphi_{Rj}}{\partial n} = n_j \tag{3}$$

Within the source formulation, the potential at any point in the fluid can be expressed in the following form:

$$\varphi = \iint_{S_B^H} \sigma G dS \tag{4}$$

where G stands for the Green function, and σ is the unknown source strength which is found after solving the following integral equation:

$$\frac{1}{2}\sigma + \iint_{S_B^H} \sigma \frac{\partial G}{\partial n} dS = V_n \tag{5}$$

It should be noted that the BVP is solved using the hydrodynamic mesh as a boundary mesh.

2.1.1 *Loading of the structural model*

As already briefly mentioned, in the present approach the calculation of the hydrodynamic coefficients (added mass, damping, hydrostatic variation, excitation) is performed after integration over the wetted part of the structural FEM model and not over the hydrodynamic model. Here below we briefly explain the basic principles of the pressure transfer and integration method.

2.1.1.1 Hydrodynamic pressure

What is proposed here is to transfer the pressure from hydrodynamic model to the structural model by recalculating the pressure at the required locations instead of interpolating it from hydrodynamic model. This is possible thanks to the particularities of the BIE method which gives the continuous representation of the potential through the whole fluid domain $Z < 0$. In this way the communication between the hydrodynamic and structural codes is extremely simplified. Indeed, it is enough for the structural code to give the coordinates of the points where the potential is required and the hydrodynamic code just evaluates the corresponding potential (from which the pressure can be easily obtained) by:

$$\varphi(x_s) = \iint_{S_B^H} \sigma(x_h) G(x_h; x_s) dS \tag{6}$$

where $\boldsymbol{x}_s = (x_s, y_s, z_s)$ denotes the structural point and $\boldsymbol{x}_h = (x_h, y_h, z_h)$ the hydrodynamic point. In the case of linear seakeeping without forward speed, this operation is sufficient because the pressure is directly proportional to the velocity potential and, within the source formulation, the potential is continuous across the body wetted surface. This is a very important point because, due to the differences in the hydrodynamic and structural mesh, the structural points might fall inside the hydrodynamic mesh. Once each pressure component has been transferred onto the structural mesh, the excitation, added mass and damping coefficients are calculated by integration over the structural mesh:

$$F_i^{DI^S} = i\omega\rho \iint_{S_B^S} \left(\varphi_I^S + \varphi_D^S\right) n_i dS \tag{7}$$

$$\omega^2 A_{ij}^S + i\omega B_{ij}^S = \rho\omega^2 \iint_{S_B^S} \varphi_{Rj}^S n_i dS \tag{8}$$

Let us also note that different ways of applying the pressure loads on the FEM model can be used depending on the particularities of the FEM packages. In the present approach the pressure is calculated at the Gauss points of the finite element and integrated (inside the interface code) in order to obtain the corresponding nodal forces which are used as a final loading of the structural model. This choice was made to ensure the compatibility with any general FEM package.

2.1.1.2 Hydrostatic pressure variations

Here we concentrate on the calculation of the hydrostatic restoring matrix which is obtained after the integration of the hydrostatic pressure variations due to the body motions. The procedure is rather similar to the one for hydrodynamic pressure, and we just need to integrate over the structural mesh. For the sake of clarity, let us first rewrite the hydrostatic pressure variations in the following compact form:

$$p^{hs} = \sum_{j=1}^{6} \xi_j p_j^{hs} \tag{9}$$

where

$$p_j^{hs} = 0, \quad j = 1, 2, 6 \tag{10}$$

$$p_3^{hs} = -\rho g \tag{11}$$

$$p_4^{hs} = -\rho g\left(Y - Y_G\right) \tag{12}$$

$$p_5^{hs} = \rho g\left(X - X_G\right) \tag{13}$$

With these notations, the first part of the hydrostatic restoring matrix becomes:

$$C_{ij}^p = \iint_{S_B^S} p_j^{hs} n_i dS \tag{14}$$

In order to obtain the complete hydrostatic restoring matrix, one additional term accounting for the change of coordinate system should be added to the above expression. In the earth fixed coordinate system, this additional term is accounted for by the change of the normal vector. However, the structural response is calculated in the body fixed coordinate system in which the normal vector does not change. It can be shown (e.g. see Malenica (2003)) that, in the body fixed coordinate system, the change of the normal vector is equivalent to the change of the gravity action, so that we can write:

$$F^g = -mg\Omega \wedge k = [C]^g \{\xi\} \tag{15}$$

where the only non zero elements of the matrix $[C]^g$ are (2, 4) and (1, 5) and they will be canceled by the contributions implicitly present in $[C]^p$. The total restoring matrix becomes:

$$[C]^S = [C]^P + [C]^g \tag{16}$$

where the superscript "S" indicates that the pressure related part was calculated by integration over the structural mesh.

2.1.1.3 Motion equation and final loading of the structural model

The rigid body motion equation, in frequency domain, is now written in the following form:

$$\left(-\omega^2\left([M]+[A]^S\right) - i\omega[B]^S + [C]^S\right) \times \{\xi\}^S = \{F^{DI}\}^S \tag{17}$$

Solving of this equation gives the body motions vector $\{\xi\}^S$ so that the total linear pressure can be written as:

$$p^S = p_I^S + p_D^S + \sum_{j=1}^{6} \xi_j^S \left(p_{Rj}^S + p_j^{hs}\right) \tag{18}$$

In summary, the final loading of the structural model will be composed of the following 3 parts:

- $-\omega^2 m_i \xi_i^S$ • Inertial loading
- p_i^S • Pressure loading
- $-m_i g\Omega^S \wedge k$ • Gravity term

The inertial loading and the gravity term are to be applied on each finite element. The pressure loading is to be applied only on wetted finite elements. It is clear that the above structural loading will be in perfect equilibrium because this equilibrium is implicitly imposed by the solution of the motion equation in which all different coefficients were calculated by using directly the information from the structural FEM model.

2.1.1.4 A few results

Figure 1 to 3 show few numerical results obtained after transferring linear seakeeping loads from HYDROSTAR to NX NASTRAN using the method described above.

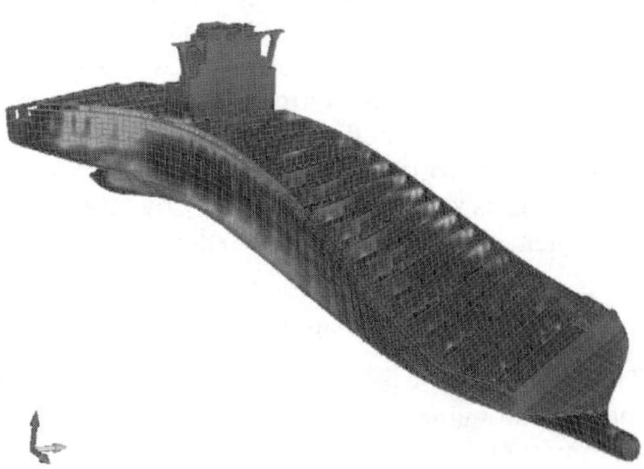

Figure 1. Von Mises stresses on Container ship.

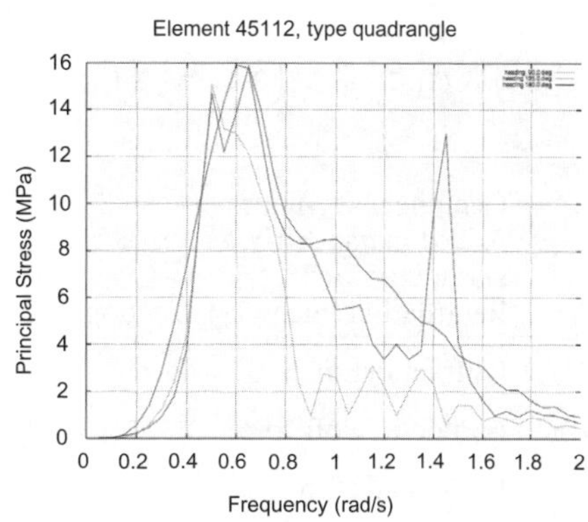

Figure 2. RAO of the stress at a particular finite element for different wave headings.

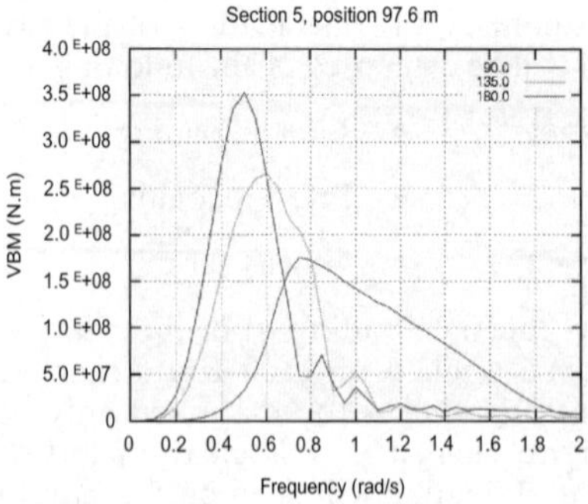

Figure 3. RAO of the vertical bending moment at midship section for different wave headings.

3 NONLINEAR SEAKEEPING

Linear seakeeping analysis can be used with fair accuracy as long as the wave height remains small. For calculation of the structural response in extreme sea states it is necessary to upgrade the linear model by the introduction of the relevant non linear effects. Due to the complexity of the hydrodynamic problem, it is impossible to include all the nonlinear effects in a fully consistent way. As far as the relatively long waves are concerned, the most important nonlinear effect is believed to be related to the nonlinearities included in the incident wave. This model is known as Froude Krylov model and its implementation is explained below.

3.1 *Non linear seakeeping model*

The non-linear seakeeping problem is solved in the time domain using the method proposed by Cummings (Cummings (1962), Ogilvie (1964)):

$$(A(\infty)+m)\cdot\ddot{\xi}+B(\infty)\cdot\dot{\xi}+\int_{-\infty}^{t}K(t-\tau)\cdot\dot{\xi}d\tau = \vec{f}_D\left(\vec{\xi},t\right)+\vec{f}_{fk}\left(\vec{\xi},t\right)+\vec{f}_h\left(\vec{\xi}\right)+\vec{f}_g\left(\vec{\xi}\right) \quad (19)$$

where:

m	Genuine mass matrix
A	Added mass matrix
B	Damping matrix
K	Retardation (memory) functions
$\vec{f}_D$	Diffraction force
$\vec{f}_{fk}$	Froude Kryloff force
$\vec{f}_h$	Hydrostatic force
$\vec{f}_g$	Gravity force
$\vec{\xi}$	Displacements

It has been shown in (e.g. Ogilvie (1964)) that the retardation functions can be calculated using the frequency domain hydrodynamic coefficients:

$$K(t)=\frac{2}{\pi}\int_0^{\infty}\left(B(\omega_e)-B(\infty)\right)\cdot\cos(\omega_e\cdot t)d\omega_e \quad (20)$$

The right hand side of the above equation consists of the different excitation forces: diffraction, Froude Krylov, hydrostatics and gravity. As mentioned above the main nonlinear part is related to the incident wave nonlinearities. Here below we explain how this nonlinearity is included in the above model.

The wave elevation of a single wave component at a location is equal to:

$$\zeta_i(t,x_w,\omega)=\zeta_a(\omega_i)\cos\left(\omega_i t+\varepsilon_\zeta(\omega_i)+x_w k(\omega_i)\right) \quad (21)$$

where:

$\zeta\alpha$	Wave amplitude
ε	Phase angle
κ	Wave number

This results in a total wave elevation of:

$$\zeta(t,x_w)=\sum_{i=1}^{Nfreq}\zeta_i(t,x_w,\omega) \quad (22)$$

The Froude-Krylov force is:

$$\begin{cases} p_{fk}=\sum_{i=1}^{Nfreq}\rho g\zeta_i(t,x_w,\omega)e^{k(\omega_i)z} & z<0 \text{ and } z<\zeta(t,x_w) \\ p_{fk}=\rho g\zeta(t,x_w)\left(1-\frac{z}{\zeta(t,x_w)}\right) & z\geq 0 \text{ and } z<\zeta(t,x_w) \\ p_{fk}=0 & z\geq\zeta(t,x_w) \end{cases} \quad (23)$$

The hydrostatic pressure is equal to:

$$\begin{cases} p_h=-\rho g z & z<0 \; and \; z<\zeta(t,x_w) \\ p_h=0 & z\geq 0 \; or \; z\geq\zeta(t,x_w) \end{cases} \quad (24)$$

The diffraction force is linear and equal to:

$$\vec{f}_D(t,x_w)=\sum_{i=1}^{Nfreq}\zeta_a(\omega_i)\left(\Re\left(\vec{F}_{D(i)}\right)\cos\left(\omega_i t+\varepsilon_\zeta(\omega_i)+x_w k(\omega_i)\right)+\Im\left(\vec{F}_{D(i)}\right)\sin\left(\omega_i t+\varepsilon_\zeta(\omega_i)+x_w k(\omega_i)\right)\right) \quad (25)$$

Finally, the gravity force is the mass times gravity acceleration vector expressed in the body fixed reference system.

3.2 *Load transfer to structural model*

In principle it would be possible to generate the structural load cases directly during the seakeeping calculations, but there are two major disadvantages using that approach:

- The retardation function and convolution integral has to be evaluated for every point at the structural mesh. This will be a significant overhead.
- Usually only a small part of the time domain calculation is selected to perform a structural analysis. This is usually the interval where the most extreme loading occurs. Using a post-processor allows to select these events.

That is why it is more efficient to create the loading cases in a post processing phase after the global body motions are known. Here below we explain how this is done within the present model.

3.2.1 *Load case*

The balance between the hydrodynamic loading and acceleration forces at the structural model is ensured by solving the dynamic motion equation again. The nodal contributions of all hydrodynamic force components are first calculated. These forces are based on the displacements and velocity calculated by the seakeeping calculation. The acceleration is (re)solved using the following expression:

$$\left(A(\infty)+m\right)\cdot \ddot{\vec{\xi}}(t)=\vec{f}_g\left(\vec{\xi}(t)\right) + \sum_{i=1}^{nnodes}\left(\vec{f}^{\,n}_{D,i}\left(t,\vec{\xi}\right)+\vec{f}^{\,n}_{R,i}\left(\dot{\vec{\xi}}(t)\right) + \vec{f}^{\,n}_{fk,i}\left(t,\vec{\xi}\right)+\vec{f}^{\,n}_{h,i}\left(t,\vec{\xi}\right)\right) \tag{26}$$

The superscript n indicates force at the nodes of the structural model. The infinite added mass is still at the left hand side to ensure that the hydrodynamic loading at the structural model is consistent with the accelerations.

The total hydrodynamic loading at the structural model is calculated after solving the motion equation:

$$\vec{f}^{\,n}_{t,i}=\vec{f}^{\,n}_{D,i}\left(t,\vec{\xi}\right)+\vec{f}^{\,n}_{R,i}\left(\dot{\vec{\xi}}(t)\right)+\vec{f}^{\,n}_{A(\infty),i}\left(\ddot{\vec{\xi}}\right) + \vec{f}^{\,n}_{fk,i}\left(t,\vec{\xi}\right)+\vec{f}^{\,n}_{h,i}\left(t,\vec{\xi}\right) \tag{27}$$

Both the diffraction and radiation pressures are integrated over the finite element in order to obtain the nodal forces. The integration is performed using the Gauss quadrature method which means that the pressures have to be evaluated at the Gauss points and not at the nodes or centroids of the finite elements. The radiation force is the most difficult force component to compute. An approach could be to create and evaluate the retardation function for every Gauss point. This approach should result in exactly the same total radiation force as computed by the seakeeping program. However the calculation effort to do this is huge. Another approach could be to use just the frequency domain RAO values. This will not be correct if it is based on the linear motion calculated in the frequency domain. Therefore a Fourier transform of the seakeeping velocity is used to obtain the amplitude and phases at different frequencies. These results in a total radiation force which is very close to the radiation force calculated by the seakeeping program.

The frequency content of the velocity in the earth reference frame is equal to:

$$\dot{\vec{\xi}}^{f}(\omega)=F\,\dot{\vec{\xi}}(t) \tag{28}$$

The nodal force in the body reference frame due to radiation, with the infinite frequency added mass subtracted is:

$$\vec{f}^{\,n}_{R,i}(t)=T_b\cdot\sum_{j=1}^{nfreq}\left\{\Re\left(\vec{f}^{\,fn}_{R,i}(i)-\vec{f}^{\,fn}_{R,i}(\infty)\right)\cdot\Re\left(\dot{\vec{\xi}}^{f}(i)\cdot e^{-i\omega(i)t}\right) + \Im\left(\vec{f}^{\,fn}_{R,i}(i)\right)\cdot\Im\left(\dot{\vec{\xi}}^{f}(i)\cdot e^{-i\omega(i)t}\right)\right\} \tag{29}$$

The nodal force by infinite frequency added mass is:

$$\vec{f}^{\,n}_{A\infty,i}(t) = \mathrm{T}_b \cdot \Re\left(\vec{f}^{\,fn}_{R,i}(\infty)\right) \cdot \ddot{\vec{\xi}}(t) \tag{30}$$

where $\mathrm{T_b}$ is the transformation matrix from the earth system to the body fixed reference system.

The nodal diffraction forces are equal to:

$$\begin{aligned}\vec{f}^{\,n}_{D,i}(t,x_w) = \mathrm{T}_b \sum_{i=1}^{Nfreq} \zeta_a(\omega_i) \Bigg(& \Re\left(\vec{F}^{fn}_{D(i)}\right)\cos\left(\omega_i t + \varepsilon_\zeta(\omega_i) + x_w k(\omega_i)\right) \\ & + \Im\left(\vec{F}^{fn}_{D(i)}\right)\sin\left(\omega_i t + \varepsilon_\zeta(\omega_i) + x_w k(\omega_i)\right)\Bigg)\end{aligned} \tag{31}$$

3.2.2 *A few results*

Figure 4 shows the time history of the stress in the deck of a particular ship, both from linear and nonlinear seakeeping load cases. As expected, the sagging stress is significantly increased in the non-linear case while the hogging stress decreases.

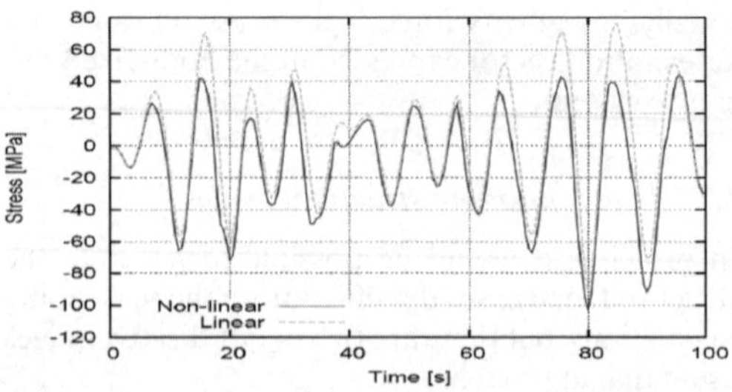

Figure 4. Stress in a deck of the ship.

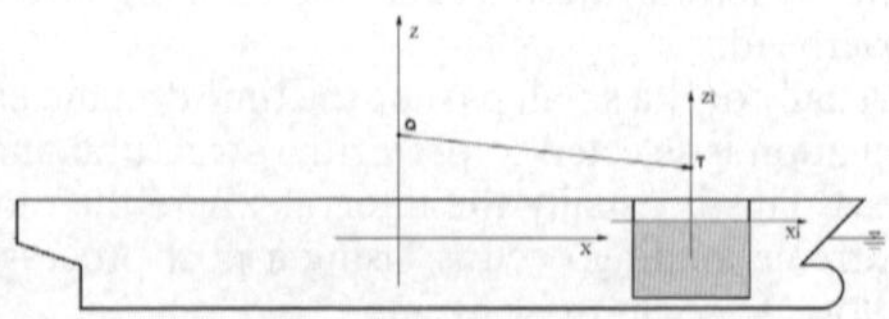
Figure 5. Basic configuration.

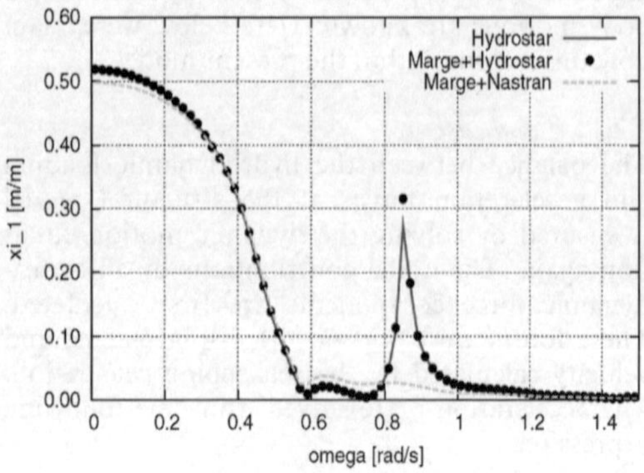

Figure 6. Surge RAO for the uncoupled and coupled problem.

4 COMBINED ACTION OF SEAKEEPING AND SLOSHING

Here below we discuss some particular applications of the quasi static hydro-structure analyses in order to show that the above described procedure can be applied for various practical purposes.

The problem of the loads resulting of the combined action of seakeeping and sloshing is relevant both for the ships transporting liquid in tanks (LNG carriers, tankers, ...) as well as for any ship sailing in ballast conditions.

4.1 *Coupled seakeeping—sloshing analysis*

The approach for sloshing is almost the same as for the simple seakeeping analysis without any tank. The Boundary Value Problems are formulated both for hull and tanks, and solved by HYDROSTAR. After that the hydrodynamic pressures are recalculated at the corresponding structural meshes. The hydrodynamic coefficients for the hull and tanks are obtained by integration of this pressure on the structural mesh and the motion equation is solved to obtain the accelerations.

The linear coupled motion equation in frequency domain is written in the following form:

$$\begin{aligned}&\left(-\omega^2\left(\left[\mathrm{M}_Q\right]+\left[\mathrm{A}_Q\right]+\left[\mathrm{A}_T\right]+\left[\mathrm{A}_{TQ}\right]\right)\right. \\ &\left.-i\omega\left[\mathrm{B}_Q\right]+\left(\left[\mathrm{C}_Q\right]+\left[\mathrm{C}_T\right]+\left[\mathrm{C}_{TQ}\right]\right)\right)\left\{\xi_Q\right\} = \left\{F_Q^{DI}\right\}\end{aligned} \tag{32}$$

- $\{\xi_Q\}$ rigid body ship motions
- $[\mathrm{M_Q}]$ genuine mass matrix of the ship
- $[\mathrm{A_Q}]$ hydrodynamic added mass matrix of the ship
- $[\mathrm{B_Q}]$ hydrodynamic damping matrix of the ship
- $[\mathrm{C_Q}]$ hydrostatic restoring matrix of the ship
- $\{F_Q^{DI}\}$ hydrodynamic wave excitation force
- $[\mathrm{A_T}]$ hydrodynamic added mass matrix of the tank

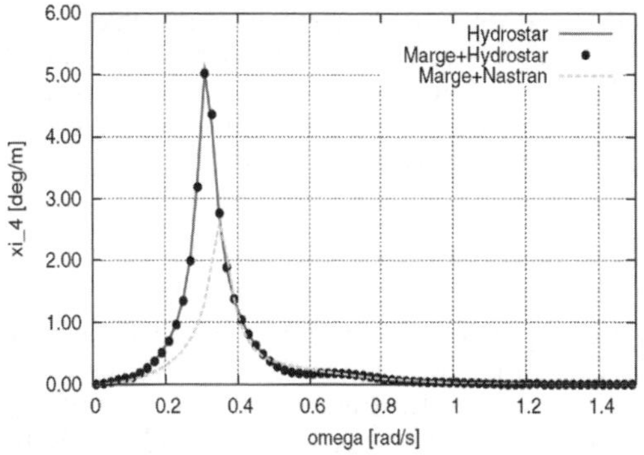

Figure 7. Roll RAO for the uncoupled and coupled problem.

$[A_{TQ}]$ added mass transfer matrix for the tank
$[B_T]$ hydrodynamic damping matrix of the tank
$[B_{TQ}]$ damping transfer matrix for the tank
$[C_T]$ hydrostatic restoring matrix of the tank
$[C_{TQ}]$ hydrostatic restoring transfer matrix of the tank

4.2 *A few results*

For the case of coupling between seakeeping and sloshing, figures 6 and 7 show the surge and roll motion RAO obtained by HYDROSTAR alone and the ones obtained through the coupling method. The surge RAO shows a peak that is due to the tank resonance. The roll resonance is shifted and the peak value is modified by the coupling. Once the seakeeping and sloshing loads are applied to the FE model, the resulting stresses will also be affected by those effects.

5 PARTIAL STRUCTURAL MODEL

Here we concentrate on the so called 3 cargo hold structural model. Indeed, in the preliminary design stage the 3 cargo hold model is often used in order to quickly perform some preliminary checks. The advantage of using 3 cargo hold models is the important reduction of the time necessary to build the FE model and the drawback is the reduced accuracy. The fact that there is a part of the structural model missing introduces some technical difficulties related to the balancing of the FE model. The method that we propose here is based on the construction of an equivalent hybrid full model. This transforms the problem of balancing of a three cargo hold model into the same problem for the corresponding full model. Once the equivalent full model is built, we are back to the case of a full FE model and the usual procedure is used.

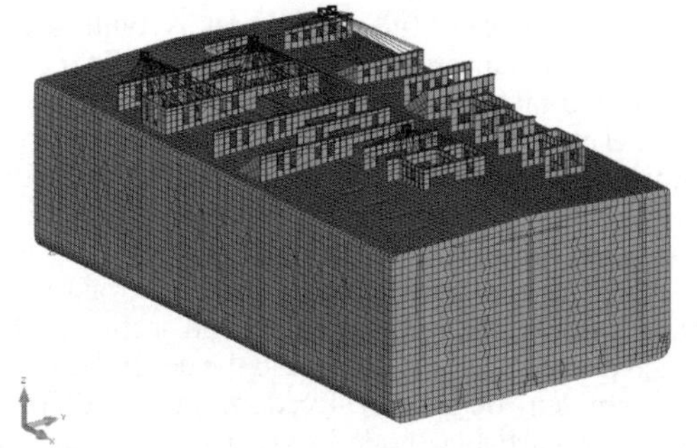

Figure 8. Partial structural model.

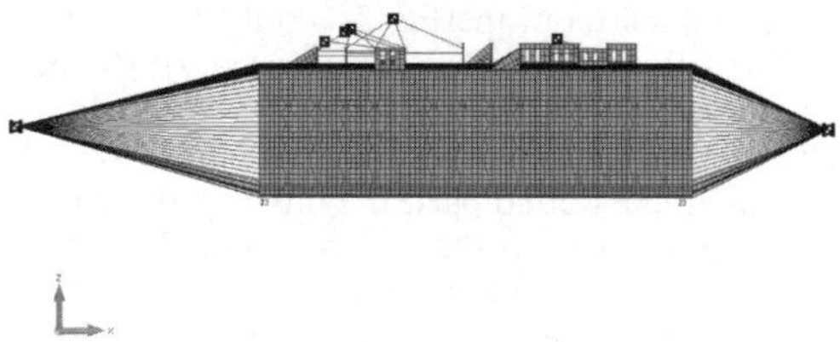

Figure 9. Equivalent full FE structural model.

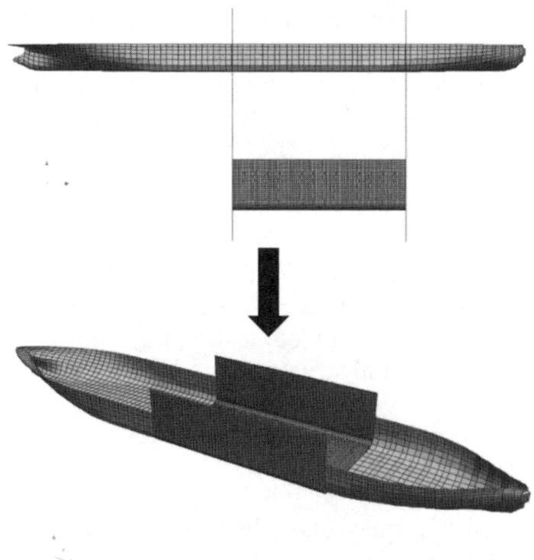

Figure 10. Hybrid wetted part of the FE model.

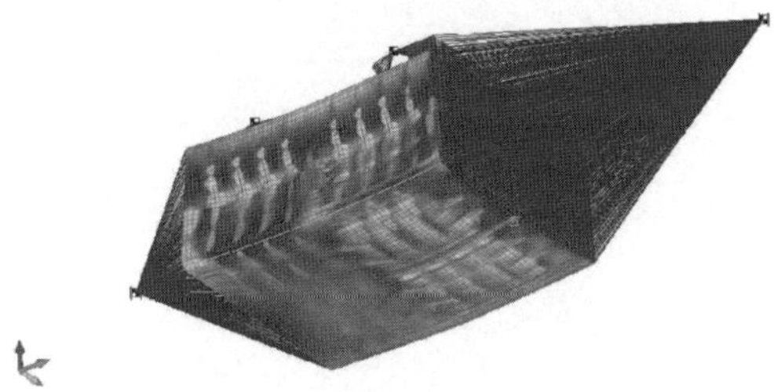

Figure 11. Stresses in the structure for the partial FE model.

The equivalent full FE model is built using concentrated masses and rigid elements. Two concentrated masses are added to the three cargo FE model at the fore and aft part. They are respectively positioned at the center of gravity of the missing parts. The mass and inertia properties of the missing parts are given to these mass points.

Each concentrated mass is linked to a set of nodes of the three cargo model end sections. The rigid elements have no mass and the dependant and independent degrees of freedom can be defined for each particular node in order to represent the physical link with the missing parts of the model. The so build equivalent full FE model has the same mass and inertia properties as a full FE model.

The FE model wetted part is used for the integration of the pressures. In order to build the equivalent full FE model wetted part from the partial model wetted part, the missing fore and aft part are taken from the hydro model. To achieve this, the hydro model is cut at the fore and aft sections of the three cargo hold FE model. The fore and aft parts of the hydro model are then added to the partial FE model wetted part. The force and moment arising from integration of the pressure on the fore and aft parts are then applied on the FEM nodes at the center of gravity of the missing parts. This allows for keeping exactly the same coupling procedure as for the full FE model.

5.1 *FE results*

Figure 11 displays the Von Mises stress on the perfectly balanced partial FE model in the case of the transfer of the coupled seakeeping and sloshing loads on a partial structural model.

6 CONCLUSIONS

A consistent quasi static hydro-structure coupling methodology has been presented. It was demonstrated on typical merchant ships, taking into account their specific loading such as sloshing for FPSOs and LNGs. Once implemented, the method is very robust and can be easily adapted to any commercial FE package.

ACKNOWLEDGMENTS

This paper presents part of the teaching material prepared and used in the scope of the course "Modelling of Environment and Environmental Loads" within the project "Advanced Ship Design for Pollution Prevention (ASDEPP)" that was financed by the European Commission through the Tempus contract JEP-40037-2005.

REFERENCES

Chen, X.B. 2004. Hydrodynamics in offshore and naval applications—Part 1., *6th Intl. Conf. on Hydrodynamics*.

Cummings, W. 1962. The impulse response function and ship motions. *Schiffstechnik*.

Malenica, S. 2003. Some aspects of hydrostatic calculations in linear seakeeping", *NAV Conference*.

Ogilvie, T.F. 1964. Recent progress toward the understanding and prediction of ship motions. *Proc. 5th Symposium on Naval Hydrodynamics*.

Tuitman J.T, Sireta F.X., Malenica S. & Bosman T.N. 2009. Transfer of non-linear seakeeping loads to FEM model using quasi static approach. *ISOPE Conference*.

Advanced Ship Design for Pollution Prevention – Guedes Soares & Parunov (eds)
© 2010 Taylor & Francis Group, London, ISBN 978-0-415-58477-7

Application of an advanced beam theory to ship hydroelastic analysis

I. Senjanović, S. Tomašević, N. Vladimir & M. Tomić
Faculty of Mechanical Engineering and Naval Architecture, University of Zagreb, Zagreb, Croatia

Š. Malenica
Bureau Veritas, Marine Division, France

ABSTRACT: Modern sea transport requires building of Very Large Container Ships (VLCS), which are relative flexible structures. Bearing in mind this fact, and taking into account the speed of VLCS, it is obvious that their natural frequencies could fall into the range of the encounter frequencies in an ordinary sea spectrum. Present Classification Rules for ship design and construction don't cover such conditions completely. This encourages scientists and engineers to develop more powerful and reliable tools for the analysis of ship behavior in seas and to improve the Rules. Hydroelastic analysis of VLCS seems to be appropriate solution for this challenging problem. Methodology of hydroelastic investigation is based on mathematical model which includes structural, hydrostatic and hydrodynamic submodels which are assembled into hydroelastic one. The hydroelastic problem can be solved at different levels of complexity and accuracy. It is obvious that the best way is to consider 3D FEM structural model and 3D hydrodynamic model, but this approach would be too expensive, especially in preliminary design stage. At this level it would be more appropriate to couple 1D FEM model of ship hull with 3D hydrodynamic model. In this paper, the emphasis is given on the advanced beam model which includes shear influence on torsion as an extension of shear influence on bending, and contribution of transverse bulkheads to hull stiffness. Beside structural model, hydrostatic and hydrodynamic submodels, as constitutive parts of hydroelastic model are briefly described. Verification of proposed numerical procedure is done by correlation analysis of the simulation results and the measured ones for flexible barge, for which the test results are available in the literature. Numerical example, which includes complete hydroelastic analysis of 7800 TEU container ship, is also given. In this case, validation of 1D FEM model is checked by correlation analysis with the vibration response of the fine 3D FEM model. The obtained results confirm that advanced thin-walled girder theory is a reasonable choice for determining wave load effects on VLCS, in preliminary design stage.

1 INTRODUCTION

Rapid increase in ship transport induces building of Very Large Container Ships (VLCS), which are relative slender, fast and quite flexible ships. Because of these features, structural natural frequencies of VLCS could fall into the range of encounter frequencies in an ordinary sea spectrum. It is very important to have reliable and powerful design tool to avoid those resonant stages during the navigation.

The classical theories for determination of ship motions and wave loads, as for example (Salvesen et al. 1970), are based on the assumption that the ship hull is a rigid body. Usually, the wave load obtained according to these theories is imposed to the elastic 3D FEM model of ship structure in order to analyze global strength, as well as local strength with stress concentrations related to fatigue analysis. Although the above approach is good enough for ships with closed cross-section and ordinary hatch openings such as tankers, bulk carriers or general cargo ships, it is not reliable as it should be for ultra large container ships due to mutual influence of the wave load and structure response (Senjanović et al. 2009a). Therefore, a more reliable solution requires analysis of wave load and ship vibration as a coupled hydroelastic problem (Bishop & Price, 1979). This is very important for impulsive loads such as ship slamming which causes whipping.

Numerical procedure for ship hydroelastic analysis requires definition of structural model, ship and cargo mass distributions, and geometrical model of ship surface (Senjanović et al. 2007, 2008a, 2009b).

In this paper, the emphasis is given on advanced numerical procedure based on the beam and thin-walled girder theories for calculation of dry natural vibrations of container ships, as an important step in their hydroelastic analysis (Senjanović et al. 2009c). This theory includes shear influence on torsion as an extension of shear influence on bending (Pavazza, 2005), as well as contribution of transverse bulkheads to hull stiffness (Senjanović et al. 2008b), similarly to the contribution of deck transverse strips to hull stiffness (Pavazza et al. 1998). Beside advanced beam theory, methodology of ship hydroelastic analysis is briefly described and illustrated. Also, short description of hydrostatic and hydrodynamic submodels, and hydroelastic model is given. Applied numerical procedure as well as developed computer codes is verified. Finally, the results of hydroelastic analysis of 7800 TEU container ship are given and properly interpreted.

2 METHODOLOGY OF SHIP HYDROELASTIC ANALYSIS

As mentioned before, structural model, ship and cargo mass distributions and geometrical model of ship surface have to be defined to make hydroelastic analysis of the ship. At the beginning of the analysis, dry natural vibrations have to be calculated, and after that modal hydrostatic stiffness, modal added mass, damping and modal wave load are determined. Finally, wet natural vibrations as well as the transfer functions (RAO) for determining ship structural response to wave excitation are obtained (Senjanović et al. 2008a, 2009b).

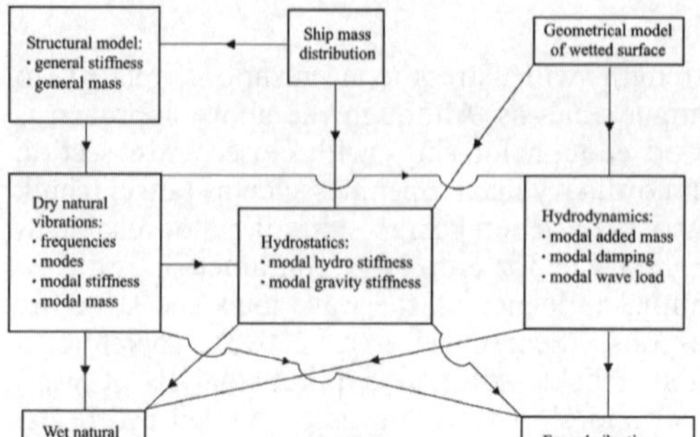

Figure 1. Methodology of the hydroelastic analysis.

3 STRUCTURAL MODEL BASED ON ADVANCED BEAM THEORY

3.1 *General remarks*

A ship hull, as an elastic non-prismatic thin-walled girder, performs longitudinal, vertical, horizontal and torsional vibrations. Since the cross-sectional centre of gravity and centroid, as well as the shear centre positions are not identical, coupled longitudinal and vertical, and horizontal and torsional vibrations occur, respectively. The distance between the centre of gravity and centroid for longitudinal and vertical vibrations, as well as distance between the former and shear centre for horizontal and torsional vibrations are negligible for conventional ships. Therefore, in the above cases ship hull vibrations can be analyzed separately. However, the shear centre in ships with large hatch openings is located outside the cross-section, i.e. below the keel, and therefore the coupling of horizontal and torsional vibrations is extremely high. The above problem is rather complicated due to geometrical discontinuity of the hull cross-section. The accuracy of the solution depends on the reliability of stiffness parameters determination, i.e. of bending, shear, torsional and warping moduli. The finite element method is a powerful tool to solve the above problem in a successful way. One of the first solutions for coupled horizontal and torsional hull vibrations, dealing with the finite element technique, is given in (Kawai, 1973, Senjanović & Grubišić, 1991). Generalised and improved solutions are presented in (Pedersen, 1985, Wu & Ho, 1987). In all these references, the determination of hull stiffness is based on the classical thin-walled girder theory, which does not give a satisfactory value for the warping modulus of the open cross-section (Haslum & Tonnessen, 1972, Vlasov, 1961). Apart from that, the fixed values of stiffness moduli are determined, so that the application of the beam theory for hull vibration analysis is limited to a few lowest natural modes only. Otherwise, if the mode dependent stiffness parameters are used the application of the beam theory can be extended up to the tenth natural mode (Senjanović & Fan, 1989, 1992, 1997).

3.2 *Outline of an advanced beam theory*

Referring to the flexural beam theory (Senjanović & Grubišić, 1991), the total beam deflection, w, consists of the bending deflection, w_b, and the shear deflection, w_s, i.e., Figure 2

$$w = w_b + w_s. \tag{1}$$

The shear deflection is a function of w_b

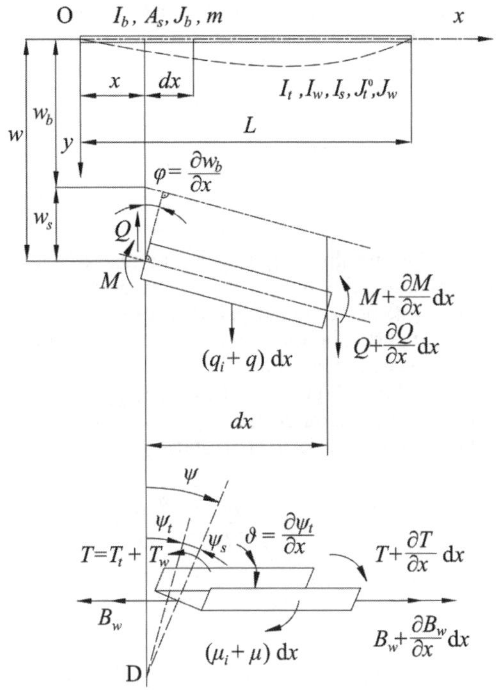

Figure 2. Beam bending and torsion.

$$w_s = -\frac{EI_b}{GA_s} \frac{\partial^2 w_b}{\partial x^2}, \tag{2}$$

where E and G are the Young's and shear modulus, respectively, while I_b, and A_s are the moment of inertia of cross-section and shear area, respectively. The angle of cross-section rotation is caused by the bending deflection

$$\varphi = \frac{\partial w_b}{\partial x}. \tag{3}$$

The cross-sectional forces are the bending moment and the shear force

$$M = -EI_b \frac{\partial^2 w_b}{\partial x^2}, \tag{4}$$

$$Q = GA_s \frac{\partial w_s}{\partial x} = -EI_b \frac{\partial^3 w_b}{\partial x^3}. \tag{5}$$

Concerning torsion, the total twist angle, ψ, consists of the pure twist angle, ψ_t, and the shear contribution, ψ_s, i.e., Figure 2.

$$\psi = \psi_t + \psi_s. \tag{6}$$

Referring to the analogy of torsion and bending (Pavazza, 2005), the shear angle depends on the twist angle, similarly to Eq. (2)

$$\psi_s = -\frac{EI_w}{GI_s} \frac{\partial^2 \psi_t}{\partial x^2}, \tag{7}$$

where I_w is the warping modulus and I_s is the shear inertia modulus. The second beam displacement, which causes warping of cross-section (similarly to the cross-section rotation due to bending) is a variation of the pure twist angle

$$\vartheta = \frac{\partial \psi_t}{\partial x}. \tag{8}$$

The sectional forces include the total torque, T, which consists of pure torsional torque, T_t, and the warping torque T_w i.e.

$$T = T_t + T_w, \tag{9}$$

where

$$T_t = GI_t \frac{\partial \psi_t}{\partial x} \tag{10}$$

$$T_w = GI_s \frac{\partial \psi_s}{\partial x} = -EI_w \frac{\partial^3 \psi_t}{\partial x^3} \tag{11}$$

and the bimoment given by

$$B_w = EI_w \frac{\partial^2 \psi_t}{\partial x^2}. \tag{12}$$

1D FEM procedure for vertical ship hull vibrations is well known in literature. Coupled horizontal and torsional vibrations are a more complex problem. Due to analogy between bending and torsion the same shape functions, represented by Hermitian polynomials, are used. The matrix finite element equation for coupled vibration yields (Senjanović, 1998)

$$\mathbf{f^e} = \mathbf{k^e}\boldsymbol{\delta}^\mathbf{e} + \mathbf{m^e}\ddot{\boldsymbol{\delta}}^\mathbf{e}, \tag{13}$$

where $\mathbf{f^e}$ is nodal forces vector, $\boldsymbol{\delta}^\mathbf{e}$ is nodal displacements vector, $\mathbf{k^e}$ is stiffness matrix, and $\mathbf{m^e}$ is mass matrix. These quantities consist of flexural and torsional parts

$$\mathbf{f^e} = \begin{Bmatrix} \mathbf{P} \\ \mathbf{R} \end{Bmatrix}, \quad \boldsymbol{\delta}^\mathbf{e} = \begin{Bmatrix} \mathbf{U} \\ \mathbf{V} \end{Bmatrix} \tag{14}$$

$$\mathbf{k}^{\mathrm{e}} = \begin{bmatrix} \mathbf{k}_{\mathrm{bs}} & 0 \\ 0 & \mathbf{k}_{\mathrm{wt}} \end{bmatrix}, \quad \mathbf{m}^{\mathrm{e}} = \begin{bmatrix} \mathbf{m}_{\mathrm{sb}} & \mathbf{m}_{\mathrm{st}} \\ \mathbf{m}_{\mathrm{ts}} & \mathbf{m}_{\mathrm{tw}} \end{bmatrix}. \tag{15}$$

Vectors of nodal forces and displacements are

$$\mathbf{P} = \begin{Bmatrix} -Q(0) \\ M(0) \\ Q(l) \\ -M(l) \end{Bmatrix}, \quad \mathbf{R} = \begin{Bmatrix} -T(0) \\ -B_w(0) \\ T(l) \\ B_w(l) \end{Bmatrix} \tag{16}$$

$$\mathbf{U} = \begin{Bmatrix} w(0) \\ \varphi(0) \\ w(l) \\ \varphi(l) \end{Bmatrix}, \quad \mathbf{V} = \begin{Bmatrix} \psi(0) \\ \vartheta(0) \\ \psi(l) \\ \vartheta(l) \end{Bmatrix} \tag{17}$$

In the above formulae symbols Q, M, T and B_w denote shear force, bending moment, torque and warping bimoment, respectively. Also, w, φ, ψ and ϑ are deflection, rotation of cross-section, twist angle and its variation, respectively. The submatrices, which are specified in (Senjanović et al. 2009c), have the following meaning:

$\mathbf{k}_{bs}$ – bending—shear stiffness matrix
$\mathbf{k}_{wt}$ – warping—torsion stiffness matrix
$\mathbf{m}_{sb}$ – shear—bending mass matrix
$\mathbf{m}_{tw}$ – torsion—warping mass matrix
$\mathbf{m}_{st} = \mathbf{m}_{ts}^{T}$ – shear—torsion mass matrix.

It is evident that coupling between horizontal and torsional vibrations is realized through the mass matrix due to eccentricity of the centre of gravity and shear centre.

Before assembling of finite elements it is necessary to transform Eq. (13) in such a way that all the nodal forces as well as nodal displacement, Eqs. (16) and (17), are related to the first and then to the second node. Furthermore, Eq. (13) has to be transformed from local to global coordinate system. The origin of the former is located at the shear centre, and of the latter at the base line.

3.3 *Contribution of transverse bulkheads to hull stiffness*

This problem for container ships is extensively analyzed in (Senjanović et al. 2008b), where torsional modulus of ship cross-section is increased proportionally to the bulkhead strain energy. The bulkhead is considered as an orthotropic plate with very strong stool (Szilard, 2004). The bulkhead strain energy is determined for the given warping of cross-section as a boundary condition. The warping causes bulkhead screwing and bending. Here, only the review of the final results is presented. The bulkhead deflection (axial displacement) is given by the following formula, Figure 3:

$$u(y,z) = -y\left\{(z-d) + \left[1-\left(\frac{y}{b}\right)^2\right]\frac{z^2}{H}\left(2-\frac{z}{H}\right)\right\}\psi', \tag{18}$$

where H is the ship height, b is one half of bulkhead breadth, d is the distance of warping centre from double bottom neutral line, y and z are transverse and vertical coordinates, respectively, and ψ' is the variation of twist angle.

The bulkhead grillage strain energy includes vertical and horizontal bending with contraction, and torsion (Senjanović et al. 2008b).

$$U_g = \frac{1}{1-\nu^2}\left[\frac{116H^3}{35b}i_y + \frac{32b^3}{105H}i_z + \frac{8Hb}{75}\nu\left(i_y+i_z\right) + \frac{143Hb}{75}(1-\nu)i_t\right]E\psi'^2 \tag{19}$$

where i_y, i_z and i_t are the average moments of inertia of cross-section and torsional modulus per unit breadth, respectively. The stool strain energy is comprised of the bending, shear and torsional contributions

$$U_s = \left[\frac{12h^2 I_{sb}}{b} + 72(1+\nu)\frac{h^2}{b^3}\frac{I_{sb}^2}{A_s} + \frac{9bI_{st}}{10(1+\nu)}\right]E\psi'^2 \tag{20}$$

where I_{sb}, A_s and I_{st} are the moment of inertia of cross-section, shear area and torsional modulus, respectively. Quantity h is the stool distance from the inner bottom, Figure 4.

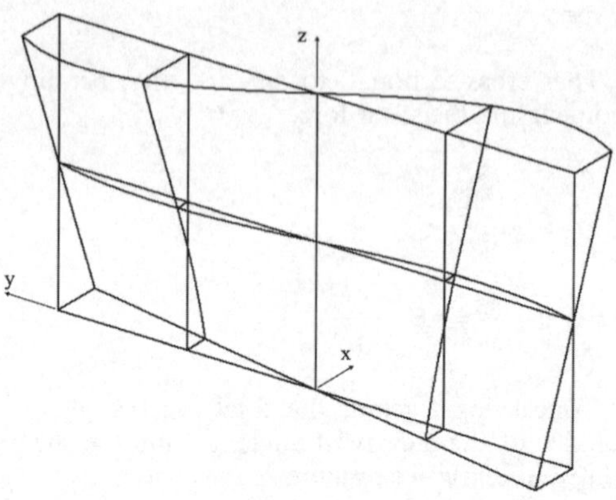

Figure 3. Shape of bulkhead deformation.

The equivalent torsional modulus yields, Figure 4

$$I_t^* = \left[1 + \frac{a}{l_1} + \frac{4(1+\nu)C}{I_t l_0}\right] I_t, \tag{21}$$

where a is the web height of bulkhead girders (frame spacing), l_0 is the bulkhead spacing, $l_1 = l_0 - a$ is the net length, and C is the energy coefficient

$$C = \frac{U_g + U_s}{E\psi'^2}. \tag{22}$$

The second term in (21) is the main contribution of the bulkhead as the closed cross-section segment of ship hull, and the third one comprises the bulkhead strain energy.

3.4 *Natural vibration analysis*

If the FEM approach is used (1D or 3D model), the governing equation of dry natural vibrations yields (Bathe, 1996)

$$(\mathrm{K} - \Omega^2\mathrm{M})\boldsymbol{\delta} = \mathbf{0}, \tag{23}$$

where $\mathbf{K}$ is stiffness matrix, $\mathbf{M}$ is mass matrix, Ω is dry natural frequency and $\boldsymbol{\delta}$ is dry natural mode. As solution of the eigenvalue problem (23) Ω_i and $\boldsymbol{\delta}_i$ are obtained for each the i-th dry mode, where $i = 1,2...N$, N is total number of degrees of freedom. Now natural modes matrix can be constituted

$$\boldsymbol{\delta} = \left[\boldsymbol{\delta}_1, \boldsymbol{\delta}_2 ... \boldsymbol{\delta}_i ... \boldsymbol{\delta}_N\right] \tag{24}$$

and the modal stiffness and mass can be determined (Senjanović, 1998)

$$\mathbf{k} = \boldsymbol{\delta}^T \mathrm{K} \boldsymbol{\delta}, \quad \mathbf{m} = \boldsymbol{\delta}^T \mathrm{M} \boldsymbol{\delta}. \tag{25}$$

Since the dry natural vectors are mutually orthogonal, matrices $\mathbf{k}$ and $\mathbf{m}$ are diagonal. Terms

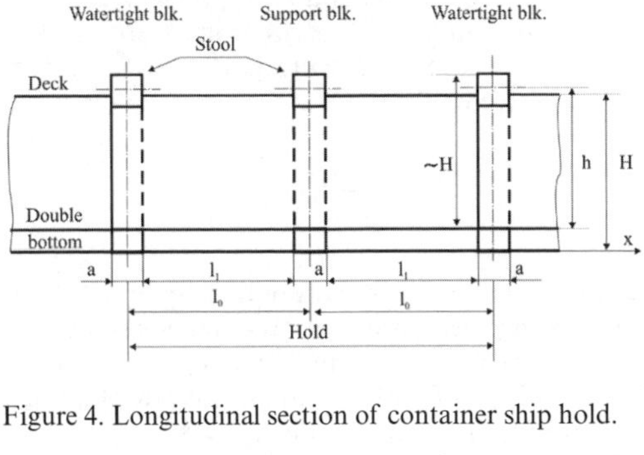

Figure 4. Longitudinal section of container ship hold.

k_i and $\Omega_i^2 m_i$ represent strain and kinetic energy of the i-th mode respectively.

Note that generally the first six natural frequencies Ω_i are zero with corresponding eigenvectors representing the rigid body modes. As a result, the first six diagonal elements of $\mathbf{k}$ are also zero, while the first three elements in $\mathbf{m}$ are equal to structure mass, the same in all directions x, y, z, and the next three elements represent the mass moment of inertia around the corresponding coordinate axes.

If 1D analysis is applied, the beam modes are spread to the ship wetted surface using the expressions for vertical vibrations (Senjanović et al. 2009a)

$$\mathbf{h}_i = -\frac{\mathrm{d}w_{vi}}{\mathrm{d}x}(z - z_N)\mathbf{i} + w_{vi}\mathbf{k}, \tag{26}$$

and for coupled horizontal and torsional vibrations

$$\mathbf{h}_i = \left(-\frac{\mathrm{d}w_{hi}}{\mathrm{d}x}y + \frac{\mathrm{d}\psi_i}{\mathrm{d}x}\bar{u}\right)\mathbf{i} + \left[w_{hi} + \psi_i(z - z_S)\right]\mathbf{j} - \psi_i y\mathbf{k}, \tag{27}$$

where w is hull deflection, ψ is twist angle, y and z are coordinates of the point on ship surface, and z_N and z_S are coordinates of centroid and shear centre respectively, and $\bar{u} = \bar{u}(x, y, z)$ is the cross-section warping intensity reduced to the wetted surface (Senjanović et al. 2009d).

4 HYDRODYNAMIC MODEL

The coupling procedure does not depend on the used hydrodynamic model, and is therefore described here for the zero speed case, as the simplest one. Harmonic hydroelastic problem is considered in frequency domain and therefore we operate with amplitudes of forces and displacements. In order to perform the coupling of structural and hydrodynamic models, it is necessary to express the external pressure forces in a convenient manner (Malenica et al. 2003). First, the total hydrodynamic force F^h has to be split into two parts: the first part F^R depending on the structural deformations, and the second one F^{DI} representing the pure excitation. Furthermore, the modal superposition method can be used. Vector of the wetted surface deformations $\mathbf{H}$ (x, y, z) can be presented as a series of dry natural modes $\mathbf{h}_i$ (x, y, z).

The potential theory assumptions are adopted for the hydrodynamic part of the problem. Within this theory, the total velocity potential φ, in the case of no forward speed, is defined with the Laplace differential equation and the given boundary values. Furthermore, the linear wave theory enables the following decomposition of the total potential (Senjanović et al. 2008a)

$$\varphi = \varphi_I + \varphi_D - i\omega\sum_{j=1}^{N} \xi_j \varphi_{Rj}, \varphi_I = -i\frac{gA}{\omega}\mathrm{e}^{\nu(z+ix)}, \quad (28)$$

where φ_I is incident wave potential, φ_D is diffraction potential, φ_{Rj} is radiation potential and A and ω represent wave amplitude and frequency respectively. Once the potentials are determined, the modal hydrodynamic forces are calculated by pressure work integration over the wetted surface, S. The total linearised pressure can be found from Bernoulli's equation

$$p = i\omega\rho\varphi - \rho gz. \quad (29)$$

First, the term associated with the velocity potential φ is considered and subdivided into excitation and radiation parts

$$F_i^{DI} = i\omega\rho\iint_S (\varphi_I + \varphi_D)\mathbf{h}_i\mathbf{n}\,\mathrm{d}S, \quad (30)$$

$$F_i^R = \rho\omega^2\sum_{j=1}^{N}\xi_j\iint_S \varphi_{Rj}\mathbf{h}_i\mathbf{n}\,\mathrm{d}S. \quad (31)$$

Thus, F_i^{DI} represents the modal pressure excitation. Now one can decompose (31) into the modal inertia force and damping force associated with acceleration and velocity, respectively

$$F_i^a = \mathrm{Re}(F_i^R) = \omega^2\sum_{j=1}^{N}\xi_j A_{ij}, \; A_{ij} = \rho\mathrm{Re}\iint_S \varphi_{Rj}\,\mathbf{h}_i\mathbf{n}\,\mathrm{d}S, \quad (32)$$

$$F_i^v = \mathrm{Im}(F_i^R) = \omega\sum_{j=1}^{N}\xi_j B_{ij}, \; B_{ij} = \rho\omega\mathrm{Im}\iint_S \varphi_{Rj}\,\mathbf{h}_i\mathbf{n}\,\mathrm{d}S. \quad (33)$$

where A_{ij} and B_{ij} are elements of added mass and damping matrices, respectively.

Determination of added mass and damping for rigid body modes is a well-known procedure in ship hydrodynamics. Now the same procedure is extended to the calculation of these quantities for elastic modes. The hydrostatic part of the total pressure,—ρgz in (29), is considered within the hydrostatic model.

5 HYDROSTATIC MODEL

Hydroelasticity is a known issue for many years, and there are few solutions for restoring stiffness (Price & Wu, 1985; Newman, 1994; Huang & Riggs, 2000; Malenica, 2003). In this study consistent formulation of restoring stiffness is used (Senjanović et al. 2009a, b), and its condensed form is given below.

The restoring stiffness consists of hydrostatic and gravity parts. Work of the hydrostatic pressure, which represents the generalized force, can be derived in the following form

$$F^h = -\rho g\iint_S \left[H_z + Z(\nabla\mathbf{H})\right]\mathbf{Hn}\mathrm{d}S, \quad (34)$$

where ∇ is Hamilton differential operator, $\mathbf{H}$ is displacement vector, $\mathrm{d}S$ is differential of wetted surface, Z is its depth and $\mathbf{n}$ is unit normal vector. According to definition, the stiffness is relation between incremental force and displacement, so it is determined from the variational equation

$$\delta F^h = -\rho g\iint_S \left[H_z + Z(\nabla\mathbf{H})\right]\delta\mathbf{Hn}\mathrm{d}S. \quad (35)$$

Furthermore, the modal superposition method is used, and the variation is transmitted to modes, i.e. modal forces and displacements

$$\delta F^h = \sum_{j=1}^{N}\delta F_j^h, \quad \mathbf{H} = \sum_{j=1}^{N}\xi_j\mathbf{h}_j, \quad \delta\,\mathbf{H} = \sum_{j=1}^{N}\mathbf{h}_j\delta\xi_j. \quad (36)$$

In that way, Eq. (35) is decomposed into the modal equations

$$\delta F_i^h = -\sum_{j=1}^{N}\left[\left(C_{ij}^p + C_{ij}^{nh}\right)\xi_j\right]\delta\xi_i, \quad (37)$$

where

$$C_{ij}^p = \rho g\iint_S \mathbf{h}_i h_z^j\mathbf{n}\mathrm{d}S, \quad C_{ij}^{nh} = \rho g\iint_S Z\mathbf{h}_i\left(\nabla\mathbf{h}_j\right)\mathbf{n}\mathrm{d}S, \quad (38)$$

are stiffness coefficients due to pressure, and normal vector and mode contributions, respectively.

Similarly to the pressure part, the generalized gravity force reads

$$F^m = -g\iiint_V \rho_s(\mathbf{H}\nabla)H_z\mathrm{d}V, \quad (39)$$

where ρ_s and V are structure density and volume, respectively. In order to obtain consistent variational equation, it is necessary to strictly follow the definition of stiffness and to vary displacement vector in (39) and not its derivatives

$$\delta F^m = -g\iiint_V \rho_s(\delta\mathbf{H}\nabla)H_z\mathrm{d}V. \tag{40}$$

Application of the modal superposition method leads to the modal variational equation

$$\delta F_i^m = -\sum_{j=1}^{N} C_{ij}^m \xi_j \delta\xi_i, \tag{41}$$

where

$$C_{ij}^m = g\iiint_V \rho_s(\mathbf{h}_i\nabla)h_z^j\mathrm{d}V, \tag{42}$$

are the gravity stiffness coefficients. Finally, the complete restoring stiffness coefficients are obtained by summing up its constitutive parts

$$C_{ij} = C_{ij}^p + C_{ij}^{nh} + C_{ij}^m. \tag{43}$$

6 HYDROELASTIC MODEL

After the definition of the structural, hydrostatic and hydrodynamic models, the hydroelastic model can be constituted. The governing matrix differential equation for coupled ship motions and vibrations is deduced

$$\left[\mathrm{k} + \mathrm{C} - i\omega(\mathrm{d} + \mathrm{B}(\omega)) - \omega^2(\mathrm{m} + \mathrm{A}(\omega))\right]\xi = \mathbf{F}, \tag{44}$$

where **k**, **d**, and **m** are structural stiffness, damping and mass matrices, respectively, **C** is restoring stiffness, $\mathbf{B}(\omega)$ is hydrodynamic damping, $\mathbf{A}(\omega)$ is added mass, $\boldsymbol{\xi}$ is modal amplitudes, **F** is wave excitation and ω is encounter frequency. All quantities, except ω and ξ, are related to the dry modes. The solution of (44) gives the modal amplitudes ξ_i and displacement of any point of the structure obtained by re-tracking to (36).

7 VERIFICATION OF PROPOSED NUMERICAL PROCEDURE

The computer software DYANA for ship hydroelastic analysis, based on the presented theory, has been developed. Both theory and code are checked by correlation analysis of the simulation results and the measured ones for a flexible segmented barge consisting of 12 pontoons, for which test results are available (Malenica et al. 2003, Remy et al. 2006), Figure 5. Good agreement between measured and calculated transfer functions of

Figure 5. Barge test in waves.

horizontal bending moment and torque, as function of wave period, T, implies that developed procedure can be used for ship hydroelastic analysis, Figures 6, 7 (Malenica et al. 2003, 2007). Also, convergence of the applied modal superposition method in hydroelastic analysis is confirmed in the case of the same flexible barge with no forward speed (Tomašević, 2007). Figure 8 shows absolute amplitude values ξ_i of normalized modes. The first three modes are related to sway, roll and yaw, while the remaining modes are elastic.

8 NUMERICAL EXAMPLE

For the illustration purposes, hydroelastic analysis of 7800 TEU VLCS is done, Figure 9.

8.1 *Particulars of the analyzed ship*

The main vessel particulars are the following:

Length overall	$L_{oa} = 334$ m
Length between perpendiculars	$L_{pp} = 319$ m
Breadth	$B = 42.8$ m
Depth	$H = 24.6$ m
Draught	$T = 14.5$ m
Displacement, full load	$\Delta_f = 135336$ t
Displacement, ballast	$\Delta_b = 68387$ t
Engine power	$P = 69620$ kW
Ship speed	$v = 25.4$ kn.

The midship section, which shows a double skin structure with the web frames and longitudinals, is presented in Figure 10. Rows and tiers of containers at the midship section are indicated in Figure 11. Vertical positions of neutral line, deformation (shear, torsional) centre, and centre of gravity are also marked in the figure. Large distance between

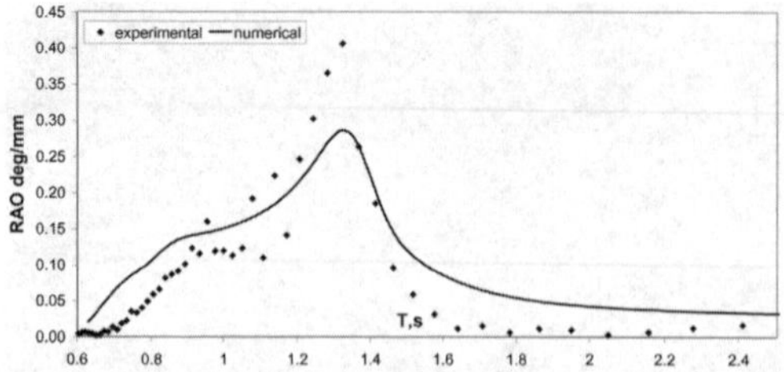

Figure 6. Transfer function of barge horizontal bending moment, $\chi = 60°$.

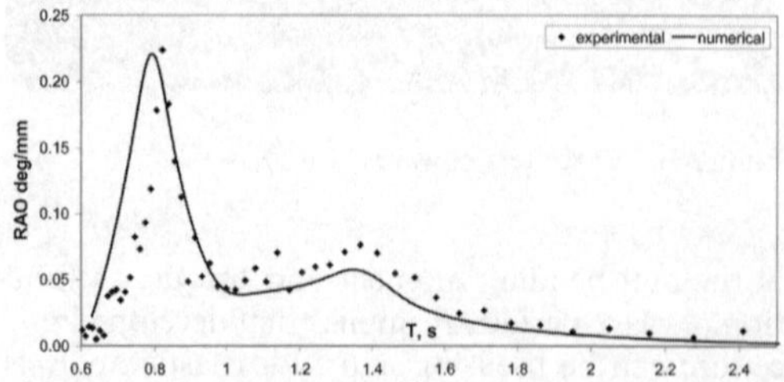

Figure 7. Transfer function of barge torque, $\chi = 60°$.

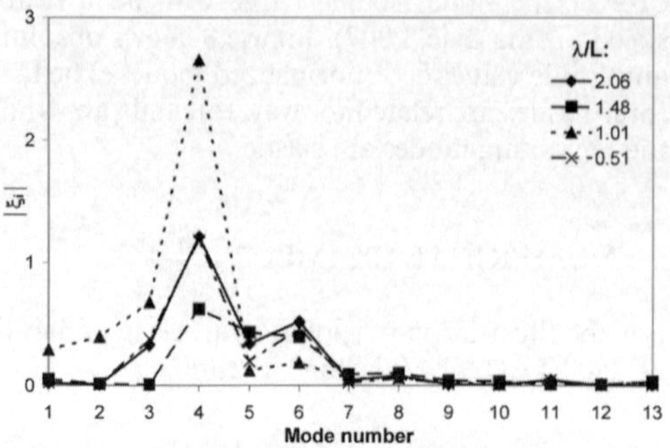

Figure 8. Modal amplitudes of coupled horizontal and torsional vibrations of flexible barge, $V = 0$ kn, $\chi = 120°$.

gravity centre and deformation centre causes high coupling of horizontal and torsional vibrations.

The ship hull stiffness properties are calculated by program STIFF, based on the theory of thin-walled girders (Senjanović & Fan, 1992, 1993). The geometrical properties rapidly change values in the engine and superstructure area due to closed ship cross-section. This is especially pronounced in case of torsional modulus, which takes quite small values for open cross-section and rather high for the closed one (Tomašević, 2007).

Influence of the transverse bulkheads is taken into account by using the equivalent torsional modulus for the open cross-sections instead of the actual values, i.e. $I_t^* = 2.4 I_t$. This value is applied for all ship-cross sections as the first approximation. The stiffness parameters of the bulkhead girders are listed in Tables 1 and 2, while the stool parameters are given in Table 3.

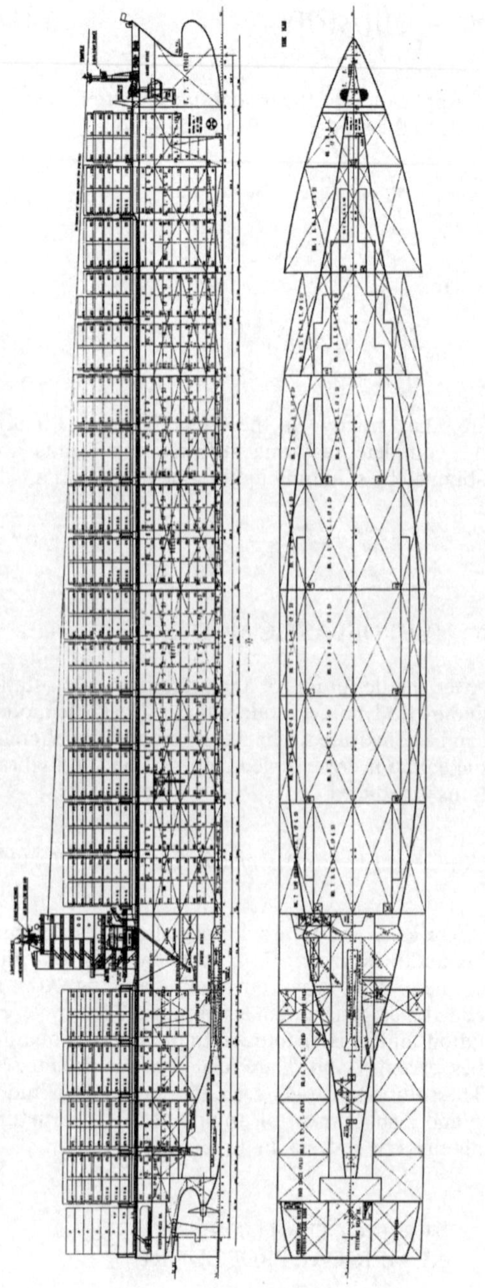

Figure 9. 7800 TEU Container Ship.

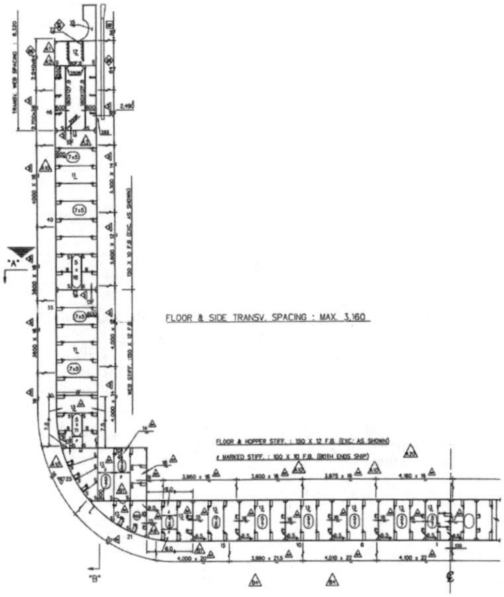

Figure 10. Midship section of the analyzed ship.

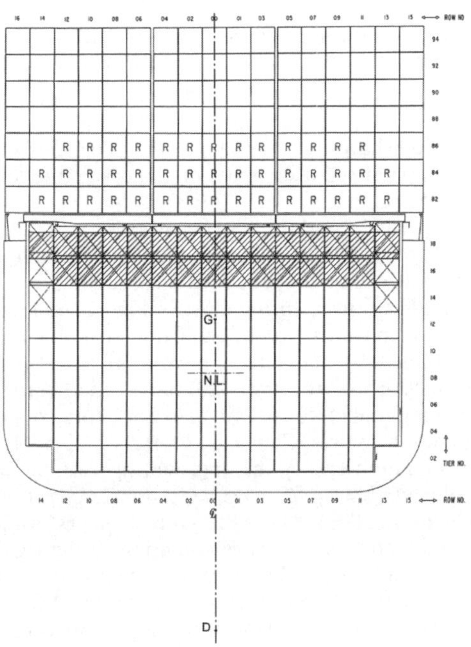

Figure 11. Container distribution at midship section.

The bulkhead strain energy, determined according to formulae presented in Chapter 3 is given in Table 4.

Table 1. Stiffness parameters of watertight bulkhead.

Girder	Moment of inertia I (m^4)	Tors. modulus I_t (m^4)	Spacing c (m)	Moment of inertia i (m^3)	Tors. modulus i_t (m^3)
Horiz.	0.02356	0.01555	2.6	0.00906	0.00493
Vert.	0.04196	0.03205	7.9	0.00531	

Table 2. Stiffness parameters of support bulkhead.

Girder	Moment of inertia I (m^4)	Tors. modulus I_t (m^4)	Spacing c (m)	Moment of inertia i (m^3)	Tors. modulus i_t (m^3)
Horiz.	0.00972	0.00486	2.6	0.00374	0.001696
Vert.	0.01944	0.01215	7.9	0.00246	

Table 3. Stool stiffness parameters.

Shear area A_s (m^2)	Moment of inertia I_s (m^4)	Tors. modulus I_{ts} (m^4)
0.45	0.07804	0.131

Table 4. Bulkhead strain energy, $U/(E\psi'^2)$.

Watertight bulkhead		Support bulkhead		Energy coefficient
Grillage	Stool	Grillage	Stool	C, Eq. (C5)
(1)	(2)	(3)	(4)	(5) = [(1) + (2) + (3) + (4)]/2
29.691	28.872	12.051	28.872	49.743

8.2 *Validation of 1D FEM model*

The reliability of 1D FEM analysis is verified by 3D FEM analysis of the considered ship. For this purpose, the light weight loading condition of dry ship with displacement Δ = 33692 t is taken into account. The lateral and bird view of the first dominantly torsional mode of the wetted surface, determined by 1D model, is shown in Figure 12. The first 3D dry coupled natural modes of the complete ship structure is shown in Figure 13, where Y and Z are vertical and transversal axis, respectively. It is similar to that of 1D analysis for the wetted surface. Warping of the transverse bulkheads, which increases the hull torsional stiffness, is evident.

The first four corresponding natural frequencies obtained by 1D and 3D analyses are compared in Table 5.

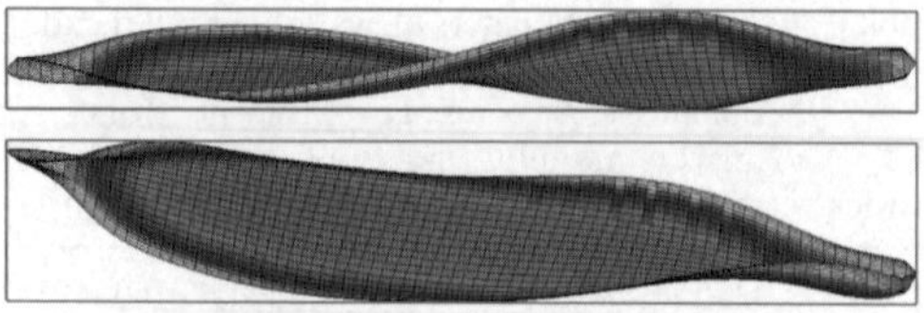

Figure 12. The first dominantly torsional mode, lateral and bird view, light weight, 1D model.

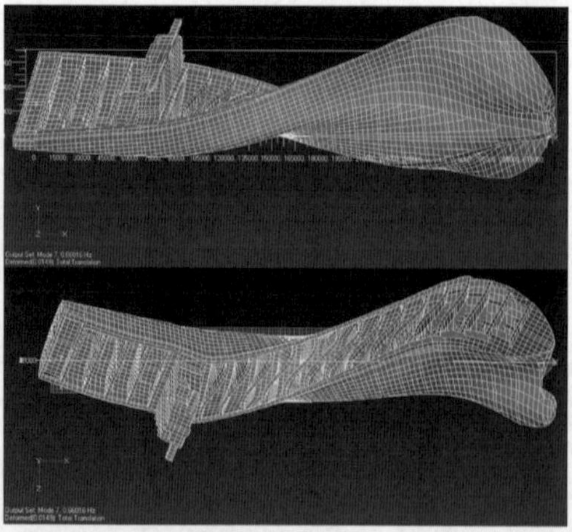

Figure 13. The first dominantly torsional mode, lateral and bird view, light weight, 3D model.

Table 5. Dry natural frequencies, light weight, ω_i [rad/s].

	Vert.		Horiz. + tors.		
Mode no.	1D	3D	1D	3D	Mode no.
1	7.35	7.33	4.17	4.15	1(H0 + T1)
2	15.00	14.95	7.34	7.40	2(H1 + T2)
3	24.04	22.99	12.22	12.09	3(H2 + T3)
4	35.08	34.21	15.02	16.22	4(H3 + T4)

Quite good agreement is achieved. Values of natural frequencies for higher modes are more difficult to correlate, since strong coupling between global hull modes and local substructure modes of 3D analysis occurs.

8.3 *Results of the ship response calculation*

Transfer functions of torsional moment and horizontal bending moment at the midship section are shown in Figures 14 and 15, respectively. They are compared to the rigid body ones determined by program HYDROSTAR. Very good agreement is obtained in the lower frequency domain, where

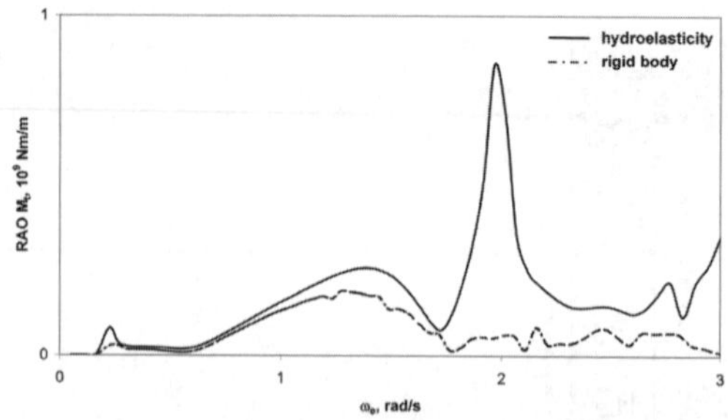

Figure 14. Transfer function of torsional moment, $\chi = 120°$, $U = 25$ kn, $x = 155.75$ m from AP.

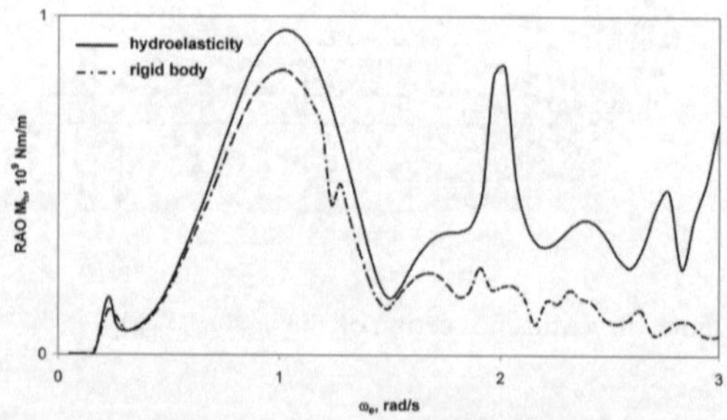

Figure 15. Transfer function of horizontal bending moment, $\chi = 120°$, $U = 25$ kn, $x = 155.75$ m from AP.

the ship behaves as a rigid body. Discrepancies are very large at the resonances of the elastic modes, as expected.

Necessary condition for convergence of sectional forces to zero value as the wave frequency approaches to zero can be used as a benchmark for validation of the restoring stiffness. Figure 16 shows the zoomed transfer function of torsional moment determined by the direct integration and three formulations of restoring stiffness in the hydroelastic approach: consistent one from this paper, symmetric matrix $\bar{C}_{ij} = \left(C_{ij} + C_{ji}\right)/2$ obtained by the minimum energy method, and hybrid matrix in which $C_{ij}^{*p} = C_{ij}^{p}$, $C_{ij}^{*n} = C_{ij}^{n}$, $C_{ij}^{*h} = C_{ji}^{h}$, $C_{ij}^{*m} = C_{ji}^{m}$, (Malenica, 2003). Only the consistent restoring stiffness satisfies the above condition as the rigid body solution does. In the case of symmetric and hybrid matrices the ship is not equilibrated. Moreover, the consistent restoring stiffness emphasizes the roll resonance at 0.23 rad/s.

Shear influence on torsion is investigated in the case of a pontoon with the cross-section equal to the midship section of the considered 7800 container ship. One end of the pontoon is fixed and another is loaded with the concentrated torque. Calculation is performed analytically by employing the advanced beam theory and numerically by 3D

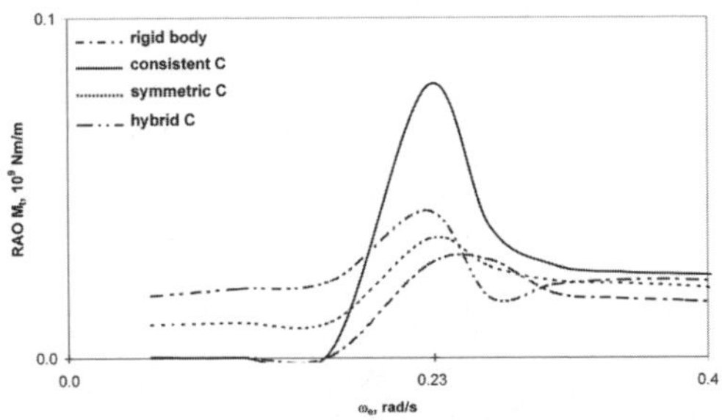

Figure 16. Zoomed transfer function of torsional moment, $\chi = 120°$, $U = 25$ kn, $x = 155.75$ m from AP.

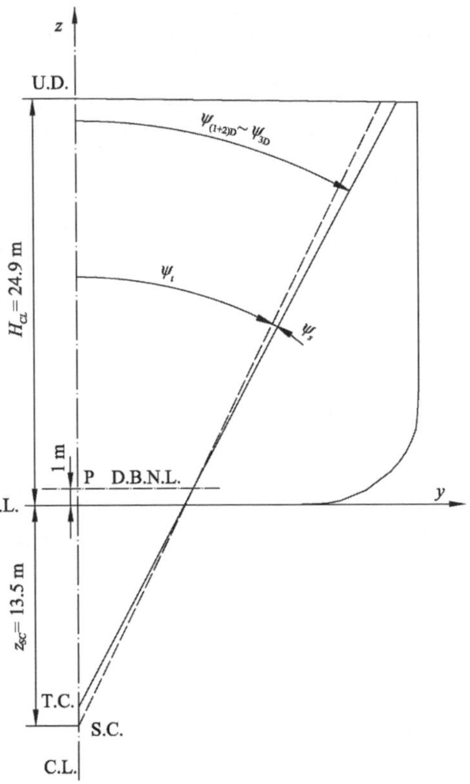

Figure 17. Twist angle at the pontoon end.

FEM model. Rotation angles of the free pontoon end are shown in Figure 17.

Pure twist angle ψ_t is realized around the shear centre, S.C., and is somewhat smaller then the twist angle determined by 3D FEM model. If the shear twist angle ψ_s is added to ψ_s around the double bottom centroid, value of the total twist angle approaches that of 3D analysis. As a result, the twist centre is determined, T.C.

9 CONCLUDING REMARKS

Ultra large container ships are quite flexible so they stretch the bounds of present classification rules for reliable structural design. Therefore, hydroelastic analysis has to be performed (Senjanović et al. 2008a, 2009a).

The illustrative numerical example of the 7800 TEU container ship shows that the developed hydroelasticity theory, utilizing the improved 1D FEM structural model and 3D hydrodynamic model, is an efficient tool for application in ship hydroelastic analyses. The obtained results point out that the transfer functions of hull sectional forces in case of resonant vibration (springing) are much higher than in resonant ship motion. Very good agreement between ship response determined by hydroelastic analysis and rigid body analysis in vicinity of zero frequency is obtained due to use of the consistent restoring stiffness. The both solutions converge to zero, as frequency approaches zero value.

The used advanced beam model of ship hull, based on advanced thin-walled girder theory with included shear influence on torsion and contribution of transverse bulkheads to stiffness, is a reasonable choice for determining wave load effects. However, stress concentration in hatch corners calculated directly by the beam model is underestimated. This problem can be overcome by applying substructure approach, i.e. 3D FEM model of substructure with imposed boundary conditions from beam response. In any case, 3D FEM model of complete ship is preferable from the viewpoint of determining stress concentration.

In order to complete hydroelastic analysis of container ships and confirm its importance for ship safety, it is necessary to proceed further to ship motion calculation in irregular waves for different sea states, based on the known transfer functions. This includes determination of global wave loads, i.e. bending and torsional moments and their conversion into stresses, stress concentration in critical areas of ship structures, especially in hatch corners due to restrained warping, and fatigue of structural details.

At the end of a complete investigation, which also has to include model tests and full-scale measurements, it will be possible to decide on the extent of the revision of Classification Rules for the design and construction of ultra large container ships.

REFERENCES

Bathe, KJ. 1996. *Finite Element Procedures*. Prentice Hall.

Bishop, RED, Price, WG. 1979. *Hydroelasticity of Ships*. Cambridge University Press.

Haslum, K., Tonnessen, A. 1972. An analysis of torsion in ship hull. *European Shipbuilding* (5/6): 67–89.
Huang, LL., Riggs, HR. 2000. The hydrostatic stiffness of flexible floating structure for linear hydroelasticity. *Marine Structures* 13: 91–106.
Kawai, T. 1973. The application of finite element method to ship structures. *Computers and Structures* (2): 1175–1194.
Malenica, Š. 2003. Some aspects of hydrostatic calculations in linear seakeeping. *Proc. 14th NAV Conf.*, Palermo.
Malenica, Š., Molin, B., Remy, F., Senjanović, I. 2003. Hydroelastic response of a barge to impulsive and non-impulsive wave load. *Proc. Hydroelasticity in Marine Technology*, Oxford, 287–314.
Malenica, Š., Senjanović, I., Tomašević, S., Stumpf, E. 2007. Some aspects of hydroelastic issues in the design of ultra large container ships. *Proc. IWWWFB*, Plitvice Lakes.
Newman, JN. 1994. Wave effects on deformable bodies. *Applied Ocean Research* 16: 47–59.
Pavazza, R. 2005. Torsion of thin-walled beams of open cross-sections with influence of shear. *International Journal of Mechanical Sciences* 47: 1099–1122.
Pavazza, R., Plazibat, B., Matoković, A. 1998. Idealisation of ships with large hatch opening by a thin-walled rod of open section on many elastic supports. *Thin-walled Structures* 32: 305–325.
Pedersen, PT. 1985. Torsional response of container ships. *Journal of Ship Research* 31: 194–205.
Price, WG, Wu, Y. 1985. *Hydroelasticity of Marine Structures*. In: Theoretical and Applied Mechanics, FI. Niordson & N. Olhoff, eds. Elsevier Science Publishers B.V.: 311–337.
Remy, F., Molin, B., Ledoux, A. 2006. Experimental and numerical study of the wave response of flexible barge. *Proc. Hydroelasticity in Marine Technology*, Wuxi, 255–264.
Salvesen, N., Tuck, EO., Faltinsen, OM. 1970. *Ship motion and sea loads*. Cambridge University Press.
Senjanović, I. 1998. *Finite Element Method in Ship Structures*. University of Zagreb, Zagreb, (in Croatian).
Senjanović, I. Fan, Y. 1989. A higher-order flexural beam theory. *Computers & Structures* 32(5): 973–986.
Senjanović, I. Fan, Y. 1992. A higher-order theory of thin-walled girders with application to ship structures. *Computers & Structures* 43(1): 31–52.
Senjanović, I. Fan, Y. 1997. A higher-order torsional beam theory. *Engineering Modelling* 32(1–4): 25–40.
Senjanović, I. Grubišić, R. 1991. Coupled horizontal and torsional vibration of a ship hull with large hatch openings. *Computers & Structures* 41(2): 213–226.
Senjanović, I., Malenica, Š, Tomašević, S, Rudan, S. 2007. Methodology of ship hydroelastic investigation. *Brodogradnja* 58(2): 133–145.
Senjanović, I., Tomašević, S., Tomić, M., Rudan, S., Vladimir, N. 2008a. Hydroelasticity of Very Large Container Ships. *Proc. Design and Operation of Container Ships*, RINA, London, 51–70.
Senjanović, I., Tomašević, S, Rudan, S, Senjanović, T. 2008b. Role of transverse bulkheads in hull stiffness of large container ships. *Engineering Structures* 30: 2492–2509.
Senjanović, I., Tomašević, S., Vladimir, N., Tomić, M., Malenica, Š. 2009a. Ship hydroelastic analysis with sophisticated beam model and consistent restoring stiffness. *Proc. Hydroelasticity in Marine Technology*, University of Southampton, Southampton, 69–80.
Senjanović, I., Tomašević, S., Vladimir, N., Malenica, Š. 2009b. Numerical procedure for ship hydroelastic analysis. *Proc. Intl. Conf. Computational Methods in Marine Eng.*, CIMNE, Barcelona, 259–264.
Senjanović, I., Tomašević, S., Vladimir, N. 2009c. An advanced theory of thin-walled girders with application to ship vibrations. *Marine Structures* 22(3): 387–437.
Senjanović, I., Malenica, Š., Tomašević, S. 2009d. Hydroelasticity of large container ships. *Marine Structures* 22(2): 287–314.
Szilard, R. 2004. *Theories and Applications of Plate analysis*. John Wiley & Sons, New York.
Tomašević, S. 2007. *Hydroelastic model of dynamic response of Container ships in waves*. Doctoral Thesis, University of Zagreb, Zagreb, (in Croatian).
Vlasov, VZ. 1961. *Thin-Walled Elastic Beams*. Israel Program for Scientific Translation, Jerusalem.
Wu, JS., Ho, CS. 1987. Analysis of wave induced horizontal and torsion coupled vibrations of ship hull. *Journal of Ship Research* 31(4): 235–252.

Advanced Ship Design for Pollution Prevention – Guedes Soares & Parunov (eds)
 ISBN 978-0-415-58477-7

Methods for estimating parametric rolling

A. Turk & J. Prpić-Oršić
Faculty of Engineering, University of Rijeka, Rijeka, Croatia

S. Ribeiro e Silva & C. Guedes Soares
Centre for Marine Technology and Engineering (CENTEC), Technical University of Lisbon, Instituto Superior Técnico, Lisboa, Portugal

ABSTRACT: This paper focuses on describing the phenomenon of parametric roll resonance with the emphasis on the physics that governs such a motion. It reviews the various approaches adopted to represent that effect, including models with 1, 1.5, 3 and 6 degrees of freedom (DOF). It mentions the parametric resonance observed in head seas in a post-Panamax, C11 class containership which proved to be particularly prone to exhibit this behaviour and presents some results of parametric roll obtained for that ship with a time-domain, non-linear, 6 DOF numerical code.

1 INTRODUCTION

Parametric roll resonance is known as one of dangerous modes of ship motions in waves. This resonance occurs mainly as a result of restoring variation, in astern or head sea. Pure loss of stability, due to exponential increase of roll in either broaching or head sea conditions, causes new losses of containers in seas and accidents of container vessels particularly for a new generation of post-Panamax ships. Under certain conditions in which a ship with unfavourable characteristics can be found, a violent rolling motion of the ship in pure head or following seas can be generated rather quickly.

Paulling and Rosenberg (1959) were probably the first to account for parametric excitation in the ship motion problem. Hua et al. (1992) have studied the roll motion in resonance with the wave excitation, showing that the *GM*-variation of a ship in waves is an important evaluation factor in that problem.

Neves et al. (1999) have concentrated their attention to study the occurrence of parametric roll in fishing boats, conducting experimental work that has shown such a phenomenon.

Ribeiro e Silva and Guedes Soares (2000) demonstrated that both linearised and nonlinear theories could be used to predict parametric rolling in regular head waves, although te later would provide better predictions.

France et al. (2003) have reported that a post-Panamax, C11 class containership lost 1/3 of her deck containers and damaged another 1/3 in a severe storm. They have studied the incident by means of numerical simulations and model tests and demonstrated that the phenomenon could be reproduced by those methods. The evidence of parametric rolling in such a large container ship received wide attention and renewed the interest of several researchers in studying parametric roll.

Since the parametric roll phenomenon is caused by the time variation of transverse stability, the numerical simulation method must be able to adequately model the changes of geometry of the immersed part of the hull due to large waves and ship motions (Belenky et al. 2003). The results confirmed that the vessel suffered from a severe case of parametric roll during the storm (Shin et al. 2004).

Since the publication of these results, ship operators and ship designers have become more aware of the fact that this phenomenon can occur for larger vessels in confused seas (Umeda et al. 2004).

Surendrana et al. (2007) conducted a full numerical procedure for a post-Panamax container vessel for predicting the onset of parametric rolling. based on the Duffing method with excitation taken from the difference in restoring force from trough to crest condition when the ship is encountering head seas proved it can push the ship into a parametric mode of rolling.

When a ship sails in head or following seas, the geometry of the underwater hull is constantly changing with time as a result of the wave surface variation along the hull as well as the pitching and heaving motions. The righting moment of a ship as a product of the righting lever and its displacement changes in longitudinal waves as both parameters oscillate, which causes a periodic variation of their product, as pointed out by Francescutto et al. (2004).

Displacement varies only within a limited range but the righting arm can undergo periodic variations of a large extent. This represents the predominant cause of possible roll excitation, hence the terms "parametric excitation" and "parametric roll". The fundamental dynamics that create this kind of behaviour is considered nowadays as reasonably clarified, namely that the frequency of encounter with waves of length similar or larger than the ship length is comparable to twice the ship's roll natural frequency.

Hull forms with pronounced bow flare, flat transom stern and non-vertical ship sides near the waterline are most vulnerable to parametric rolling (Prpić-Oršić et al. 2007a). This applies especially to large container ships and passenger vessels, which show greatly different shape forward and aft. In the forebody the breadth is sharply reduced whereas in the afterbody there is pronounced flare and the hull's full breadth extends up to the transom. Such features contribute to the variation of the ship's stability characteristics, namely to the development of righting levers owing to the flare of the frames and superstructures (e.g. the forecastle).

The established understanding of this phenomenon has led some classification societies to propose the guidelines for the assessment of parametric roll resonance in the design of container carriers such as ABS (2004).

This paper will describe various methods to model parametric roll, starting from 1DOF models (Francescutto, 2001) and moving to 1.5DOF (Bulian et al. 2003), 3DOF (Neves & Rodríguez 2007) and 6DOF models (Ribeiro e Silva et al. 2005).

2 THEORETICAL BACKGROUND FOR PARAMETRIC ROLLING OF SHIPS

2.1 *Physics of parametric roll resonance*

Excitation of roll motion caused by wave slope is well known and comparatively easily explained, especially when the ship is under way or drifting in beam seas (Prpić-Oršić et al. 2007b). Nevertheless, under certain conditions of encounter period, roll motion can be excited in longitudinal seas, via a different phenomenon. With monohull merchant ships a periodic variation of the righting lever in longitudinal waves is caused by the combined effects of two processes: waves moving along the hull, and periodic oscillatory change in stability at approximately twice the natural roll period. Although these processes always occur simultaneously, they will be explained individually in order to provide a clear description of the different effects.

Consider the static case of a ship in calm water. Any disturbance in the transverse direction (as from a wind gust) will lead to roll motions dominated by the "natural roll period".

If the crest of a wave with a length similar to the ship passes the midships part of the hull (Fig. 1), both ends of the ship emerge at the troughs of the waves. Righting levers at both ends of the ship are smaller, with a minimum when the crest is approximately amidships, since the waterplane at the immersed portions of the bow and stern are narrower than in calm water.

In contrast to the above, if a ship is located in a wave trough (Fig. 2), the flared parts of the bow and stern are more deeply immersed than in calm water, making the instantaneous waterplane wider than in calm water with the result that the metacentric height (*GM*) is increased over the calm water value, subsequently leading to the condition of improving or increasing stability.

In pure longitudinal seas condition, the first order roll wave excitation is zero. However, the ship may experience a very small roll disturbance from some external or internal cause. Nevertheless, if the period of wave encounter is approximately one-half the natural period of roll, a rolling motion can exist even in the absence of a direct roll exciting moment, but from the periodic variation of ship's righting arms with the ship's longitudinal position relative to wave profile (Chang 2008) (Fig. 3).

Thus in parametric resonance, one can observe that the roll motion and periodic variation of stability coincide in a certain distinctive way. This relationship between the two is enforced by the fact that the ship will incline when stability is reduced, i.e. when the ship encounters a wave crest. As a result inclining takes place under the condition of reduced stability and as a result, the ship rolls more than it would in calm water with the same roll disturbance. Before righting can begin, the ship's stability will improve in the course of its periodic variation. During righting, stability continues to

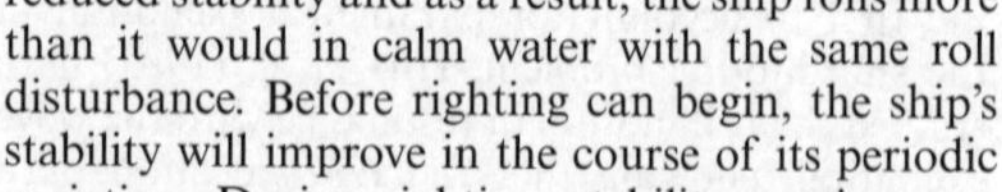

Figure 1. Profile of waterline in wave crest.

Figure 2. Profile of waterline in wave trough.

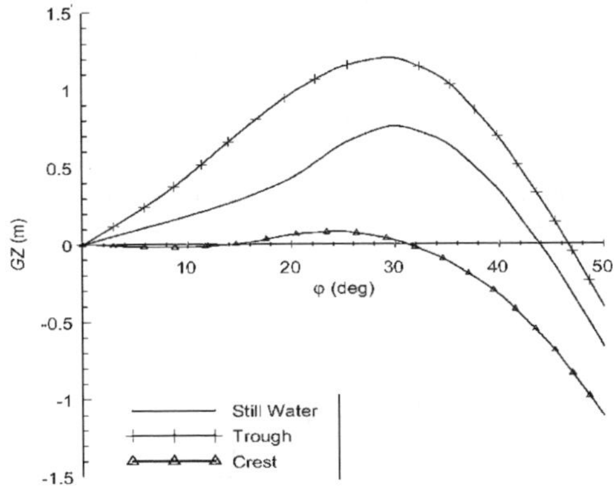

Figure 3. The righting arm curves of ship in seas.

grow towards its maximum. Thus righting takes place under the condition of increased stability i.e. when the ship encounters a wave trough, consequently "pushing" the ship back to the vertical position with a larger-than-calm-water moment.

This combination of restoring (with a larger-than-calm-water) and resisting the roll (with less-than-calm-water) can cause the roll angle to progressively increase to a large and possibly dangerous level. The fact that inclining at reduced stability alternates with righting at increased stability can only lead to excitation of roll if this alternation is repeated regularly and for sufficiently long time. Firstly, the ship gains additional momentum during each righting, whic occurs twice during a complete roll motion. Secondly, stability changes at twice the rate relative to the roll motion. This causes stability to shift closer to its maximum prior to and during righting. The difference between effective stability during inclining on one hand and during righting on the other is intensified. This increases the additional momentum gained during each half roll.

Then, during a number of consecutive rolls, the momentum inherent in the rolling ship at the end of each righting is greater than it had been at the beginning, will make roll angles increase as long as it is greater than the loss of momentum due to roll damping. Actually, roll damping has a relatively minor effect in regular waves, and in the majority of speed range the build-up of roll motion is mainly limited by restoring nonlinearities. To summarize, parametric excitation of roll is enhanced with increasing variation of stability and with the extent of this variation being large in comparison with the ship's still water stability.

2.2 *Influence of roll damping*

Roll damping is an important parameter in parametric rolling (Turk et al. 2009). In calm water, roll damping decreases roll amplitudes owing to the ship generated waves, eddies and viscous drag. Normally, the damping coefficients can be obtained from free decay experiments (Fig. 4), in which the model is released for a free roll from a given inclination.

Usage of such an empirical roll damping assessment, established from the free-decay model tests led to a few applicable analytical methods (Miller 1974, Ikeda et al. 1978).

There is a term associated with the previously mentioned effect of roll damping called roll damping threshold for parametric roll resonance. The parametric roll resonance can take place if the roll damping moment is below the threshold, meaning that the energy "gain" per cycle caused by the changing stability in longitudinal seas is more than the energy "loss" due to damping. This fact can simply be explained by the energy balance between damping and change of stability (Taylan 2007). To put it another words, if the reduction in roll damping is larger than the roll damping from the hull and appendages an unstable situation occurs and the vessel is subject to parametric roll.

Nonlinear damping tends to increase with the rolling velocity of ships. Actually linear damping increases with speed due to lift effect, while nonlinear damping due to hull vortex shedding tends to reduce. Nonlinear damping due to bilge keels tends to be quite unaffected by speed. However, since parametric roll inception is associated with linear damping, the increase of speed increases also the threshold for the inception of parametric rolling. Thus, the damping will increase, some time or later, up to a certain extent. At that time, the dissipated energy from damping is more than the extra input energy from parametric resonance. As a result, nonlinearity of damping has the effect of stabilizing the parametric rolling motion of ships.

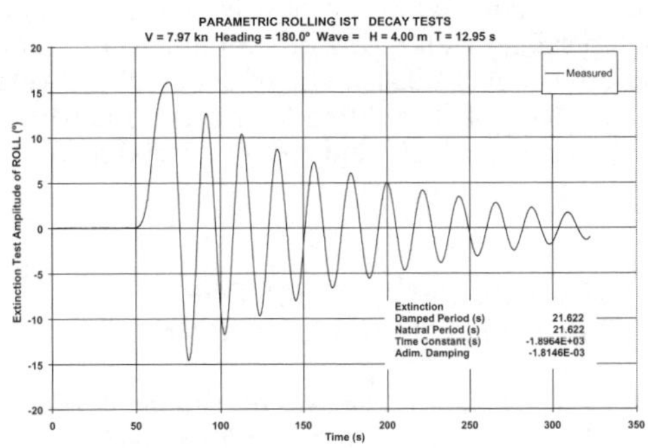

Figure 4. Successively decreasing roll amplitudes due to roll damping in calm water.

2.3 *Amplitude of parametric roll*

When considering the amplitude of parametric rolling one of the most important factors that determines it is the shape of the *GZ* curve. Apart from the above condition of frequencies to build-up a parametric rolling, meaning that the ship gains additional momentum during each righting motion and thus twice during a complete roll motion, a threshold wave height must be determined as well.

While the *GZ* curve usually is practically linear in the first 10–12 degrees of heel angle (at least for sufficiently large *GM*), the *GM* does not change, so both natural roll period and frequency remain constant for small values of roll angle (up to about 10–12 degrees). Once the roll angle increases beyond the linear portion of the *GZ* curve, the instantaneous *GM* value changes as the *GZ* curve bends (Fig. 3). This in turn affects instantaneous restoring energy in rolling motion and eventually total stability qualities. One can no longer treat metacentric height of the ship as constant. This complicates the evaluation of ship stability. Standard ship stability curve (*GZ*) is dependent on wave crest position along the ship length and must be determined instantaneously. It can be concluded that periods of encounter which are half the period of roll at dangerous roll amplitudes are a function of *GM* as well.

Because of the pronounced tendency of the roll period being adapted to the encounter period, parametric resonance will exist for a range of periods of encounter. Since the wave encounter frequency remains the same, the roll natural frequency may no longer be close to twice the encounter frequency. As a result, parametric resonance conditions no longer exist and roll motions no longer receive additional energy at each cycle. Therefore, after certain angles of roll, conditions for parametric resonance cannot be established and the growth stops.

One of the typical realisations of the development of parametric rolling for a model test of a C11 post-Panamax containership is presented in Figure 5. For example, a post-Panamax containership sailing with a *GM* of 1.90 m has a natural period of roll at small angles of about 29 s. Owing to the shape of its righting lever curve, the period at angles between 30° and 40° is only about 25 s.

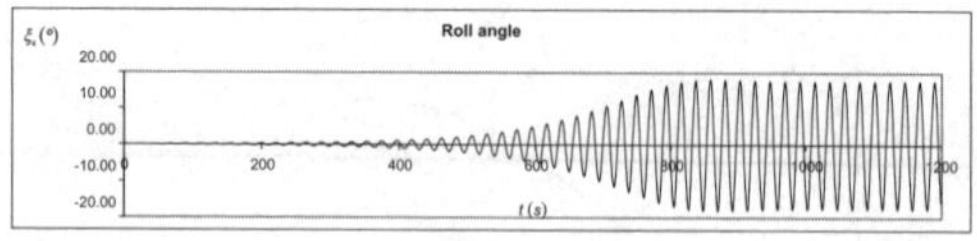

Figure 5. Development of parametric rolling during tests in regular waves (Turk et al. 2009).

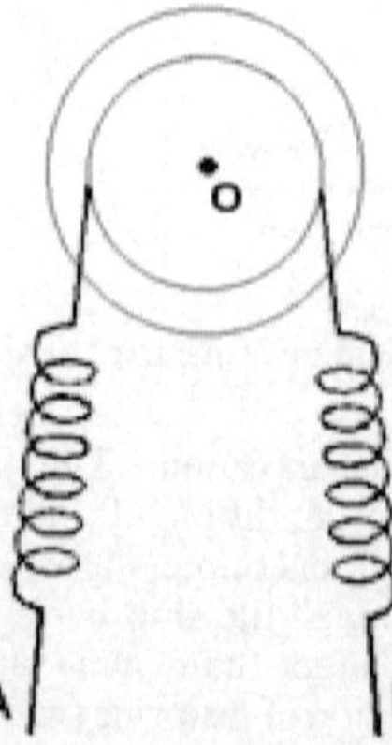

Figure 6. The rotational oscillator.

3 ASSESSMENT MODELS FOR PARAMETRIC ROLLING ON SHIPS

3.1 *Prediction of occurrence of parametric rolling in regular waves (susceptibility criteria)*

Some essential elements of roll motion are reviewed first. At a fundamental level, the equation of linear rolling motion is that of an excited ($M_x(t)$) or non excited rotational oscillator (Fig. 6).

Oscillatory motion is periodic motion where the displacement from equilibrium varies from a maximum in one direction to a maximum in the opposite or negative direction.

Considering a single degree of freedom equation for roll motion in head seas, and taking into account the changing *GM* due to wave encounter Shin et al. (2004) adopted:

$$\ddot{\eta}_4 + 2\zeta\dot{\eta}_4 + \frac{W \cdot GM(t)}{I_{XX} + A_{44}}\eta_4 = 0 \tag{1}$$

where η_4 represents the roll amplitude, ζ is the linearised damping coefficient, W is the displacement of a ship, I_{XX} is the transverse moment of inertia, and A_{44} is the added mass moment of inertia in roll. The single degree of freedom roll motion of a ship in head or following seas may then be described by an equation of motion similar to that for still water. However, the restoring moment is not only a function of angle of heel but it also varies sinusoidally with time, so we may use the small amplitude moment expression with a time varying metacentric height, which may result in parametric resonance,

$$GM(t) = GM_m + GM_a \cos(\omega t) \tag{2}$$

Here, GM_m is the mean value of the GM and GM_a is the amplitude of the GM changes in waves. Mean roll frequency and its variation amplitude:

$$\omega_m = \sqrt{\frac{W \cdot GM_m}{I_{XX} + A_{44}}}; \quad \omega_a = \sqrt{\frac{W \cdot GM_a}{I_{XX} + A_{44}}} \tag{3}$$

can be included in (1), with the following equation derived:

$$\ddot{\eta}_4 + 2\zeta\dot{\eta}_4 + \left(\omega_m^2 + \omega_a^2 \cos(\omega t)\right)\eta_4 = 0 \tag{4}$$

To check if parametric resonance is possible, the roll equation (4) must be transformed to the form of a Mathieu type varying restoring coefficient of the form,

$$\frac{d^2\eta_4}{d\tau^2} + (p + q\cos\tau)\eta_4 = 0 \tag{5}$$

where p is a function of the ratio of forced and natural frequency, q is the parameter that dictates the amplitude of parametric excitation, and τ represents nondimensional time If ship rolling is Mathieu-type then there is a possibility of parametric resonance. It will be manifested by an oscillatory build-up of roll, despite the absence of direct excitation. The physical phenomenon is based on successive alterations of the restoring rolling moment lever between crests and troughs, exhibited by many ships in steep longitudinal waves with a clear analogy with a simple oscillator governed by the Mathieu equation with damping.

To obtain the standard form of the Mathieu equation (5) it is necessary to transform equation (4) by introducing a dimensionless time: $\tau = \omega t \Rightarrow t = \tau/\omega$. By doing that and dividing both terms by the square of the wave frequency ω^2, the dimensionless quantities:

$$\mu = \frac{\zeta}{\omega}; \quad \bar{\omega}_m = \frac{\omega_m}{\omega}; \quad \bar{\omega}_a = \frac{\omega_a}{\omega} \tag{6}$$

are obtained, where ω_a is the roll variation frequency and ω_m is the mean value of roll frequency. The damping part is given by:

$$\eta_4(\tau) = x(\tau)\exp(-\mu\tau) \tag{7}$$

This finally expresses roll in the form of a Mathieu equation (4) with coefficients,

$$p = \left(\bar{\omega}_m^2 - \mu^2\right); \quad q = \bar{\omega}_a^2. \tag{8}$$

Solution of the Mathieu equation may be found in many references and depends strictly on the values of p and q. Thus, the solution may be periodic, increasing or decreasing in nature. Figure 7 is the stability diagram for this equation. The shaded regions are stable corresponding to (p,q) pairs where motion cannot exist and the unshaded regions which are unstable, i.e., motion can exist. If (p,q) lie in an unstable region, an arbitrarily small initial disturbance will trigger an oscillatory motion that tends to increase indefinitely with time. In a stable region, the initial disturbance will die out with time (Turk et al. 2008).

A linear damping in the Mathieu equation does not limit the amplitude of a solution. The effect of linear damping only raises the boundary of unstable zones by creating a threshold for the amplitude of GM variation in waves (q value). The unstable rolling motions will still happen if the p value is larger than the threshold value. This is why it is possible to use Mathieu equation for modelling the occurrence of amplified motions caused by parametric excitation, but not for evaluating how large the parametric oscillations might develop. To do so, nonlinear damping and/or stiffness terms must be added to "stabilize" the rising oscillations.

The transformation into the Mathieu type equation was a prerequisite in order to use the Ince-Strutt diagram to examine the properties of the solutions. An example on C11 post-Panamax is given. As shown in Figure 8, the operational conditions fall into a Mathieu type zone of instability. The ship may be susceptible to parametric rolling and the severity criterion has to be checked.

To summarize, the most comprehensive mathematical method to assess susceptibility to parametric rolling is the solution of the Mathieu equation in order to see if parametric rolling occurs in numerical simulations. Actually, it would be better to use a

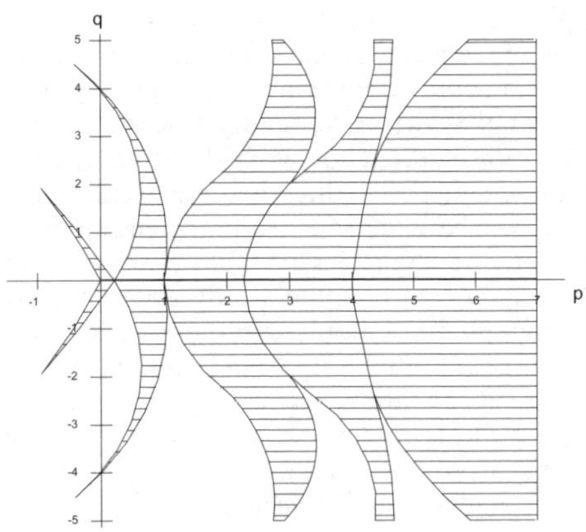

Figure 7. Stability diagram for the Mathieu equation.

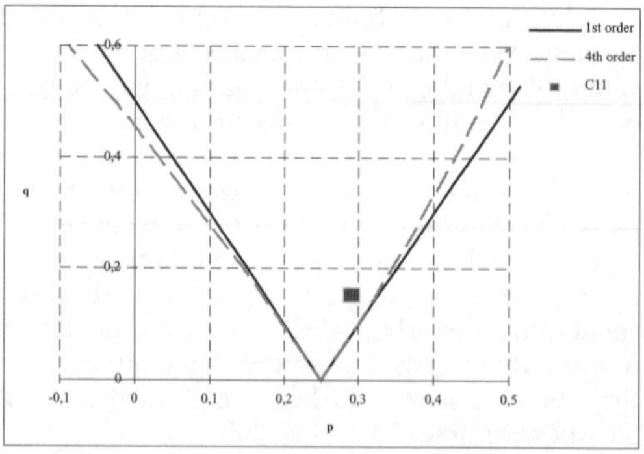

Figure 8. Linear (solid) and high-order (dashed) approximations for the boundary of the first instability zone (Turk et al. 2009).

linearised (with respect to roll) 3-DOF system coupling pitch heave and roll. This is what Neves et al. (2006) has been doing and Nabergoj et al. (1994) as well. The "pure Mathieu" approach is reasonable when the amplitude of the parametric oscillation (and also the remaining parameters) takes heave and pitch into account. To conclude, parametric rolling occurs when the following requirements are satisfied:

1. natural period of roll is equal to approximately twice the wave encounter period
2. wave length is on the order of the ship length (between 0.8 and 2 times LBP)
3. wave height exceeds a critical level
4. roll damping is low

3.2 *Prediction of amplitude of parametric roll (severity criteria)*

Applicability of any solution based on the Mathieu equation is limited because it is linear: it can only indicate the conditions when parametric roll can be generated, but it is unable to predict the roll amplitude. Such an answer is not enough for engineering practice: the solution must determine how large the parametric rolling might develop if conditions satisfy the susceptibility criteria.

A chain of events between the suitable combination of ship speed, direction, metacentric height and wave characteristics leads to parametric roll resonance. In the real world, however, parametric rolling is limited to finite steady state amplitude, even though the motion angles develop rapidly and up to large amplitudes. Two major nonlinear effects that stabilize parametric rolling are the shape of the righting arm curve and the roll damping. Nonlinearity of the righting arm curve at larger heeling angles leads to variations of natural roll frequencies of the ship, as mentioned before. Since the encounter frequency between ship and waves remains invariable at the moment, the condition which triggers the parametric rolling motion disappears. So, once the roll of ship reaches a certain angle, an extra energy gain during parametric rolling ceases to exist.

Researchers dealing with problems of nonlinear dynamics usually distinguish between two approaches. The first one involves applications of classic mathematical analyses of nonlinear dynamic equations, i.e. modelling in one degree of freedom. The second is based on as a 6-degrees-of-freedom system travelling at a given mean angle relative to the dominant direction of a stationary seaway.

3.2.1 *Models of one degree of freedom*

So far it can be said that the probability of parametric roll has not been yet "controlled" completely and eliminated at the design stage. There is another way to express this variation of GM that can be reflected in the equation of roll motion as follows (ITTC 2006):

$$\ddot{\eta}_4 + 2\zeta\omega_{\eta_4}\dot{\eta}_4 + \omega_{\eta_4}^2\left(1 - h\cos\left(\omega_e t - \varepsilon_e\right)\right)\eta_4 = M_x(t) \tag{9}$$

where ζ is the roll damping ratio, ω_e is the encounter frequency and ε_e is an appropriate phase shift. The variation of the roll natural frequency:

$$\omega_{\eta_4} = \sqrt{mg \cdot \overline{GM}/(I_X + A_{44})} \tag{10}$$

The ITTC procedures for parametric roll implicitly address the presence of multiple steady states when providing the analytical solution of the steady rolling amplitude for a nonlinear 1-DOF model.

If the amplitude of parametric roll is moderate to large, a fifth order polynomial is likely to be required for modelling the restoring moment. It is known that nonlinear terms in the roll equation stabilize parametric roll. The nonlinear equation of roll in a longitudinal sea that could be used for predicting the steady roll amplitude in the vicinity of principal resonance would be like:

$$\begin{aligned}\ddot{\eta}_4 + 2\zeta\omega_{\eta_4}\dot{\eta}_4 + \omega_{\eta_4}^2\left(1 - h\cos\left(\omega_e t - \varepsilon\right)\right)\eta_4 \\ - c_3\omega_0^2\eta_4^{\,3} = 0\end{aligned} \tag{11}$$

which yields

$$\begin{aligned}\ddot{\eta}_4 + 2\zeta\omega_{\eta_4}\dot{\eta}_4 + \omega_{\eta_4}^2\left(1 - h\cos\left(\omega_e t - \varepsilon\right)\right)\eta_4 \\ - c_3\omega_{\eta_4}^2\eta_4^{\,3} - c_5\omega_{\eta_4}^2\eta_4^{\,5} = 0\end{aligned} \tag{12}$$

where c_3 and c_5 are nonlinear stiffness coefficients, corresponding respectively to the third and fifth order restoring terms.

It is essential to ensure that the amplitude of parametric roll oscillation A, which might be generated in an extreme seaway, is kept small. The amplitude A may be expressed by the following solution of (9), (ITTC 2006):

$$A^2 = \frac{4}{3c_3}\left[\left(1-\frac{1}{a}\right) \pm \sqrt{\frac{h^2}{4} - \frac{4\zeta^2}{a}}\right] \tag{13}$$

where $a = 4\omega_{\eta_4}^2 / \omega_e^2$.

For a parametric roll Dunwoody (1989a and 1989b) proposed such a method, which is based on a 1 DOF equation of motion for roll, using a time varying restoring coefficient. This motion equation is known as the Mathieu equation, which was presented earlier.

The restoring moment or static stability GM of the ship when sailing in waves will vary in time and the variation is a function of the actual wetted surface contour and thus depends on the hull lines around the calm water line. A formula for the time varying restoring force, for the upright ship, in regular waves is given by (2), and damped Mathieu equation from (4), where the threshold value for parametric roll in regular waves is according to the 1-DOF model of Francescutto (2001):

$$\frac{\delta GM}{GM} = \frac{4\zeta}{\omega_n} \tag{14}$$

The method proposed by Dunwoody (1989a) is based on the following assumptions (Levadou et al. 2007):

1. The roll motion can be expressed by the differential equation for a single degree of freedom motion method with parametric excitation of the stiffness.
2. There is a linear relation between roll stiffness excitation (GM fluctuations) and wave height.
3. A relation can be established between the spectrum of the stiffness fluctuations with the incident wave height spectrum and the speed of the ship.
4. GM fluctuations produce an effect analogous to a reduction in the roll damping.
5. If the reduction in roll damping is larger than the roll damping from the hull and appendages an unstable situation occurs and the vessel is subject to parametric roll.

According to Dunwoody (1989b) the non dimensional damping reduction follows from:

$$\Delta\xi = \frac{\pi g^2 S_{GM}}{4\omega_\Phi^3 k_{xx}^4} \tag{15}$$

where S_{GM} denotes the spectral density of the GM fluctuations, which can be expressed as the product of the spectral density of the wave encounter spectrum and the square of the transfer function of the GM fluctuations. Parametric roll will occur if the non dimensional roll damping reduction exceeds the total roll damping (B_{total}):

$$\frac{B_{total}}{B_{crit}} - \Delta\xi \le 0 \qquad B_{crit} = 2\sqrt{(A_{xx} + I_{xx})\rho g \nabla GM} \tag{16}$$

There are various other methods of this kind with different complexities and nonlinearities involved. Umeda et al., (2004) and Hashimoto & Umeda, (2004) use time domain numerical simulations of a nonlinear 1-DOF parametric roll model showing the possible coexistence of a stable upright position and a resonant solution in some ranges of parameters.

The other model, due to Bulian et al. (2003), is a 1.5 DOF nonlinear mathematical model developed in order to have the possibility of obtaining an approximate solution in analytical form, preserving on the other hand the fully nonlinear and stochastic features of the phenomenon. The model is based on a 1.5-DOF system: the roll motion is modelled dynamically using a single degree of freedom, whereas the additional half DOF indicates that the coupling with heave and pitch is taken into account by means of hydrostatic calculations without considering dynamic effects. The equation of motion for the 1.5-DOF roll motion can be written as:

$$\ddot{\eta}_4 + d(\eta_4, \dot{\eta}_4) + \omega_0^2 \frac{\overline{GZ}(\eta_4, t)}{\overline{GM}} = 0 \tag{17}$$

A first attempt to analyse the equation (17) is based on the analysis of the fluctuation of the righting arm around the still water (SW) value.

$$\overline{GZ}(\eta_4, t) = \overline{GZ}_{SW}(\eta_4) + \delta\overline{GZ}(\eta_4, t) \tag{18}$$

where $\delta\overline{GZ}(\eta_4, t)$ represents the variation of the righting arm with respect to the still water condition. The final derivative of the 1.5-DOF roll motion formulation depends on two parameters, one of which represents the relative variation of the metacentric height, whereas the second one is used in order to take into consideration some nonlinear aspects of the righting arm fluctuation. The procedure for the analytical approximation of the GZ curve and its use in time domain simulations is conducted as well (Bulian 2005), basically by changing from "crest position" domain (shape of the *GZ* curve) to time domain. An example of roll time history from Bulian (2005) method is presented in figure 9.

While reviewing the methods proposed there is one more aspect of parametric roll that is of concern related to the parametric rolling in irregular waves. Considering a the threshold value for parametric roll different types of stability requirements can be imposed in the case of stochastic parametric excitation (Francescutto & Bulian 2004):

- stability of the mean and asymptotic stability of the mean;
- stability of mean square and asymptotic stability of mean square;
- almost sure asymptotic stability.

They lead to quite different estimates of the threshold. The effective linear damping possesses a stochastic component, which is to be subtracted from the hydrodynamic damping in the same way as it was done in the case of parametric roll in a regular sea.

3.2.2 *Models of three degrees of freedom*

In the literature it is possible to find the numerical evidence of possible multiple steady states, as, e.g., in the application of a 3-DOF model by Neves & Rodríguez (2007), where the simulated steady state roll amplitude was shown to be dependent of the initial conditions in some range of waves and ship speeds. A 3-DOF analytical model for roll, heave and pitch is employed. In this case the equations of motions are fully analytic. The stability boundaries for the loss of stability of the upright position are determined analytically in approximate form, however the amplitude of motions are determined by time domain simulations, since an analytical approach with 3-DOF for the steady state amplitude of motions is mostly prevented by the complexity of the system of equations.

3.2.3 *Models of six degrees of freedom*

Another approach is a simulation based on a 6-degrees-of-freedom system where the application of a full numerical procedure with different possible analytical methods can be applied; see (Umeda et al. 2004, Neves et al. 2003, Ribeiro e Silva et al. 2005, Ribeiro e Silva and Guedes Soares, 2008). As compared with other methods, a numerical simulation method will be more comprehensive and flexible. A fully numerical approach is able, in principle, to address the problem taking into account all the necessary coupling between motions (e.g. France et al. 2003, Shin et al. 2004). In this approach the problem is addressed in time domain only.

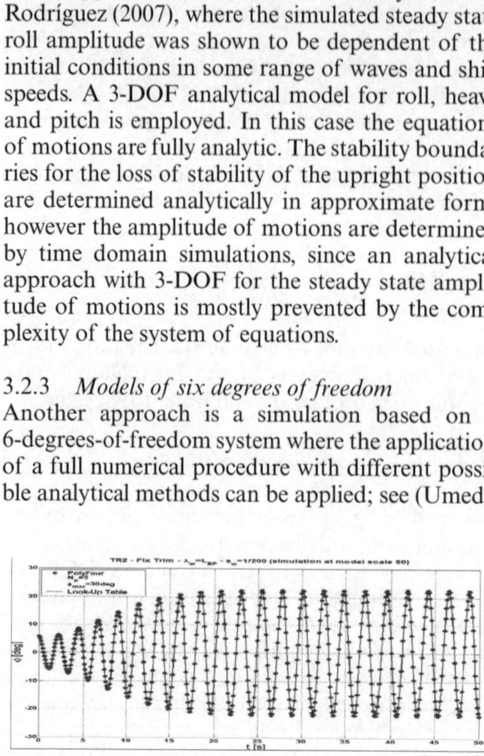

Figure 9. Roll time histories comparison between analytical and numerical method (Bulian 2005).

During the simulation, the chosen mean ship speed and mean wave encounter angle remain constant, whereas the instantaneous ship speed and heading are influenced by the ships motions, which are simulated in all 6 degrees of freedom. All simulation methods are based on assumptions and simplifications in their mathematical model in order to reduce computing time. In large wave conditions with large amplitude motions the assumptions behind both the seakeeping and manoeuvring theory are violated since large variations in wetted surface are not accounted for when the basic coefficients in the models are calculated. Model tests are an essential guidance for the user of non-linear time domain simulation tools to gain experience in the use of a unified model.

For this paper the brief background of the theory behind the code implementation from Ribeiro e Silva and Guedes Soares, (2008) is presented. Forces due to wave excitation (incident wave forces and diffraction forces) and reaction (restoring and radiation) forces due to wave-induced ship motions have to be taken into account. With the model adopted for this case study, radiation and wave excitation forces are calculated at the equilibrium waterline evaluated with a standard strip theory, where the two-dimensional frequency-dependent coefficients of added mass and damping are computed by the Frank's close fit method and the sectional diffraction forces are evaluated using the Haskind-Newman relations.

The non-linear restoring coefficients in heave, roll, and pitch motions in waves are calculated using a quasi-static approach. In the time domain simulations, maintaining the hypothesis of the linear hydrodynamic model of Ribeiro e Silva et al. (2005), the underwater part of the hull is calculated at each time step, together with its geometric, hydrostatic and hydrodynamic properties, represented as follows

$$\Phi = \Phi_U + \Phi_I + \Phi_D + \Phi_R \tag{19}$$

where Φ_U denotes the potential of the steady motion in still water, Φ_I is the incident wave potential, Φ_D is the diffraction potential and Φ_R is the forced motion potential.

Owing to complex interactions between the hull and ship generated waves, the governing equations can be written in the form of integro-differential equations. Combining all hydrodynamic forces with the mass forces, using subscript notations, it is possible to obtain six linear coupled differential equations of motion, in abbreviated form, such as,

$$\sum_{J=1}^{6}\left\{\left(M_{kj}+A_{kj}\right)\ddot{\eta}_j+B_{kj}\dot{\eta}_j+C_{kj}(t)\,\eta_j\right\}=F_k e^{i\omega_e t} \tag{20}$$

where the subscripts k, j are associated with forces in the k-direction due to motions in the j-mode (k = 1, 2, 3 represent the surge, sway and heave directions, and 4, 5, 6 represent roll, pitch and yaw directions). M_{kj} are the components of the mass matrix for the ship, A_{kj} and B_{kj} are the added mass and damping coefficients, $C_{kj}(t)$ are the hydrostatic (time dependent) restoring coefficients, and F_k are the amplitudes of the exciting forces, where the forces are given by the real part of $F_k e^{i\omega_e t}$.

The forces and moments, calculated using these instantaneous properties, are used to derive the resulting translational and rotational motions. These motions are applied to the hull and the time step incremented. Forces acting on the ship body are estimated from the stance of ship at this moment, so that roll and surge motions of the ship for the next time step can be calculated, using nonlinear ship motion equations. This process is cyclic, the previous time step providing conditions for the current time step. As shown in equations (20), a quasi-static approach is adopted to calculate the non-linear restoring coefficients in heave, roll, and pitch motions in waves, in which calculations of significant variations on these restoring coefficients are calculated over the instantaneous waterline.

As illustrated in Figure 10, the two-dimensional hydrostatic force and moment calculations are made using the pressure integration technique along each segment (C_x) of each transverse section of the ship hull, rather than using area and volume integration of the ship offsets. The original theoretical approach to the pressure integration technique outlined by Schalck & Baatrup (1990) has been adopted in conjunction with a practical method to generate the segments required to calculate the hydrostatic pressure distribution under either a regular or irregular wave profile.

Following the theory a computer code was developed implementing the features necessary to predict ship motion instabilities, particularly utilized for a design wave and forward speed that could lead to the occurrence of parametric rolling. In order to run simulations the program is set up by tuning the parameters so that parametric rolling can be expected, such as, the natural period of roll

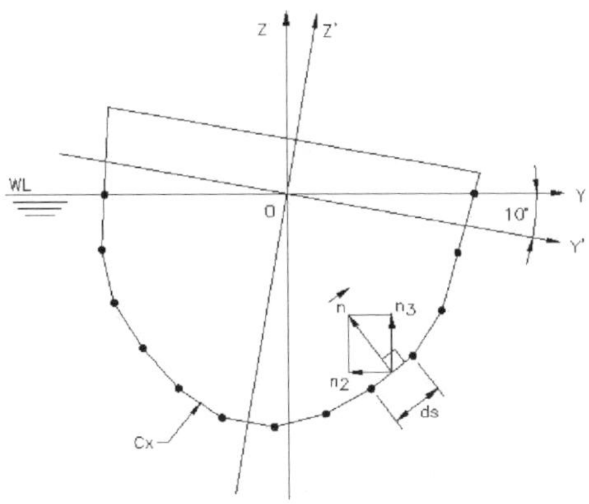

Figure 10. Definition of geometric properties of a transverse section of the underwater part of the hull.

is equal to approximately twice the wave encounter period and the wave length is in the order of the ship length (between 0.8 and 2 times LBP). Wave height needs to exceed a critical level and roll damping is small. The duration of each regular wave simulation record should be long enough for steady state ship motions to be established. The initial conditions for each simulation should be chosen from the steady state regime corresponding to the initial phase of the waves.

Figure 11 shows one of the realisations of the resulting translational and rotational motions observed as parametric resonance phenomena on the C11 post-Panamax containership (Fig. 12).

For the simulations in irregular waves, both randomly distributed amplitude and phase components are utilised to generate an incident wave realization. More specifically, the modelling of the incident wave spectrum is made by a finite number of harmonic waves where a lower and an upper limit for the wave frequency, ω_{min} and ω_{max} are defined. The continuous incident wave spectrum is therefore discretised by a number of harmonic wave components *Nwc* of frequency and amplitude (Ribeiro e Silva et al. 2005).The phase angle of the regular waves are also randomly distributed in the entire range although the energy of the discretised wave systems resulting from the above approach equals the energy of the incident irregular seaway adopted.

To conclude, such simulations are needed in conjunction with the polar diagrams in a written version of the operational guidelines for the Master to mitigate or avoid parametric rolling. Polar diagrams can be produced for different sea states and loading conditions. Figure 13 shows a sample polar diagram where the shaded regions present conditions for rolling resonance and therefore to be avoided.

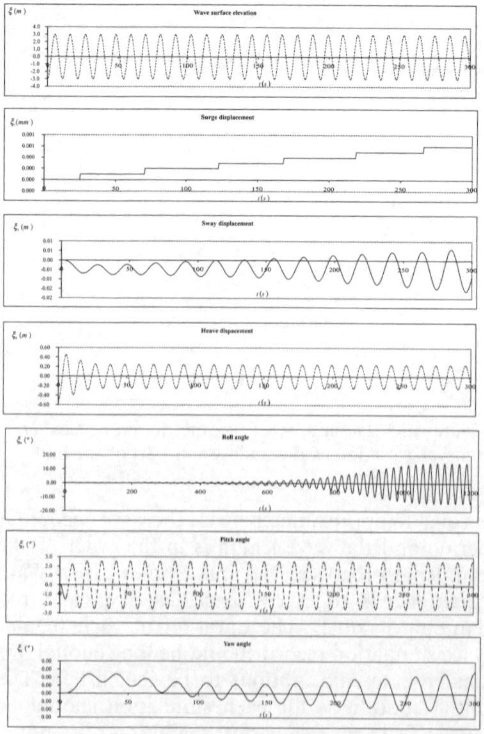

Figure 11. Development of parametric rolling in regular waves (H_W = 6 m, T_W = 12.95 s, speed 8 knots) in numerical calculations. (Turk et al. 2009).

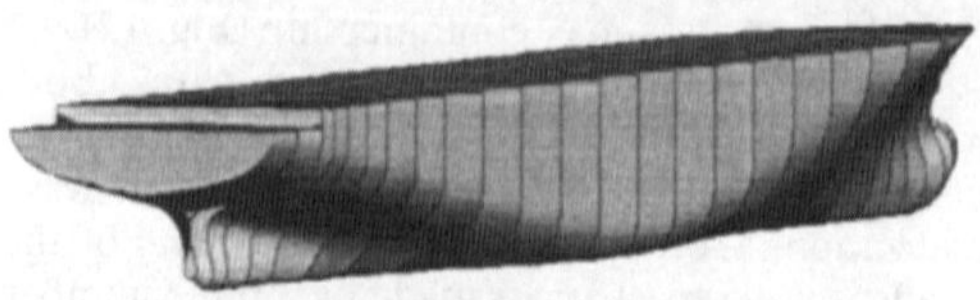

Figure 12. C11 hull form.

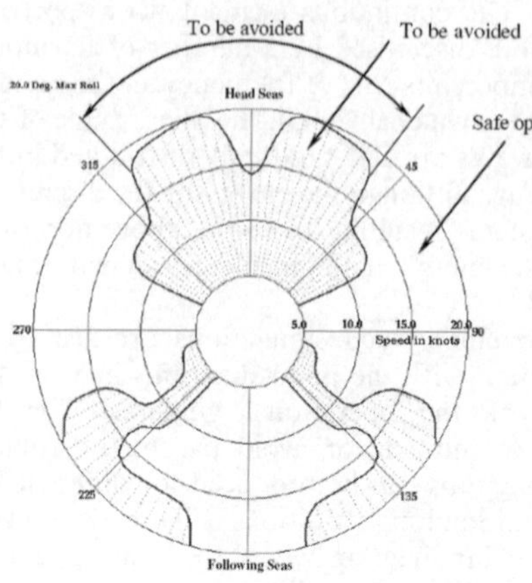

Figure 13. Sample polar diagram (ABS 2004).

4 CLASSIFICATION GUIDES REGARDING PARAMETRIC ROLLING

The American Bureau of Shipping (ABS) was the first society that issued in September 2004 a guide for the assessment of parametric roll resonance in the design of container carriers. This was a supplement to the Rules and the other design and analysis criteria that ABS issues for the classification of container carriers. The Guide contains a brief description of the physical phenomenon of parametric roll resonance, which may cause an excessive roll of a containership in longitudinal (head and following) waves. The Guide also contains a description of criteria used to determine if a particular vessel is vulnerable to parametric roll (susceptibility criteria) and how large these roll motions might be (severity criteria).

In March 2008 ABS has awarded the first class notation specific to parametric roll to three ships in the Hyundai Merchant Marine fleet. The optional class notation was issued against criteria contained in the ABS Guide for the Assessment of Parametric Roll Resonance in the Design of Container Carriers, which provides design and analysis measures to determine if a particular vessel is vulnerable to parametric roll and the potential magnitude of the roll motions. The "PARR C1" notation has been granted to the 4,700 TEU Hyundai Forward, and 8,600 TEU vessels Hyundai Faith and Hyundai Force.

Det Norske Veritas issued a container ship update, in addition to traditional class services during the design, construction and operational phases. They provide owners and operators with increasingly 'popular' services, such as Active Operator Guidance, advice on extreme roll motions (parametric rolling) and how to avoid these for a sea state in which the vessel's hull, given the wave length and height as well as the distance between waves, may be subject to extreme roll motions.

Lloyd's Register of Shipping (LRS 2003) supports initiatives made to introduce guidance on avoidance of parametric roll. At present the IMO sub-committee on stability and load lines and on fishing vessel safety (IMO SLF) is tasked with addressing this issue. In response to concerns voiced in the industry, Lloyd's Register has investigated its container securing requirements. LR suggests simplified Susceptibility Assessment Method, which considers the roll motions of the vessel. Any seakeeping software can be used to determine the ship motions in chosen wave conditions, for a range of speeds and headings. However linear seakeeping codes cannot predict parametric rolling. From the resulting data, the variation in vessel stability examined through the waves will help to avoid the possibility of parametric rolling.

Bureau Veritas (BV 2005) is recommending appropriate solutions of the simplest mathematical model of parametric roll considering the one degree of freedom roll motion equation, in which the restoring coefficient is made time dependent. Calculated hydrostatic variations are approximated by the sinusoidal function, which leads to the Mathieu type equation for roll, from which the regions where the roll instability takes place can be identified. To summarize, BV approach defines a two step procedure,

- preliminary checks using the simplified semi-analytical model
- fully non-linear simulations for critical cases, after which it is possible to produce the polar plots which represents the maximum expected roll motion for a given sea state with respect to ship speed and heading.

Present IMO-SLF work towards the creation and implementation of performance based criteria specifically addressing parametric roll are calling for a deeper understanding of the quite subtle issues that could be associated, in the framework of the approval of simulations/experiments, with the presence of coexisting multiple steady states (IMO 2005).

5 CONCLUSIONS

The theoretical background of the parametric rolling phenomena has been described with emphasis on prediction of the occurrence of parametric rolling (susceptibility criteria) as well as the prediction of the amplitude of parametric rolling (severity criteria).

This paper has described methodologies for the prediction of parametric roll which is sufficiently fast to identify the effect, and thus to avoid or to reduce the effects of that phenomenon.

This paper showed the usefulness of the numerical model as a tool to study the influence of vessel operational restrictions on the tendency of containerships towards parametric roll but mostly to provide understanding of the physical phenomenon that triggers such a highly nonlinear motions and what are the tools to understand it.

Classification guides regarding parametric rolling phenomena are also outlined.

ACKNOWLEDGMENTS

This paper presents material that is in the scope of the course "Modelling of Environment and Environmental Loads" within the project "Advanced Ship Design for Pollution Prevention (ASDEPP)" that was financed by the European Commission through the Tempus contract JEP-40037-2005.

The work was initiated during visits of the first two authors to the Center of Marine Technology and Engineering of Instituto Superior Técnico, which were funded by the ASDEPP project.

REFERENCES

ABS, 2004, Guide for the Assessment of Parametric Roll Resonance in the Design of Container Carriers, *American Bureau of Shipping*, Houston, USA.

Belenky, V.L., Weems, K.M., Lin, W.M., & Paulling, J.R. 2003, Probabilistic Analysis of Roll Parametric Resonance in Head Seas. *Proc. 8th Int. Conf. on Stability of Ships and Ocean Vehicles (STAB)*, Madrid, 325–340.

Bulian, B. 2005, Nonlinear parametric rolling in regular waves—a general procedure for the analytical approximation of the GZ curve and its use in time domain simulations. *Ocean Engineering*, 32, 309–330.

Bulian, G. Francescutto, A. & Lugni, C. 2003, On the nonlinear modeling of parametric rolling in regular and irregular waves. *Proc. 8th International Conference on the Stability of Ships and Ocean* (STAB), Madrid.

Bureau Veritas. 2005, Parametric roll—Bureau Veritas approach, Technical note.

Chang, B.C. 2008, On the parametric rolling of ships using a numerical simulation method. *Ocean Engineering* 35, 447–457

Dunwoody, A.B. 1989a, "Roll of a ship in astern Seas – Metacentric height spectra", *Journal of Ship Research vol* 33, No. 3, pp. 221–228.

Dunwoody, A.B. 1989b, Roll of a ship in astern Seas – Response to GM fluctuations, *Journal of Ship Research vol* 33, No. 4, pp. 284–290.

France, W.N., Levadou, M., Treakle, T.W., Paulling, J.R., Michel, R.K. & Moore, C. 2003, An investigation of head-sea parametric rolling and its influence on container lashing systems. *Marine Technology,* 40, 1 1–19.

Francescutto, A. 2001, "An experimental investigation ofparametric rolling in head waves", *Journal of OffshoreMechanics and Arctic Engineering*, Vol. 123, pp. 65–69.

Francescutto, A. Bulian, G. & Lugni, C., 2004, Nonlinear and stochastic aspects of parametric rolling modeling. *Marine Technology* 41 (2), 74–81.

Hashimoto, H., Umeda, N. 2004, "Nonlinear analysis ofparametric rolling in longitudinal and quartering seas with realistic modeling of roll-restoring moment", *Journal ofMarine Science and Technology, Vol.* 9, pp. 117–126.

Ikeda, Y., Himeno, Y. & Tanaka, N. 1978. *A prediction method for ship roll damping.* Report No. 405, of Department of Naval Architecture, University of Osaka.

IMO Document SLF48/4/12, "*On the development of performance-based criteria for ship stability in longitudinal waves*", Submitted by Italy, 11 July. 2005.

ITTC, 2006, *Recommended Procedures and Guidelines; Predicting the Occurrence and Magnitude of Parametric Rolling.*

Levadue, M. & van 't Veer, R. 2007, Parametric Roll and Ship Design. *Proceedings of the 9th International Conference on Stability of Ships and Ocean Vehicles,*

Lloyd's Register, 2003, Head-sea parametric rolling of container ships, *Marine Services Information Sheet,*

Miller, E. R.; Roll Damping, *Technical Report 6136-74-280*, NAVSPEC. 1974.

Nabergoj, R., Tondl, A., Virag, Z. 1994, Autoparametric Resonance in an Externally Excited System, *Chaos, Solutions & Fractals,* 4, 2, 263–273.

Neves, M.A.S., Perez, N.A., & Valerio, L. 1999, Stability of small fishing vessels in longitudinal waves. *Ocean Engineering*. 26(12), 1389–1419.

Neves, M., Pérez, N., Lorca, O. & Rodriguez, C. 2003, Hull design considerations for improved stability of fishing vessels in waves. *Proc. of STAB'03 8th International Conference on Stability of Ships and Ocean Vehicles,* Madrid.

Neves, Marcelo, A.S. & Rodriguez, Claudio A. 2006 On unstable ship motions resulting from strong non-linear coupling. *Ocean Engineering*. 33(14–15), 1853–1883.

Neves, M.A.S., Rodríguez, C.A. 2007, "Nonlinear Aspects of Coupled Parametric Rolling in Head Seas", *Proc. 10th International Symposium on Practical Design of Ships and Other Floating Structures*, Houston.

Paulling, J.R. & Rosenberg, R.M. 1959, "On Unstable Ship Motion Resulting from Non-linear Coupling", *Journal of Ship Research*, Vol. 3, No.1.

Prpić-Oršić, J. Čorić, V. & Turk, A. 2007, Parametric rolling estimation on high waves // *Proceedings of the 5th International Conference on Computer Aided Design and Manufacturing CADAM 2007*/71–73.

Prpić-Oršić, J. Čorić, V. & Turk, A. 2007, The influence of large amplitudes on the accuracy of parametric rolling estimation, *International Journal Advanced Engineering.* 2, 1; 87–97.

Ribeiro e Silva, S. & Guedes Soares, C. 2000, Time Domain Simulation of Parametrically Excited Roll in Head Seas, *Proceedings of the 7th International Conference on Stability of Ships and Ocean Vehicles (STAB'2000)*; Renilson, M. (ed) Launceston, Tasmania, Australia. 652–664.

Ribeiro e Silva, S., Santos, T.A., & Guedes Soares, C. 2005, Parametrically excited roll in regular and irregular head seas. *Int. Shipbuild. Progr.,* 52, 29–56.

Ribeiro e Silva, S. & Guedes Soares, C. 2008, Non-Linear Time Domain Simulation of Dynamic Instabilities in Longitudinal Waves, *Proceedings of the 27th International Conference on Offshore Mechanics and Arctic Engineering;(OMAE),* Estoril, Portugal, Paper OMAE2008-57973.

Schalck, S. & Baatrup, J. 1990, "Hydrostatic Stability Calculations by Pressure Integration", *Ocean Engineering*, 17, 155–169.

Shin, Y., Belenky, V.L., Paulling, J.R., Weems, K.M. & Lin, W.M. Criteria for parametric roll of large containerships in head seas. *Transactions of SNAME,* Vol. 42, 2004.

Surendrana, S.; Leeb, S.K.; Sohn K.H. 2007, Simplified model for predicting the onset of parametric rolling. *Ocean Engineering* 34, Pages 630–637.

Taylan, M. 2007, On the parametric resonance of container ships. *Ocean Engineering,* 34, 1021–1027.

Turk, A. Matulja, D. Prpić-Oršić, J. & Čorić, V. 2008, Simple methods for parametric rolling estimation, *Proc. Zbornik Sorta 2008* 457–471.

Turk, A. Ribeiro e Silva, S. Guedes Soares, C. & Prpić-Oršić, J. 2009 An investigation of dynamic instabilities caused by parametric rolling of C11 class containership *Proc the 13th Congress IMAM 2009*/Istanbul 167–175.

Umeda, N. Hashimoto, H. Vassalos, D. Urano, S. & Okou, K., 2004, Nonlinear Dynamics on Parametric Roll Resonance with Realistic Numerical Modelling, *International Shipbuilding Progress,* 51:2/3, 205–220.

Advanced Ship Design for Pollution Prevention – Guedes Soares & Parunov (eds)
© 2010 Taylor & Francis Group, London, ISBN 978-0-415-58477-7

Effect of water depth and forward speed on ship dynamic behaviour in waves

I. Kolacio
Faculty of Engineering, University of Rijeka, Rijeka, Croatia

C. Guedes Soares
Centre for Marine Technology and Engineering (CENTEC), Technical University of Lisbon, Instituto Superior Técnico, Lisboa, Portugal

J. Prpić-Oršić
Faculty of Engineering, University of Rijeka, Rijeka, Croatia

ABSTRACT: The aim of this study was to examine the behaviour of a vessel in shallow water and to compare it to the behaviour in deep water. The analysis of dynamic behaviour of tanker in different sea depths is performed for several heading angles. The response amplitude operators for surge, heave and pitch are calculated at different water depths and compared. For solving the problem of the velocity potential and fluid pressure on the submerged surfaces of the bodies the boundary integral equation method, also known as the panel method, is used. A convergence study is made to analyse the influence of the number of panels on results. As the traditional approach to calculate ship motions is using strip theories, the results of the panel code are compared with strip theory results. Finally, to have an idea of the relative importance of the changes in ship dynamic behaviour with depth with the changes with speed, strip theory results are obtained for various ship speeds.

1 INTRODUCTION

The effect of water depth on the dynamic behaviour of a ship in waves is important for shallow water areas such as happens in port approaches. Furthermore the behaviour of ships and ship like offshore platforms I shallow water is relevant as often they are moored in this type of water depths.

Strip theories have been developed mainly in late 60's and several are available, based on somewhat different assumptions and adopting different calculation methods. Guedes Soares (1990) presented a comparison of the results of various codes, which may give an idea of the type of variability that one can expect among strip theory results.

Strip theory has also served as the basis to include some of the most important non-linear effects as done by Fonseca and Guedes Soares (1998) or by Prpić-Oršić, et al. (1999). Several other similar types of theories are available and comparisons between several of them can be found in Watanabe and Guedes Soares (1999).

While the influence of forward speed and ship heading with respect to the waves is usually accounted for by strip theories, the effect of water depth is seldom considered. Some work was done in the late 70's as for example by Van Oortmerssen, (1976), Andersen, (1979), and Svendsen, and Madsen, (1981), but not many more references are found at later times.

Developments of panel methods have allowed a better representation of three dimensional effects than strip theory allows and they have become popular to apply for ships but in particular for offshore structures, in which speed effects do not require being modelled. One such code is WAMIT, which is based on theoretical aspects described for example in Lee and Newman (2004). WAMIT incorporates several features including the ability of determining motions of stationary floaters in deep and shallow water. Therefore this code has been adopted in the present study to assess the effect of shallow water on the dynamic behavior of a tanker.

Another aspect considered in the present study is the comparison of the performance of the panel code compared with a strip theory code, in conditions applicable to both, i.e. deep water and zero forward speed. The linear strip theory code adopted is one based on the theory of Salvesen et al. (1970).

Finally, the results of strip theory predictions are obtained for different ship speeds are compared to establish the effect of the forward speed and to compare it with the relative change that shallow water induces on a stationary ship. These comparisons will allow and understanding of the relative order of magnitude of the two effects. In this way the less studied effect of shallow water can be related to the well known effect of ship speed on wave induced motions.

2 MATHEMATICAL MODEL

2.1 *Introduction*

The dynamic response of the ship advancing in a seaway is complicated phenomenon involving the effect of several distinct hydrodynamic forces. The ship advancing at a steady mean forward speed with arbitrary heading in the train of regular waves will oscillate in six degrees of freedom. Consequently, six non-linear equations of motion, with six unknowns, must be set up and solved simultaneously. However, for slender vessels in low or moderate sea states it is possible to assume that the ship motions will be small and hence to apply a linear theory. With these assumptions, the linear theory can be applied to predict ship motions.

Experimental and theoretical investigations have shown that a linear analysis of ship motions gives excellent predictions over a wide variety of sea conditions and vessel types. The exceptions are the large amplitude ship oscillations in the resonant range. The linear theory neglects the variation of the submerged part of the ship hull. The velocity potential is computed for the mean hull position in still water, what is acceptable assumption for the small amplitudes, but on the higher waves the nonlinearity can not be neglected.

The most often used tool for seakeeping prediction of slender ships is the strip method. Strip theory considers the ship as being made from finite number of transverse two-dimensional slices which are rigidly connected. The hydrodynamic problem of each slice is solved by formulating the hydrodynamic problem of an oscillating infinitely long cylinder with the cross sections identical to slice cross section (Figure 1).

Thus in essence the three-dimensional problem is reduced to a set of two dimensional boundary value problems. Fundamentally, strip theory is valid only for a long and slender body but it has been shown that it could be successfully applied for ships with a length to beam ratio larger than three.

Strip theory is known as a high frequency theory but due to the fact that at lower frequencies the problem becomes quasi- static, the hydrodynamic effects are not so important and the theory also

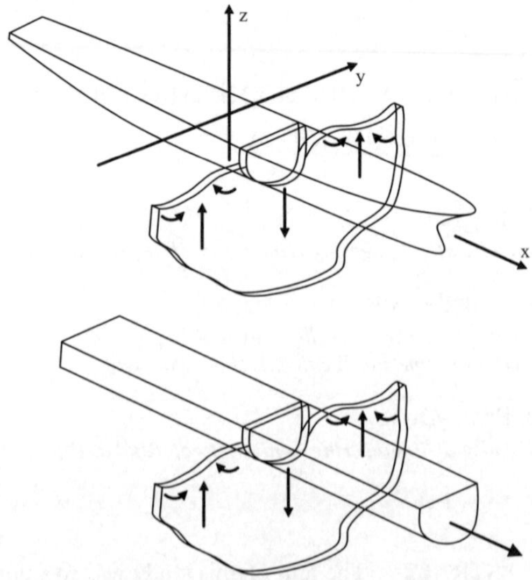

Figure 1. Principle of strip method.

gives good approximations for lower encounter frequencies.

WAMIT is a computer program based on the linear and second-order potential theory for analyzing floating of submerged bodies, in the presence of ocean waves (WAMIT, 2006). The boundary integral equation method (BIEM), also known as the panel method, is used to solve the velocity potential and fluid pressure on the submerged surfaces of the bodies. The second-order module, Version 6S, provides complete second-order nonlinear quantities in addition.

WAMIT includes several unique options to facilitate its application in the most effective manner. In addition to the conventional low-order method, where the geometry is represented by small quadrilateral panels and the velocity potential is assumed constant on each panel, a powerful higher-order method is also available based on the representation of the potential by continuous B-splines and with a variety of options to define the geometry of the body surface approximately or exactly.

The WAMIT software includes a special utility F2T which is intended to transform first-order frequency-domain outputs to time-domain impulse response functions.

The main difference between the strip method and the panel method approaches is the fact that the panel method, due to its three-dimensional nature, can be applied to arbitrary shaped bodies, despite being restricted to zero mean forward speed. On the other side, the strip method is suitable for seakeeping analysis of long slender ships with forward speeds.

All the following analyses have been performed by a linear strip code based on the theory of Salvesen et al. (1970) and by WAMIT (2006), in a conventional low-order method.

2.2 *The boundary value problem*

The objective is to evaluate the unsteady hydrodynamic pressure, loads and motions of the body, as well as the induced pressure and velocity in the fluid domain. The free-surface and body-boundary conditions are linearized, the flow is assumed to be potential, free of separation or lifting effects. Harmonic time dependence is adopted. Figure 2 illustrates a three-dimensional body interacting with plane progressive waves in water of finite water depth H.

The flow is assumed to be potential so the Laplace equation can be applied in the fluid domain

$$\nabla^2\Phi = \frac{\partial^2\Phi_w}{\partial x^2} + \frac{\partial^2\Phi_w}{\partial y^2} + \frac{\partial^2\Phi_w}{\partial z^2} = 0 \tag{1}$$

The velocity potential of wave defined by amplitude ζ_a and angular frequency ω is defined as

$$\Phi_0(x,y,z;t) = \frac{\zeta_a g}{\omega}\frac{\cosh k(H+z)}{\cosh kH} \cdot \sin(kx\cos\beta + ky\sin\beta - \omega t) \tag{2}$$

where g stands for gravitational acceleration and the heading angle β is the angle between the wave propagation vector and the x axis. The wave number k is defined by the dispersion relation

$$\frac{\omega^2}{g} = k\tanh kH \tag{3}$$

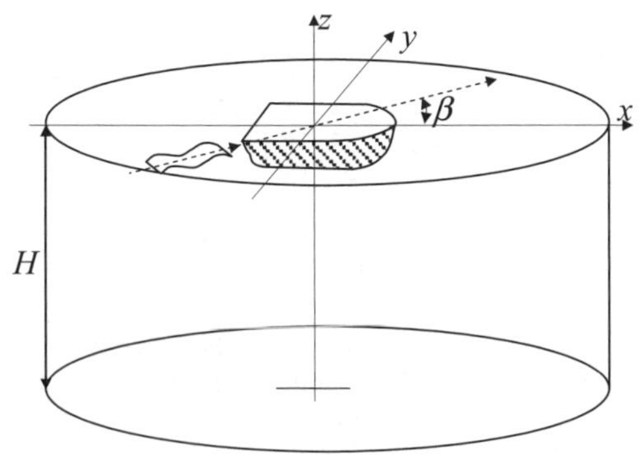

Figure 2. Coordinate system.

The harmonic dependence allows the definition of the complex velocity potential ϕ_0, related to Φ_0 by

$$\Phi_0 = \mathrm{Re}\left[\zeta_a\phi_0 \mathrm{e}^{-i\omega t}\right] \tag{4}$$

The complex velocity potential of the incident wave is defined as

$$\phi_0 = -i\frac{g}{\omega}\frac{\cosh\left[k(H+z)\right]}{\cosh(kH)}\mathrm{e}^{i(kx\cos\beta + ky\sin\beta)} \tag{5}$$

The linearization of the problem permits decomposition of the velocity potential on radiation and diffraction part

$$\Phi(x,y,z;t) = \Phi_0(x,y,z;t) + \sum_{j=1}^{6}\Phi_j(x,y,z;t) + \Phi_7(x,y,z;t) \tag{6}$$

or

$$\Phi(x,y,z;t) = \mathrm{Re}\left\{\left(\begin{array}{l}\zeta_a(\phi_0(x,y,z) + \phi_7(x,y,z)) \\ +\sum_{j=1}^{6}\delta_j\phi_j(x,y,z)\end{array}\right)e^{-i\omega t}\right\} \tag{7}$$

With the same assumptions adopted in the progressive harmonic wave model (ideal fluid, potential flow, dynamic and kinematic boundary condition on free surface and of impermeability of sea bottom), the diffraction wave component has to satisfy some additional conditions: boundary condition on the wetted body surface (Figure 3) and radiation condition in infinity.

The first of the two additional conditions is the consequence of impermeability of the body

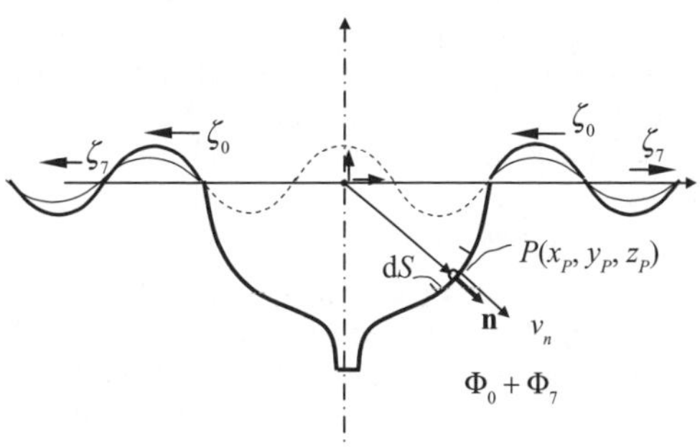

Figure 3. Boundary condition on wetted surface.

wetted surface S and can be expressed by the normal component of the resulting normal flow velocity v_n at point P

$$v_n = \mathbf{n}\,\mathrm{grad}\,\Phi = \frac{\partial \Phi}{\partial n} = 0 \text{ on } S, \tag{8}$$

or by the appropriate flow velocities potentials of each component

$$v_n = \frac{\partial}{\partial n}(\Phi_0 + \Phi_7) = 0 \text{ on } S, \tag{9}$$

from which follows the condition to determine the unknown function Φ_7

$$\frac{\partial \Phi_7}{\partial n} = -\frac{\partial \Phi_0}{\partial n} \text{ on } S, \tag{10}$$

as well as its complex part ϕ_7

$$\frac{\partial \phi_7}{\partial n} = -\frac{\partial \phi_0}{\partial n} \text{ on } S, \tag{11}$$

For the complex radiation potential the impermeability condition can be expressed as

$$\mathbf{n}\,\mathrm{grad}\phi_j = \frac{\partial \phi_j}{\partial n} = v_{nj}^P, \quad j = 1 \ldots 6, \tag{12}$$

where v_{nj}^P is the normal velocity component of point P because of motion of jth degree of freedom.

3 ANALYSIS OF THE RESULTS

The vessel used in this study is a tanker (Figure 4) with the following characteristics:

$L_{OA} = 275$ m
$L_{PP} = 264$ m
Breadth = 45.1 m
Mean draught = 17.1 m
Displacement = 173164 t

Throughout the analysis Response Amplitude Operators (RAOs) were determined and their features are shown as a function of the number of panels and water depth in the model.

3.1 *Convergence study*

A comparison was performed for six different sets of panel grid, starting with 129 panels as the smallest number, then 214 by increasing the number of transverse panels, 246 by increasing the number of longitudinal panels, 310, 480 and 1680. The comparison was performed to reduce the run time for each case that is to be examined.

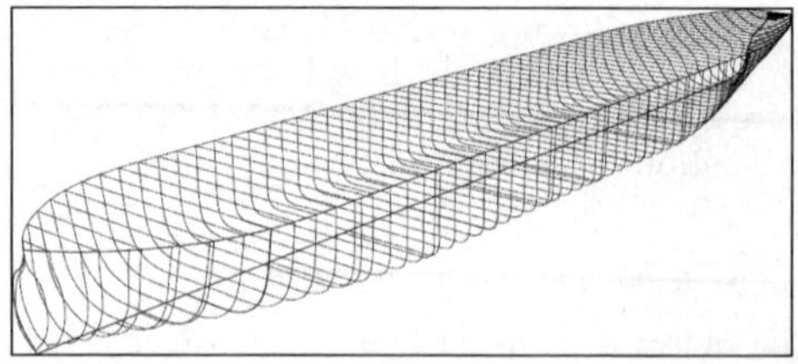

Figure 4. View of the vessel geometry from the bow.

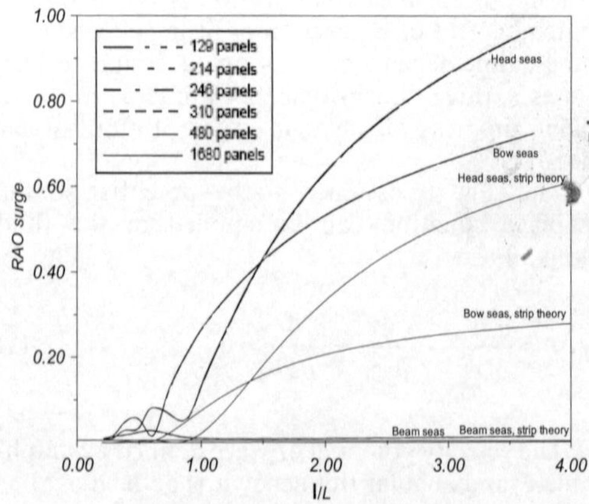

Figure 5. RAO for surge for WAMIT meshes with 129 to 1680 panels and for the strip theory.

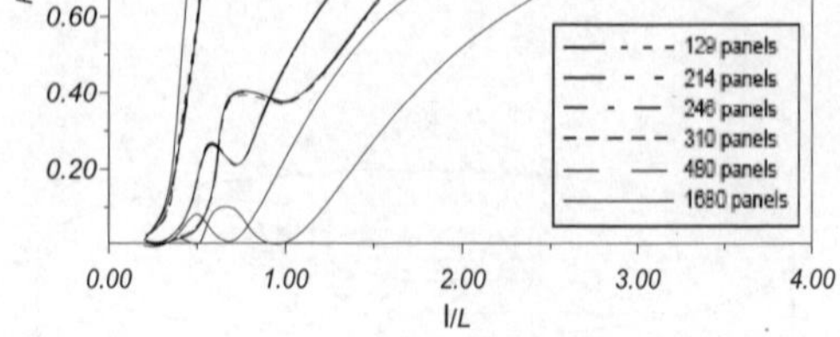

Figure 6. RAO for heave for WAMIT meshes with 129 to 1680 panels and for the strip theory.

For each number of panels, the cases of three response amplitude operators—heave, pitch and surge, and three wave headings—head, beam and bow seas are examined. Surge response amplitude operators are shown in Figure 5, heave in Figure 6 and pitch in Figure 7.

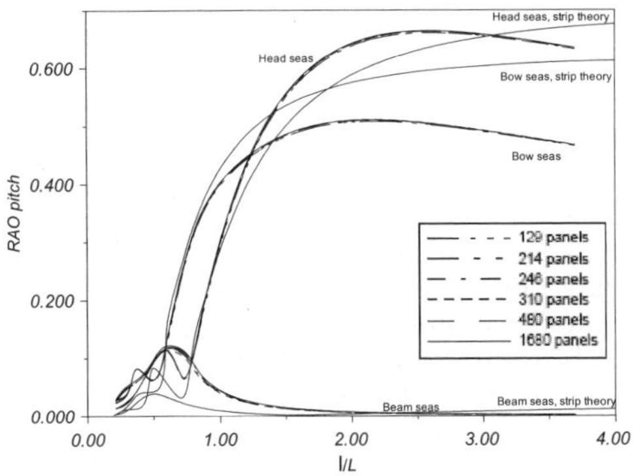

Figure 7. RAO for pitch for WAMIT meshes with 129 to 1680 panels and for the strip theory.

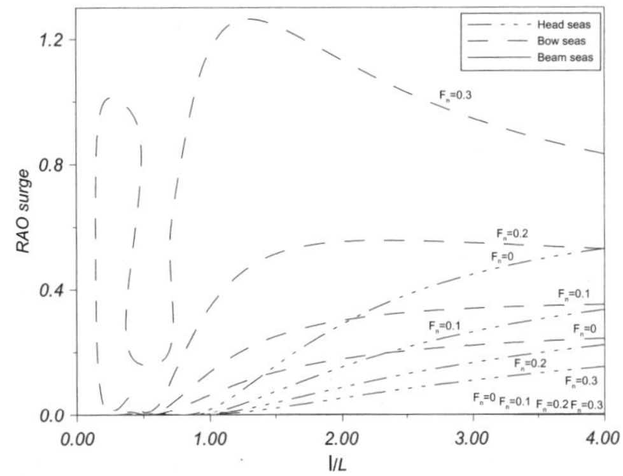

Figure 8. Comparison of ship speed effect on surge RAO.

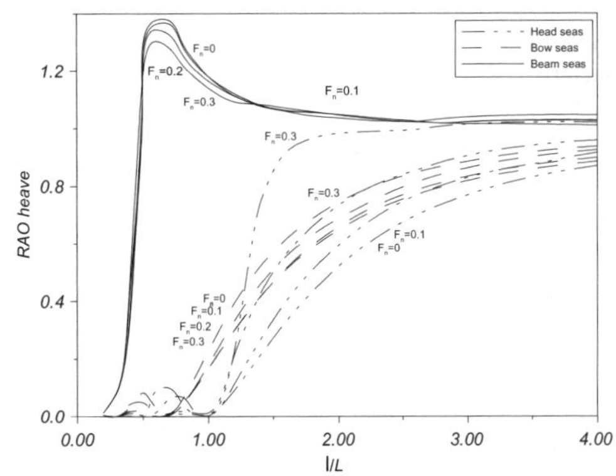

Figure 9. Comparison of ship speed effect on heave RAO.

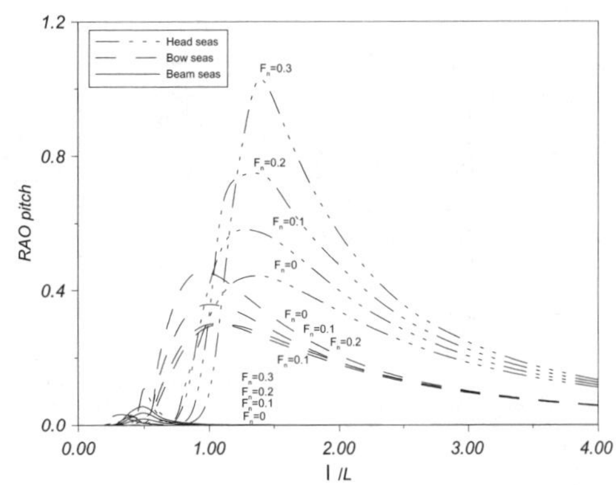

Figure 10. Comparison of ship speed effect on pitch RAO.

When the results for different number of panels are compared, it can be noted the overall results do not differ from each other. The conclusion is therefore that the method performed in a stable manner and can give good results even with the smaller number of panels.

However, since the run time necessary to perform the analyses does not increase significantly when the 480 number of panels case is run, compared to the 129 number of panels one, the number of panels for which the analyses were performed was chosen to be 480. The cases compared gave no significant difference, but it was also possible that some other results given by the same analysis could be affected more by a smaller number of panels.

For the analysis performed by the linear strip theory method, 23 strips were defined, and the mass inertia moments were approximated by simplified expressions.

The results obtained by the panel code are compared with values calculated with strip code in Figures 5 to 7. It can be noted that generally for zero speed the values obtained by strip theory are smaller then those obtained by panel method.

3.2 *Comparison of results for different ship speed*

Figures 8 to 10 present results obtained for the four different Froude numbers: $F_n = 0.0$, 0.1, 0.2, and 0.3, and for head, bow and beam wave seas heading. Figure 8 represents the effect of ship speed for surge, Figure 9 for heave and Figure 10 for pitch response amplitude operators. The values of RAOs for head, bow and beam wave seas heading are presented in Table 1, Table 2 and Table 3 respectively.

Ship speed has important influence on ship motions especially at head wave seas. The values of the amplitude of all three motions in head wave seas increase as the speed increases. For heave and

Table 1. RAO values for head seas and different ship speed.

Froude number	λ/L	Surge	Heave	Pitch
0.0	**0.5**	0.3336	0.5208	0.3376
	1.0	0.0218	0.0083	0.3207
	2.0	0.0025	0.0044	0.1093
0.1	**0.5**	0.1768	0.5948	0.3990
	1.0	0.0084	0.0034	0.4167
	2.0	0.0005	0.0006	0.0199
0.2	**0.5**	0.1020	0.7336	0.4734
	1.0	0.0034	0.0134	0.3289
	2.0	0.0001	0.0002	0.0041
0.3	**0.5**	1.8837	11.0300	3.2535
	1.0	0.7430	1.8752	1.7714
	2.0	0.3172	1.1590	0.9508

Table 2. RAO values for bow seas and different ship speed.

Froude number	λ/L	Surge	Heave	Pitch
0.0	**0.5**	0.0015	1.0360	0.0000
	1.0	0.0007	1.1811	0.0001
	2.0	0.0003	1.0485	0.0059
0.1	**0.5**	0.0015	1.0483	0.0000
	1.0	0.0007	1.1806	0.0006
	2.0	0.0003	1.0640	0.0190
0.2	**0.5**	0.0015	1.0478	0.0001
	1.0	0.0007	1.1651	0.0016
	2.0	0.0003	1.1162	0.0368
0.3	**0.5**	0.0015	1.0466	0.0003
	1.0	0.0007	1.1406	0.0027
	2.0	0.0003	1.1680	0.0554

Table 3. RAO values for beam seas and different ship speed.

Froude number	λ/L	Surge	Heave	Pitch
0.0	**0.5**	0.2063	0.7440	0.2066
	1.0	0.0751	0.2283	0.4528
	2.0	0.0006	0.0826	0.0182
0.1	**0.5**	0.3478	0.6881	0.1877
	1.0	0.1589	0.1796	0.3584
	2.0	0.0019	0.0487	0.0163
0.2	**0.5**	0.6368	0.6595	0.1793
	1.0	0.3967	0.1653	0.2999
	2.0	0.0101	0.0188	0.0072
0.3	**0.5**	1.3072	0.6481	0.1865
	1.0	1.3397	0.1673	0.2928
	2.0	0.1852	0.0104	0.0057

pitch motions at beam and bow wave seas the influence of ship speed is not so strong. For surge the influence is noticeable for head and bow wave seas while for beam sea is almost negligible.

3.3 *Comparison of results for different water depths*

Figures 11 to 13 present results obtained with WAMIT for the head wave seas heading. The values of RAOs for surge, heave and pitch are presented in Table 4. Figure 11 represents water depth comparison for surge, Figure 12 for heave and Figure 13 for pitch response amplitude operators, head wave heading.

All curves for motions of the vessel follow the same pattern of behaviour; they increase with the period. Surge and pitch motions increase gradually with the period. The bumps that appear on the curves are greater for smaller water depths, and come into sight at greater periods. Heave motions for all water depths are similar, although it can be noticed that the values for infinitive water depths are somewhat greater than smaller water depths.

Figures 14, 15 and 16 present results obtained for the bow seas wave heading. The values of RAOs for surge, heave and pitch are presented in Table 5.

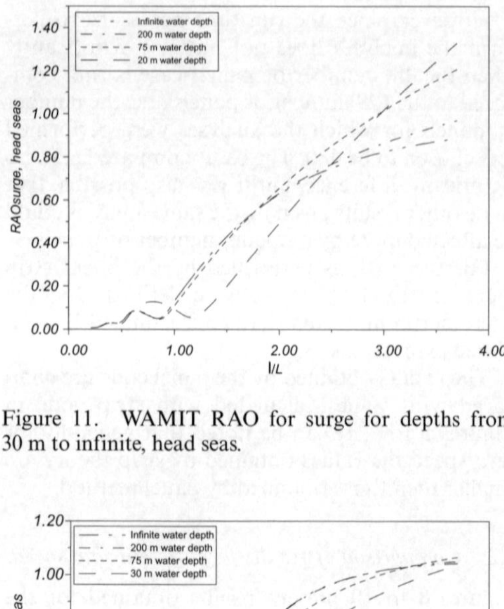

Figure 11. WAMIT RAO for surge for depths from 30 m to infinite, head seas.

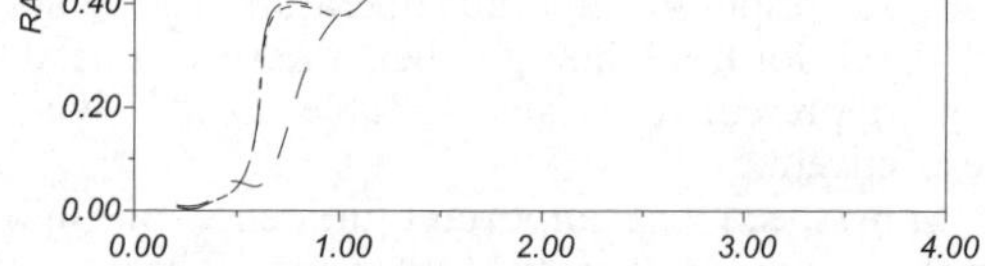

Figure 12. WAMIT RAO for heave for depths from 30 m to infinite, head seas.

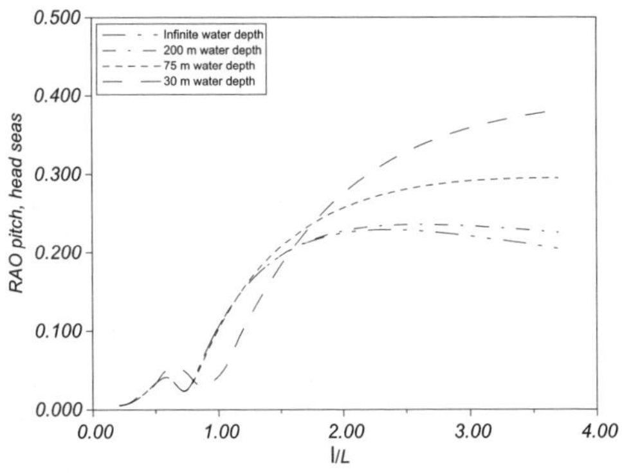

Figure 13. WAMIT RAO for pitch for depths from 30 m to infinite, head seas.

Table 4. WAMIT RAO values for head seas.

Water depth	λ/L	Surge	Heave	Pitch
30 m	**0.5**	0.0222	0.0571	0.0006
	1.0	0.0932	0.3798	0.0009
	2.0	0.4444	0.6714	0.0053
75 m	**0.5**	0.0298	0.0373	0.0006
	1.0	0.1088	0.3763	0.0020
	2.0	0.6177	0.7683	0.0050
200 m	**0.5**	0.0299	0.0368	0.0006
	1.0	0.1297	0.3736	0.0021
	2.0	0.6038	0.8249	0.0045
Infinite	**0.5**	0.0299	0.0368	0.0006
	1.0	0.1299	0.3738	0.0021
	2.0	0.6009	0.8271	0.0045

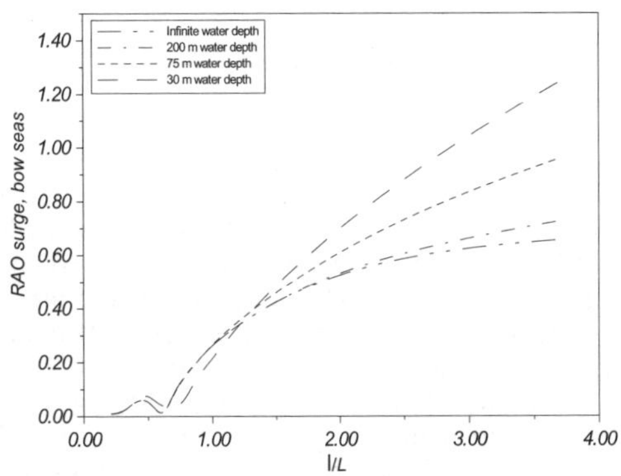

Figure 14. RAO for surge for depths from 30 m to infinite, bow seas.

Figure 14 represents water depth comparison for surge, Figure 15 for heave and Figure 16 for pitch response amplitude operators, bow wave heading.

All curves for motions of the vessel follow the same pattern of behaviour; they increase with the period.

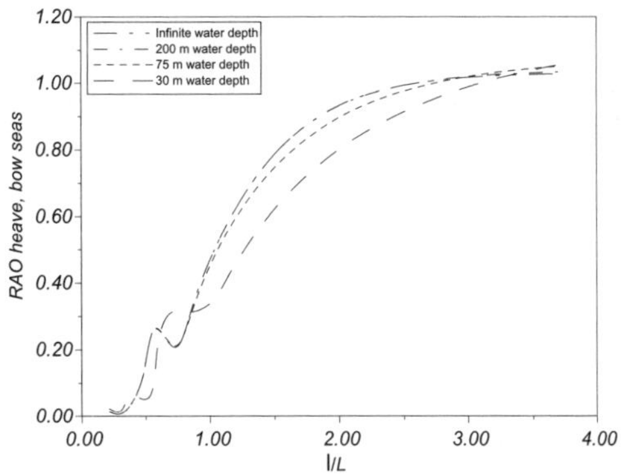

Figure 15. RAO for heave for depths from 30 m to infinite, bow seas.

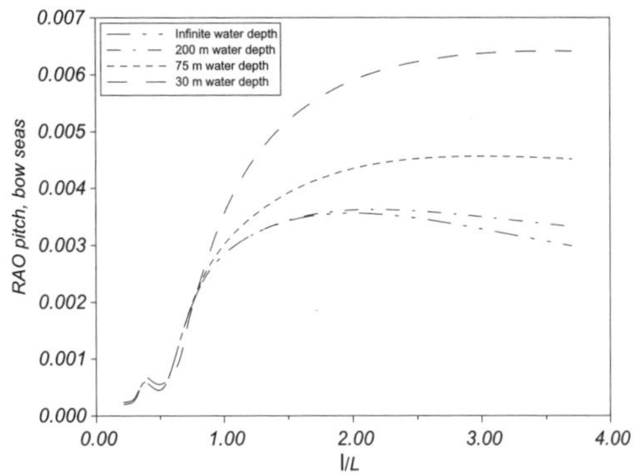

Figure 16. RAO for pitch for depths from 30 m to infinite, bow seas.

Table 5. WAMIT RAO values for bow seas.

Water depth	λ/L	Surge	Heave	Pitch
30 m	**0.5**	0.0763	0.0515	0.0006
	1.0	0.2155	0.3382	0.0035
	2.0	0.6657	0.7770	0.0058
75 m	**0.5**	0.0586	0.1248	0.0004
	1.0	0.2660	0.4508	0.0030
	2.0	0.5865	0.8775	0.0043
200 m	**0.5**	0.0587	0.1266	0.0004
	1.0	0.2629	0.4686	0.0028
	2.0	0.5169	0.9150	0.0036
Infinite	**0.5**	0.0587	0.1267	0.0004
	1.0	0.2631	0.4686	0.9158
	2.0	0.5118	0.0028	0.0036

The results for bow seas are similar to those for head seas, and the conclusions drawn are consequently similar.

Figures 17 to 19 present results obtained for the beam seas wave heading. The values of RAOs for surge, heave and pitch are presented in Table 6.

Figure 17 represents water depth comparison for surge, Figure 18 for heave and Figure 19 for pitch response amplitude operators, beam wave heading.

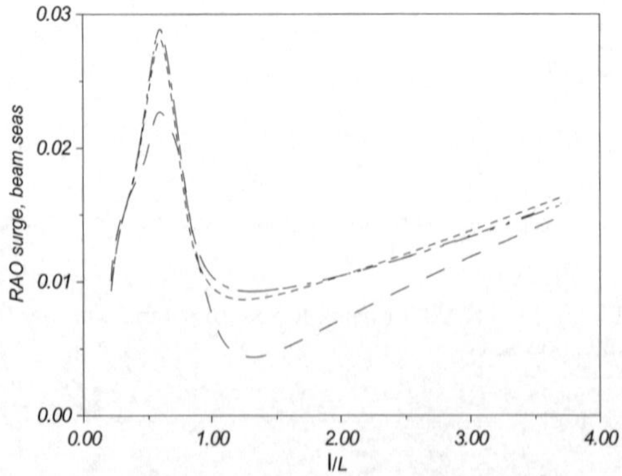

Figure 17. WAMIT RAO for surge for depths from 30 m to infinite, beam seas.

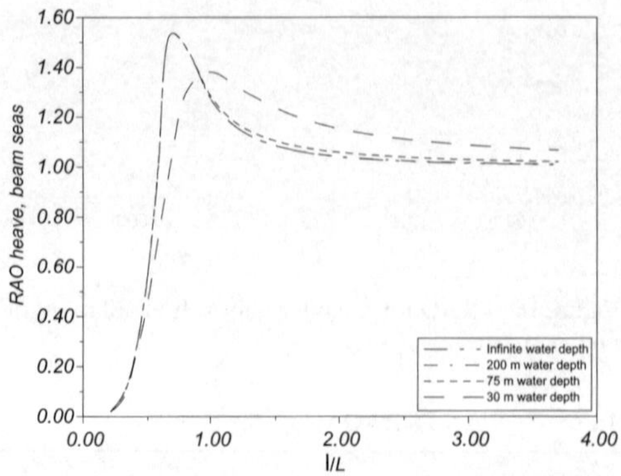

Figure 18. WAMIT RAO for heave for depths from 30 m to infinite, beam seas.

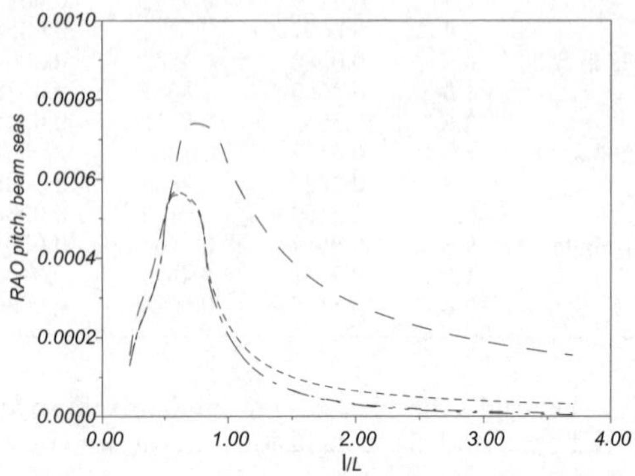

Figure 19. WAMIT RAO for pitch for depths from 30 m to infinite, beam seas.

Table 6. RAO values for beam seas.

Water depth	λ/L	Surge	Heave	Pitch
30 m	**0.5**	0.0193	0.4070	0.0007
	1.0	0.0077	1.3817	0.0010
	2.0	0.0067	1.1627	0.0004
75 m	**0.5**	0.0227	0.4932	0.0006
	1.0	0.0097	1.2868	0.0004
	2.0	0.0102	1.0631	0.0001
200 m	**0.5**	0.0231	0.4894	0.0006
	1.0	0.0105	1.2760	0.0003
	2.0	0.0102	1.0455	0.0001
Infinite	**0.5**	0.0231	0.4894	0.0006
	1.0	0.0105	1.2757	0.0003
	2.0	0.0102	1.0445	0.0000

Compared to all other results for different wave headings, the behaviour of the vessel at wave heading of 90° differs significantly. All curves for motions of the vessel follow the same pattern of behaviour; they increase with the period.

Surge and pitch response amplitude operators show a great growth of the values as water depth reduces, so for 30 m water depth it can be noted that the values are much greater than those for infinite water depths. The values for pitch motions are very small, as expected, since small motions are expected to occur with this wave heading.

Heave response amplitude operators for all water depths are similar, although it can be noticed that the peaks at infinitive water depth is somewhat greater than the one at smaller water depths; and it occurs at a smaller period.

4 DISCUSSION OF RESULTS

The study has been performed for water depths of 30 m, 40 m, 50 m, 75 m, 100 m, 150 m, 200 m and infinite. Since the behaviour of the vessel changes gradually with the reduction of the water depth, the results that have been shown are only the ones for extreme water depths—30 m and infinite, as well as 75 m and 200 m, in order to establish the limits of behaviour of the vessel without overcrowding the figures.

Values for all quantities evaluated in this study can be divided into two groups—one is related with the sectors of head and bow wave headings, and the other for beam wave heading. Due to extreme behaviour of the vessel in beam waves, the final conclusions was also be derived separately.

For head and bow seas, response amplitude operators for surge motions have very similar behaviour for each water depth at different headings. They all reach a maximum value at approximately a non dimensional wave length of 1.40, and for smaller periods the values are almost identical (up to approximately T = 10 s).

As the period increases, curves still behave correspondingly, but the values for smaller water depths are greater than those for infinite water depth; the peaks at bumps also occur with greater periods and values for smaller water depths. The curves for infinite water depth have a tendency of slowing its growth at high periods.

All the values are somewhat smaller for smaller water depths, in the range of about 30 per cent, for smaller λ/L, and for greater ones, they increase up to 50 per cent.

Heave motion curves all behave in a similar way, and the values for lower water depths increase substantially compared to those for infinite water depth (up to 60 per cent). At smaller periods the curves have similar values.

Values for pitch motions for each of the water depths at different headings are very similar. Also, among different water depths, the values for smaller periods are very similar. After the maximal value has been reached at higher water depths for infinite water depth, the curves start slowing its growth, while the maxima are still to be reached for low water depths.

It is important to note that the pitch motions for 30 m water depths are much greater (around 70 per cent) than those for infinite water depths, and their values increase double at maximum period $T = 25$ s.

For beam seas, response amplitude operators for surge at infinite and 30 m water depths have a similar behaviour. However, all the values for surge motions at 90° wave heading are quite small, and can therefore be neglected. This is as expected, since no high surge motions are expected to appear at this wave heading.

For heave motions the values for 30 m water depth are about 20 per cent higher than those for infinite water depth.

For pitch motions, response amplitude operators are greater for 30 m water depth by about 40 per cent. This shows that the pitch motions are much greater at small water depths with beam heading waves.

5 CONCLUSIONS

The comparison of the results of the WAMIT panel code for various water depths and the strip theory code for various speeds has allowed interesting quantitative results about the effect of water depth and ship speed on the wave induced motions. It was also possible to compare the predictions of both codes for infinite water depth and zero speed.

The comparison of the motion RAOs for different water depths show that if the water depth is less than twice the draught of the vessel, the wave-induced motion becomes significantly larger than in deep water. Furthermore the peak of the frequency response function of the motions shifts towards lower frequencies with decreasing water depth and thus the influence of the water depth also depends on the shape of the wave spectrum.

The results obtained by panel code compared with values calculated with strip code shows that for zero speed the values obtained by strip theory are smaller then those obtained by panel method by about 30% in average.

The influence of ship speed on motions is significant, especially for head seas, where the percentages can increase for up to 100 per cent for head, 80 per cent for bow and 50 per cent for beam seas when Froude number is increased from 0 to 0.3.

ACKNOWLEDGEMENTS

The work reported was performed during a visit that the first author made to the Centre for Marine Technology and Engineering at Instituto Superior Técnico, which was funded by the Portuguese Foundation of Science and Technology.

REFERENCES

Andersen, P. 1979. Ship Motions and Sea Loads in Restricted Water Depth, *Ocean Engineering*, Vol. 6, pp. 557–569.

Fonseca, N. & Guedes Soares, C. 1998. Time-Domain Analysis of Large-Amplitude Vertical Ship Motions and Wave Loads, *Journal of Ship Research,* 42(2):139–153.

Guedes Soares, C. 1990. Comparison of Measurements and Calculations of Wave Induced Vertical Bending Moments in Ship Models, *International Shipbuilding Progress.* 37(412):353–374.

Lee, C.-H. & Newman, J.N., 2004. Computation of Wave Effects Using the Panel Method, *in Numerical Models in Fluid-Structure Interaction*, S. Chakrabarti, Ed., WIT Press, Southampton.

Prpić-Oršić, J., Čorić, V. & Dejhalla, R. 1999. Nonlinear Heave and Pitch Motion of Ship in Head Waves, *Brodogradnja,* Vol. 47, Br. 3, pp. 239–245.

Salvesen, N., Tuck, E.O., & Faltinsen, O. 1970. Ship motions and sea loads, *Transactions Society Naval Architects Marine Engineers*, Vol. 78, pp. 250–287.

Svendsen, I.A. & Madsen, P.A. 1981. The Dynamics of Wave Induced Ship Motions in Shallow Water, *Ocean Engineering,* Vol. 8, No. 5, pp. 443–479.

Van Oortmerssen, G. 1976. The Motions of a Ship in Shallow Water. *Ocean Engineering*, Vol. 3, pp. 221–255.

Watanabe, I. & Guedes Soares, C. 1999. Comparative Study on Time Domain Analysis of Non-Linear Ship Motions and Loads, *Marine Structures.* 12(3):153–170.

WAMIT. 2006. *WAMIT Manual*, Wamit Incorporated and Massachusetts University of Technology, Boylston, USA.

Advanced Ship Design for Pollution Prevention – Guedes Soares & Parunov (eds)
© 2010 Taylor & Francis Group, London, ISBN 978-0-415-58477-7

Comparison of various classification societies requirements regarding bow area subject to wave impact loads

D. Bajič
Lloyd's Register EMEA, Trieste, Italy

J. Prpić-Oršić
Faculty of Engineering, University of Rijeka, Rijeka, Croatia

ABSTRACT: Bowflare slamming describes dynamic wave impact on the bow side shell structure above the design waterline. During water entry the bow structure is subject to high pressure loads which sometimes lead to local damages that usually do not affect the survivability of the ship. For practical purposes, the values of these pressures are usually obtained by simplified procedures suggested by classification societies. In this paper the comparison of bowflare slamming pressures estimated by three classification societies are presented. The results suggest significant difference in estimated values of pressure as well as values of plating thickness.

1 INTRODUCTION

In order to increase area available for cargo loading, in particular on container and Ro-Ro ships, the recent trend is to widen deck forward. Consecutively bow flare angles are lowered. Bow flare angle is intended as an angle in a transverse section between side shell and horizontal axes. This raises the flare slamming pressure excessively and may cause structural damages. It has been observed that such ships often suffer damages in the fore end region while sailing in heavy weather conditions. It has been realized that damages are caused by dynamic wave impacts in these regions due to excessive ship motions. The problem is recently exacerbated by modern trend to "drive through" bad weather particularly when in conjunction with high speeds.

The major problem in this respect is to accurately predict values of these pressures in order to determine required reinforcements in the bow area. Various classification societies have developed in the past various sets of Rules. These Rules have been created in a different ways combining analytical techniques with results of direct calculations all calibrated with damages experienced in service. Intention of this paper is to investigate differences in their approaches in terms of predicted pressures in a bow flare area as well as sensitivity to different input parameters. Lloyd's Register (LR), American Bureau of Shipping (ABS) and Det Norske Veritas (DNV) have been chosen here for the purpose of this investigation.

2 BOWFLARE SLAMMING PHENOMENA

As shown in Figure 1 the problem of bowflare slamming is the problem of a symmetrical body penetrating the water surface. The y and z axes are taken along the undisturbed free surface and along the body's centerline pointing upward, respectively. The problem corresponds to the vertical motion of a hull section in heading or following waves, in addition, the vertical speed $V(t)$ in the figure can be considered the body's relative velocity with the water surface of incoming waves at the target section. The fluid domain is surrounded by the boundaries consisting of the free surface, the body surface, the centerline, the side walls, and the

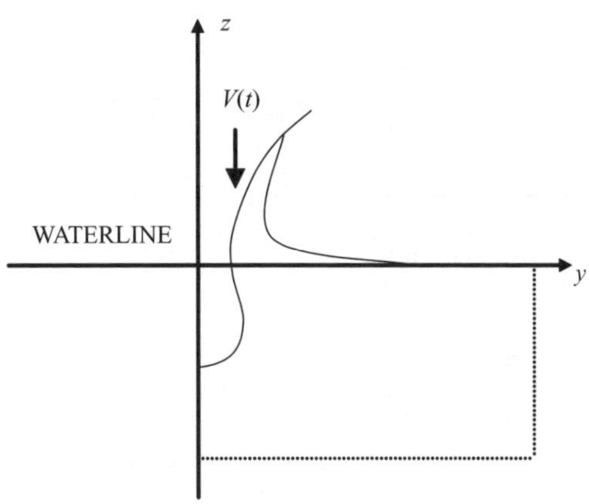

Figure 1. Bowflare slamming.

bottom. Assuming the fluid is incompressible and the flow is irrotational, the fluid motion is specified by the velocity potential.

Traditionally the water impact problem is applied to the slamming analysis. So far enormous works have been done in this respect (e.g. Korobkin, (1996), Faltinsen (2005)). The water impact approach is classified roughly into analytical and numerical. The analytical approach is based on mathematical models. Although the Wagner theory is well known, the recent studies are devoted to deal with the slamming problem of the arbitrary shaped body (Korobkin (2005), Takagi et al. (2007)). The numerical approach is represented by the boundary element method (BEM). A certain level of success was already achieved previously, but the important progress has been actuated when the numerical simulation of the impact jet is realized, Sun et al. (2007).

Moreover, the problem additionally become more complex because of stochastic sea nature and the problem of estimation of worst weather condition which the ship can suffer during exploitation.

However, in practice the problem of design slamming pressures are usually obtained by empirical formulae given by classification societies. These formulas are simplified and take into account simplified form of ship defined by basic parameters (length, beam, draft, block coefficient etc.). The bow form is defined by relevant angles α (angle between axes x and section inclination in vertical longitudinal plane), β (angle between axes y and section inclination in vertical transversal plane) and γ(angle between axes x and section inclination in horizontal plane) shown at Figure 2.

The procedures proposed by classification societies are based on performed direct calculations and correlated by service experience. Consequently, these formulas are not necessary valid for novel designs.

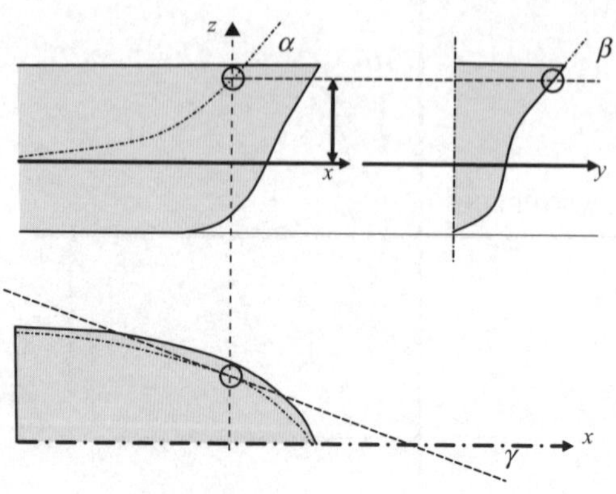

Figure 2. Relevant angles.

3 CLASSIFICATION SOCIETIES REQUIREMENTS

3.1 *Applicable ship type*

LR has two sets of relevant Rules (Lloyd's Register, (2009)). First one contemplated in the Part 3, Chapter 5 are intended for ships with moderate block and moderate speed hull forms. Since these requirements were not adequate for high speed hull forms with more significant bow flare angles a revised bow flare slamming pressure requirements were introduced in the Part 4, Chapter 2 for passenger ships, Ro-Ro ships, ferries and container ships.

ABS has different sets of Rules (American Bureau of Shipping, (2006)) for oil tankers contemplated in the Part 5-1-3, bulk carriers contemplated in the Part 5-3-3 and Container ships contemplated in the Part 5-5-3.

DNV has unified Rules (Det Norske Veritas, (2004)) contemplated in the Part 3, Chapter 1, Section 7, para E300 intended for all ship types. Normally only ships with well rounded bow lines and or flare will need strengthening.

3.2 *Applicable bow region*

ABS requires investigating bowflare slamming in the region above the waterline between 0.0125 and 0.25 Rule length (L) from the forward perpendicular (FP). As per DNV requirements, the effect of bow impact loads is in general to be evaluated in the region forward of a position 0.1 L from aft FP and above the summer load waterline. On the other hand LR requires strengthening shell envelope against bow flare typically over the fore end side and bowing structure above the waterline in the areas where the hull exhibits significant flare.

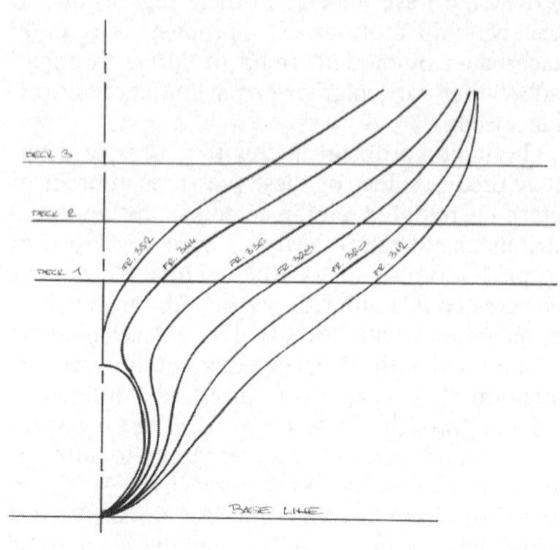

Figure 3. Bow form.

3.3 *Selected hull form*

Considering chapter 3.1., the container ship has been chosen as a referent ship type for this comparison. The main characteristics of chosen hull form are the following:

$L = 240$ m
Breadth = 36 m
Draught = 8.5 m
Block coefficient = 0.71
Service speed = 22 kn

Relevant bow sections are shown at the Figure 3.

Considering chapter 3.2 this investigation has been limited to the area forward of 0.1 L from FP above waterline. Selected frames 312, 320, 328, 336, 344 and 352 are located at approximately 91, 93, 96, 98, 100 and 103% of length from aft end respectively. Pressures have been calculated at each frame for three different deck levels, i.e at 11 m (D1), 13.8 m (D2) and 16.6 m (D3) above base line.

4 BOWFLARE SLAMMING PRESSURE ESTIMATION

For selected positions three different Rule calculation procedures have been applied and following values of bowflare slamming pressures, shown in Table 1 and Figure 4 have been obtained.

Table 1. Bowflare slamming pressure.

		Bowflare slamming pressure, kN/m²		
		D3	**D2**	**D1**
Fr.312	ABS	110	130	130
	DNV	240	380	600
	LR	40	230	510
Fr.320	ABS	120	140	150
	DNV	320	500	680
	LR	150	410	670
Fr.328	ABS	130	150	200
	DNV	410	670	750
	LR	270	610	800
Fr.336	ABS	140	150	310
	DNV	580	830	780
	LR	530	860	880
Fr.344	ABS	150	160	520
	DNV	800	920	720
	LR	840	1140	900
Fr.352	ABS	160	280	980
	DNV	860	830	470
	LR	1010	1160	740

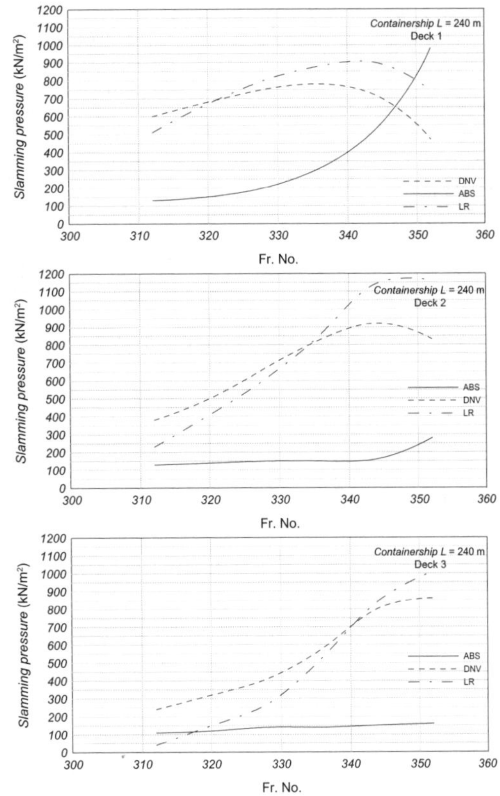

Figure 4. Pressure distribution over decks.

As can be seen from Figure 4 LR and DNV values are generally close. Compared to them ABS values are significantly lower. The same trend can be seen also when calculating required shell plating thickness as shown in Table 2 and Figure 5.

Calculation of required plating thickness is based on the spacing of side shell longitudinals of 700 m and span of 2.8 m between two side shell transverses in way.

5 SENSITIVITY STUDY

Differences found in the comparison are significant. Analyzing the different empirical formulas as proposed in the Rules by each classification society it is not possible to get entirely insight into the background of theirs procedures. For example LR requirements originally were developed from a comparative study carried out on cargo ships which experienced fore end damages from heavy weather and on undamaged ships having a similar configuration and service speed (to ensure that strengthening for bow slamming was not carried out unnecessarily). This method was further

Table 2. Required shell plating thickness.

		Required shell plating thickness, mm		
		D3	**D2**	**D1**
Fr.312	ABS	9	9	10
	DNV	10	13	16
	LR	4	11	16
Fr.320	ABS	9	10	10
	DNV	12	15	18
	LR	9	14	18
Fr.328	ABS	10	10	12
	DNV	14	17	18
	LR	12	17	20
Fr.336	ABS	10	10	15
	DNV	16	19	19
	LR	16	21	21
Fr.344	ABS	10	11	19
	DNV	19	20	18
	LR	20	24	21
Fr.352	ABS	11	14	26
	DNV	20	19	15
	LR	22	24	19

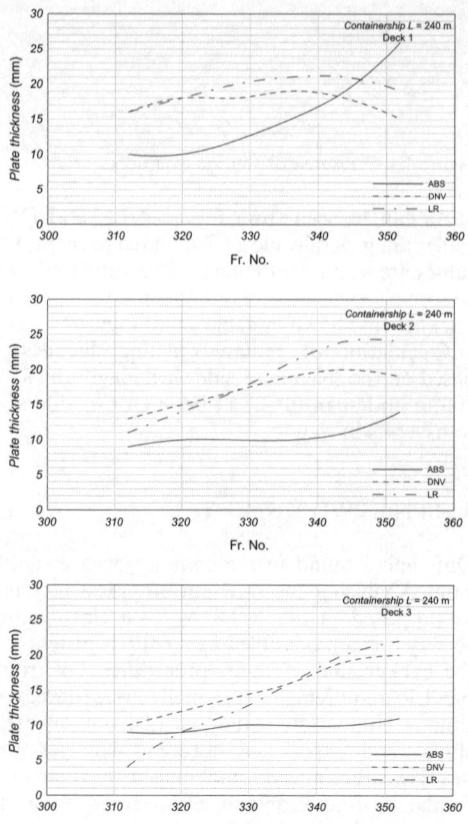

Figure 5. Shell plating thickness.

calibrated using general cargo ship type hull forms, i.e. moderate block and moderate speed hull forms. Later, it was found that these requirements were not adequate for high speed hull forms with more significant bow flare angles. For this reason a revised bow flare slamming pressure requirement was introduced for passenger ships, Ro-Ro ships, ferries and container ships. This method is based on the theoretical approach using ship motions correlated with service experience.

Looking at the basic formulas of all three classification societies difference in the formula structure and approach can be easily noted:

ABS basic formula for calculating the value bowflare slamming pressure P_{ij}:

$$P_{ij} = P_{0ij} \text{ or } P_{bij} \tag{1}$$

whichever is the greater, where:

$$P_{0ij} = k_1 (9M_{Ri} - h_{ij}^2)^{0.5} \text{ kN/m}^2 \tag{2}$$

where k_1 is constant value while M_{Ri} depends on position along the ship length L, ship speed V, ship length and block coefficient. Value h_{ij} is the vertical distance of chosen point form waterline. The second term is expressed as

$$P_{bij} = k_2 [c_2 + K_{ij} M_{Vi} (1 + E_{ni})] \text{ kN/m}^2 \tag{3}$$

where k_2 and c_2 are constant values while K_{ij} depend on bow form, M_{Vi}, h_{ij}, position along ship length, ship speed and block coefficient.

DNV basic formula for calculating the value bowflare slamming pressure P_{sl}:

$$P_{sl} = C (2.2 + C_f) (0.4 V \sin\beta + 0.6\sqrt{L})^2 \text{ kN/m}^2 \tag{4}$$

where C depends on wave coefficient and vertical distance of considered point from waterline while C_f depends on bow form and ship motions.

LR basic formula for calculating the value bowflare slamming pressure P_{bf}:

$$P_{bf} = 0.5 (K_{bf} V_{bf}^2 + K_{rv} H_{rv} V_{rv}^2) \text{ kN/m}^2 \tag{5}$$

where K_{bf} and K_{rv} depend on hull form, V_{bf} wave impact velocity, V_{rv} relative forward speed and H_{rv} depends on heading angle.

Therefore, it was decided to examine the differences in the results at this stage through influence of various input parameters. Only parameters required by all three societies are considered. Among them ship speed, distance of considered point from waterline and angle between side shell and waterline in a transverse section have been selected as the most influencing ones. The results are shown in the Figures 6, 7 and 8 respectively.

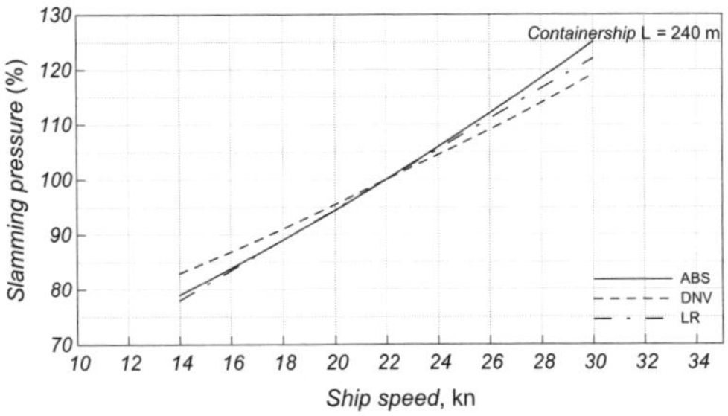

Figure 6. Ship speed influence on slamming pressure.

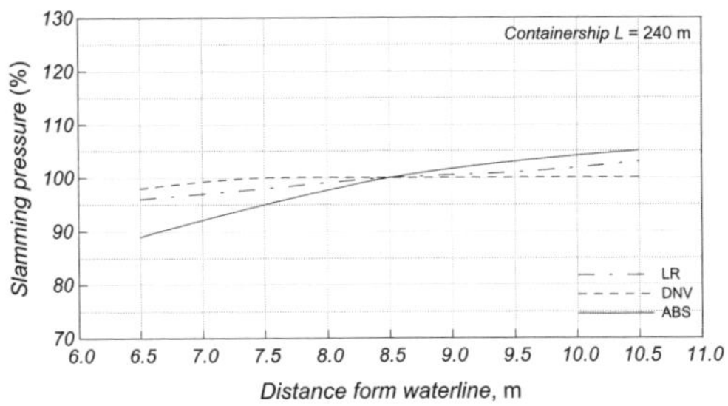

Figure 7. Ship draught influence on slamming pressure.

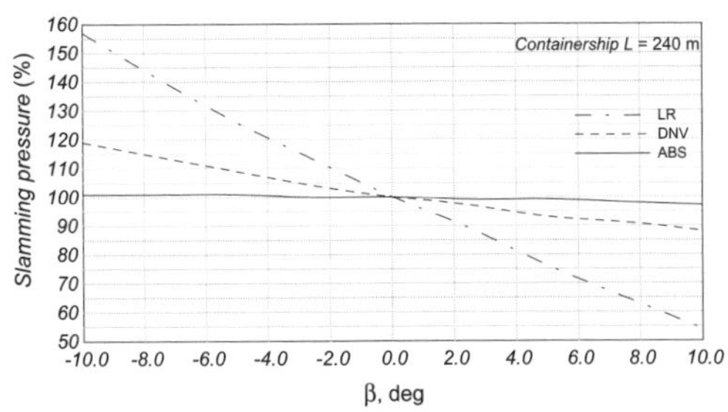

Figure 8. Angle β influence on slamming pressure.

6 CONCLUSION

This analysis has shown that while DNV and LR requirements for bowflare pressures are generally close, ABS estimates significantly lower values. In terms of scantlings, ABS would require approximately 10% thicker plating for the same pressure but this finally does not lead to similar scantlings as those required by DNV and LR for the same pressure.

All three classification societies have similar influence of speed. Increment of speed of approximately 20% leads generally to a medium increment of pressure in order of 10%. It is important to point out that medium increment of pressure is calculated as an average for entire bow region.

As regards distance of the point where bowflare slamming pressure calculated from the waterline it can be noted that it has almost no influence in the DNV requirements while the most influencing in the ABS requirements. Sensibility of LR requirements is between them. Generally can conclude bowflare pressures for all classification societies are not significantly dependent on the distance from the waterline.

On the other hand the most influencing parameter appears to be angle between side shell and horizontal axes. In LR requirements this dependence is the highest and decrease of this angle for only 1 degree leads to increment of pressure for about 2%. Similarly DNV Rules for 1 degree lower angle shows increment of about 1%. Only exception in this respect is ABS requirements where resulting bowflare pressures is nearly constant.

The further investigation will be directed toward analyzing novel forms extreme bowflare slamming pressures by direct calculation and comparing with values estimated by empirical formulas.

REFERENCES

American Bureau of Shipping, 2006, *Rules for Building and Classing Steel Vessels*, Houston.

Det Norske Veritas, 2004, *Rules for Ships*, Hovik.

Faltinsen, O.M., 2005, *Hydrodynamics of High-Speed Marine Vehicles*, Cambridge University Press.

Korobkin, A.A., 1996, *Water impact problem in ship hydrodynamics*, Advance in Marine Hydrodynamics, M. Ohkusu, Ed (Chap.7), Computational Mechanics Publishing, Southampton, Boston.

Korobkin, A.A., 2005, *Analytical Models of Water Impact*, Euro. J. Applied Mathematics, 16, pp. 1–18.

Lloyd's Register, July 2009, *Rules and Regulations for the Classification of Ships*, London.

Sun, H. and Faltinsen, O.M., 2007, *Water Impact of Horizontal Circular Cylinder and Cylindrical Shells*, Applied Ocean Research, Vol. 28, pp. 299–311.

Takagi, K. and Ogawa, Y., 2007, *Flow Models of the Flare Slamming*, Proc. of International Conference on Violent Flows, pp. 173–179.

Advanced Ship Design for Pollution Prevention – Guedes Soares & Parunov (eds)
© 2010 Taylor & Francis Group, London, ISBN 978-0-415-58477-7

Dynamic analysis of structures partially submerged in water

D. Sedlar, Ž. Lozina & D. Vučina
Faculty of Electrical Engineering, Mechanical Engineering and Naval Architecture, University of Split, Split, Croatia

ABSTRACT: Natural frequencies of cantilever beam partially submerged in water are investigated. Influence of water on the beam is numerically simulated through effect of virtual (added) mass. The difference between natural frequencies in the air and in the water, which is obtained this way, is compared with experimental one. Experimental natural frequencies are obtained by performing impact test and measuring frequency response functions. Possibility of determining the depth of immersion beam is shown using residual force method.

1 INTRODUCTION

The dynamic interaction between the deformable structure and fluid is often in engineering practice. The fluid-structure interaction of a partially submerged cantilever beam significantly helps understanding naval architecture issues, e.g. vibrations of rudders and periscope. Therefore, in this paper change of natural frequencies of cantilever beam partially submerged in still water are shown, Figure 1. Natural frequencies of beam are changed under influence of different level of water. One way to simulate cantilever beam submerged in water is virtual mass principle. This principle enables modeling the submerged beam like the simple beam with increased mass. In this case complicated modeling of fluid structure interaction is avoided. The virtual mass is a portion of the surrounding fluid that is accelerated as though it where rigidly attached to the structure.

Since virtual mass can be consider like structural change, the residual force vector, Ge (2005), can be implemented in detection of beam submergence depth, as shown later on.

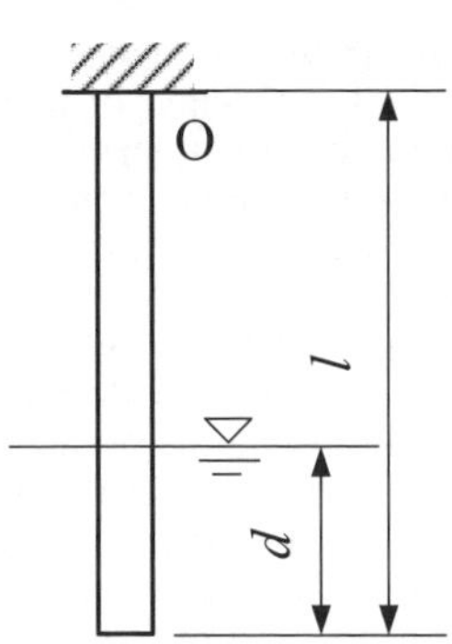

Figure 1. Cantilever beam partially submerged in water.

The experiments are performed under laboratory conditions. The natural frequencies and corresponding damping for cantilever beam in air and partially submerged in still water are extracted from the measured data. The impact test is applied to obtain frequency response functions and rational fraction polynomial method (RFP) is used to calculate natural frequencies and damping coefficients, Ewins (2000).

2 VIRTUAL MASS EFFECT

The governing equation for viscous damping and single degree of freedom system in the absence of the fluid is

$$m\ddot{x} + c\dot{x} + kx = f(t) \tag{1}$$

where m = mass; c = damping; and k = stiffness. The equation that governs the coupled fluid-structure behavior is especially complex when the response of the structure modifies significantly the flow field leading into large vibration amplitudes. When the response of the structure does not affect the flow field, the fluid-structure interaction affects only the structure terms of the motion equation. In this case mass, damping and stiffness are modified. So, for structure in still fluid (as far as fluid flow does not significantly affect natural frequencies) equation of motion can be written

$$(m + m_a)\ddot{x} + (c + c_a)\dot{x} + (k + k_a)x = f(t) \tag{2}$$

where m_a = virtual mass; c_a = virtual damping; and k_a = virtual stiffness. These virtual members are the consequence of the submergence in still fluid. This virtual mass effect is important when the density of the fluid compared with density of structures cannot be neglected, Rodriguez (2006).

Therefore, if the fluid is the air interacting with steel structure the virtual mass can typically be neglected and the values of m_a, c_a and k_a are ignored, as done in Equation (1). If the fluid is water the virtual mass cannot be neglected. Obviously the natural frequencies of the structure in water will be different from the frequencies of structure in air. In general, the virtual mass is the function of the geometry, submergence and vibration amplitude. If the structure vibrates in the flowing fluid then the virtual mass can also be the function of flow condition. Vibration amplitudes can affect the virtual mass if they are large enough to separate the fluid from the structure. In that case, the motion induced force will change from the inertial force (virtual mass) to the damping force. The virtual stiffness can be important in some cases like in floating structures.

2.1 *System and modal parameters identification*

The modal parameters, like damping ratio and natural frequency can be expressed using system parameters (mass, damping and stiffness). For single degree of freedom damping ratio is

$$\zeta = \frac{c}{2\sqrt{km}} \tag{3}$$

natural frequency

$$\omega_n = \sqrt{\frac{k}{m}} \tag{4}$$

and damped natural frequency

$$\omega_d = \omega_n\sqrt{1-\zeta^2} \tag{5}$$

The solution of the equation of motion can be written as

$$x = Xe^{st} \tag{6}$$

where s = complex value; and X = maximum value of x. Obtaining system parameters from modal parameters is inverse process. This process is more complicated. It requires measuring and modal analysis as first step. Excitation and response is measured to obtain frequency response function. Frequency response function is used in modal analysis to determine modal parameters. Frequency response function can be written in the form

$$H = \frac{X}{F} = \frac{1/k}{\sqrt{\left(1-\omega^2/\omega_n^2\right)^2 + \left(2\zeta\omega/\omega_n\right)^2}} \tag{7}$$

For low damping ratio $\omega_n = \omega_d$ can be assumed and for maximum frequency response function amplitude where $\omega = \omega_d$ it is possible to determine stiffness in terms of measured data

$$k = \frac{1}{2\zeta H_{(\omega=\omega_d)}} \tag{8}$$

Now from Equations (3), (4) and (8) it is possible to determine mass and damping

$$m = \frac{1-\zeta^2}{2\omega_d^2\zeta H_{(\omega=\omega_d)}} \tag{9}$$

$$c = \frac{\sqrt{1-\zeta^2}}{\omega_d H_{(\omega=\omega_d)}} \tag{10}$$

The above procedure can be repeated for multi degree of freedom system. Similar to the single degree of freedom the modal stiffness, modal mass and the modal damping of the multi degree of freedom can be written

$$k_r = \frac{{}_rA_{jk}}{2\zeta_r\, {}_rH_{jk(\omega=\omega_d)}} \tag{11}$$

$$m_r = \frac{{}_rA_{jk}\left(1-\zeta_r^2\right)}{2\omega_{dr}^2\zeta_r\, {}_rH_{jk(\omega=\omega_d)}} \tag{12}$$

$$c_r = \frac{{}_rA_{jk}\sqrt{1-\zeta_r^2}}{\omega_{dr}\, {}_rH_{jk(\omega=\omega_d)}} \tag{13}$$

where r = matches particular mode; ${}_rH_{jk}$ = frequency response function for N degree of freedom system at response point j due to applied force at point k; k_r = modal stiffness of mode r; and ${}_rA_{jk}$ = modal constant. Modal constant corresponds to the normalized mode shape of mode r. If the modes of the system are well separated, which means that the behavior of a system is dominated by the single mode in vicinity of the resonance, then the frequency response function in this tight band is controlled by just one of the term in the series.

From Equations (11), (12) and (13) it is possible to find values of mass, damping and stiffness. One has to be careful because the damping ratio is hard to obtain with high accuracy.

2.2 *Virtual mass analysis*

The virtual changes of the structure parameters due to fluid-structure interaction can be decoupled from the free structure parameters using values of modal mass, modal damping and modal stiffness obtained for the system in the air and for the system submerged in the fluid. Comparing the characteristic equations of motion for the system in the air and the system in the fluid we have

$$\left(1+\frac{k_{ar}}{k_r}\right)\left(1+\frac{m_{ar}}{m_r}\right)^{-1}=\frac{1-\zeta_r^2}{1-\zeta_{wr}^2}\left(\frac{\omega_{wdr}}{\omega_{dr}}\right)^2 \tag{14}$$

where index w = fluid (water). If the stiffness effect of the fluid is neglected, which is reasonable assumption for most structures; ratio of mass can be found

$$\frac{m_{ar}}{m_r}=\frac{1-\zeta_{wr}^2}{1-\zeta_r^2}\left(\frac{\omega_{dr}}{\omega_{wdr}}\right)^2-1 \tag{15}$$

The ratio m_{ar}/m_r is less sensitive to the errors in the determination of the damping and it is sensitive to the frequency ω_d that can be more accurately obtained. A negative value of Equation (15) means that water subtracts mass to the system, a zero value means that added mass is negligible and a positive value means that water virtually adds mass to the system.

3 NUMERICAL SIMULATION

Simulation of partially submerged beam is done to obtain natural frequencies and mode shapes which are necessary to calculate residual force vector. Influence of water is modeled like added mass to the beam. Own code is written for simulation. Also, motion of water and beam is simulated in ADINA software.

3.1 *Residual force vector*

For the multi degree of freedom system, neglecting damping, under free vibration the equation of motion in matrix form is given by

$$\mathbf{m}\ddot{\mathbf{x}}+\mathbf{k}\mathbf{x}=\mathbf{0} \tag{16}$$

where $\mathbf{k}$ = stiffness matrix; and $\mathbf{m}$ = mass matrix. Eigen-value equation can be derived and written as

$$\left(\mathbf{k}-\omega_r^2\mathbf{m}\right)\boldsymbol{\varphi}_r=0 \tag{17}$$

where ω_r = r-th natural frequency; and $\boldsymbol{\varphi}_r$ = r-th mode shape. If only virtual mass in fluid is considered, eigen-value equation for structure in the fluid can be written as

$$\left(\mathbf{k}-\omega_{wr}^2\left(\mathbf{m}+\mathbf{m}_a\right)\right)\boldsymbol{\varphi}_{wr}=0 \tag{18}$$

The above equation can be rearranged in the way that left and right side has unit force

$$\mathbf{R}_r=\left(\mathbf{k}-\omega_{wr}^2\mathbf{m}\right)\boldsymbol{\varphi}_{wr} \tag{19}$$

$$\mathbf{R}_r=\omega_{wr}^2\mathbf{m}_a\boldsymbol{\varphi}_{wr} \tag{20}$$

Equation (19) and (20) represent residual force vector for mode r. If the natural frequency and mode shape for just one mode are available, for the structure in fluid, the residual force vector can be calculated from Equation (19). Also, this equation requires the knowing of stiffness and mass matrices for the structure in air. These matrices can be obtained with finite element model of structure. The ith entry of the residual force vector represents the ith degree of freedom. It is evident from Equation (20), that the particular elements of residual force vector will be different from zero only if virtual (added) mass belongs to the associated degrees of freedom. If there is no virtually added mass in considered degree of freedom, corresponding element of the residual force vector will be zero.

Therefore, the depth of the structure submergence can be detected by identifying the non-zero elements of residual force vector. Theoretically, the residual force vector approach is possible if only one mode parameters is available. However, in practice mode shapes are contaminated by noise and mode shapes accuracy is not always satisfactory. Typically modal parameters for more then one mode are available. If modal parameters for p modes are available pseudo residual force vector can be defined as

$$\mathbf{R}=\left\{R_1,R_2,\ldots,R_n\right\}^T \tag{21}$$

where R_i is calculated from equation

$$R_i=\left[\prod_{r=1}^{p}\left|R_i\right|_r\right]^{\frac{1}{p}} \tag{22}$$

in which $|R_i|_r$ is absolute value of the ith element of vector $\mathbf{R}_r$ calculated from Equation (19). In this way, the influence of the noise is reduced and the depth of submergence is more accurately detected.

3.2 *Numerical example*

In this section the numerical example of application of the pseudo residual force vector employed in the detection of the cantilever submergence depth is given. The cantilever beam is 712.6 mm long, rectangular cross section is 25.4 × 5.25 mm, density 7800 kg/m^2 and elastic modulus 210 GPa. The cantilever beam is divided into seven elements where the last two are in the fluid. The fluid influence to governing equation is simulated by increasing mass of the finite elements for estimated 20%. Each finite element has two nodes with one translation and one rotation degree of freedom per node. The consistent mass matrix was used. Table 1 shows natural frequencies for simulated model of cantilever beam in the air and in the fluid.

The calculated pseudo residual force vector is shown in Figure 2. The first six modes are used to calculate pseudo residual force vector. Pseudo residual force vector consists of sixteen members, eight points and two degrees of freedom in every point. It is evident that the last six values of the pseudo residual force vector are different from zero. These values correspond to the degrees of freedom that belong to the finite elements which are submerged in water. The translational degrees of freedom compared with rotational are much more sensitive to the virtual (added) mass change what is confirmed with the residual force vector.

Figure 3 shows first four mode shapes for cantilever beam in the air and in the fluid.

3.3 *ADINA simulation*

ADINA enables a simulation of fluid structure-interaction problems. Unfortunately there is no possibility to calculate natural frequencies of structures which are submerged in fluid. Modeling this type of problem is done in two steps. First, the structure is modeled independently of fluid. Then fluid is modeled independently of structure. It is necessary to define fluid-structure interaction boundary for structure and fluid model. These two models are used in dynamic analysis. Impact test simulation is done. So, the only load is concentrated force which acts very short time period.

Figure 4 shows model of beam partially submerged in water. Dynamic analysis gives displacement, velocities and pressures in fluid. Figures 5–6 show velocities and pressures in fluid for second step.

Table 1. Natural frequencies of cantilever beam—nume-rical.

Num.	In air	In fluid
1	54.44	50.75
2	341.27	332.53
3	956.41	930.13
4	1879.19	1822.91
5	3123.73	3046.82
6	4705.28	4584.85

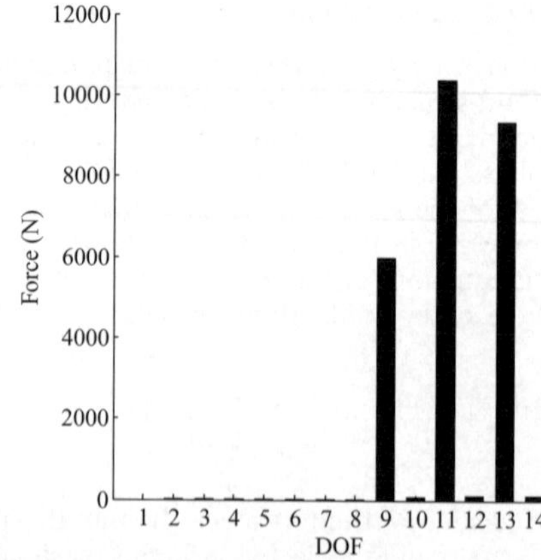

Figure 2. Pseudo residual force vector.

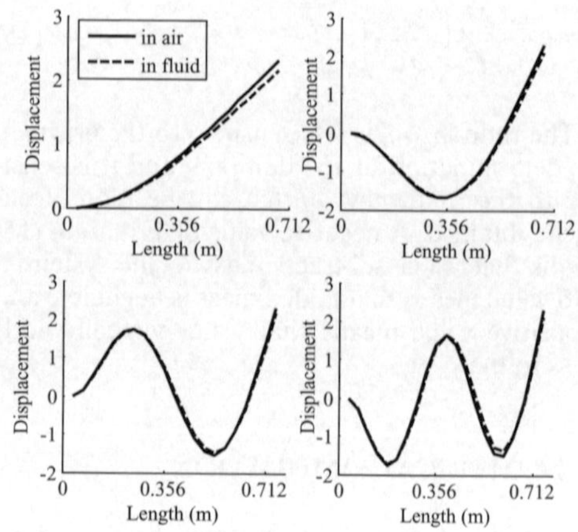

Figure 3. First four mode shapes.

4 EXPERIMENTAL RESULTS

The experiment is performed with cantilever beam that has the same geometrical and material properties like beam in numerical example, paragraph 3.1. Again, beam is divided in seven elements. The lower beam ends that correspond to the last two elements are submerged in the water.

Impact test was performed. Reference accelerometer is mounted on node two and hammer is moved from node to node to obtain set of frequency response functions.

Figure 7 shows comparison between the frequency response functions of the cantilever beam

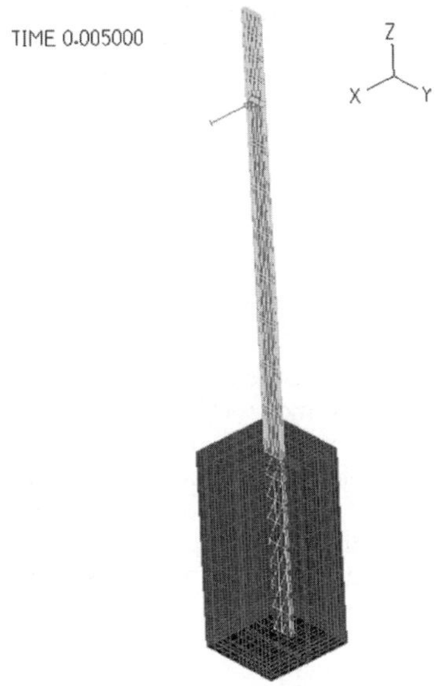

Figure 4. Fluid-structure interaction model.

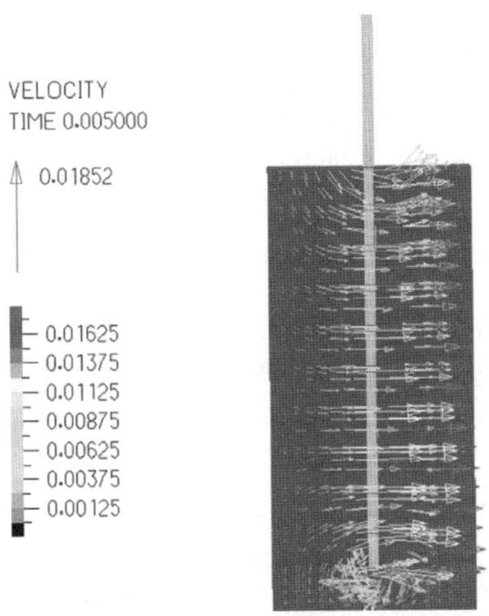

Figure 5. Velocities distribution in fluid.

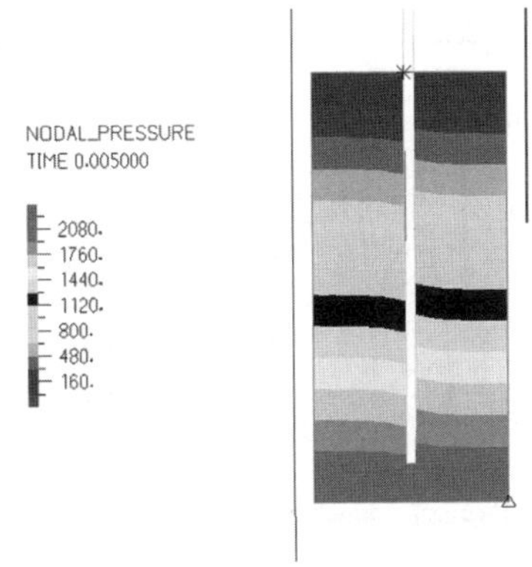

Figure 6. Pressure distribution in fluid.

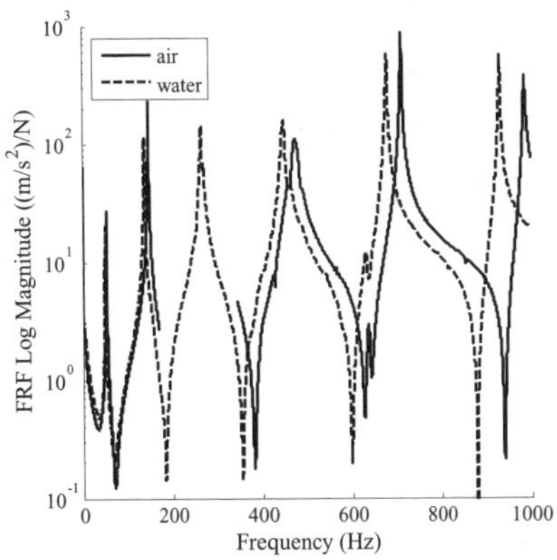

Figure 7. Frequency response function for drive point 2–2, air-water.

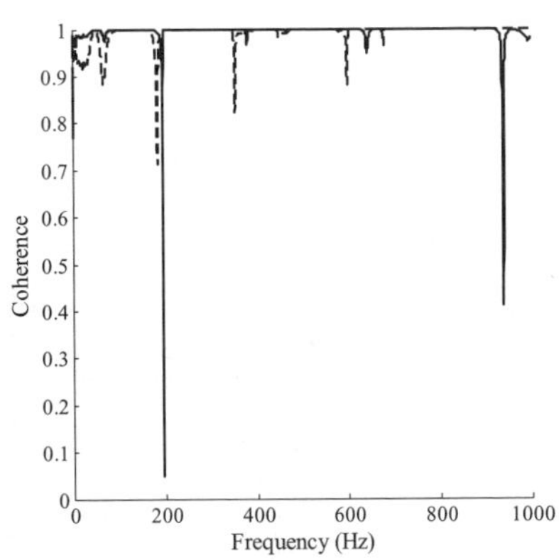

Figure 8. Coherence function for drive point air-water.

in air and the beam partially submerged in water. Figure 8 shows comparison between the coherence functions of the cantilever beam in air and the beam partially submerged in water.

Figures 9–10 show two other comparisons for frequency response function 3–2 and 5–2.

Obtained natural frequencies for cantilever beam in the air and in the water are shown in Table 2. Obtained damping ratios for cantilever beam in the air and in the water are shown in Table 3.

Comparing experimental, Table 2, and numerical, Table 1, natural frequencies for cantilever beam in air difference can be noted. These differences arise from mismatching of the numerical and real model. If higher similarity is required, the finite element model update procedure has to be performed. However, in both cases (numerical

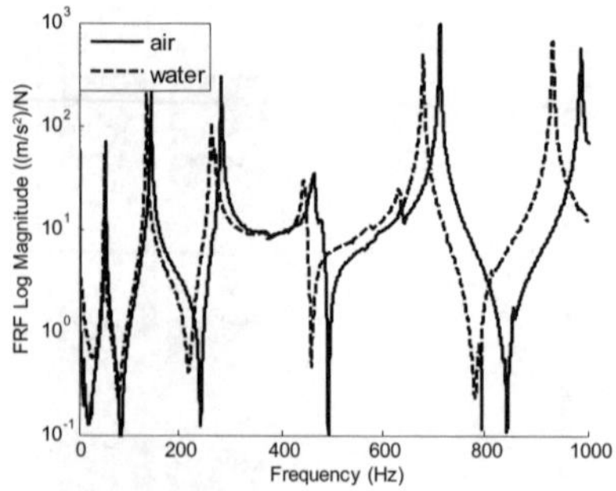

Figure 9. Frequency response function 3–2, air-water.

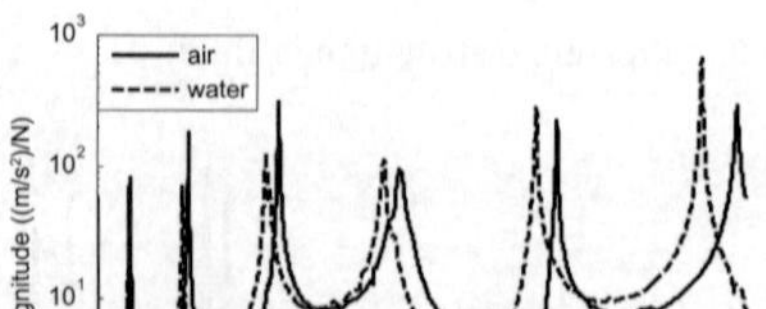

Figure 10. Frequency response function 5–2, air-water.

Table 2. Natural frequencies of cantilever beam—experimental.

Num.	In air	In fluid
1	53.30	45.53
2	328.56	309.89
3	914.45	856.20
4	1779.33	1660.39
5	2959.50	2790.80
6	4462.00	4258.55

Table 3. Damping ratio of cantilever beam—experimental.

Num.	In air	In fluid
1	8.65E-2	3.37E-2
2	7.48E-2	2.25E-1
3	1.59E-1	4.41E-1
4	9.71E-1	5.21E-1
5	9.21E-2	1.39E-1
6	1.46E-1	9.16E-2

and experimental) there is significant difference between natural frequencies for cantilever beam in the air and in the water. The natural frequencies of the partially submerged beam are lower comparing them with the frequencies of the beam in the air.

Table 4. Mass ratio.

Mode	Mass ratio
1	0.1241
2	0.1407
3	0.1484
4	0.1246
5	0.0978
6	0.1201

The approximate frequency decrease is about 6% except for the first mode where decrease is 13%.

Using experimentally obtained data from Table 2 and Table 3 it is possible to find mass ratio by Equation (15). Mass ratios for first six modes are shown in Table 4.

It is evident that virtual mass uniformly distributes across modes.

5 CONCLUSION

The influence of the fluid-structure interaction to the natural frequencies of the partially submerged cantilever beam in still water is analyzed. The still water causes the added mass effect, which means that part of surrounding water is virtually rigidly attached to the beam. Numerical simulation and experimental impact test were performed to obtain natural frequencies of beam partially submerged in water. As a result natural frequencies for partially submerged beam are decreased approximately for 6%. This is confirmed by numerical and experimental test. The pseudo residual force vector was defined and the numerical simulation used to detect depth of submergence.

REFERENCES

Ergin, A. & Ugurlu, B. 2003. Linear vibration analysis of cantilevers plates partially submerged in fluid. *Journal of Fluids and Structures* 17: 927–939.

Ewins, D.J. 2000. *Modal Testing: Theory, Practice and Application*. Hertfordshire: Research Studies Press.

Ge, M. & Lui, E.M. 2005. Structural damage identification using system dynamic properties. *Computers and Structures* 83: 2185–2196.

Kimber, M., Lonergan, R. & Garimella, S.V. 2009. Experimental study of aerodynamic damping in arrays of vibrating cantilevers. *Journal of Fluids and Structures* Article in Press.

Liang, C.C., Liao, C.C., Tai, Y.S. & Lai, W.H. 2001. The free vibration analysis of submerged cantilever plates. *Ocean Engineering* 28: 1225–1245.

Rodriguez, C.G., Egusquiza, E., Escaler, X., Liang, Q.W. & Avellan, F. 2006. Experimental investigation of added mass effects on a Francis turbine runner in still water. *Journal of Fluids and Structures* 22: 699–712.

Ship structural reliability with respect to ultimate strength

Advanced Ship Design for Pollution Prevention – Guedes Soares & Parunov (eds)
© 2010 Taylor & Francis Group, London, ISBN 978-0-415-58477-7

Reliability assessment of intact and damaged ship structures

A.P. Teixeira & C. Guedes Soares
Centre for Marine Technology and Engineering (CENTEC), Technical University of Lisbon, Instituto Superior Técnico, Lisboa, Portugal

ABSTRACT: This paper reviews the approaches that have been adopted to assess the implicit safety levels of intact and damaged ships and presents some results of the reliability analysis of damaged tankers due to grounding and collision events. The paper starts by reviewing recent work published on the reliability of intact and damaged ships and presents the relevant considerations that should be addressed when formulating the reliability problem of damaged ships. Particularly, the effect of the damage on the ship ultimate strength and the relevant changes on the still water and wave induced loads following damage, including the environmental conditions and duration of exposure that are applicable for the damaged ship, are discussed.

1 INTRODUCTION

During the last decades there have been considerable developments of methods and tools for structural reliability assessment, as reviewed by Rackwitz, (2001). The structural reliability methods are now well accepted tools capable of representing in a rational manner the uncertainties in the design of structures.

Currently, Structural Reliability Analysis (SRA) is performed mainly to evaluate the implicit reliability level of ship structures for different failure modes with the state-of-the-art models and representative uncertainty models (Guedes Soares et al., (1996), Paik and Frieze, (2001), Bitner-Gregersen et al., (2002)), and to calibrate design formats where a consistent reliability level is required (Spencer et al., (2003), Teixeira and Guedes Soares, (2005), Hørte et al., (2007b)).

An important application of the structural reliability methods is on the assessment of the notional probability of structural failure that result from different ship types as well as from different actual concepts of the same ship (Guedes Soares and Teixeira, (2000)). Also, the time dependent degrading effect of fatigue cracking and corrosion on the ultimate moment has also been taken into account in the reliability assessment of different ship types including FPSO structures (e.g. Guedes Soares and Garbatov, (1996a), Garbatov et al., (2004)).

More recently, structural reliability analysis has been used to assess the safety level of damaged ships structures due to accidental events (e.g. Luís et al., (2009), Hussein and Guedes Soares, (2009), Hørte et al., (2007a), Fang and Das. (2005)).

Luís et al., (2009) performed a reliability analysis of a Suezmax double hull tanker accidentally grounded. The loading of the ship was defined based on the extremes that the ship could find during one voyage of one week to dry-dock through European coastal areas. It was shown that in spite of the reliability being lower in the intact condition for sagging, in the damaged condition it is possible to find lower values for hogging, depending on the location and size of the damaged areas.

Hussein and Guedes Soares, (2009) have studied the residual strength and the safety levels of three double hull tankers designed according to the new International Association of Classification Societies (IACS) common structural rules (CSR). Different damage scenarios at side and bottom were considered and the residual strength of the ship in each scenario was calculated. The reliability of the damaged ships was then calculated considering the changes in the still water bending moment and the decrease in the ultimate strength due to the damage.

Hørte et al., (2007a) have performed a cost benefit analysis to identify the optimum safety level of an intact and damaged VLCC and LNG ship considering the cost of marginally increasing the scantlings and the risk reduction effect this would have in term of improved safety, reduced risk of environmental damage and reduced risk of property loss. By comparing the results for the intact and the damaged case it was shown that the intact criteria is dimensioning, i.e the scantlings at the cost optimum target safety level for the damaged case are lower than those for the intact case, even with a number of rather pessimistic assumptions related to the damage scenario. Therefore, on the basis of the case study reported, deck strengthening is not a cost effective risk control option concerning the damaged condition, since the ship is not likely to break after a collision even.

Fang and Das (2005) have studied the risk level of a ship damaged in different grounding and collision scenarios in different service conditions. They found that the failure probability of the ship damaged due to grounding is far less than due to collision. Moreover it was shown that the ship damaged by grounding or collision is at high risk unless necessary operational precautions are taken in order to reduced the expected loads to which the ship is subjected.

In order to be able to assess the hull girder structural safety of a damaged ship, it is necessary to identify what are the critical scenarios that potentially may lead to hull girder failure (i.e. damage location, and extent together with its likelihood) and evaluate the impact of the damage in the ship ultimate strength. Additionally, it is necessary to properly take into account the relevant changes the still water and wave induced bending moments following damage. Furthermore, the environmental conditions and duration of exposure that are applicable for the damaged ship should be specified as these will have a important impact in the probabilistic models of the load effects.

This paper reviews the approaches that have been adopted to assess the implicit safety levels of intact and damaged ships and presents some results of the reliability analysis of damaged tankers due to grounding and collision events.

2 STRUCTURAL RELIABILITY ANALYSIS

In general, the structural reliability problem is characterized by an n–vector X of basic (directly observable) random variables and a subset Ω of their outcome space, which defines the "failure" event. The probability of failure, is given by:

$$P_f = \int_{\Omega} f_X(x)\, dx \qquad (1)$$

where $f_x(x)$ is the joint probability density function (*pdf*) of X and Ω is the failure domain. For a structural component with a single failure mode, Ω can formally be written as $\Omega = \{x : g(x) \leq 0\}$, where $g(x)$ is the limit state (or failure) function for the failure mode considered. A failure domain is defined when $g(x) < 0$, a safe domain is defined when $g(x) > 0$ and a failure surface is defined when $g(x) = 0$ (Figure 1). The vector of basic random variables comprises physical variables describing uncertainties in loads, material properties, geometrical data and calculation modelling.

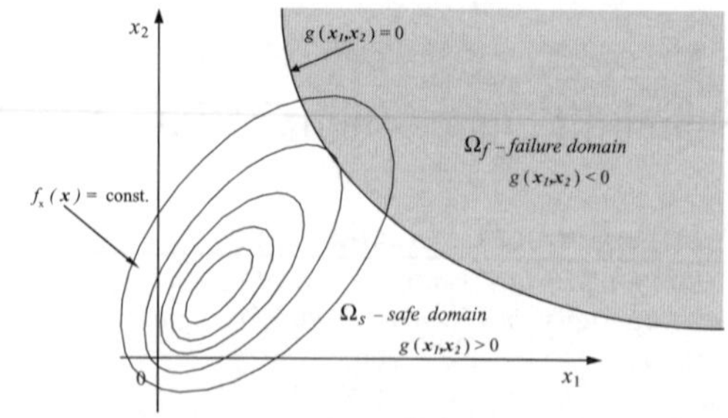

Figure 1. Generalized reliability problem.

The difficulty in computing the failure probability P_f directly from the integral given by equation 1 led to the development of approximate methods (Cornell, (1969) Hasofer and Lind, (1974)). Such methods are based on approximations of the failure surface to some simple forms, such as hyperplane or quadratic surfaces, at some locations, which are so-called design points. The methods dealing with this calculation algorithm are called as the level II (approximate) methods, in which the multi-dimensional integral given by equation 1 is calculated after the transformation of the basic random variables (the vector X) onto a set of independent normal random variables denoted by the (U) vector (Ditlevsen, (1981a), Hohenbichler and Rackwitz, (1981) and Der Kiureghian and Liu, (1986)) and the approximation of the limit state (failure) function in the (U) space to a linear or a second order (quadratic) functions at the failure surface to form a hyperplane or a quadratic failure surface.

If a linear approximation of the limit state function is used, then the method is called as the First Order Reliability Method (FORM) (Hasofer and Lind, (1974), Rackwitz and Fiessler, (1978), Rackwitz and Fiessler, (1978), Liu and Der Kiureghian, (1991)). However, if a second order approximation of the limit-state function is used, then it becomes the Second Order Reliability Method (SORM) (Fiessler et al., (1979), Der Kiureghian et al., (1987) and Tvedt, (1990), Breitung, (1984)).

First and Second order reliability methods (FORM/SORM), developed mainly to deal with explicit limit state functions, are well established techniques that have been successfully used for reliability assessment of structures, as reviewed by Rackwitz, (2001).

There are a number of textbooks on the subject that describe the material in more detail. Thoft-Christensen and Baker, (1982) still remains a good introductory book on the subject. Other textbooks include: Ang and Tang, (1984), Ditlevsen, (1981b), Ditlevsen and Madsen, (1996), Madsen et al., (1986), Melchers, (1999), Thoft-Christensen and

Mourotsu, (1986), Haldar and Mahadevan, (2000) and Nikolaidis et al., (2005). The DNV, (1992) Classification Note 30.6 provides guidance on the practical application of reliability methods, and ISO 2394, (1998) gives more formal information on the general principles of structural reliability.

3 RELIABILITY FORMULATION

The reliability formulations that have been proposed for reliability assessment of ship structures started off by closely following established approaches in the design codes. The main difference is that the basic variables were modelled as being random while the Rules of the Classification Societies specify their nominal value as function of ship parameters.

For many years the classification society rules were specifying the minimum section modulus, which resulted from considering the elastic behaviour of the hull under longitudinal bending induced by still water and wave induced load effects and considering a failure when the stresses at the deck or bottom would reach the yield stress. Although being conservative, formulations based on this check were adopted in the initial studies of ship reliability of Mansour, (1972) and Mansour and Faulkner, (1973), as well as in other subsequent studies reviewed by Guedes Soares, (1998a).

In recent reliability assessments of the primary ship structure more correct descriptions of the hull girder failure based on the ultimate collapse moment have been adopted. This moment, in general between the elastic and the plastic moment, is the sum of the contribution of all elements, taking into account their load deflection characteristics, including their post-collapse strength, (Gordo et al., (1996)). The limit state equation considered corresponds to the hull girder failure under vertical bending under combined vertical and horizontal bending:

$$1-\left(\frac{M_s+M_{wv}}{M_{uv}}\right)^{\alpha}-\left(\frac{M_{wh}}{M_{uh}}\right)^{\alpha}\leq 0 \qquad (2)$$

where M_s is the vertical still water bending moment; M_{wv}, M_{wh} are the wave induced vertical and horizontal bending moments, respectively; M_{uv}, M_{uh} are the ultimate vertical and horizontal bending moments; and α is the exponent of the interaction equation.

When the levels of horizontal bending moments are very small it may be appropriate to deal only with the vertical bending moments. In these cases the corresponding failure equation used in the reliability analysis is given by (Guedes Soares et al., (1996); Guedes Soares and Teixeira, (2000)):

$$g=X_uM_u-\left(X_sM_s+X_wX_{nl}M_w\right) \qquad (3)$$

where M_u is the random ultimate longitudinal bending strength of the ship, M_s and M_w are the stochastic still water and wave induced bending moments, respectively. X_u, X_s, X_w, and X_{nl} are model uncertainties that have already been used by Guedes Soares and Teixeira, (2000) to introduce in the reliability analysis the uncertainty associated with the predictions of the longitudinal strength of the ship and of the induced load effects. In particular, X_u and X_s are model uncertainties on the on ultimate capacity of the ship and on the values of the still water bending moment, respectively. X_w is a model uncertainty on the linear wave load calculations and X_{nl} reflects the uncertainty associated with the non-linear effects that are particularly significant for ships with a low block coefficient leading to differences between sagging and hogging bending moments.

Several reliability analyses of ships have been performed using this approach. Guedes Soares et al., (1996) assessed the reliability of the primary hull structure of several tankers and containerships using the ultimate strength of the ships as well as new probabilistic models for the still water and wave induced bending moment. The results of the reliability analysis were the basis for the definition of a target safety level, which was used to assess the partial safety factors for a new design rules format.

Teixeira and Guedes Soares, (1998) quantified the changes in notional probability of structural failure that result from ships being subjected to different wave environments and also from different ships of the same type (Teixeira and Guedes Soares, (2005)), different ship types and different concepts of the same ship (Guedes Soares and Teixeira, (2000)).

More recently, similar reliability formulations have been adopted to assess the safety level of damaged ships structures due to accidental events (e.g. Fang and Das, (2005), Hørte et al., (2007a), Luís et al., (2009)).

4 STOCHASTIC MODELLING OF ULTIMATE STRENGTH OF THE SHIPS

In order to be able to determine structural reliability of a ship it is necessary to evaluate the longitudinal strength of the hull girder and to define probabilistic models which can characterise the variability expected from the structural strength estimates.

Although the elastic bending moment has been widely treated as a measure of longitudinal bending strength of the ships, it does not provide with information concerning their resistance in extreme conditions. This can be achieved by the evaluation of the hull girder ultimate strength (maximum bending moment a hull can carry), which becomes an important parameter in ship structural rational design and in the reliability analysis of ships.

4.1 *Hull girder ultimate strength*

One of the first methods for calculating the ultimate strength of a midship section was suggested by Caldwell, (1965). However, Caldwell's method did not account for the post-collapse strength of the structural members which significantly influence the collapse strength. This problem was addressed by Smith, (1977) who proposed a method which assumes that each element of the cross section, made up of a longitudinal stiffener and the respective associated plate, behaves in its pre- and post-collapse phase, independently from the neighbouring components and that the contribution of the various components is summed up to produce the bending moment which makes the transverse section collapse. In this way, the ultimate moment of the transverse section, M_u, is given as:

$$M_u = \sum_i^n \frac{\sigma_{ui}}{\sigma_y} d_i\ \sigma_y \qquad (4)$$

where d_i is the central distance of the structural element to the neutral axis, σ_{ui} is the ultimate strength of each element which can be the yield stresses σ_y if it is in tension or the elasto-plastic buckling stress, σ_c, if it is in compression.

A number of simplified progressive collapse methods that adopt this type of approach have been proposed and they were compared in a ISSC study (Yao, (2000)). Amongst the various methods the one of Gordo et al., (1996) must be highlighted which was satisfactorily compared with experimental results (Gordo and Guedes Soares, (1996)).

Another type of method is the Idealized Structural Unit Method (ISUM) proposed by Ueda and Yao, (1982). This method adopts the finite element terminology which relates the nodal point force increments to the nodal point displacement increments through an incremental stiffness matrix. The main difference between the FEM and the ISUM is the size of the elements. The ISUM uses large elements where one structural member is considered to be one element. The number of degrees of freedom is thus considerably decreased and a large amount of computer time is saved. The ISUM is continuously developed by other researchers (e.g. Ueda and Rashed, (1984), Paik, (1993) and Bai et al., (1993)).

The FEM can also be a powerful method to calculate the ultimate strength of ships. However, the hull girder is too complex to perform progressive collapse analysis by the ordinary FEM and, therefore, most of the studies on the probabilistic modelling of the ultimate strength of hull girders have been based on predictions of Smith-type simplified progressive collapse methods.

4.2 *Ultimate strength of damaged ships*

Some studies on the impact of structural damages in the ship ultimate strength have been made. Zhang et al., (1996) proposed a semi-analytical method of assessing the residual longitudinal strength of damaged ship hulls. Paik et al., (1998b) developed a fast method for assessing the collapse of the hull girder in the damaged condition using the formulation of the American Bureau of Shipping (ABS, (1995)).

Wang et al., (2002) reviewed the state-of-the-art research on collision and grounding. The focus was on the three issues that a standard for design against accidents needs to address: definition of accident scenarios, evaluation approaches, and the acceptance criteria. Later, Wang et al., (2002) investigated the longitudinal strength of damaged ship hulls for a broad spectrum of collision and grounding accidents. Simple relations to assess residual hull girder strength were obtained, which may be used as handy and reliable tools to help make timely decisions in the event of an emergency. Kalman and Manta, (2002) investigated the effect of different damage modes of a hull section on the residual longitudinal strength of an impaired ship based on elastic theory, fully plastic resistance moment theory and ultimate bending moment approach.

Gordo and Guedes Soares, (2000), Ziha and Pedisic, (2002), and Fang and Das, (2004) have studied the ability of simplified methods for the calculation of the vertical ultimate bending moment to predict the ultimate longitudinal strength of damaged ships. The approach generally adopted in these studies considers that the elements within the damaged area are removed and the ultimate strength of the ship is recalculated using the simplified methods. It was found that the width of the damaged area influenced considerably the ultimate strength of the ship. However, accidental damages of ships can occur in any number of ways being the two most concerning ones the collision with other ships and grounding on rocky seabed.

Guedes Soares et al. (2008), reported the results of a Benchmark study in which the strength of a damaged ship hull was calculated with 3D nonlinear finite elements and was compared with the strength predicted by various codes based on the Smith method showing in general a good correlation.

Luís et al., (2009) have assessed the effect of damages around the keel area due to an accidental grounding of a Suezmax tanker. Figure 2 illustrates the adopted rectangular damaged areas. The basic damage has a height of $3/4H$ (where H is the double bottom height). A major damage with a higher penetration (H), was also considered to simulate a major accident that would cause cargo spill. The width is equal to $B/6$ (were B is the breadth), which was varied ±20%. Table 1 presents the reduction on the vertical ultimate strength of the ship due to the grounding damages in terms of residual strength index (RIF) defined as:

$$RIF = M_{u,int}/M_{u,dam} \tag{5}$$

where $M_{u,dam}$ is the vertical ultimate moment of the damaged section and $M_{u,int}$ is the vertical ultimate moment of the intact section both calculated by the simplified progressive collapse method developed by Gordo et al., (1996).

The vertical bending moment is indeed the most important load effect when considering the hull girder collapse. However, in many types of ships, the combined effect of the vertical and the horizontal bending moments is important especially after the ship is damaged, as the ship is likely to be in a heeled condition. As in the case of biaxial compressive strength of the plates, the nature of the

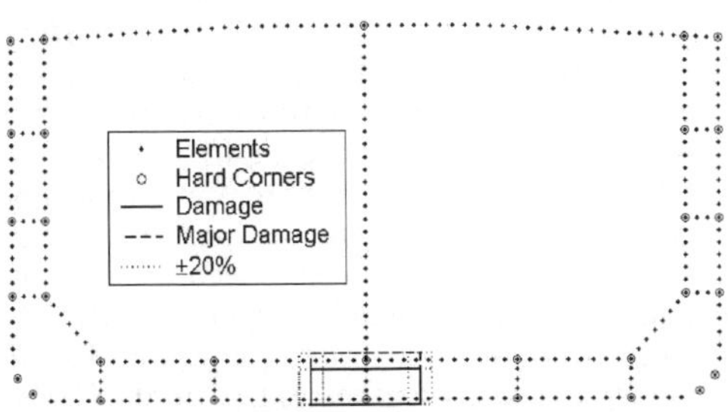

Figure 2. Damage dimensions.

Table 1. Residual strength, RIF.

	Sagging	Hogging
Intact	1.000	1.000
Damage (GCD) – 20%	0.985	0.957
Damage (GCD)	0.982	0.948
Damage (GCD) + 20%	0.979	0.935
Major Damage (GCMD) – 20%	0.983	0.925
Major Damage (GCMD)	0.979	0.903
Major Damage (GCMD) + 20%	0.970	0.880

interaction problem requires the solution of two issues. One is the collapse load in each individual mode, which may be used as normalizing factor in the interaction formula. The second problem is the interaction formula itself in order to adequately describe the combined effect of vertical and horizontal collapse moments (Gordo and Guedes Soares, (1995), Mansour et al., (1995), Gordo and Guedes Soares, (1997)).

The combined effect of vertical and horizontal bending moments in a damaged VLCC ship has been studied by Hørte et al., (2007a) using simplified progressive collapse methods by rotating the neutral axis around which the curvature is applied. The study considered grounding damages at the keel (GCMD) and at bilge area (GBMD) and also a collision damage that occurred at the upper part of the side shell (CSMD). Figure 3 illustrates the models of the intact and damaged cross sections of the VLCC. Figure 4 illustrate the interactions between the horizontal and vertical bending

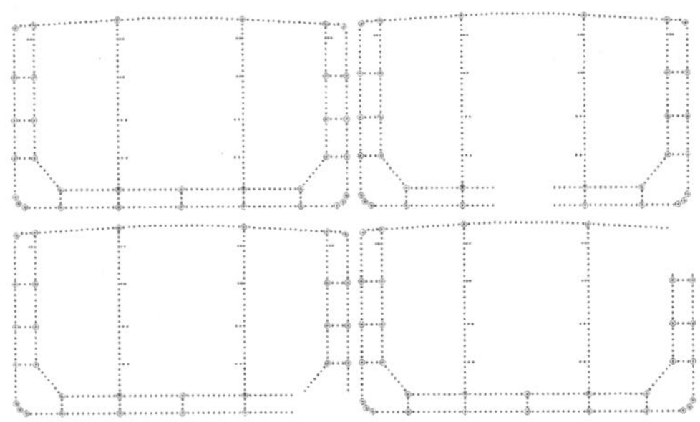

Figure 3. VLCC intact and damaged sections (intact, GCD & GBD and CSD.

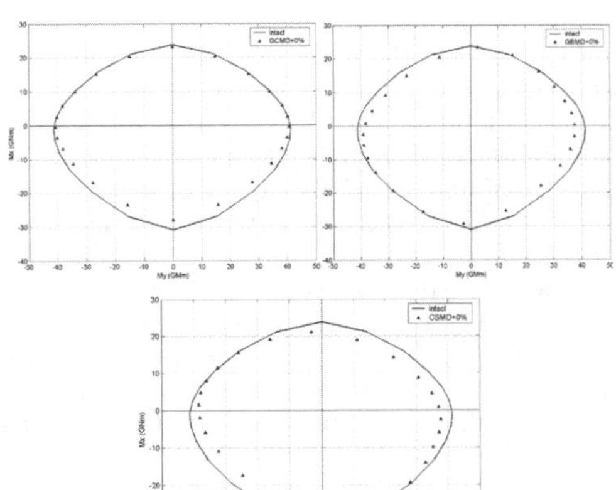

Figure 4. Interactions between the horizontal and vertical moments for the VLCC.

moments of the ship with major damages due to collision and groundings. Positive vales on the vertical axis represent sagging whereas negative are hogging.

For the typical damages studied, it was shown that side damage due to a collision can reduce the longitudinal strength of the VLCC tanker under sagging bending moments by approximately 12%. In hogging the strength of the ship can be reduced by up to the maximum of 11% when the ship is damaged due to a grounding accident.

4.3 *Probabilistic models of the ultimate strength*

In several reliability analyses the ultimate bending strength calculated by the deterministic methods based on the characteristic values of the material and of the geometrical parameters of the hull girder has been considered to be the expected value of the ultimate strength of the mid-ship section, and all uncertainty are typically concentrated in a model uncertainty random variable. A log-normal distribution with a mean value of unity is usually selected to describe this model uncertainty. It takes into account both the uncertainty in the yield strength and the model uncertainty of the method to assess the ultimate capacity of the mid ship section. Since the coefficient of variation of the yield strength of the steel normally ranges from 8% to 10% it has been assumed that the additional uncertainty will bring the overall coefficient of variation to 0.15 (Guedes Soares et al., (1996), Guedes Soares and Teixeira, (2000)).

This result has been demonstrated by Hansen, (1996) who quantified the uncertainty of the predictions of hull collapse, showing that it is very small and dominated by the uncertainty of the yield stress of the material, slightly increased by the uncertainty of the collapse strength.

Alternatively, First Order Second Moment approaches, Monte Carlo simulation or Response Surface Methods can be used to construct the probabilistic models of the ultimate strength of the ships based on the probabilistic models of the geometrical and material properties of the mid-ship cross section.

More recently, Teixeira and Guedes Soares, (2009) have constructed a probabilistic model of the longitudinal strength of the ship based on an inverse application of the First Order Reliability Method. The technique consists in fitting a three-parameter lognormal distribution to three fractiles of the ship longitudinal strength obtained by inverse FORM (iFORM) (Sadovský and Páles, (1999), Teixeira and Guedes Soares, (2007)). In this approach the collapse strength of the hull girder was calculated numerically by a progressive collapse program developed by Gordo et al., (1996), for a given realization of the vector of random variables of the problem. The set of random variables included the uncertainty on the material properties of elements of the cross section of the ship and also the uncertainty in the geometry of all elements grouped by their thicknesses represented by a normal distribution with mean value equal to the nominal value of the thickness and a coefficient of variation 0.05 (one variable per plate or web or flange thickness). Table 2 presents the main dimensions of the ship, i.e., length (L), beam (B), depth (D) and block coefficient (C_b) as well as the probabilistic model of the ultimate strength of the hull girder obtained for sagging and hogging bending moment (BM).

As indicated previously, the hull collapse moment is a result of the contribution of the different plate elements that make up the cross-section. Therefore the basic structural element is the plate and the next step in the level of complexity is the stiffened panel. It has been shown by different authors that the strength of the plate elements depend of the design formulation used to predict the plate collapse strength (Guedes Soares, (1988), on the shape and amplitude of initial imperfections (Guedes Soares and Kmiecik, (1993); Kmiecik and Guedes Soares, (2002)), on the level of residual stresses and on the boundary conditions. Therefore, the uncertainty on the prediction of the strength of these basic structural elements would be reflected in the ultimate strength of the ship.

The reduction of plate thickness due to corrosion is another important factor that affects the strength of plate elements under in-plane compression. The first studies have assumed a constant corrosion rate, leading to a linear relationship between the material loss and time, and a uniform reduction of plate thickness due to corrosion. This approach has been adopted by several authors for probabilistic modelling of the collapse strength of corroded plates and for the reliability analysis of plates (Hart et al., (1986), Guedes Soares, (1988), White and Ayyub,

Table 2. Ultimate vertical strength of the ship (M_u).

$L = 236\,m$, $B = 42\,m$, $D = 19.2\,m$, $C_b = 0.81$	Sagging	Hogging
M_e *Elastic. BM*$(MN \cdot m)$		8161
Mean value of M_u/M_e	1.014	1.195
Std. dev. of M_u/M_e	0.067	0.081
COV	0.066	0.068
Skewness	−0.128	−0.096
5.0% Fractile $(M_{u\,5\%}/M_e)$	0.794	0.933
0.1% Fractile $(M_{u\,0.1\%})$	0.901	1.060

(1992), Shi, (1993), Guedes Soares and Garbatov, (1999)). However, in addition to the general wastage that is reflected in the generalized decrease of plate thickness, the microscopic variations on the surface of the metal tend to cause different forms of corrosion and also variations in the corrosion rate over wide or small areas (Melchers, (2003)).

Recently Teixeira and Guedes Soares, (2007) have considered the spatial distribution of the corrosion in the probabilistic modelling of the collapse strength of plates by representing the thickness of the corroded plate by stochastic simulation of random fields. The non-linear corrosion model proposed by Guedes Soares and Garbatov, (1999) was used to define the probabilistic characteristics of the random fields based on corrosion data measured in plate elements at different locations of several bulk carriers reported by Paik et al., (1998a). The random fields of corrosion were discretized using the Expansion Optimal Linear Estimation method proposed by Li and Der Kiureghian, (1993). The collapse strength of the plate is then assessed by nonlinear finite element analysis as done by Teixeira and Guedes Soares, (2006). This study has shown the importance of the spatial representation of corrosion by random fields in alternative to the traditional approach based on a uniform reduction of plate thickness.

4.4 *Model uncertainty*

The uncertainty associated with the predictions of the simplified methods used to calculate the hull girder ultimate strength can be assessed by comparing the results obtained by different methods. For this purpose different simplified progressive collapse methods developed by several organizations have been used to calculate the ultimate bending strength of a RoRo vessel in the project MARSTRUCT (Guedes Soares et al. (2008)). Table 3 presents the collapse bending moments obtained by each organization for hogging and sagging conditions. "Method 1" corresponds to the simplified progressive collapse method developed by Gordo et al., (1996), which is based on the load-shortening curves derived by Gordo and Guedes Soares, (1993). The other methods differ mainly on the "average stress-average strain" curves used to define the behaviour of the stiffened plate elements. The results in the sagging condition show a larger coefficient of variation than the ones in hogging, 17.7% and 8.7%, respectively. This study gave an idea of the range of variation of the perditions of the various methods that can be incorporated in the reliability formulation by an additional model uncertainty random variable affecting the

Table 3. Results for the collapse bending moment in sagging and hogging conditions without residual stresses (Guedes Soares et al 2008).

	Hogging		Sagging	
Method	(GN · m)	Diff.	(GN.m)	Diff.
Method 1	3935	−1.3%	2071	−25.6%
Method 2	4222	5.9%	2492	−10.4%
Method 3	4052	1.7%	3008	8.1%
Method 4	3613	−9.4%	2344	−15.7%
Method 5	3296	−17.3%	2656	−4.5%
Method 6	3942	−1.1%	2530	−9.1%
Method 7	4163	4.5%	3526	26.8%
Method 8	4406	10.5%	3455	24.2%
Method 9	4243	6.5%	2954	6.2%
Mean	3990	0%	2780	0%
St.Dev.	345		493	
COV	0.087		0.177	

stochastic model of the ultimate strength of the hull girder.

5 STOCHASTIC MODEL OF WAVE INDUCED LOAD EFFECTS

5.1 *Intact condition*

The stochastic model of wave induced load effects is typically constructed based on the evaluation of the wave induced load effects that occur during long-term operation of the ships in a seaway. The procedure for calculating long-term cumulative probability distributions of the wave induced responses is a well defined and accepted one (e.g. Guedes Soares, (1998b)). To standardize the procedure for computation of long-term distributions, IACS, (2002) has issued Recommendation note No. 34 suggesting that the IACS North Atlantic scatter diagram (ATLN) covering areas 8, 9, 15,16 given by Global Wave Statistics (Hogben et al., (1986)) should be used for intact ships (Figure 5). This scatter diagram was considered representative of a North Atlantic crossing although other assumptions about the route could obviously be made. The effect of choosing alternative routes and sea areas was dealt for example by Guedes Soares and Moan, (1991) and Teixeira and Guedes Soares, (1998).

A Weibull model is usually fitted to the long-term distribution, which describes the distribution of the peaks at a random point in time. However one is normally interested in having the probability distribution of the maximum amplitude of wave induced effects over a time period T that is given by a Gumbel distribution:

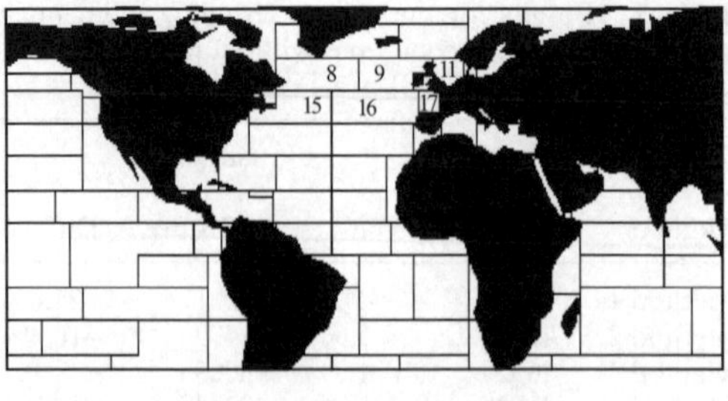

Figure 5. Global wave statistics ocean areas.

$$F_e(x_e) = \exp\left[-\exp\left(-\frac{x_e - x_n}{\sigma}\right)\right] \tag{6}$$

The parameters x_n and σ of the Gumbel distribution can be estimated from the initial Weibull distribution using the following equations:

$$x_n = w \cdot \left[ln(n)\right]^{\frac{1}{k}} \quad \text{and} \quad \sigma = \frac{w}{k}\left[ln(n)\right]^{\frac{1-k}{k}} \tag{7}$$

where w and k are the Weibull parameters and n corresponds to the mean number of load cycles expected over the time period T (typically, one year of operation). The number of wave cycles corresponding to the return period is calculated considering the average wave period equal to 7 s, which is the one that is applicable for the areas of the North Atlantic. However the mean value of the Gumbel distribution does not change significantly when changing the average wave period. In fact, considering an average wave period of 10 s, the mean value of the Gumbel distribution only decreases by about 2.7% Guedes Soares and Teixeira, (2000).

5.2 *Damaged condition*

For a damaged ship a model similar to the intact case can been applied. However, the applicable environmental conditions are now taken to be less severe than the North Atlantic model since groundings and collisions are likely to occur in coastal areas where the traffic density is higher. In this case scatter diagrams of European costal areas (Figure 6) given by Global Wave Statistics (Hogben et al., (1986)) can be adopted, as done by Luís et al., (2009).

In alternative to the scatter diagrams of European costal areas, the data from the HARDER project (Rusaas, (2003)) can be used to define the distribution of the significant wave height at the time of collision, as done by Hørte et al., (2007a). Figure 7 shows the data and a Weibull (cumulative) distribution fitted to the data.

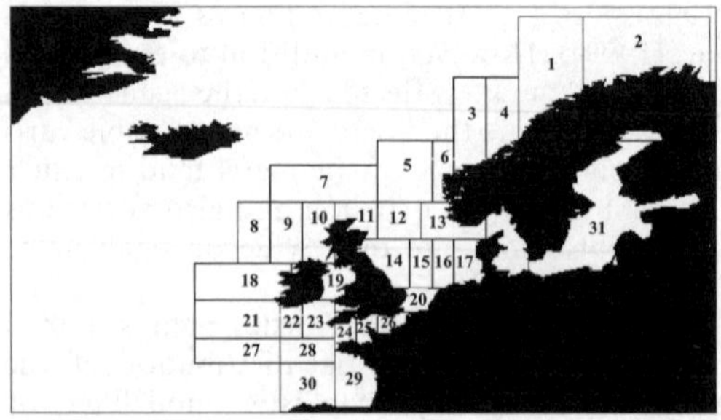

Figure 6. Global wave statistics coastal areas of Europe.

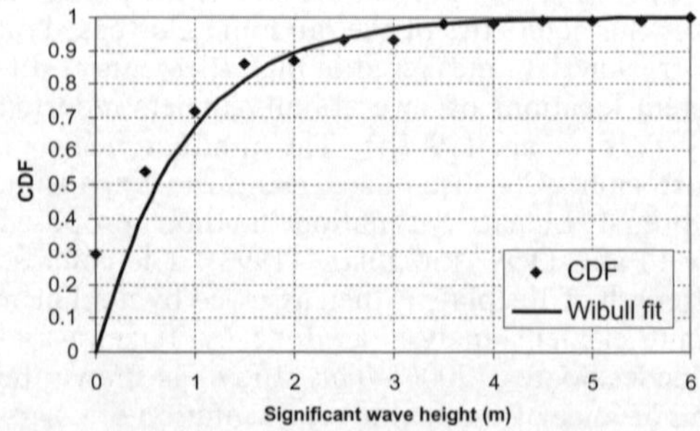

Figure 7. Distribution of the significant wave height at the time of collision (Rusaas, (2003)).

$$F_{H_s}(h_s) = 1 - \exp\left[-\left(\frac{h_s - \gamma}{\alpha}\right)^{\beta}\right] \tag{8}$$

The corresponding distribution parameters are: $\alpha = 0.9$ (scale), $\beta = 1.0$ (shape) and $\gamma = 0.0$ (location). The assumption regarding heading and distribution of the zero crossing wave period T_Z can be assumed to follow the same relationship as used for the North Atlantic model.

Furthermore, a reduced exposure time to the environmental conditions after damaged and before the ship is taken to a safe location should be considered when establishing the stochastic model of the extreme wave induced bending moment. Hørte et al., (2007a) and later Luís et al., (2009) have considered a time period T of one week as the voyage duration of the damaged ship to dry-dock.

Table 4 shows that the mean value of the distribution of the extreme values of the vertical wave induced bending moment can reduce in around 15% for Suezmax tanker when reducing exposure time from one year in the North Atlantic to one week under the environmental conditions of European coastal areas (areas 27, 28 and 30 of Figure 6).

Table 4. Stochastic model of extreme wave induced loads in North Atlantic and European coastal areas.

Gumbel distribution ($F_e(x_e)$)	North Atlantic	European coastal areas (27, 28 and 30)
Exposure time (T)	$T = 1$ year	$T = 1$ week
Mean value	6199.1	5291.0
St. Dev.	476.1	491.3

The traditional approach for assessing the wave induced loads in intact ships structures assumes that the sea states are dominated by wave systems generated by local winds. However, marine structures are subjected to all types of sea states that can occur, which in many situations are a result of the combination of more than one wave system, and in this case, the frequency spectrum exhibits two peaks. Double-peaked wave spectra can be observed whenever a swell system that is typically confined to a very narrow range of directions combines with a locally wind-driven system.

Teixeira and Guedes Soares, (2009) have demonstrated that, for a trading ship of non-restricted operation, the long-term distributions of the wave induced vertical bending moment for combined sea states do not change significantly when compared with the ones obtained from sea states of a simple component. However, it has been recognized that double-peaked wave spectra can have a significant impact on design and operability of fixed (e.g. Bitner-Gregersen et al., (1992)) and offshore platforms (e.g. Ewans et al., (2003)) and therefore it would be important to assess its impact on damaged ships since collisions and groundings may occur in sea areas with swell dominated sea states and the manoeuvrability of the ship may be affected as a consequence of the accident.

6 STOCHASTIC MODEL OF STILL WATER BENDING MOMENT

6.1 *Intact condition*

The stochastic model of the still water loads that is frequently adopted in most of the reliability analyses of intact ships assumes that the values of the still water vertical bending moment follow a normal distribution. The statistical parameters can be defined on the basis of the statistical analysis performed by Guedes Soares and Moan, (1988), which covered different types of ships. Guedes Soares and Dias, (1996) have conducted more detailed studies in order to find out if it was possible to identify different probabilistic models applicable to data from different ship routes and also the effect of operational conditions that truncate the distributions of the still water bending moments (Guedes Soares, (1990), Guedes Soares, (1990)).

Rizzuto, (2006) has examined various aspects of the loading procedure of tankers and bulk carriers in order to identify the sources of uncertainties in the prediction of still water bending loads. Rizzuto, (2006) suggested that all loading modes should be treated separately (as done by Guedes Soares), in order to get coherent statistics of bending moments and found that considerable part of the variability in the corresponding bending moments is related to the approximations in the realization of the selected loading plan.

Recently, characteristic values of the loading manual (such as its maximum value) have been adopted to define the probabilistic models of the still water bending moment used in the reliability analysis of individual ships. Hørte et al., (2007b) found that the mean value of the still water bending moment in sagging was between 49% and 85% of the maximum value in the loading manual. They also found that when the mean value of still water bending moment is large the standard deviation is relatively small. Based on the analysis of 8 test tankers, Hørte et al., (2007b) proposed a stochastic model that describes the still water vertical bending moment by a normal distribution with mean value and standard deviation of 70% and 20% of the maximum value in the loading manual, respectively.

Since there is no truncation in this model, there is a chance that the still water moment may attain a value that exceeds the maximum value in the loading manual, although the probability of this is small; i.e. less than 7%. Furthermore, since the model is based on data from 8 ships one wonders how representative it is.

6.2 *Damaged condition*

The still water bending moment in the damaged conditions (M_s^D) is typically based on the distribution for the intact ship (M_s), and the change in still water loads as a consequence of the damage is added or represented by a still water load combination coefficient (K_{us}) that affects stochastic model for the intact condition.

$$M_s^D = K_{us} M_s \qquad (9)$$

The ABS, (1995) guide for damaged ships recommends the use of $K_{us} = 0.9$ for hogging and $K_{us} = 1.1$ for sagging. In this context, ABS also recommends that the North Atlantic wave induced bending moment should be reduced by 50% when calculating the total bending moment for the damaged condition.

Luís et al., (2009) have used the value of 1.0 for K_{us} to calculate the one year reliability of the intact ship and 1.1 and 1.5 to analyze the impact of increased still-water loads on the reliability of the damaged ship.

Santos and Guedes Soares, (2007) have developed a method that is able to calculate the changes in still-water loads as a result of the flooding that follows hull rupture and this approach can be used to determine appropriate values of the still-water load combination coefficient (K_{us}).

Typically, after a grounding or a collision event the ship gains weight (it embarks sea water) while losing buoyancy. Depending on the longitudinal position and extension of the damage and of the initial load condition, the still water bending moment may increase or decrease.

In most cases, flooding of ballast compartments in the midship region is most critical, and this causes the sagging moment to increase. Examples of this effect for a VLCC are presented in Figure 8 (Hørte et al., (2007a)). The basis for the calculation for the intact ship is a homogenous full load condition (that is not the most severe one), and the label in the figure indicates which ballast compartment is flooded, 1 being the foremost and 5 the aftmost hold. E.g. "dam3" represents flooding of the ballast compartment outside cargo hold 3, and implies water ingress in the double bottom from the centreline up to the still waterline in the double side. A significant increase in the sagging bending moment is seen as a consequence of flooding of ballast tanks near the centre of the ship.

Hussein and Guedes Soares, (2009) have also calculated the effect the flooding of ballast compartments on maximum value the still water bending moment (SWBM) of a double hull tanker. It was shown that when one of the compartments in the midship region (at one sine) is damaged, he SWBM will increase 30%. If two compartments (starboard and portside) are damaged the increase will be 46%.

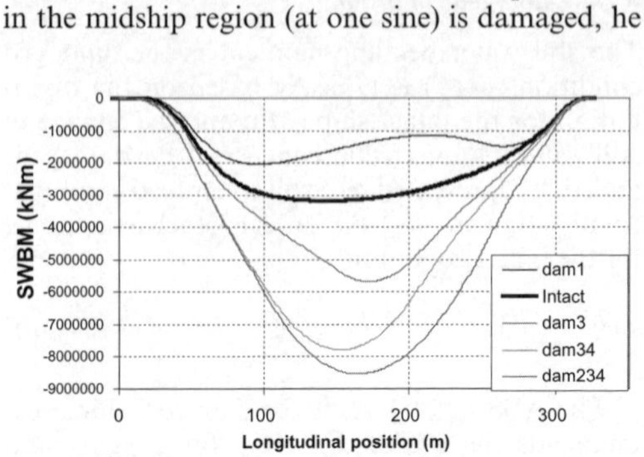

Figure 8. Still water bending moment for intact and various damaged conditions (negative values are sagging).

7 RELIABILITY ANALYSIS

In the damaged condition the annual probability of failure $P_{f\,damage}$ can be defined as the product of the probability of a critical damage scenario $P_{damage\,i}$ times the conditional probability of failure given this scenario $P_{f|damage\,i}$, and accumulated over the number of relevant scenarios.

$$P_{f\,damage} = \sum_{i=1}^{all\,scenarios} P_{f\,damage\,i} = \sum_{i=1}^{all\,scenarios} P_{damage\,i} \cdot P_{f|damage\,i} \tag{10}$$

In practice a very limited number of scenarios need to be considered since for most of them $P_{damage\,i}$ or $P_{f|damage\,i}$ are not relevant.

In fact, the most critical damage scenario regarding hull girder failure causes the most unfavourable combination of the following effects:

- a reduction in the ultimate strength,
- an increase in the still water bending,
- unfavourable wave loading.

The reduction in the hull girder strength is most critical for the region where the loading, both wave and still water loads, is high; i.e. in the midship region.

An increase in the still water loads depends on the loading condition of the ship. For a fully loaded ship in sagging, flooding of the ballast compartments in the midship region will cause an increase in the still water bending moment, whereas damages at the ship ends will reduce the sagging moment. Severe damages that penetrate into the cargo hold in the midship region are, of course, critical from an environmental point of view, but may be less critical concerning subsequent hull girder failure, since the oil outflow will tend to reduce the still water bending moment.

The identification of the critical damage scenarios has been studied by the use of a risk-based approach implemented in a Bayesian Network model (Garre et al., (2006)). A few scenarios were identified as the critical ones. Garre et al., (2006) have focused on using a risk based approach to identify those scenarios that most likely will cause a given increase in the still water loads.

As concerns damage size and location, significant collision damages in the upper part of the side is considered the most critical concerning ultimate hull girder strength. This type of damages reduce the sagging strength of the hull girder, whereas grounding damage (unless very extensive) generally has less impact on the sagging capacity due to the double bottom. According to Kjellstrom and Johansen, (2004) the probability of occurrence of such a damage is close to $1 \cdot 10^{-2}$ for double hull tankers above 10,000 dwt. More severe damages may also occur, however the probability of such events is so small that it becomes irrational to attempt control the associated risk by structural means.

Table 5 illustrates the results of the reliability analysis of a VLCC in the intact condition and with a significant collision damage using rather conservative assumptions (Hørte et al., (2007a)). It can be seen that the benefit of the reduced magnitude of wave loads due to calmer environmental conditions exceeds the disadvantage of increased still water loads and reduced strength of the damaged ship.

The analysis of the damaged case has also been carried out assuming that the collision event may happen at any time worldwide, and that ship is exposed to worldwide environmental conditions for a week after the collision. In this case the annual probability of failure is about 2.5 times higher than the intact case; however, when the probability of the collision event ($\approx 1 \cdot 10^{-2}$) is multiplied to this result, the hull girder failure probability of the damaged ship becomes significantly lower than the chance of failure for an intact ship.

As concerns structural damages resulting from grounding accidents, Table 7, and Figure 9 and Figure 10, present the reliability indices of a Suezmax tanker under sagging and hogging, respectively. The reliability indices are plotted as a function of increasing damages around the keel area of the tanker, illustrated in Figure 2. Two operational conditions, ATLN and ECA, have been considered. ATLN refers to the failure of the hull girder in severe weather in the North Atlantic (ATLN), which corresponds to the conventional conditions used to assess the structural reliability of intact ships. ECA condition is the one proposed to asses the reliability of the damaged ship, which considers that the ship operates in costal areas for one week that is a more likely scenario for a damaged ship. ECA condition also considers an increase of 50% of the still water bending moment as a result of the damage (k_{us} = 1.5 in Eqn. 9). Table 6 summarizes characteristics of the two operational profiles considered in the reliability calculations.

It can be seen that in general the reliability in sagging is smaller than in hogging and the reduced wave induced bending moment compensates in most of the cases the reduction in the strength of the ship and the increased still water loads due to the damaged. However, since grounding damages affect considerably the strength of the ship under hogging bending moments, the hogging reliability of the ship with large damages around the keel area (e.g. "GCMD" and "GCMD + 20%") may be lower than the one of the intact ship in sagging

Table 5. Annual probabilities of failure of a VLCC in intact and damaged conditions.

Case	Reference period	P_f	β
Intact failure, (North Atlantic)	Annual	7.2×10^{-4}	3.19
Intact failure, (world wide)	Annual	2.5×10^{-5}	4.06
Failure of damaged ship, $P_{fldamage}$ (world wide environment)	1 week	1.8×10^{-3}	2.91
Failure of damaged ship, $P_{fldamage}$ ("Collison" environment)	1 week	8.3×10^{-5}	3.77
Prob. of collision event, P_{damage}	Annual	1×10^{-2}	
Fail. damaged, ship, $P_{damage} \cdot P_{fldamage}$ ("world wide" environment)	Annual	1.8×10^{-5}	4.13
Fail. damaged ship, $P_{damage} \cdot P_{fldamage}$ ("Collison" environment)	Annual	8.3×10^{-7}	4.79

Table 6. Operational profiles of the Suezmax tanker.

Condition	Sea zone	Ref. period	k_{us}
ATNL	ATLN	Annual	1.0
ECA	European Coastal areas (27, 28 and 30 of Figure 6)	1 week	1.5

Table 7. Reliability indices of a Suezmax in sagging and hogging.

	Sagging			Hogging		
	RIF	ATLN	ECA	RIF	ATLN	ECA
Intact	1.0	1.77	2.15	1	2.12	2.15
GCD – 20%	0.985	1.67	1.88	0.957	1.84	1.88
GCD	0.982	1.65	1.82	0.948	1.78	1.82
GCD + 20%	0.979	1.63	1.74	0.935	1.69	1.74
GCMD – 20%	0.983	1.66	1.67	0.925	1.61	1.67
GCMD	0.979	1.63	1.52	0.903	1.46	1.52
GCMD – 20%	0.970	1.57	1.36	0.88	1.28	1.36

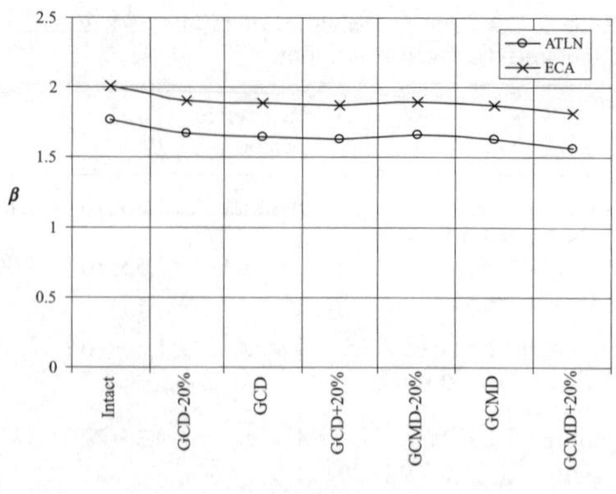

Figure 9. Sagging reliability indices of a Suezmax.

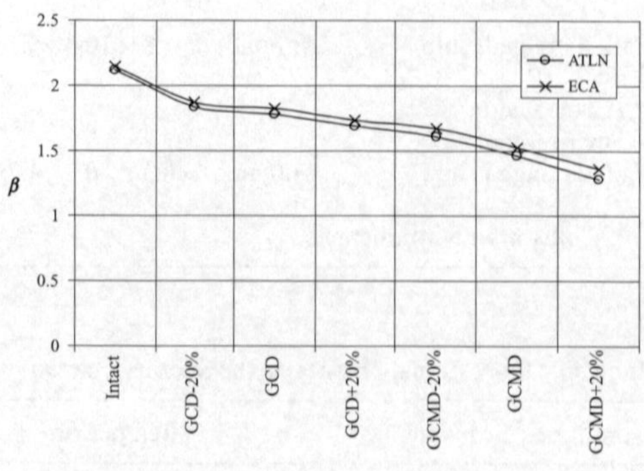

Figure 10. Hogging reliability indices of a Suezmax.

under the ATLN conditions. This suggests that the hogging failure of the ship in ballast and partial loading conditions should be analysed carefully when assessing the structural safety of ships with grounding damages.

8 CONCLUSIONS

The structural reliability methods have been successively applied to assess the implicit safety levels of intact ships and can be easily extended to damaged ships due to grounding and collision events.

In order to asses the structural safety of damaged ships it is necessary to identify the critical damage scenario, or scenarios, with the associated annual probability of occurrence. The use of Bayesian Network modelling may be applied for this purpose. In practice a very limited number of scenarios regarding hull girder failure need to be considered since the occurrence probability of most of them are very small and only a limited number of scenarios are relevant for the hull girder failure.

It should be noted that reliability analysis must account for the effect of the damage on the ship ultimate strength and should properly take into account the relevant changes the still water and wave induced loads following damage. Furthermore, the environmental conditions and duration of exposure that is applicable for the damaged ship should be specified as these will have an important impact in the probabilistic models of the load effects.

The studies already performed indicate that, although hull girder failure of an intact ship in severe weather conditions is a critical factor in dimensioning of the midship sagging capacity, a strength criterion for damaged conditions appears to be unnecessary since the probability of failure of the damaged ship is typically lower than the one of the intact ship. However, additional studies are need with improved descriptions of damage scenarios and better predictions of the effect of the damage on the loads an on the strength of the ships in order the support this conclusion.

ACKNOWLEDGMENTS

This paper presents part of the teaching material prepared and used in the scope of the course “Ship structural reliability with respect to ultimate strength” within the project “Advanced Ship Design for Pollution Prevention (ASDEPP)” that was financed by the European Commission through the Tempus contract JEP-40037-2005.

REFERENCES

ABS, (1995), *Guide for assessing hull-girder residual strength for tankers*, American Bureau of Shipping.

Ang, A.H.S. and Tang, W.H., (1984), *Probability concepts in engineering planning and design. Vol II, Decision, risk and reliability*, John Wiley & Sons, New York.

Bai, Y., Bendiksen, E. and Pedersen, P.T., (1993), Collapse analysis of ship hulls, *Marine Structures*, 6, pp. 485–507.

Bitner-Gregersen, E.M., Haver, S., and Lřseth, R., (1992), Ultimate Limit States with Combined Load Processes, *Proceeings of the 2nd Offshore and Polar Engineering Conference*, Vol. IV, pp. 515–522.

Bitner-Gregersen, E.M., Hovem, L., and Skjong, R., (2002), Implicit Reliability of Ship Structures, *Proc. of the Int. Conf. on Offshore Mechanics and Arctic Engineering (OMAE2002)*, Oslo, Norway, June 23–28, 2002, OMAE2002-28522.

Breitung, K., (1984), Asymptotic approximations for multi-normal integrals, *J. Eng. Mech. Div. ASCE*, No. 3, Vol. 110, pp. 357–366.

Caldwell, J.B., (1965), Ultimate longitudinal strength, *Transactions of RINA*, 107, 411–430.

Cornell, C.A., (1969), Structural Safety Specifications Based on Second Moment Reliability Analysis, *Symp. on Concepts of Safety of Structures and Methods of Design*, IABSE, London, pp. 235–246.
Der Kiureghian, A., Lin, H.-Z., and Hwang, S-J., (1987), Second-order reliability approximations, *J. Eng. Mech. Div. ASCE*, N. 8, Vol. 113, pp. 1208–1225.
Der Kiureghian, A. and Liu, P.-L., (1986), Structural reliability under incomplete probability information, *Journal of Eng. Mechanics*, No. 1, Vol. 112, pp. 85–104.
Ditlevsen, O., (1981a), Principle of normal tail approximation, *J. Eng. Mech. Div. ASCE*, N. 6, Vol. 107, pp. 1191–1209.
Ditlevsen O., (1981b), *Uncertainty modelling*, McGraw-Hill, New York.
Ditlevsen, O. and Madsen, H.O., (1996), *Structural Reliability Methods*, John Wiley & Sons, England.
DNV, (1992), Classification Note 30.6, *Structural Relaibility Analysis of Marine Structures*, Det Norske Veritas.
Ewans, K., Valk, C., Shaw, C., Haan, J., Tromans, P., and Vandershuren, L., (2003), Oceanographic and motion response statistics for the operation of a weathervaning LNG FPSO, *Proceedings of the 22nd International Conference on Offshore Mechanics and Artic Engineering (OMAE)*.
Fang, C. and Das, P.K., (2004), Hull Girder Ultimate Strength of Damaged Ship, *9th Symposium on Practical Design of Ships and Other Floating Structures (PRADS2004)*.
Fang, C. and Das. P.K, (2005), Survivability and reliability of damaged ships after collision and grounding, *Ocean Engineering*, 32, pp. 293–307.
Fiessler, B., Neumann, H., and Rackwitz, R., (1979), Quadratic Limit States In Structutal Reliability, *J. Eng. Mech. Div., ASCE*, No. 4, Vol. 105, pp. 661–676.
Garbatov, Y., Teixeira, A.P., and Guedes Soares, C., (2004), Fatigue Reliability Assessment of a Converted FPSO Hull. *Proceedings of the OMAE Specialty Conference on Integrity of Floating Production, Storage & Offloading (FPSO) Systems*, ASME, Houston, paper n. FPSO'0035.
Garre, L., Friis-Hansen, P., Hørte, T., Austefjord, H.N., and De Francisco, V.F., (2006), Probabilistic Models for Load Effects: Uncertainty modelling of load and responses; Identification of most critical collision damage scenarios, *SAFEDOR report no SAFEDOR-D-2.2.1-2006-10-15-DTU-loadrev-3*.
Gordo, J.M. and Guedes Soares, C., (1993), Approximate Load Shortening Curves for Stiffened Plates Under Uniaxial Compression, *Proc. Integrity of Offshore Structures—5*, D. Faulkner, M.J. Cowling, A. Incecik and P.K. Das, EMAS, U.K., pp. 189–211.
Gordo, J.M. and Guedes Soares, C., (1995), Collapse of Ship Hulls Under Combined Vertical and Horizontal Bending Moments, *Proceedings of the 6th International Symposium on Practical Design of Ships and Mobile Units (PRADS'95)*, Korea, Vol. II, pp. 808–819.
Gordo, J.M. and Guedes Soares, C., (1996), Approximate Method to Evaluate the Hull Girder Collapse Strength, *Marine Structures*, Vol. 9, pp. 449–470.
Gordo, J.M. and Guedes Soares, C., (1997), Interaction Equation for the Collapse of Tankers and Containerships under Combined Bending Moments, *Journal of Ship Research*, Vol. 41, No. 3, pp. 230–240.
Gordo, J.M. and Guedes Soares, C., (2000), Residual strength of damaged ship hulls, *Proceedings of the 9th international congress of international maritime association of the mediterranean*, Ischia, Italy.
Gordo, J.M., Guedes Soares, C., and Faulkner, D., (1996), Approximate Assessment of the Ultimate Longitudinal Strength of the Hull Girder, *Journal of Ship Research*, N. 1, Vol. 4, pp. 60–69.
Guedes Soares, C., (1988), Uncertainty Modelling in Plate Buckling, Structural Safety, Vol. 5, pp. 17–34.
Guedes Soares, C., (1990), Stochastic Modeling of Maximum Still-Water Load Effects in Ship Structures, *Journal of Ship Research*, N°. 3, Vol. 34, pp. 199–205.
Guedes Soares, C., (1998a), Ship Structural Reliability, *Risk and Reliability in Marine Structures*, C. Guedes Soares (Ed.), A.A. Balkema, pp. 227–244.
Guedes Soares, C., (1998b), Stochastic Modelling of Waves and Wave Induced Loads, *Risk and Reliability in Marine Technology*, Guedes Soares, C., (Ed.), A.A. Balkema, Rotterdam, pp. 197–211.
Guedes Soares, C. and Dias, S., (1996), Probabilistic Models of Still-Water Load Effects in Containers, *Marine Structures*, n. 3–4, Vol. 9, pp. 287–312.
Guedes Soares, C., Dogliani., M., Ostergaard, C., Parmentier, G., and Pedersen, P.T., (1996), Reliability Based Ship Structural Design, *Transactions of the Society of Naval Architects and Marine Engineers (SNAME)*, New York, Vol. 104, pp. 357–389.
Guedes Soares, C. and Garbatov, Y., (1996), Fatigue Reliability of the Ship Hull Girder, *Marine Structures*, 9, pp. 495–516.
Guedes Soares, C. and Garbatov, Y., (1999), Reliability of Maintained Corrosion Protected Plate Subjected to Non-Linear Corrosion and Compressive Loads, *Marine Structures*, Vol. 12, pp. 425–446.
Guedes Soares, C. and Kmiecik, M., (1993), Simulation of the Ultimate Compressive Strength of Unstiffened Rectangular Plates, *Marine Structures*, Vol. 6, pp. 553–569.
Guedes Soares, C.; Luís, R.M.; Nikolov, P.I.; Modiga, M.; Quesnel, T.; Dowes, J.; Toderan, C., and Taczala, M., (2008), Benchmark Study on the use of Simplified Structural Codes to Predict the Ultimate Strength of a damaged ship hull. *International Shipbuilding Progress.* 55(1–2):87–107.
Guedes Soares, C. and Moan, T., (1988), Statistical Analysis of Still Water Load Effects in Ship Structures, *Transactions of the Society of Naval Architects and Marine Engineers*, New York, No. 4, Vol. 96, pp. 129–156.
Guedes Soares, C. and Moan, T., (1991), Model Uncertainty in the Long Term Distribution of Wave Induced Bending Moments for Fatigue Design of Ship Structures, *Marine Structures*, Vol. 4, pp. 295–315.
Guedes Soares, C. and Teixeira, A.P., (2000), Structural Reliability of Two Bulk Carrier Designs, *Marine Structures*, Vol. 13, pp. 107–128 .
Haldar, A. and Mahadevan, S., (2000), *Reliability Assessment using Stochastic Finite element Analysis*, John Wiley & Sons, Inc. New York.
Hansen, A.M., (1996), Strength of Midships Sections, *Marine Structures*, 9, pp. 471–494.
Hart, D.K., Rutherford, S.E., and Wichham, A.H.S., (1986), Structural reliability analysis of stiffened

panels, *Trans Roy Inst Nav Architects (RINA)*, Vol. 128, pp. 293–310.

Hasofer, A.M. and Lind, N.C., (1974), An Exact and Invariant First-Order Reliability Format, *J. Eng. Mech. Div. (ASCE)*, Vol. 100, pp. 111–121.

Hogben, N., Da Cuna, L.F., and Ollivier, H.N., (1986), Global Wave Statistics, *British Marine Technology*, Publishing Urwin Brothers Limited, London.

Hohenbichler, M. and Rackwitz, R., (1981), Non-normal dependent vectors in structural safety, *J. Eng. Mech. Div. ASCE*, Vol. 107, pp. 1227–1238.

Hørte, T., Skjong, R., Friis-Hansen, P., Teixeira, A.P., and Viejo de Francisco, F., (2007a), Probabilistic Methods Applied to Structural Design and Rule Development, *RINA Conference on Developments in Classification & International Regulations*, 24–25 January 2007, London, UK.

Hørte, T., Wang, G., and White, N., (2007b), Calibration of the Hull Girder Ultimate Capacity Criterion for Double Hull Tankers, *10th International Symposium on Practical Design of Ships and Other Floating Structures*, © 2007 American Bureau of Shipping, Houston, Texas, United States of America, vol. 1, pp. 553–564.

Hussein, A.W. and Guedes Soares, C., (2009), Reliability and residual strength of double hull tankers designed according to the new IACS common structural rules, *Ocean Engineering*, 36 (17–18), pp. 1446–1459.

IACS, (2002), Recommendation No. 34, *Standard Wave Data*.

ISO 2394, (1998), *General Principles on reliability for structures*, Geneve.

Kalman, Z. and Manta, P., (2002), Tracing the Ultimate Longitudinal Strength of a Damaged Ship Girder, *International Shipbuilding Progress*, Vol. 49, No 3, pp. 161–176.

Kjellstrom, S. and Johansen, C.B., (2004), FSA Generic Vessel Risk—Double hull tanker for oil, *DNV Report No. 2003-0425*, Hovik, Norway.

Kmiecik, M. and Guedes Soares, C., (2002), Response Surface Approach to the Probability Distribution of the Strenght of Compressed Plates, *Marine Structures*, Vol. 15, N. 2, pp. 139–156.

Li, C.-C. and Der Kiureghian, A., (1993), Optimal discretization of random fields, *Journal of Eng. Mechanics*, Vol. 119, N. 6, pp. 1136–1154.

Liu, P.-L. and Der Kiureghian, A., (1991), Optimization algorithms for structural reliability, *Structural Safety*, Vol. 9, pp. 161–177.

Luís, R.M., Teixeira, A.P., and Guedes Soares, C., (2009), Longitudinal strength reliability of a tanker hull accidentally grounded, *Structural Safety*, vol. 31, 3, pp. 224–233.

Madsen, H.O., Krenk, S., and Lind, N.S., (1986), *Methods of Structural Safety*, Prentice-Hall, Eaglewood Cliffs.

Mansour, A.E., (1972), Probabilistic Design Concepts in Ship Structural Safety and Reliability, *Transactions of the Society of Naval Architects and Marine Engineers*, Vol. 80, pp. 64–97.

Mansour, A.E. and Faulkner, D., (1973), On Applying the Statistical Approach to Extreme Sea Loads and Ship Hull Strength, *Transactions Royal Institution of Naval Architects (Rina)*, Vol. 115, pp. 277–314.

Mansour, A.E., Lin, Y.H., and Paik, J.K., (1995), Ultimate Strength of Ships Under Combined Vertical and Horizontal Moments, *Proceedings, International Symposium on Practical Design in Shipbuilding, (PRADS 95)*, pp. 844–856.

Melchers, R.E., (1999), *Structural Reliability and Analysis Prediction*, 2nd Edition, John Wiley & Sons.

Melchers, R.E., (2003), Probabilistic Models for Corrosion in Strutural Reliability Assessment—Part 1: Empirical Models, *Journal of Offshore Mechanics and Arctic Engineering*, ASME, Vol. 125, pp. 264–271.

Nikolaidis, E, Ghiocel, D.M., and Singhal, S. eds, (2005), *Engineering Design Reliability Handbook*, ISBN 0-8493-1180-2, New York, CRC Press.

Paik, J.K., (1993), Advanced idealized structural units considering excessive tension-deformation effects, *Trans. of Society of Naval Architects of Korea*, (6), 30, 100–115.

Paik, J.K. and Frieze, P.A., (2001), Ship structural safety and reliability, *Progress in Structural Engineering and Materials*, John Wiley & Sons, Vol. 3, 2, pp. 198–210.

Paik, J.K., Kim, S.K., and Lee, S.K., (1998a), Probabilistic corrosion rate estimation model for longitudinal strength members of bulk carriers, *Ocean Engineering*, Vol. 25, 10, pp. 837–860.

Paik, J.K., Thayamballi, A.K., and Yang, S.H., (1998b), Residual Strength Assessment of Ships after Collision and Grounding, *Marine Technology*, Vol 35, No. 1, pp. 38–54.

Rackwitz, R., (2001), Reliability analysis—a review and some perspectives, *Structural Safety*, Elsevier, vol. 23, pp. 365–395.

Rackwitz, R. and Fiessler, B., (1978), Structural reliability under combined random load sequences, *Computers & Structures*, Vol. 9, pp. 486–494.

Rizzuto, E., (2006), Uncertainties in still water loads of tankers and bulkers, *International Conference on Ship and Shipping Research (NAV 2006)*, Genova, Italy.

Rusaas, S., (2003), Final Publishable Report, *HARDER FP5 Project, Contract No. G3RD-CT-1999–00028*, 1999–2003, Doc. Ref. N. 0-00-X-2003-03-0, 2003–08–31.

Sadovský, Z. and Páles, (1999), Third-moment identification of resistance by FORM, *Building Research Journal*, N. 3, Vol. 47, pp. 197–213.

Santos, T. and Guedes Soares, C., (2007), Time domain simulation of ship global loads due to progressive flooding, *Advancements in marine structures*, Guedes Soares, C. and Das, P.K. (Eds.), Taylor & Francis Group, London, pp. 79–88.

Shi, W.B., (1993), In-service Assessment of Ship Structures: Effects of General Corrosion on Ultimate Strength, *Transactions of the Society of Naval Architects and Marine Engineers*, Vol. 135, pp. 77–91.

Smith, C., (1977), Influence of local compressive failure on ultimate longitudinal strength of a ship's hull, *Proc. Int. Symposium on Practical Design in Shipbuilding (PRADS'77)*, Tokyo, pp. 73–79.

Spencer, J.S., Wirsching, P.H., Wang, X., and Mansour, A.E., (2003), Development of Reliability-Based Classification Rules for Tankers, *Transactions of the Society of Naval Architects and Marine Engineers*, SNAME, Vol. 111.

Teixeira, A.P. and Guedes Soares, C., (1998), On the Reliability of Ship Structures in Different Coastal Areas, *Structural Safety and Reliability*, Shiraishi, Shinozuca & Wen (Eds), A.A. Balkema, pp. 2073–2076.

Teixeira, A.P. and Guedes Soares, C., (2005), Assessment of partial safety factors for the longitudinal strength of tankers, *Maritime Transportation and Exploitation of Ocean and Coastal Resources,* Guedes Soares, C., Garbatov, Y., Fonseca, N., (Eds), Francis and Taylor, Lisbon, pp. 1601–1610.

Teixeira, A.P. and Guedes Soares, C., (2006), Ultimate strength of plates with random fields of corrosion, *Advances in Reliability and Optimization of Structural Systems*, Sorensen & Frangopol (eds), Taylor & Francis Group, London, pp. 179–186.

Teixeira, A.P. and Guedes Soares, C., (2007), Probabilistic modelling of the ultimate strength of plates with random fields of corrosion, *Proc. of the 5th International Conference on Computational Stochastic Mechanics*, G. Deodatis & P.D. Spanos (eds), Millpress, Rotterdam, pp. 653–661.

Teixeira, A.P. and Guedes Soares, C., (2009), Reliability analysis of a tanker subjected to combined sea states, *Probabilistic Engineering Mechanics*, vol. 24, n. 4, pp. 493–503.

Thoft-Christensen, P. and Baker, M.J., (1982), *Structural Reliability Theory and its Applications,* Springer-Verlag, Berlin.

Thoft-Christensen, P. and Mourotsu, Y., (1986), *Application of structural systems reliability theory*, Springer-Verlag, Berlin.

Tvedt, L., (1990), Distribution of quadratic forms in normal space, *J. Eng. Mech. Div. ASCE*, Vol. 116, N. 6, pp. 1183–1197.

Ueda, Y. and Rashed, S.M.H., (1984), The idealized structural unit method and its application to deep girder structures, *Computer and Structures*, Vol. 18, 2, pp. 227–293.

Ueda, Y. and Yao, T., (1982), The plastic node method—a new method of plastic analysis, *Computer Methods in Applied Mechanical Engineering*, 34, pp. 1089–1104.

Wang. G., Chen, Y., Zhang, H., and Peng, H., (2002), Longitudinal strength of ships with accidental damages, *Marine Structures*, 15, pp. 119–138.

Wang, G., Spencer, J., and Chen, Y., (2002), Assessment of a Ship's Performance in Accidents, *Marine Structures*, Vol 15, pp. 313–333.

White, G.J. and Ayyub, B.M., (1992), Determining the effects of corrosion on steel structures: a probabilistic approach, *Proc. of the 11th Int. Conf. on Offshore Mechanics and Arctic Engineering (OMAE'92)*, In: Guedes Soares C et al., editors, ASME, New York, USA, vol. II, p. 45–52.

Yao, T. et. al, (2000), Ultimate Hull Girder Strength, *Proceedings of the 14th International Ship and Offshore Structutures Congress (ISSC)*, Nagasaki, Japan, pp. 321–391.

Zhang, S., Yu, Q., and Mu, Y.A., (1996), Semi-analytical Method of Assessing the Residual Longitudinal Strength of Damaged Ship Hull, *International Offshore and Polar Engineering Conference*, pp. 510–516.

Ziha K. and Pedisic M., (2002), Tracing the ultimate longitudinal strength of a damaged ship hull girder, *International Shipbuilding Progress*, vol. 49, n. 3, pp. 161–76.

Advanced Ship Design for Pollution Prevention – Guedes Soares & Parunov (eds)
© 2010 Taylor & Francis Group, London, ISBN 978-0-415-58477-7

Improving the ultimate strength of ship hull

K. Žiha, J. Parunov, M. Ćorak & B. Tušek
Faculty of Mechanical Engineering and Naval Architecture, University of Zagreb, Zagreb, Croatia

ABSTRACT: The paper first briefly introduces the practical methods for ultimate strength analysis of the ship hull girder proposed by Inernational Association of Classification Societies (IACS) Common Structural Rules (CSR). Then it shortly presents the home-made computer programme ULTIS based on the IACS recommendations. Subsequently, benchmark studies of ultimate strength of a bulk carrier built in Croatian shipyard employing program ULTIS and program MARS 2000 are attached. The examples illustrate the stepwise yielding and buckling scenario of hull elements both in deck and bottom structures under sagging and hogging conditions. Finally, the paper investigates the possible improvement of ship's hull girder ultimate strength by strengthening of longitudinal structural elements following their failure sequences under bending.

1 INTRODUCTION

Since the early proposals of Caldwell, the recent practical recommendation of the classification societies the ultimate ship hull strength is of growing interest (Rutherford & Caldwell 1990, Hughes 1988). Yao et al. (2006) have suggested the implementation of ultimate strength assessments in structural design and hull maintenance based either on sophisticated but time consuming finite element methodology or on simplified and idealized more practical structural methods. The implementation of sufficiently accurate and computationally efficient procedures for ultimate strength assessment challenge engineers to optimize their structures. Therefore the aim of this paper is to investigate the possibilities for ultimate ship hull strength improvement. In contrast to the optimization of the ship hull scantlings with respect to fully elastic and fully plastic section module, it became evident that the improvement of the ultimate bending capacity of the hull, before optimization attempts, requires understanding of the ship hull collapsing scenario under imposed vertical bending for each particular case.

2 SHIP HULL ULTIMATE STRENGTH PRACTICAL ANALYSIS

For now, classification societies rely on incremental-iterative analysis procedure based on stress-strain (load-end shortening) curves derived from unique IACS' longitudinal strength regulations implemented as common structural rules for bulk carriers and oil tankers, (IACS 2006 a,b).

Incremental part of the procedure manifests in gradual increase of imaginary curvature of the ship hull girder. After each increment of the curvature a new neutral axis position is iteratively defined. At the end of each step, the total hull girder bending moment is calculated by summing the contributions of all internal forces of each effective individual longitudinal element of a transverse section to the total bending moment. The result of the procedure is the bending moment M versus curvature χ and the ultimate bending moments are selected as maxima of the curves, negative sign for sagging and positive sign for hogging condition.

By using six types of curves that represent the stress-strain relationship $\sigma - \varepsilon$ known as "load—end shortening curves" the behavior of hard corners, transversely stiffened plates and longitudinally stiffened panels are investigated, (IACS 2006 a,b). The element responses depend on tension or on compressive loads and on the position of elements regarding the neutral axis of the ship hull transverse section. The six failure modes are the elasto-plastic collapse, the beam column buckling, the torsional buckling, the web local buckling of stiffeners of flanged profiles, the web local buckling of stiffeners of flat bars and plate buckling.

The principal steps of incremental—iterative approach are as follows:

1. Subdivision of the hull transverse section into stiffened plate elements
2. Definition of the neutral axis for non-deformed structure
3. Definition of stress—strain relationships for all elements
4. Increment with respect to the initial curvature

5. Calculation of corresponding stresses for each element (here intermediate outputs are required)
6. Finding the new position of the neutral axis
7. Calculation the total bending moment by summing the contribution of each element

Particularly for tankers the single step procedure (HULS-1) is provided for calculation of the sagging hull-girder ultimate bending capacity applying a simplified method based on reduced hull-girder bending stiffness also accounting for buckling of the main deck (IACS 2006a, Guedes Soares & Parunov 2008).

The problem becomes more complex for damaged ship hull, e.g. (Žiha & Pedišić 2000) .

2.1 *The computer program*

The computer program MARS 2000, with implemented code for ultimate strength computation developed at Technical University in Szczecin, Poland, is taken for benchmarking the final results (Bureau Veritas 2003). The program MARS 2000 applies the formulation for the ultimate strength assessment according to the requirements of Bureau Veritas and also from IACS CSR.

For the purposes of this study a computer program under the working name ULTIS was developed in order to check the intermediate results during hull bending process leading to collapse of the ship structure due to violation of the ultimate strength. The computer program ULTIS employs the incremental—iterative approach for ultimate bending moment evaluation using the load-end shortening curves as it is recommended by the IACS CSR. The program is developed in the Visual Basic 5 programming language ans also uses spreadsheet utilities for tabular and graphical interpretations of results (Hergert 1998, Žiha et al. 2007).

2.2 *Example of ship hull ultimate strength improvement*

Ultimate strength is checked for a bulk carrier built in the Brodosplit shipyard in the year 1999 as the yard 409 with following main particulars (Žiha et al. 2007):

Class: ✠ 100 A1 Bulk Carrier Strengthened for Heavy Cargoes
Length overall: $L_{OA} = 187.6$ m
Length between perpendiculars: $L_{PP} = 179.37$ m
Scantling length: $L_{LR} = 177.95$ m
Breadth: $B = 30.8$ m
Design draft: $T = 10.1$ m
Block coefficient: $C_B = 0.823$ m
Trial speed: v_{PP} = 14.5 kt at 10.1 m draft and 6435 kW/127 o/min

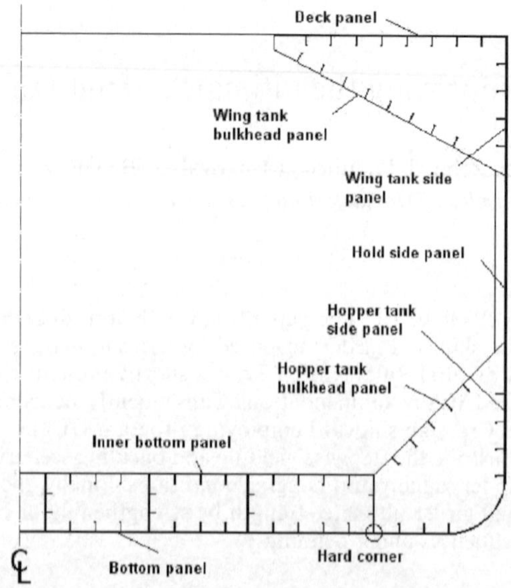

Figure 1. Transverse section division on to stiffened panels.

According to the shipyard data for the maximum still water bending moments and IACS recommendations for wave bending moments, the total calculated bending moments are:

Total hogging bending moment:

$M_{TH} = M_{SWH} + M_{WH} = 914783 + 1433907$
$M_{TH} = 2348690$ kNm

Total sagging bending moment:

$M_{TS} = M_{SWS} + M_{WS} = -1014354 - 1536242$
$M_{TS} = -2550596$ kNm

In order to follow the process of ship hull collapse under vertical bending the hull transverse section is divided into stiffened panels and hard corners, Fig.1.

Hull elements are of grade A MS steel with yielding stress 235 N/mm^2, while the deck panels are of grade AH 32 HT steel with the yielding stress of 315 N/mm^2.

3 RESULTS OF CALCULATION

The bending moments for imposed hull curvatures obtained by ULTIS are presented in Fig. 2:

The results are checked by the computer program MARS 2000 and presented in Fig. 3. The two programs show tolerable deviations mostly due to differences in modeling and in convergence criteria, Table 1.

4 ANALYSIS OF THE COLLAPSE PROCESS USING ULTIS

In contrast to the commonly adopted longitudinal strength analysis in the elastic domain, the analysis of the ship hull girder collapse requires information on the collapsing sequence of the cross sectional elements contributing to the irreversible accumulation of damages leading to collapse. Therefore the primary intention in preparing the home-made program ULTIS was to provide intermediate results for individual elements' stresses as a function of the imposed hull girder curvature (Žiha et al. 2007).

Of particular interests in the investigation of the collapsing sequence in this paper are the states of hull elements before (point 1-pre-critical), during (point 2-critical) and after (point 3-post-critical) during bending of the ship hull for both hogging (H) and sagging (S), Fig. 2.

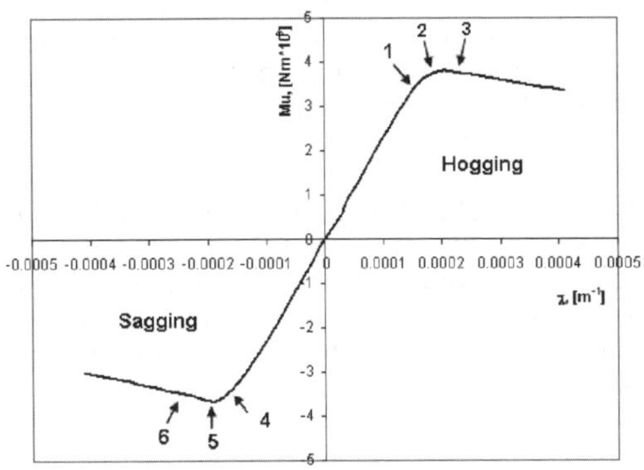

Figure 2. Bending moment versus hull curvature for hogging and sagging obtained by ULTIS.

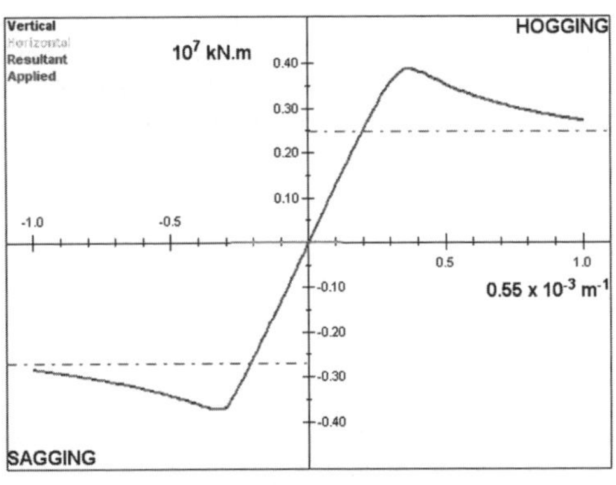

Figure 3. Bending moment versus hull curvature for hogging and sagging according to MARS 2000.

4.1 *Hogging condition*

1H-pre-critical-point-hogging: Deck of high tensile steel is under tensile loads but still below the HT material yielding stress of 315 N/mm^2 so yielding of the deck doesn't take place. However few stiffeners of the wing tank lower bulkhead built from mild steel are subjected to initiation of plastic deformations after exceeding the MS yielding stress of 235 N/mm^2. The bottom structure built of MS steel under compressive loads doesn't attain the yielding nor critical buckling stresses as yet.

2H-critical-point-hogging: The tensile stresses in deck panel of HT steel are still below the HT steel yield stress. But the tensile stresses in more and more stiffeners of the wing tank lower bulkhead of MS steel exceed 235 N/mm^2 yield stress of MS inducing plastic deformation, Fig. 4.

Compressive stresses in the bottom structure built from MS steel exceed the working stresses of 175 N/mm^2 and then the critical buckling stresses which reduce the load carrying capacity of the bottom panel. At the same time compressive stresses in the hard corners (bottom side girders) exceed the MS yield stress of 235 N/mm^2 what leads to plastic deformations in bottom longitudinal girders, Fig. 4.

3H-post-critical-point-hogging: Deck panels continue to yield under tensile loads inducing further plastic deformations. In the same time almost all stiffeners of the wing tank lower bulkhead reach their yielding point resulting in intense plastic deformations.

Table 1. Comparison of results obtained from ULTIS and MARS 2000.

	Initial neutral axis [m]	Elastic section modulus [m^3]	Ultimate hogging bending moment [MNm]	Hogging safety factor	Ultimate sagging bending moment [MNm]	Sagging safety factor
ULTIS	6.706	13.21	3789	1.613	–3646	1.430
MARS 2000	6.745	13.13	3886	1.655	–3730	1.463
Difference,[%]	0.58	0.58	2.5	2.54	2.24	2.30

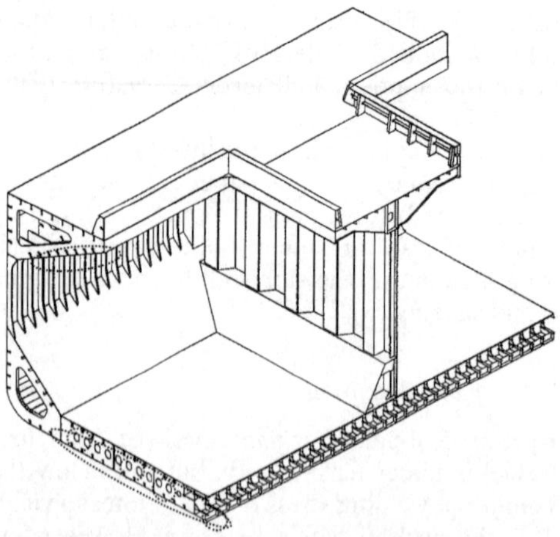

Figure 4. Ship's hull element failures in critical point under bending in hogging condition.

Compressive stresses in the double bottom structure induce further buckling due to violation of critical stresses and additional plastic deformations in girders due to exceeding of MS steel yielding stress. However, the post-critical conditions are out of the scope of the ultimate hull girder strength in this study.

4.2 *Sagging condition*

4S-pre-critical-point-sagging: Deck of high tensile steel is under compressive loads and since the stresses exceed the working stress for HT steel of 225 N/mm^2 buckling starts in the deck panel. But since the stresses are below the HT material yielding point of 315 N/mm^2 the plastic deformation of the deck doesn't take place as yet. The compressive stresses in wing tank lower bulkhead and in side shell stiffeners built from MS steel exceed the working stress of 175 N/mm^2 closing to the yield stress 235 N/mm^2 what induce buckling prior yielding followed by reduction of load carrying capacity of the upper ship hull strake.

The stresses in the bottom panel structure built of MS steel (excluding the double bottom, tanktop and bilge structure) under tensile loads exceeds the yielding point of 235 N/mm^2 what causes plastic deformations in the bottom panel. The tensile stresses in the bottom longitudinal girders (hard corners) are still below the MS yield stress 235 N/mm^2 and no plastic deformations occur as yet.

Tensile stresses in the wing tank stiffeners get higher of the hopper tank stiffeners since the position of the neutral axis gets lower because of the reduction of load carrying capacity due to buckled deck scantlings.

5S-critical-point-sagging: Deck panel and wing tank bulkhead and side shell stiffeners suffer further buckling under increasing compressive loads resulting in stresses over the working stresses of 225 N/mm^2 and 175 N/mm^2, although not exceeding the yield points 315 N/mm^2 and 235 N/mm^2 for HT and MS steel, respectively, as yet, Fig. 5.

The tensile stresses in the bottom structure, flat keel, garboard strake and the bilge (but not yet in the tank-top of the double bottom) exceed the yielding point of MS steel of 235 N/mm^2 what leads to elasto–plastic collapse, Fig. 5.

The stresses in the bottom longitudinal girders (hard corners) exceed the MS steel yield stress 235 N/mm^2 inducing progression of plastic deformations, Fig. 5.

6S-post-critical-point-sagging: The continuous growth of compressive stresses over 225 N/mm^2 in the deck structure leads to reduction in load carrying capacity due to further buckling under compressive loads although not exceeding the HT steel yielding point of 315 N/mm^2.

The growth of tensile stresses in the bottom structure, flat keel, garboard strake and the bilge (but not yet in the double bottom) under tensile loadings over the yielding point (235 N/mm^2) induces the progression of plastic deformations leading to elasto—plastic collapse.

In the same time the tensile stresses in the bottom longitudinal girders (hard corners) growth above the yield stress 235 N/mm^2 inducing further plastic deformations leading to elasto-plastic collapse.

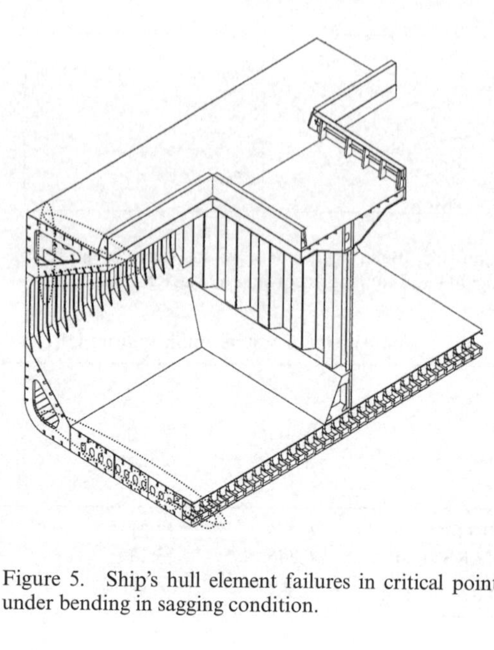

Figure 5. Ship's hull element failures in critical point under bending in sagging condition.

4.3 *Influence of deck panel material during sagging*

The compressive loads during sagging induce buckling prior yielding in the deck panel of high tensile steel soon after exceeding the working stress of 225 N/mm², a lot before reaching the yielding stress of 315 N/mm². Therefore the collapsing scenario for deck of MS steel instead of HT steel is investigated. The equivalent dimensions of the deck panel made of grade A MS steel are determined on the basis of material factor *K*, as follows:

Thickness of deck plate:

$$t_{P235} = \frac{t_{P315}}{\sqrt{K}} = \frac{26.5}{\sqrt{0.78}} = 30 \text{ mm},$$

Stiffener section modulus:

$$W_{235} = \frac{W_{315}}{K} = \frac{1450}{0.78} = 1860 \text{ cm}^3.$$

(HP 430 × 17, W = 1950 cm³).

After calculating the ultimate strength for the alternative deck material it appears that the deck panel, now of adequately scantlings built from MS grade A steel, will yield prior buckling. Critical buckling stress of the panel built of MS steel is slightly higher than the working stress of the HT steel (225 N/mm²), so it greatly exceeds the working stress of the grade A steel (175 N/mm²), Fig. 6. Altering deck materials from HT to MS steel, the overall ultimate hull bending moment has not been changed significantly (the ultimate bending moment is reduced for 3% for hogging and for about 1% for sagging) but the character of the failure is not the same. This observation supports the experience of careful usage of HT steels in building the ship's hull.

4.4 *Bottom panel in hogging*

The repeated application of the program ULTIS confirmed that the bottom panel built of MS steel in the case of hogging (compression load) suffers buckling prior yielding soon after exceeding the working stress (175 N/mm²) and far from the yielding stress (235 N/mm²), Fig. 7.

4.5 *One step method*

Assuming that yielding occurs continuously without buckling during the imposed ship hull bending process, it is possible to determine the upper edge of ultimate bending strength M_U in one step using vertical plastic section modulus M_p:

$$M_U = M_p = R_e^h \cdot \frac{A}{2} d = R_e^h \cdot Z_p \qquad (1)$$

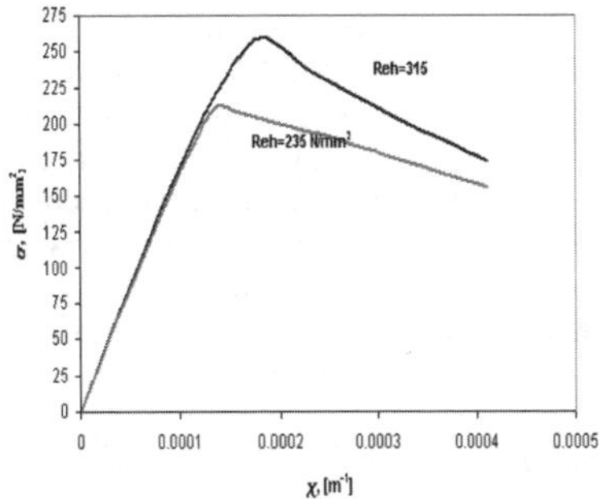

Figure 6. Deck panel stress versus hull curvature during sagging.

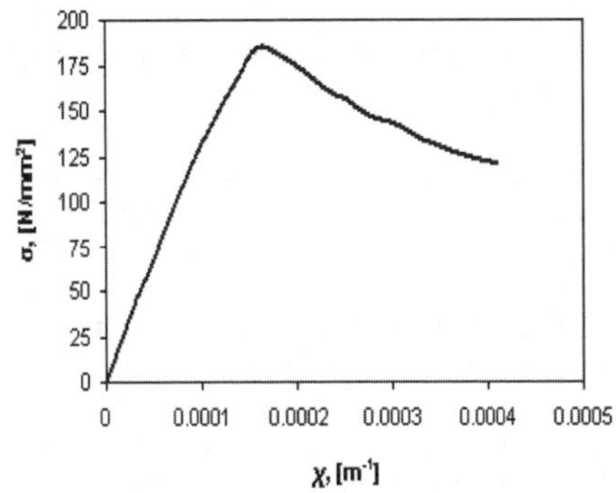

Figure 7. Bottom panel stress versus hull curvature for the case of hogging.

The shape factor ν, is defined as the coefficient between the fully plastic M_p and the fully elastic M_e bending moment as follows:

$$\nu = \frac{M_p}{M_e} = \frac{R_e^h \cdot \frac{A}{2} \cdot d}{R_e^h \cdot \frac{I}{y_{max}}} = \frac{R_e^h \cdot Z_p}{R_e^h \cdot Z_e} = \frac{Z_p}{Z_e} \qquad (2)$$

A is the total cross-sectional area and *I* is he cross-sectional moment of inertia about horizontal axis
y_{max} is the greatest distance of a section from horizontal neutral axis
d is the distance between the centroids of upper and lower half of the resisting area
R_e^h is the material upper yield stress
Z_P is the vertical fully plastic section modulus and
Z_E is the vertical fully elastic section modulus

The vertical plastic bending moment for the example bulk carrier is:

$$M_U = M_p = Z_P \cdot R_h^e \cdot 10^3$$
$$M_U = 13.21287 \cdot 315 \cdot 10^3 = 4162000 \text{ kNm} \qquad (3)$$

The vertical plastic bending moment in (3) exceeds the result for ultimate bending moments obtained from ULTIS by 12%, and the result obtained from MARS 2000 by 10%. The differences between the ultimate and fully plastic bending moments are the consequence of imperfections in structural arrangement since some weaker elements buckle or yield prior their anticipated normal sequences.

The value of shape factor is

$$\nu = M_p/M_e = 4162000/3728000 = 1.12$$

which points at safety regarding exceeding the plastic section modulus in relation to elastic section modulus.

5 IMPROVEMENT OF THE SHIP HULL ULTIMATE STRENGTH

The optimization of the ship hull longitudinal strength is normally driven by objectives either of weight or of cost reduction subjected to constrained optimization procedure in order to satisfy the required elastic section modulus of the midship section that can employ different optimization techniques. The same reasoning can be applied to the optimization of the ultimate strength represented by the fully plastic section modulus. However, the iterative-incremental method for assessment of the ultimate strength that is recommended by IACS does not operate with synthetic measures of ultimate strength that can be simply taken for objective in the optimization procedure. At this point it became clear that the optimization of the limit strength within the framework of traditional optimization techniques is not an easy task as yet and therefore the investigation has been focused on possible improvements.

The improved procedure described in this study uses the collapsing scenario of the hull girder obtained by a preliminary run of the program ULTIS that identified the sequence of failures. The elements that failed first are strengthened first and then the calculation is repeated. Then the elements that failed next are strengthened and the calculation is repeated again, and so on. Neither convergence criterion nor redistribution of materials has been applied in order to avoid the violation of local strength conditions since the study recognized the complexity of the integral optimization procedure.

For considered bulk carrier the plate and stiffener elements at the wing tank bulkhead as well as at the bilge and bottom plating and stiffening that have first undergone yielding and/or buckling both in sagging conditions have gradually been strengthened as it is presented next.

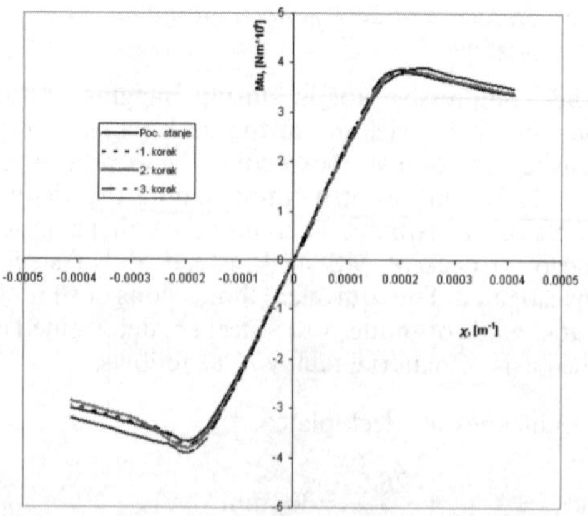

Figure 8. Curves of bending moments versus hull curvature for various optimization steps.

In the first step the wing tank bulkhead stiffeners which buckled first under compression in sagging have been strengthened from the profile HP 320 × 12 to HP 340 × 12 and from the shell plate thickness of 16 to 17 mm. Along the increase of the transverse section unit mass of 0.8%, the ultimate strength has been increased for 1.13% for hogging and 2.24% for sagging.

Furthermore, the stiffeners have been strengthened to HP 370 × 13 and shell plating thickness has been increased to 18 mm. Transverse section unit mass has been increased for 2% and the ultimate strength for 1.91% for hogging and 3.6% for sagging.

In the last step, the bottom and the bilge shell plates that collapsed under tension loads during sagging have been strengthened from 15 to 17 mm thickness while keeping the wing tank bulkhead panel dimensions from previous step. Along with 4% increase of transverse section unit mass, the ultimate strength has increased by 2.56% for hogging and 6.4% for sagging.

The curves of bending moment versus hull curvature for all steps are shown in Fig. 8.

6 DISCUSSION

According to IACS (2006b) the vertical hull girder ultimate bending capacity is to satisfy the following criteria:

$$M = \gamma_s M_{sw} + \gamma_w M_{wv-sag} \leq \frac{M_u}{\gamma_R} \tag{4}$$

where in (4) $\gamma_s = 1.0$ and $\gamma_w = 1.20$ are the partial safety factors for still water and wave induced bending momemnts (IACS 2006b).

Table 2. Relations of bending moments for bulk-carriers.

Items		B1	B2	B3	B ex
Lpp	m	228	277	266	179
B	m	32.2	45	42	30.8
H	m	18.3	24.1	23	15
d	m	12.7	17.7	16.9	10.1
M_{uH}/M_p	λ_{pH}	0.93	0.94	0.92	0.91
M_{uS}/M_p	λ_{pS}	0.77	0.75	0.79	0.88
M_{uH}/M	γ_{Rh}	1.44	1.51	1.49	1.52
M_{uS}/M	γ_{Rs}	1.31	1.20	1.27	1.35
M_p/M_e	ν	1.15	1.21	1.13	1.12
M_{uH}/M_{uS}	λ_{hs}	1.20	1.25	1.17	1.13

The ultimate bending moment in (15) is $M_u = M_{uH}$ for hogging and $M_u = M_{uS}$ for sagging, Figs. 2 and 3. The ruled based partial safety factor $\gamma_R = 1.10$.

Recent investigations indicated that the ultimate strength of typical tankers and bulk carriers satisfy this classification requirement as illustrated in Table 2 (Ćorak et al. 2009) both for sagging and hogging condition with partial safty factor $\gamma_R = 1.10$. The results obtained for the bulk carrier considered in the study also confirm this result, Table 2. However, this observation deserves a comment in the conclusion of this study.

7 CONCLUSION

The investigations brought forward in this paper aimed to add impetus for improvement of the ship's hull ultimate strength. In contrast to the optimization of the elastic or plastic section modulus the iterative-incremental approach recommended by new IACS CSR cannot be easily implemented in commonly applied optimization techniques as yet. The optimization of the elastic and plastic section modulus is normally attainable by traditional optimization techniques considering the cross sectional properties since the improvement of the ultimate strength could be additionally achieved by intervening on spacings and bracketing of element ends.

The improvement of the ultimate strength in this study was achieved by strengthening the cross sectional elements in the sequence in which they failed during imposed bending. Since no other material redistribution was performed, the improvement was followed by hardly defendable increase in the structural weight.

The study encourages the investigation of the collapsing sequences even if no improvement will be undertaken since it provides a better understanding of the ship hull behavior under extreme loads.

The information on failure sequences may at least point to weakest or critical elements of the hull that later can be subjected to more intense inspection and maintenance. The differences between the ultimate and fully plastic bending moments are the consequence of inevitable and unavoidable imperfections in structural arrangement since some weaker elements buckle or yield prior their anticipated normal sequences. However, the principal question that arouse during the study was not how but why to optimize the ultimate strength that in most cases abundantly fulfills the requirements.

It appears that for common merchant ships the optimization of the ultimate strength is not at present of high priority. Moreover simultaneous satisfying the local strength criteria and the working stress criterion in elastic region normally provides satisfactory ultimate strength. The material redistribution during iterative incremental-procedure that also retains all the other local and global constraints appears too complex at this starting stage of the investigation.

But nevertheless, the engineering challenge based on enginering ethics is to improve the improvable of the ship's hull ultimate strength.

REFERENCES

Bureau Veritas 2003. MARS 2000, Paris.

Ćorak, M., Parunov, J. & Žiha, K. 2009. The relation of working and ultimate longitudinal strength of a ship hull girder in service, *13th Congress of Intl. Maritime Assoc. of Mediterranean, IMAM, İstanbul, Turkey, 12–15 Oct. 2009.*

Guedes Soares, C. & Parunov, J. 2008. Structural Reliability of a Suezmax Oil Tanker Designed According to New Common Structural Rules, *Journal of Offshore Mechanics and Arctic Engineering*, 130 (2): 17–27.

Hergert, D. 1998. *Visual Basic 5 Bible*, IDG Books WorldWide, Foster City.

Hughes, O.F. 1988. *Ship Structural Design*, SNAME, New Jersey.

IACS, 2006a. *Common Structural Rules for Double Hull Oil Tankers.*

IACS, 2006b. *Common Structural Rules for Bulk Carriers.*

Rutheford, S.E. & Caldwell, J.B. 1990, Ultimate Longitudinal Strength of Ships: A Case Study, *Transactions SNAME*, 98; paper no.14: 1–26.

Yao et al. 2006. Ultimate strength, *Proceedings of 16th International Ship and Offshore Structures Congress, Southampton, 20–25 August 2006.*

Žiha, K., Parunov, J. & Tušek, B. 2007. Ultimate strength of ship hull (in Croatian), *Brodogradnja,* 58(1): 29–41.

Žiha, K., Pedišić, M.: Sposobnost preživljavanja broda pri oštećivanju trupa, *Brodogradnja,* 48(1):

Advanced Ship Design for Pollution Prevention – Guedes Soares & Parunov (eds)
© 2010 Taylor & Francis Group, London, ISBN 978-0-415-58477-7

Extended IACS incremental-iterative method for calculation of hull girder ultimate strength in analysis and design

S. Kitarović, J. Andrić & V. Žanić
Faculty of Mechanical Engineering and Naval Architecture, University of Zagreb, Zagreb, Croatia

ABSTRACT: Theoretical background and operational aspects are given for the proposed extension of the IACS incremental-iterative method. Method is used for the assessment of the hull girder ultimate bending moment as one of the design attributes within optimization based decision making process and is implemented in OCTOPUS design environment. Examples of application are given and the general conclusion based on the results obtained is presented.

1 INTRODUCTION

Since the ultimate strength might be perceived as the most meaningful safety measure of the ship's hull girder structure, prediction of the ultimate bending moment becomes essential and unavoidable part of the ship structural concept design process. Methods employed should support multiple failure modes and their interactions, while giving precise prediction of collapse and post-collapse behavior of the structural members involved (particularly those under compression). On the other hand, multiple executions within design loop demand utilization of stable, robust and sufficiently fast algorithms.

Consideration of the above stated demands resulted in development of the improved incremental-iterative method for longitudinal ultimate strength assessment based on IACS prescribed incremental-iterative method. Incorporated method particularities include contemporary advances which improve the accuracy during multi-deck ship application, as well as the ability to consider vertical shear force influence on the ultimate hull girder strength.

Purpose of this paper is to give an insight into theoretical background and operational aspects of this method. It is to be used for the assessment of the ultimate bending moment as one of the design attributes within optimization based decision making process, as implemented in OCTOPUS design environment (Zanic et al. 2007).

2 EXTENSION OF THE BASIC METHOD

2.1 *General remarks*

Ultimate longitudinal strength is defined herein as the value of bending moment at which the flexural stiffness of the hull girder (i.e. the slope of moment to curvature curve) assumes the value of zero, as illustrated by Figure 1.

The general approach for assessment of the moment to curvature relationship used in IACS prescribed incremental-iterative method is similar to the one originally proposed by well known and widely spread Smith's method (Smith 1977) and will not be further discussed here. Modifications of the basic method are introduced in effort to enable inclusion of the effects disregarded by the basic method and thus improve the overall accuracy of the analysis. Influence of the shear stress and deck efficiency is incorporated into basic method as illustrated by the Figure 2, which represents general flowchart of the proposed method.

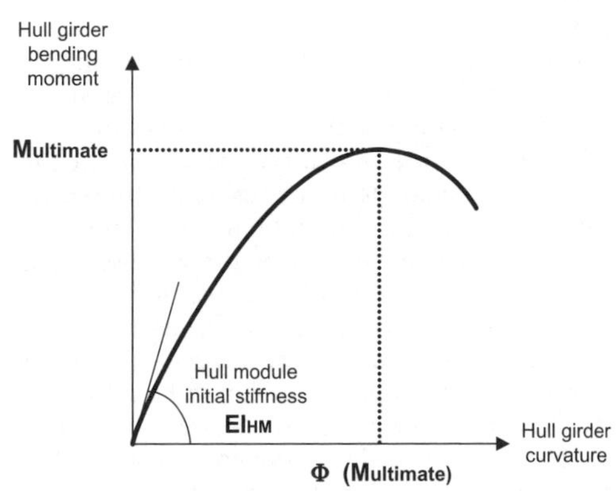

Figure 1. Qualitative hull module moment to curvature (M-Φ) response curve, obtained by utilization of the Bernoulli-Euler beam idealization of the hull girder.

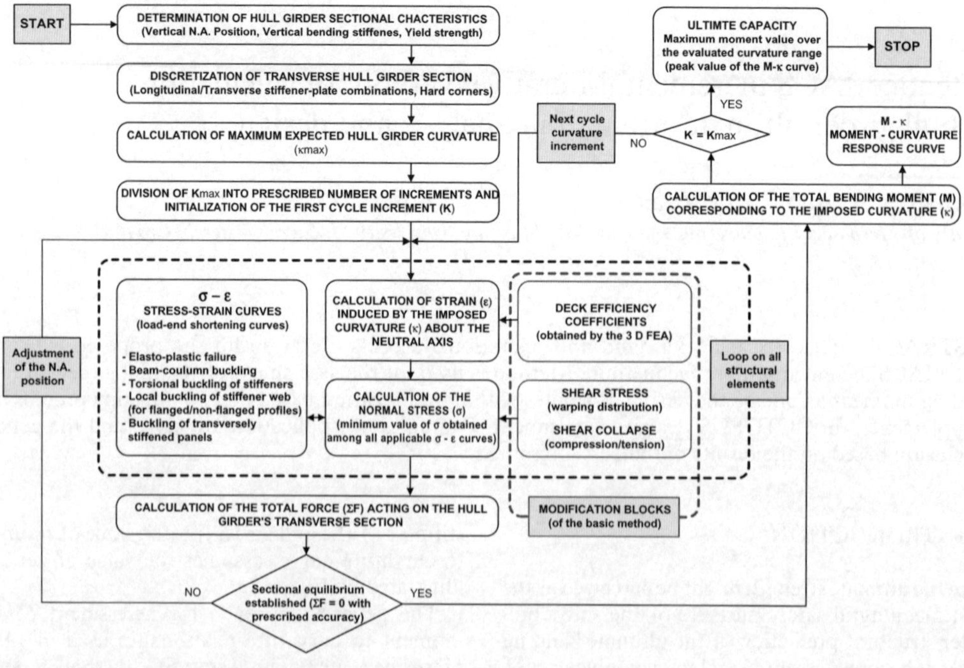

Figure 2. Flowchart of the proposed method.

2.2 *Inclusion of the shear stress influence*

Longitudinal ultimate strength is usually analyzed at the hull girder cross-section with maximum bending moment, where the shear force is negligible. Accounting for shear might be interesting when there is a cross-section along the hull girder with less then maximum value of bending moment, but with significant value of the shear force (cases of alternate loading conditions). Since in this case it is not so obvious whether the hull girder will collapse at the section with the maximum value of the vertical bending moment or at the sections with high vertical bending moment and shear force values, both scenarios deserve the due consideration.

The effect of the vertical shear force on the hull girder ultimate strength is considered trough the influence of the warping induced shear stress distribution of the hull module on the collapse (buckling, yield) of the principal structural members.

For the structural evaluation of primary response at the concept design level the beam idealization is often used. Since this evaluation is based on the extended beam theory which needs cross-sectional characteristics usually obtained using analytical methods, this can be very complicated for the realistic combinations of interconnected open and closed thin-walled (un)symmetric cross sections.

Application of the energy based numerical methods gives an opportunity for alternative approach based on decomposition of a cross section into the line finite elements between nodes *i* and *j* (Figure 3) with coordinates (y_i, z_i), (y_j, z_j); element thickness t^e; material characteristics (Young's modulus *E*, shear modulus *G*) and material efficiency *RN* and *RS* (due to hatches, cutouts, lightening holes, etc.) with respect to normal/shear stresses respectively.

The methodology (Zanic & Prebeg 2005) is based on application of the principle of minimum total potential energy with respect to the parameters which define the displacement fields of the structure.

Primary displacement field (following classical beam theory) is defined by displacements and rotations of the cross section as a whole. Secondary displacement field $\mathbf{u}_2(x,y,z) \equiv \mathbf{u}(x,y,z)$ represents warping (deplanation) of the cross section. For piecewise-linear FEM idealization of the cross section, divided into *n* elements, with shape functions **N** in element coordinate system (x,s), the warping field reads:

$$u(s)^e{}_{x=x_0} = \mathbf{N}^{\mathrm{T}} \cdot \mathbf{u}^e = \begin{Bmatrix} 1-\dfrac{s}{l^e} & \dfrac{s}{l^e} \end{Bmatrix} \begin{Bmatrix} u_i \\ u_j \end{Bmatrix} \qquad (1)$$

Element strain and stress fields ε and σ are obtained from strain-displacement and stress-strain relations:

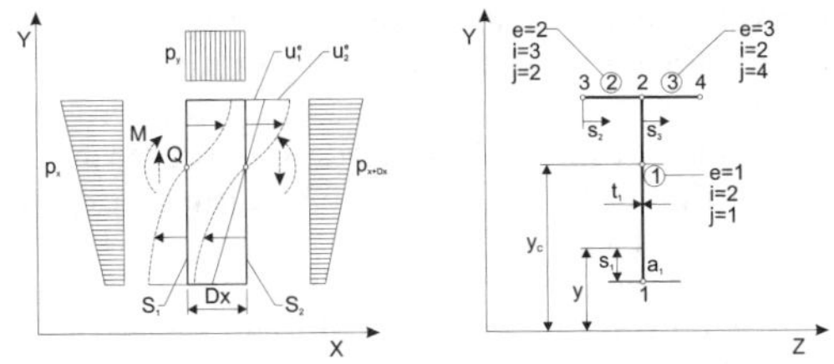

Figure 3. Cut-out from the longitudinal structural member considered as the transverse strip ($S_1 - S_2$) with external loading p and warping fields u.

$$\varepsilon \rightarrow \gamma_{xs} = \frac{\partial u}{\partial s} = \mathbf{B}^T \mathbf{u}^e = \left\{ -\frac{1}{l^e} \quad \frac{1}{l^e} \right\} \begin{Bmatrix} u_i \\ u_j \end{Bmatrix} \tag{2}$$

$$\sigma \rightarrow \tau_{xs} = G^e \gamma_{xs} = G^e \mathbf{B}^T \mathbf{u}^e \tag{3}$$

Total potential energy of the Δx-long transverse strip, or segment, of the beam (with section divided into n elements):

$$\begin{aligned} \Pi &= \sum_n \left[\int_{V^e} \sigma^T \varepsilon \, dV - \int_{S^e} p(x,s) u(s) dS \right] \\ &= \sum_e \left[\int_{V^e} \frac{1}{2} \mathbf{B}^T \mathbf{B} dV - \int_{S^e} F(s) u(s) dS \right] \\ &= \Delta x \sum_e \left[\frac{1}{2} \mathbf{u}^{eT} \mathbf{K}^e \mathbf{u}^e - \mathbf{u}^{eT} \mathbf{F}^e \right] \end{aligned} \tag{4}$$

where $p(x,s)$ = external loading on two cross sections (S_1 and S_2) of the strip. The minimization of total potential energy leads to classical FEM matrix relation $\mathbf{K}_{2D}\mathbf{u}_{2D} = \mathbf{F}_{2D}$ (shortened to $\mathbf{Ku} = \mathbf{F}$). The element stiffness matrix for the proposed linear displacement distribution along the line element is:

$$\mathbf{K}^e = \frac{G^e \cdot t^e \cdot RS^e}{l^e} \begin{bmatrix} 1 & -1 \\ -1 & 1 \end{bmatrix} \tag{5}$$

Normal stress is varying with x and s. In the beam segment of length Δx, at a certain point of the cross section, the resultant stress is given by:

$$\begin{aligned} \Delta p(x,s) \equiv \Delta\sigma(x,s) &= \sigma(x+\Delta x,s)_{S2} - \sigma(x,s)_{S1} \\ &= Q(x) \frac{\xi_C(s)}{I(x)} \Delta x = Q(x) \bar{F}(x,s) \Delta x \end{aligned} \tag{6}$$

where $\xi_c(s)$ = distance from the considered point to the neutral axis. Vector of the nodal loads for the element e of the unsymmetrical cross section, in bending about the Z axis, reads:

$$\begin{aligned} \mathbf{F}_z^e(x) &= Q_y(x) \cdot \bar{\mathbf{F}}_z^e(x) \\ &= \frac{E^e Q_y(x) t^e RN^e}{E\left(I_Y I_Z - I_{YZ}^2\right)} \\ &\times \left[I_Y \cdot \begin{Bmatrix} \frac{y_{ic} l^e}{2} + \frac{l^{e2} \sin\alpha^e}{6} \\ \frac{y_{ic} l^e}{2} + \frac{l^{e2} \sin\alpha^e}{3} \end{Bmatrix} - I_{YZ} \cdot \begin{Bmatrix} \frac{z_{ic} l^e}{2} + \frac{l^{e2} \sin\alpha^e}{6} \\ \frac{z_{ic} l^e}{2} + \frac{l^{e2} \sin\alpha^e}{3} \end{Bmatrix} \right] \end{aligned} \tag{7}$$

For the case of bending about the Z axis the matrix relations $\mathbf{Ku} = \mathbf{F}$ with $\mathbf{u} = Q(x) \cdot \bar{\mathbf{u}}$, can be converted into expressions $\mathbf{K}\,\bar{\mathbf{u}} = \bar{\mathbf{F}}$ for the warping due to unit load $\bar{\mathbf{F}}$. For the node warping $u_i(x)$, unit warping $\bar{\mathbf{u}}(x)$ must be multiplied with Q(x).

Embedment of the presented methodology into the overall incremental-iterative procedure for the ultimate strength assessment rests on the following rational assumption. Each cross section of the hull girder has particular ratio of the vertical bending moment and the vertical shear force derived from their respective distributions for the considered loading condition. This ratio is approximated as constant during the incrementation of the hull girder curvature/moment, enabling the calculation of the respective shear force within each increment. This, in a turn, implies that additional iterative loop should be placed over the balancing loop for each incremental step, since the value of the moment used for the calculation of the shear force should be equal to the one obtained by the summation of the contributions of the balanced hull module principal structural members.

Since the instantaneous hull module neutral axis vertical position is changed during the iteration and/or incrementation, the respective equivalent hull module moment of inertia, as well as the instantaneous load vector for hull module vertical bending should be updated accordingly, in order to produce valid unitary warping distribution used for the calculation of the hull module shear stress distribution. For this purpose, 'efficiency' of the principal structural members is represented by the ratio of their Young's modulus and their instantaneous secant modulus, where instantaneous secant modulus is calculated for each principal structural member as the ratio between the stress and strain corresponding to the last valid equilibrium state.

Furthermore, unitary warping distribution calculation methodology assumes higher resolution discretization of the cross sectional model when compared to the incremental-iterative method used for the ultimate strength assessment. Therefore, principal structural members are assigned with their respective line elements (with averaged nodal values of the unitary warping displacements), and the value of the unitary warping displacement is calculated for each of them in the following manner:

- Stiffener-plate combinations are assigned with the five line elements in a case of the T-profile stiffener with the attached plating. Two for the both parts of the flange, two for the both parts of the attached plating and one for the web. If the stiffener cross section is of angle or flat bar type (bulbs are represented as equivalent T-profiles), the assigned number of line elements is either four or three respectively. Since the plating line elements are always characterized by the considerably greater value of the unitary warping displacement than the stiffener line elements, the greater value among those two is chosen as representative.
- Transversely stiffened plates are represented by one line element, since only plating is longitudinally relevant, and the average value of that line element's nodal unitary warping displacements is accepted as the representative one.
- Knuckles are assigned with two line elements, while multiple panel intersections are assigned with the number of line elements equal to the number of intersecting panels. Representative value for each of those principal structural elements is calculated as the average value of the group.

When distribution of the unitary warping displacements of the hull module is known, the shear stress distribution can be calculated using the following expression for the shear stress of the considered principal structural member:

$$\tau^{PSM} = Q \cdot MAX\left(G^e \frac{\overline{u}^e}{l^e}\right)^p \tag{8}$$

where τ^{PSM} = shear stress of considered principal structural member; Q = instantaneous shear force value acting on the hull module; e denotes properties of the considered line element; p denotes plating.

In effort to quantify the influence of the calculated shear stress on the stress obtained by the means of the valid stress-strain curves for every principal structural member, the following elliptic interaction formula (Yao et al. 2005) is used, both for compression and tension:

$$\left(\frac{\sigma_M^{PSM}}{\sigma_L^{PSM}}\right)^2 + \left(\frac{\tau^{PSM}}{\tau_U^{PSM}}\right)^2 = 1 \tag{9}$$

where τ^{PSM} = calculated shear stress of considered principal structural member; $\tau_U{}^{PSM}$ = ultimate shear strength of the plating in compression/tension; $\sigma_M{}^{PSM}$ = longitudinal stress modified with the influence of the shear stress; $\sigma_L{}^{PSM}$ = longitudinal stress derived from valid stress-strain curves.

If the considered principal structural member is in compression, then the ultimate edge shear strength of the proprietary plating is considered according to the criteria given by (Paik 2003):

$$\frac{\tau_U}{\tau_Y} = \begin{cases} 1.324\dfrac{\tau_E}{\tau_Y} & \text{for } 0 < \dfrac{\tau_E}{\tau_Y} \le 0.5 \\ 0.039\left(\dfrac{\tau_E}{\tau_Y}\right)^3 - 0.274\left(\dfrac{\tau_E}{\tau_Y}\right)^2 + 0.676\dfrac{\tau_E}{\tau_Y} + 0.388 & \\ & \text{for } 0.5 < \dfrac{\tau_E}{\tau_Y} \le 2.0 \\ 0.956 & \text{for } \dfrac{\tau_E}{\tau_Y} > 2.0 \end{cases} \tag{10}$$

where τ_Y = yield stress of the material under pure shear loading according to von Mises criteria:

$$\tau_Y = \frac{\sigma_Y}{\sqrt{3}} \tag{11}$$

where σ_Y = specified yield stress of the material; τ_E = elastic shear buckling stress of the simply supported plates:

$$\tau_E = k_\tau \frac{\pi^2 E}{12(1-\nu^2)}\left(\frac{t}{b}\right)^2 \tag{12}$$

where k_τ = buckling coefficient for the shear loading, dependant on the aspect ratio of the considered plate:

$$k_\tau \approx \begin{cases} 4.0\left(\dfrac{b}{a}\right)^2 + 5.34 & \text{for } \dfrac{a}{b} \ge 1 \\ 5.34\left(\dfrac{b}{a}\right)^2 + 4.0 & \text{for } \dfrac{a}{b} < 1 \end{cases} \tag{13}$$

$\sigma_M{}^{PSM}$ is calculated using (9):

$$\sigma_M^{PSM} = \sigma_L^{PSM}\sqrt{1-\left(\frac{\tau^{PSM}}{\tau_U^{PSM}}\right)^2} = \sigma_L^{PSM} K_\tau^C \tag{14}$$

If the considered principal structural member is in tension, then the ultimate edge shear strength of the proprietary plating is considered according to the von Mises criteria, which gives $\tau_U = \tau_Y$. In this case σ_M^{PSM} is calculated according to the following expression:

$$\sigma_M^{PSM} = \sigma_L^{PSM} \sqrt{1-3\left(\frac{\tau^{PSM}}{\sigma_Y}\right)^2} = \sigma_L^{PSM} K_\tau^T \tag{15}$$

K_τ^C and K_τ^T represent shear influence coefficients of the considered principal structural member in compression or tension, respectively. According to their definition given in equations (14) and (15) it can be seen that calculated shear stress value can not exceed the considered value of the ultimate shear stress, which is therefore considered as the limiting value. In this boundary case (when calculated shear stress is equal to the ultimate shear strength in compression or yield stress in tension) the shear influence coefficients will assume the value of zero, which will effectively exclude the considered principal structural member from the balancing procedure and annihilate it's contribution to the sectional vertical bending moment.

It should also be noted that if the used ratio of the hull module vertical bending moment and the shear force is small (actual limit value is dependant of the structural model layout), a large values of the shear force will be imposed throughout the incrementation sequence. This might cause a very large values of the shear stress in a considerable number of the hull module principal structural members and the balancing procedure might fail to find the equilibrium state of the section due to large longitudinal stress reductions induced by the influence of the shear stress.

2.3 *Deck efficiency coefficients*

Important drawback of the basic method is that it assumes linear strain distribution over the hull module height. This constrains the consideration of the shear lag effects and limits the evaluation of the hull girder longitudinal strength significantly, especially in the case of multi-deck ships with complex transverse sections, where some discontinuous longitudinal hull girder components are imposed with reduced deformation. Here, linear distribution assumed by the basic method, and given by the equation (16), can not be considered as accurate.

$$\varepsilon_e = \Phi_{HM} y_e \tag{16}$$

where ε_e = strain imposed on considered principal structural element by hull module curvature Φ_{HM}; y_e = distance from the centroid of the considered principal structural element to the hull module effective horizontal neutral axis.

An approximate procedure using linear-elastic 3D FEM analysis is used for prediction of the efficiency of each principal structural element in order to obtain more accurate average strain as input for the stress-strain curves.

Stress of each principal structural element can be obtained from the rapid 3D FEM analysis of the generic coarse mesh hull girder model (Zanic et al. 2009). The ratio of this stress and the stress calculated by the means of the Bernoulli-Euler beam idealization can be interpreted as structural efficiency coefficient (Biot et al. 2006):

$$k_e = \frac{(\sigma_{FEM})_e}{(\sigma_{EB})_e} = \frac{E_e(\varepsilon_{FEM})_e}{E_e(\varepsilon_{EB})_e} = \frac{(\varepsilon_{FEM})_e}{(\varepsilon_{EB})_e} \tag{17}$$

where $(\varepsilon_{FEM})_e$ = primary strain imposed on considered principal structural element obtained by the means of the 3D FEM analysis; $(\varepsilon_{EB})_e$ = strain imposed on considered principal structural element obtained according to Bernoulli-Euler beam equation; E_e = Young's module of the considered principal structural element; $(\sigma_{FEM})_e$ = primary stress in considered principal structural element obtained by the means of the 3D FEM analysis; $(\sigma_{EB})_e$ = stress in considered principal structural element obtained according to Bernoulli-Euler beam equation.

Strain due to instantaneous hull module curvature is multiplied by the calculated structural efficiency coefficient and this product is used as input into relevant stress-strain curves. If contribution of the decks to the hull girder strength is to be evaluated, an average structural efficiency or deck efficiency coefficient valid for every principal structural element of the particular deck can be defined as an average value of the structural efficiency coefficients of all deck components. Since the total deck axial force is equal whether calculated as the sum of the forces on the deck components considering their individual structural efficiency coefficients, or by overall approach using the average structural efficiency coefficient, the definition of the deck efficiency coefficient is given by:

$$k_D = \frac{\sum_{i=1}^{N} (A_e k_e \sigma_e)_i}{\sum_{i=1}^{N} (A_e \sigma_e)_i} \tag{18}$$

where N = number of primary structural elements of the deck; A_e = area of the cross section of the considered primary structural elements of the deck; σ_e = stress of the considered structural elements of the deck.

This is valid since strain imposed on each principal structural element is not contributed by the direct interaction of the adjacent elements. Due to the decoupled nature of the principal structural elements, the imposed strain is determined solely by the distance of the considered element from the horizontal neutral axis.

Since the same value of the k_e is used throughout the curvature incrementation sequence, implementation of this modification has some obvious limitations regarding overall accuracy, but relatively simple and not so time consuming nature of the procedure enables better structural response assessment and renders this modification of the basic method as convenient for the application within the optimization based concept design loop.

3 EXAMPLES OF APPLICATION

3.1 *Chemical tanker*

The work on example of application presented herein was originally performed by the UZ-FMENA within the scope of the FP6 project IMPROVE (Naar at al. 2008), where longitudinal ultimate strength evaluation was performed for the hull girder of the 40 000 DWT ocean-going chemical tanker (Figure 4) designed to carry large variety of different cargoes in thirty cargo tanks, as one of the three specific products considered by the project. The main particulars of the vessel are as follows (SSN 2008):

Length overall:	182.88 m;
Length between perpendiculars:	175.25 m;
Beam molded:	32.20 m;
Depth to main deck:	15.00 m;
Scantling Draught:	11.10 m;
Cargo tanks capacity (total):	44 000 m^3;
Capacity of Duplex cargo tanks:	26 800 m^3;
Service speed:	15.0 Knots.

Non-linear FEM analysis of the prismatic structural model of the vessel was performed (Naar et al. 2008) for two extreme loading scenarios (briefly described in section 3.1.2) and the results obtained were used in section 3.1.3 for comparison only.

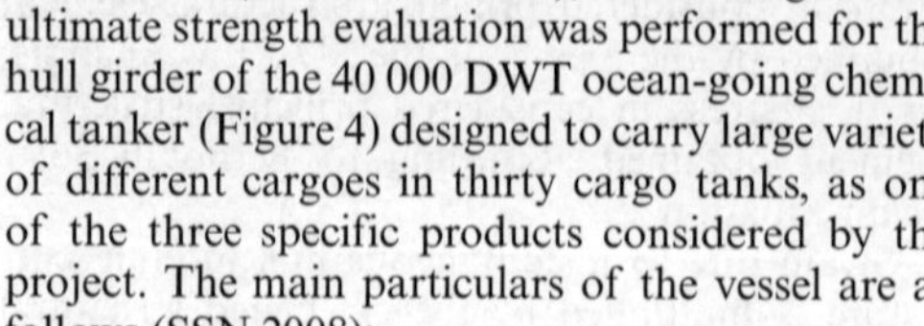

Figure 4. Chemical tanker designed by the SSN, Szczecin, Poland (SSN 2008).

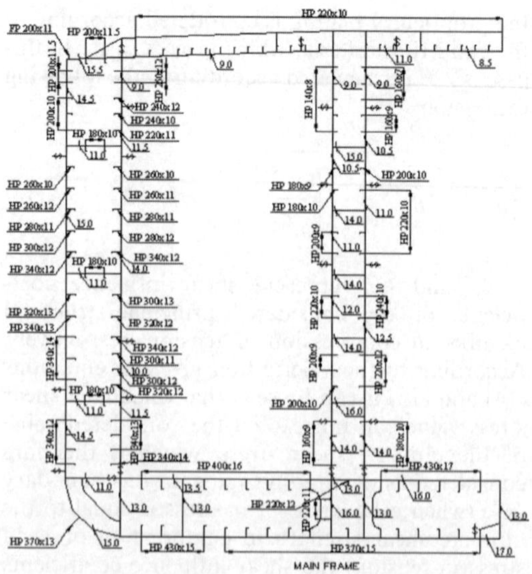

Figure 5. Midship section drawing of the analyzed structure (Naar et al. 2008).

3.1.1 *Structural model*

Since the structure was considered as prismatic, only the product's midship section (one-bay) model was produced and analyzed. Structural layout, main structural dimensions, as well as the stiffener scantlings used are given by the Figure 5.

Two different materials were used for structural elements, namely: high tensile steel (AH36) and stainless steel. Stainless steel is used only for the cargo tank plating (inner plating of the double sides and double bottom, cofferdam plating and strength deck plating), while high tensile steel is used for the rest of the structure. Relevant material properties of the materials used are specified by Table 1.

Span of the considered one bay model is 3560 mm, while the unsupported lengths of the transversely stiffened and cross stiffened panels of the cofferdam are: 890 mm (for the panels below the height of 5100 mm) and 1780 mm (for the panels above the height of 5100 mm). One bay model of longitudinally relevant structure produced using MAESTRO modeler and used as input for the longitudinal ultimate strength assessment using OCTOPUS software is given by the Figure 6.

Structural model of the chemical tanker midship section discreticised with the structural elements supported by the IACS incremental-iterative method (stiffener—plate combinations,

Table 1. Properties of the structural material.

Material property	High tensile steel	Stainless steel
Youngs modulus (N/mm^2)	210 000	210 000
Poisson ratio (–)	0.3	0.3
Yield stress (N/mm^2)	355	455

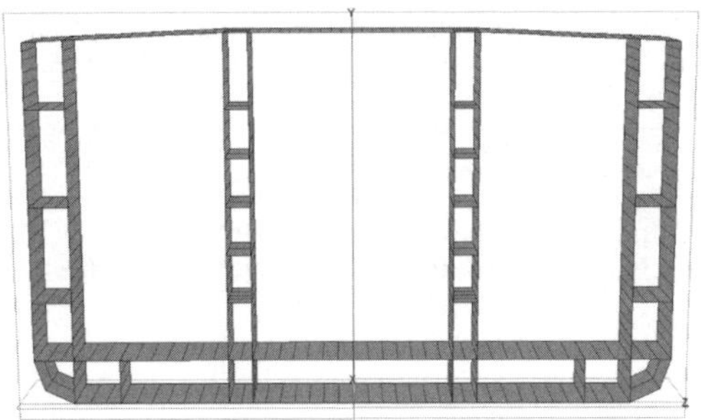

Figure 6. MAESTRO one bay structural model of the analyzed midship section.

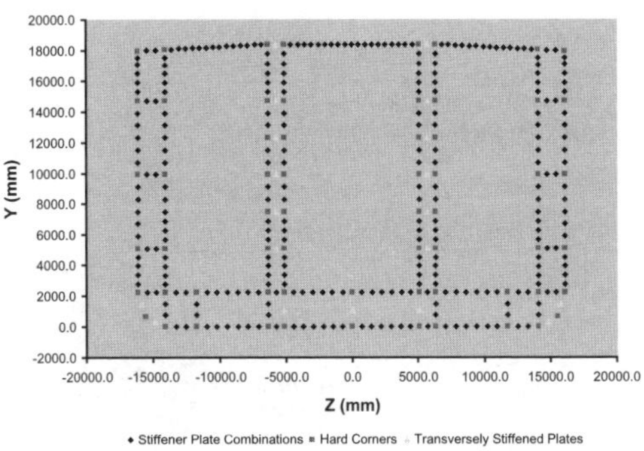

Figure 7. Longitudinal ultimate strength model of the analyzed midship section.

hard corners, transversely stiffened plates) is given by the Figure 7.

3.1.2 *Loading scenarios*

Two loading conditions of the vessel were identified and considered as extreme ones (Naar et al. 2008). In both cases the realistic loading distributions were determined using hydrostatic analysis. First considered loading scenario is the one giving the maximum hull girder vertical bending (at corresponding frame station), which constitutes the usual approach for the longitudinal ultimate strength assessment. Relevant distributions for this loading scenario are given by Figure 8.

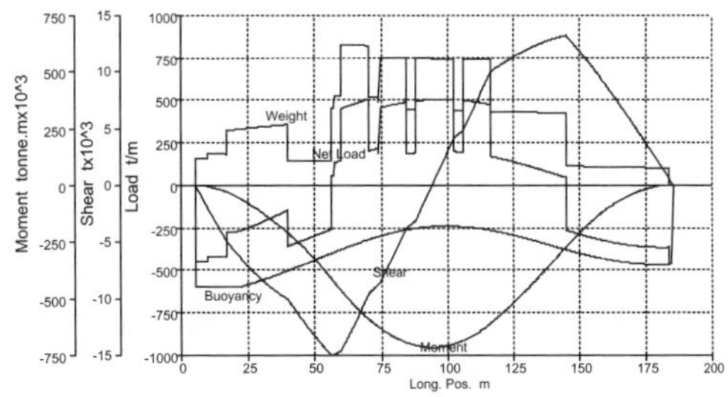

Figure 8. Loading distributions for the first extreme loading scenario (Naar et al. 2008).

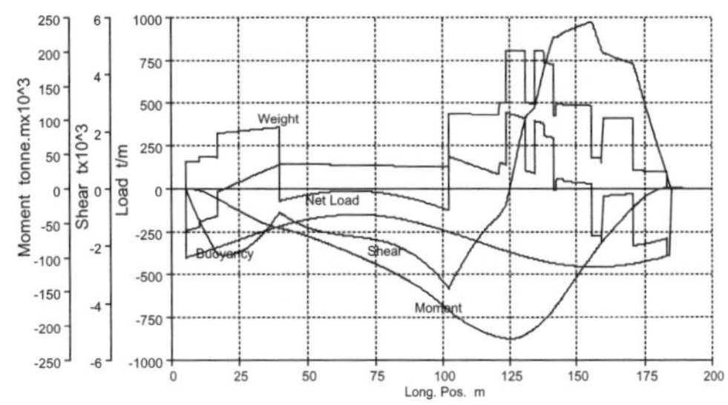

Figure 9. Loading distributions for the second extreme loading scenario (Naar et al. 2008).

The second extreme loading scenario corresponds to the loading condition with the large value of shear force and the hull girder section where the value of the shear force to vertical bending moment ratio is relatively high and might have an impact on the collapse of the hull girder. Relevant distributions for this loading scenario are given by Figure 9.

3.1.3 *Results*

For the given particular example of application consideration of deck efficiency coefficients was disregarded due to the structural configuration of the analyzed hull girder (single-deck structure) and only shear stress influence was included into analysis.

Since the first loading scenario (LC1) is analyzed at the section with maximum vertical bending moment where the shear force is negligible, the moment to curvature relationship for sagging and hogging of the hull girder indicated by the Figure 10 is obtained by the means of the original (unmodified) IACS incremental-iterative method. The plot corresponding to second extreme loading scenario (LC2) is obtained with included shear stress influence, and is also given in Figure 10. Values of the ultimate vertical bending moments obtained by extended IACS incremental-iterative

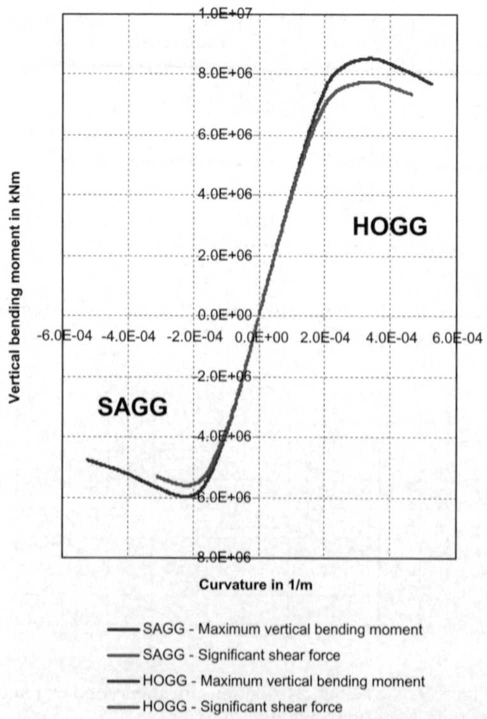

Figure 10. Hull girder vertical bending moment to curvature relationships obtained for LC1 and LC2.

method (as implemented in OCTOPUS software) for both loading scenarios in sagging and hogging are given by Table 2.

Nonlinear FEM analysis performed showed that the analyzed hull girder structure will collapse at the section of the maximum vertical bending moment for both loading scenarios considered, rendering the comparison of the results obtained by two methods employed not valid for the LC2 (since the structural collapse occurred at the different hull girder stations). Results obtained by the NL FEM are summarized in the Table 3.

Generally, based on the results obtained, it can be argued that neglecting of the shear induced effects for hull girder transverse sections with significant shear loading leads to overly optimistic and possibly unsafe estimate of the ultimate bending capacity.

3.2 *Multi-deck structure*

Application of the procedure given in the section 2.3 is exemplified by the analysis of the generic structure of the simplified passenger ship, namely the 'ISSC benchmark' (ISSC 2006). Specification of the structural layout, dimensions and loading used by the various authors/contributors involved in analysis of this reference structure by various proprietary methods (including the nonlinear FEM analysis) is given by the same reference. Here, only results of the nonlinear FEM analysis will be used for comparison purposes. Figure 11 depicts the procedure of determination of the deck efficiency coefficients for this particular example of application (Andric 2007).

Table 2. Results obtained by the extended IACS incremental-iterative method for both loading scenarios (values in kNm).

	SAGG	HOGG
LC1 (max. M_V; $Q_V \approx 0$)	-5.983×10^6	$+8.538 \times 10^6$
LC2 (max. Q_V; $M_V \neq 0$)	-5.608×10^6	$+7.753 \times 10^6$

Table 3. Results obtained by the nonlinear FEM analysis for both loading scenarios considered (values in kNm).

	SAGG	HOGG
LC1 (max. M_V; $Q_V \approx 0$)	-5.830×10^6	$+8.630 \times 10^6$
LC2 (max. M_V; $Q_V \approx 0$)	-6.100×10^6	$+7.420 \times 10^6$

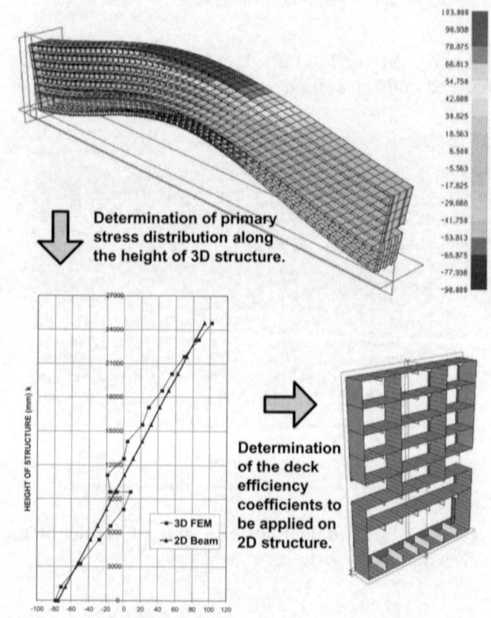

Figure 11. Procedure for determination of the deck efficiency coefficients using the prismatic ISSC benchmark model.

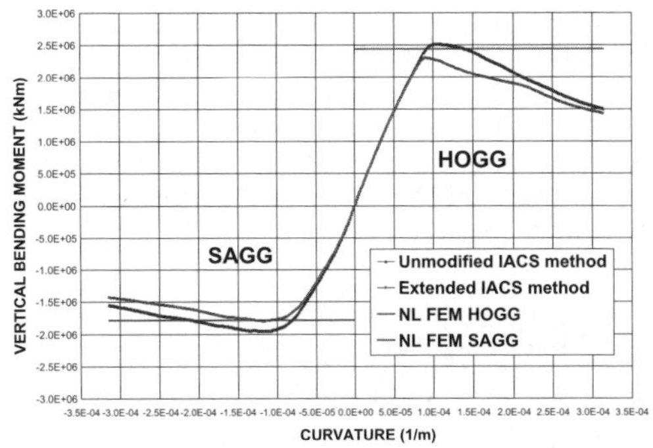

Figure 12. Resulting moment to curvature plots obtained by both unmodified and extended IACS incremental-iterative methods superimposed with the results of the nonlinear 3D FEM using the prismatic ISSC benchmark model.

Table 4. Results obtained by the three methods available for hogging and sagging loading scenarios (values in kNm).

	Unmodified IACS	Extended IACS	NL 3D FEM
HOGG	$+2.52 \times 10^6$	$+2.30 \times 10^6$	$+2.44 \times 10^6$
SAGG	-1.95×10^6	-1.80×10^6	-1.78×10^6

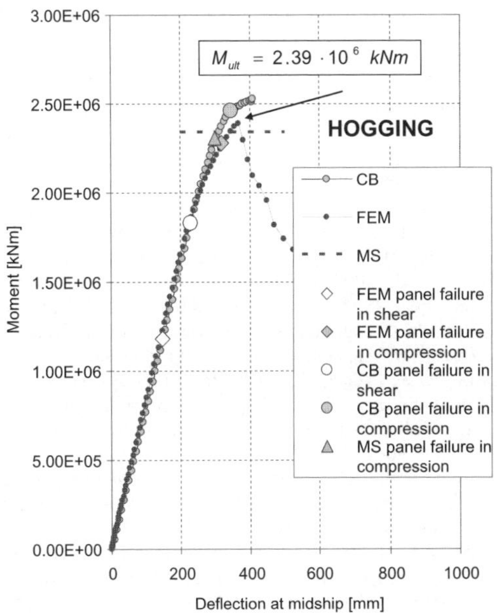

Figure 13. Moment to deflection curves at midship section for prismatic multi-deck structure with deck openings in hogging (Naar et al. 2008).

The results of the hull girder ultimate strength analyses performed by all three methods available (original IACS incremental-iterative, extended IACS incremental-iterative and nonlinear FEM) are presented superimposed in Figure 12, and are summarized in terms of the obtained ultimate vertical bending moments in Table 4.

The results presented show excellent agreement with the reference results of the nonlinear FEM for the sagging case, while for the hogging case reference results are approached from the safe side.

Extensive validation of the implemented methodology for ultimate strength evaluation of multi-deck structures trough comparison of results obtained by nonlinear FEM and Coupled Beam (CB) method (Naar et al. 2004) analysis was performed. Multiple variants (structural openings, deletion of decks, replacement of pillars with longitudinal bulkheads, etc.) of the reference ISSC multi-deck structure were analyzed within the scope of the FP6 project IMPROVE (Naar et al. 2008).

Very good agreement of the results was accomplished and reconfirmed (Figure 13).

The value $M_{ult} = 2.39 \times 10^6$ kNm is the ultimate moment obtained with NL FEM (Naar et al. 2008). For modified IACS method (denoted MS), the moment value is $M_{ult} = 2.34 \times 10^6$ kNm.

4 APPLICATION IN DESIGN

Incremental nature of the described concept for the ultimate strength assessment enables prediction of the structural collapse dynamics and establishment of the collapse sequence of the principal structural members of the hull module.

The knowledge about that sequence of the substructures (such as decks) composing a hull girder can give a very useful information regarding the weakest or critical structural areas. Collapse sequence for the second application case is presented in the Figure 14 for sagging case, as an example.

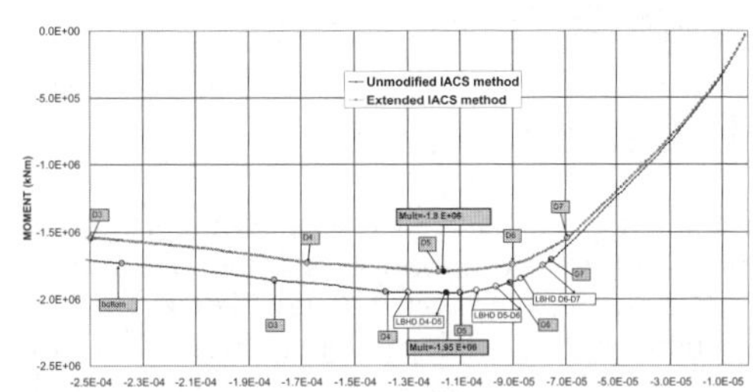

Figure 14. Collapse sequence for the sagging case.

Furthermore, identified collapse scenario can be helpful in structural optimization with the ultimate vertical bending moment as a design attribute. This enables subsequent redesign of the critical components resulting in a globally safer structure, especially if the methodology is employed within the optimization based concept design loop.

5 CONCLUSIONS

Considering the results of nonlinear FEM analysis for the first application case it can be concluded that the structural collapse, of the particular tanker structure considered, is not influenced by the vertical shear force induced effects. This can be generally explained by sufficient hull girder webbing (double-sides) able to effectively sustain imposed shear loading. Consequently, the structural collapse of the single-sided hull girder structure (i.e. bulk carrier) might prove more prone to shear effects considered.

Results of the hull girder ultimate strength analyses of the second application case (multi-deck) show very good agreement with the results of the nonlinear FEM analysis, especially for the sagging case.

Generally, it can be noted that the results obtained by extended IACS incremental-iterative method are in better agreement with the reference results of the nonlinear 3D FEM analysis when compared to the results of the original IACS method, enabling safer estimate of the hull girder ultimate bending capacity.

Further research has to be performed to fully understand the limitations of presented method and how it could be applied and improved for application to multi-deck structures. Detailed investigation will be performed to identify changes of the deck coefficients through design cycles and other possibilities of additional updates and improvements of the presented method.

Ultimate strength safety measure as a design objective could be very useful in multi-criteria decision making since it drives the design process towards the most rational material distribution and safer ship designs.

ACKNOWLEDGMENTS

Thanks are due to the Croatian Ministry of Science, Education and Sport for long-term support of the national project 120-1201829-1671. Considerable part of work presented in this paper was funded by the European Commission through the EU FP6 STREP project IMPROVE (Contract No. TST5-CT-2006-031382). Special thanks are due to Prof. H. Naar who led MEC team from Tallinn for extensive proprietary work on loading determination, non-linear FEM modeling/analysis and the pleasant collaborative work performed on the IMPROVE project.

REFERENCES

Andric, J. 2007. Decision support methodology for concept design of the ship structures with hull-superstructure interaction. PhD dissertation (in Croatian), University of Zagreb, Croatia.

Biot, M. et al. 2006. Collapse Analysis of a Modern Cruise Ship Hull Girder. San Francisco: ISOPE 2006.

Hughes, O.F. 1988. *Ship Structural Design—A Rationally-Based Computer-Aided Optimization Approach*. Jersey City: The Society of Naval Architects and Marine Engineers.

IACS 2006. Common Structural Rules for Double Hull Oil Tankers.

IACS 2008. Common Structural Rules for Bulk Carriers.

ISSC Technical Committee III.1 2006. Ultimate Strength. *Proceedings of the 16th International Ship and Offshore Structures Congress Vol 1*. Southampton.

MAESTRO v8.7, Software documentation, DRS-C3 Advanced Technology Center, Stevensville, MD, USA.

Naar, H. et al. 2004. A theory of coupled beams for strength assessment of passenger ships. *Marine Structure, Vol 17, Issue 8*: 590–611.

Naar, H. et al. 2008. Deliverable D3.2—Report on Assessment of ultimate strength at the early design stage. FP6 STREP Project IMPROVE (Contract No. TST5-CT-2006-031382).

Paik, J.K. 2003. *Ultimate Limit State Design of Steel-Plated Structures*. Chichester: John Wiley and Sons Ltd.

Smith, C.S. 1977. Influence of Local Compressive Failure on Ultimate Longitudinal Strength of a Ship's Hull. Tokyo: PRADS 1977.

SSN 2008. Information on Stability and Longitudinal Strength: Chemical Tanker 40000 DWT, Szczecin: IMPROVE/042-3, Version 09/12/2008.

Yao, T. et al. 2004. Influence of Warping due to Vertical Shear Force on Ultimate Hull Girder Strength. Luebeck-Travemuende: *9th Symposium on Practical Design of Ships and Other Floating Structures*.

Zanic, V. & Prebeg, P. 2005. Primary response assessment method for concept design of monotonous thin-walled structures. *Acta Polytechnica Vol 45, No 4/2005*: 96–103.

Zanic, V. et al. 2007. Decision Support Problem Formulation for Structural Concept Design of Ship Structures. *Proceedings of MARSTRUCT 2007*: 499–509.

Zanic, V. et al. 2009. Design Environment for Structural Design: Application to Modern Multideck Ships. *Proceedings of IMechE, part M, Journal of Engineering for the Maritime Environment, Vol.223, No.M1*: 105–120.

Advanced Ship Design for Pollution Prevention – Guedes Soares & Parunov (eds)

 ISBN 978-0-415-58477-7

Performance of the Common Structural Rules design formulations for the ultimate strength of uniaxially loaded plates and stiffened panels

M. Ćorak & J. Parunov
Faculty of Mechanical Engineering and Naval Architecture, University of Zagreb, Zagreb, Croatia

A.P. Teixeira & C. Guedes Soares
Centre for Marine Technology and Engineering (CENTEC), Technical University of Lisbon, Instituto Superior Técnico, Lisboa, Portugal

ABSTRACT: The paper deals with application of the nonlinear finite element analysis for collapse assessment of uniaxially loaded plates and stiffened panels of ship structures. Consideration is given to models extension, boundary conditions, mesh size and initial imperfections appropriate for collapse analyses. The load-end shortening curves of plates and stiffened panels computed by nonlinear finite element analysis are compared with those proposed in Common Structural Rules.

1 INTRODUCTION

The paper is concerned with the basic structural elements used in ships and marine structures, that is, the stiffened panels. In ship structures, stiffened panels are subjected to loads normal to its plane due to local water pressure, liquid cargo pressure, deck loads, etc., and to in-plane forces due to global bending of the vessel. When in-plane forces are compressive, the buckling of the panel may occur due to the loss of elasto-plastic stability, which may be very dangerous situation. As buckled panels loose rapidly their load-carrying capability, adjacent panels may become overloaded and collapse, too. Eventually, collapse of the hull girder may take place, being the worst situation that could happen to the structure of an oceangoing vessel.

The behavior of plates and stiffened panels under compressive loads has been studied for many years dating back to Faulkner (1975). Since then, many theoretical, experimental and numerical studies have been carried out to assess structural capacity of stiffened panels.

Nonlinear finite element methods (NLFEM) have been used for many years to calculate the ultimate strength of plates and stiffened panels with different characteristics. Initial efforts were reported by several authors, among whom one can refer Crisfield (1975), Frieze et al. (1977), Soreide et al. (1977), Ueda and Yao (1979) and Carlsen (1980). Many other studies have been done to determine the influence of different initial distortions and residual stresses as well as their variability in the strength assessments (Guedes Soares and Kmiecik, 1993).

Since the early work of Faulkner (1975) for plates and Faulkner et al. (1973) for stiffened panels many design formulations have been used (e.g Guedes Soares 1997a). It has been shown that the main governing parameter of the compressive strength is the plate slenderness (Faulkner, 1975) and thus the simplest design methods include only plate slenderness. However, if more accurate results are desired, other important variables must be included. In fact, Guedes Soares (1988a, 1992) recognized and incorporated the influence of initial imperfections in Faulkner's analytical formula to predict the collapse strength of plates.

Depending on the mathematical formulation adopted they incorporate different types of model errors, which need to have their bias and uncertainty assessed (Guedes Soares, 1988b, 1997b). Finite element methods are a good tool together with experimental results to calibrate design methods and to determine their model uncertainty as has also been done by various authors (e.g. Guedes Soares and Gordo, 1997).

The new Common Structural Rules (CSR) for double-hull Tankers (ABS et al. 2006) have adopted similar type of formulations for the ultimate strength of plates and stiffened panels, together with the requirement of using the ultimate hull girder collapse strength as design strength.

Qi et al. (2005), Paik et al. (2008c), and Hussein et al. (2008) have studied the accuracy of the method stipulated in the CSR for assessing the ultimate hull girder collapse strength, by using nonlinear finite element analysis (NLFEA) and other methods, while the effect of the ultimate strength formulations on reliability was examined by Parunov & Guedes Soares (2008) and Hussein and Guedes Soares (2009).

However studies addressing specifically the accuracy of the plate and stiffened panel formulations seem to be still lacking.

The present study addresses this topic and compares the CSR formulations of plate and stiffened panel strength with NLFEA, using the ANSYS code (ANSYS 2009). It deals with uniaxially loaded panels, primarily those that one can find on the main deck of an oil tanker.

Recent casualties of oil tankers "Erica" and "Prestige" that resulted in disastrous environmental pollution, huge economic losses and enormous negative publicity to the shipbuilding industry are typical examples of hull girder failures due to the instability collapse of main deck panels.

Although lateral pressure is important parameter that can affect ultimate strength it is outside the scope of the present study. However, there are design equations which can predict the reduction in the longitudinal strength of plates due to lateral pressure (Teixeira & Guedes Soares 2001).

As the collapse of stiffened panels are usually initiated by the buckling of the plate between stiffeners, the first part of the paper deals with the collapse of unstiffened plates. After that, stiffened panels are analyzed and load-end shortening curves are computed. In all cases, comparison of load-end shortening curves with those proposed in CSR is performed.

2 COLLAPSE OF UNIAXIALLY LOADED PLATES

The first part of this paper deals with the behavior of an unstiffened plate supported by longitudinals in the longitudinal direction and by transverse girders in the transverse direction (Figure 1).

In the NLFEA performed there are crucial factors that were accounted for, such as geometrical nonlinearity, material nonlinearity, type and magnitude of initial imperfections, boundary conditions, loading conditions and mesh size in the analysis.

The length of the considered plate is $a = 4300$ mm while the spacing between longitudinals is $b = 815$ mm. Three different thicknesses are assumed: 12, 15 and 18 mm, corresponding to plate slenderness β of 2.66, 2.13 and 1.77 respectively, where β is given by as follows (Faulkner 1975):

$$\beta = b/t \cdot \sqrt{(\sigma_y / E)} \tag{1}$$

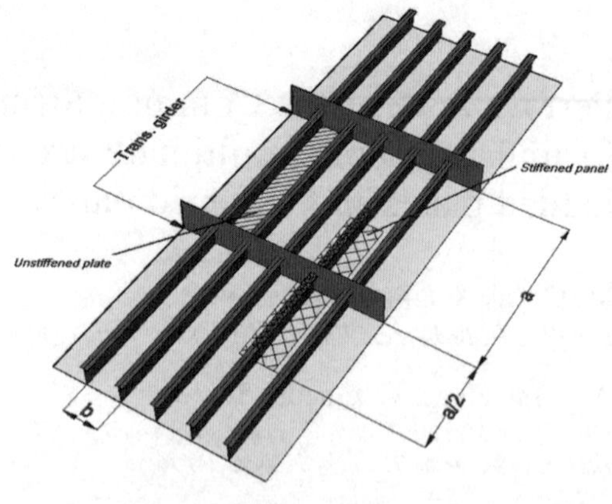

Figure 1. Proposed model of unstiffened and stiffened plate.

The values chosen are considered to cover the range of main deck thicknesses for as built state as well as for corroded state of oil tankers. For the plate material, 32 AH high tensile steel is used with yield stress $\sigma_Y = 315$ MPa, Young's modulus $E = 205.8$ GPa and Poisson's ratio $\upsilon = 0.3$.

The typical mesh size used in this analysis is 50×50 mm. This corresponds to about 16 elements in the shorter (transverse) direction and to about 80 elements in longer (longitudinal) direction. Elastic–perfectly plastic material is applied using von Mises yield criteria. Residual stresses are neglected in the present study although they could affect significantly the ultimate strength (Ueda & Tall 1967, Guedes Soares 1988a,).

2.1 *Boundary conditions and initial deflections*

The boundary conditions can affect considerably the plate strength. Guedes Soares (1988a) has shown that in the range of plate slenderness between 2.5 and 3.5 clamped plates are between 15% and 30% stronger than simply supported ones.

For the present analysis, one bay plate model is considered, where the plate is simply supported at the longitudinal edges and clamped at the transverse edges. One bay model may be appropriate for uniaxially loaded rectangular plates, while two bay models may be necessary when dealing with bi-axial load and lateral pressure (Luis et al. 2008, Paik et al. 2008a).

In most of NLFEA of plate collapse, it was assumed that the pattern of the initial plate deflection is equivalent or very similar to the plate buckling mode which may give the lowest resistance against the actions. That pattern is such as to have one half-wave of the buckled shape in the transverse direction, while the number of half-waves in

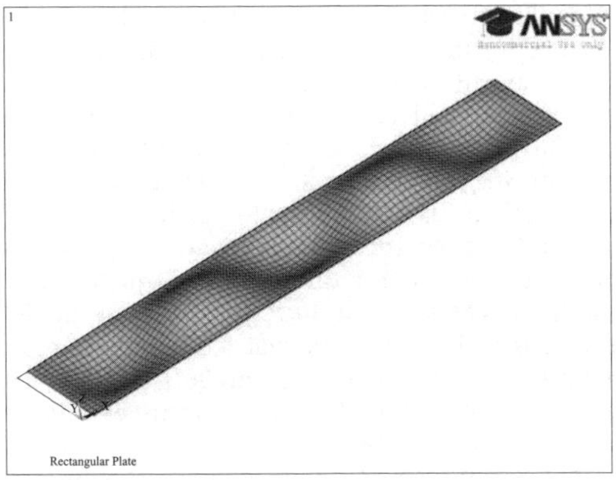

Figure 2. Plate initial imperfection.

the longitudinal direction is normally an integer ratio of the longer and shorter side of the plate. That pattern may be expressed by the following equation which is based on the Fourier series:

$$w_p = w_0 \cdot \sin\left(\frac{m \cdot \pi \cdot x}{a}\right) \cdot \sin\left(\frac{n \cdot \pi \cdot y}{b}\right) \tag{2}$$

The amplitude of the buckling waves is assumed to be $w_0 = 0.1 \cdot \beta^2 \cdot t$ (Guedes Soares 1992). A plate with initial imperfections is shown in Figure 2.

The loading of the plate is imposed as forced displacement at one of two short edges. The displacement is increased gradually resulting in concentrated force in the corresponding prevented degree of freedom. The total force is then divided by the area of the short plate edge and thus average stress is obtained.

2.2 *Results*

The results from the NLFEA are compared with the formulations from CSR. The load-end shortening curves from CSR are those specified in CSR, APPENDIX A, Chapter 2.

Load end shortening curves are presented in Figures 3–5 for different plate thicknesses together with CSR results.

The ultimate strength of uniaxially loaded plates is the maximum value of the relative stresses on the load-end shortening curves. Comparison of ultimate strengths for the three considered cases is given in the Table 1 (the ultimate strength is presented as the ratio of stresses and yield stress of the plate material).

From figures 3–5 and from Table 1, one may conclude that agreement between the load-end shortening curves is favourable.

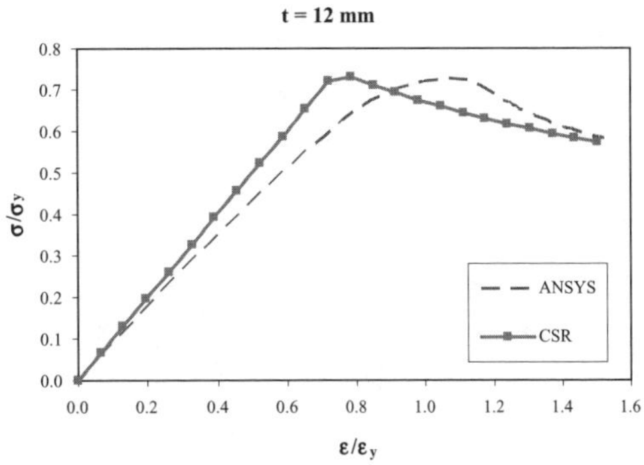

Figure 3. Load end shortening curves for t = 12 mm.

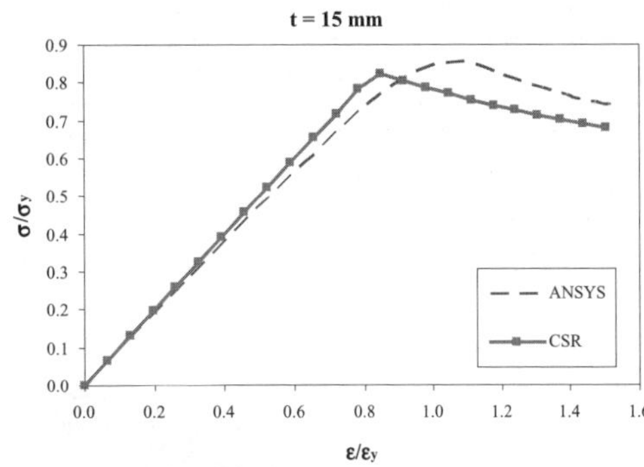

Figure 4. Load end shortening curves for t = 15 mm.

The ultimate strength calculated by NLFEM shows better agreement to CSR results for thinner plates, while load-end shortening curves agree better for thicker plates.

The deformation of the model at the collapse state is presented in Figure 6 for a plate thickness of 12 mm. It can be seen that collapse mode follows the buckling pattern due to uniaxially compressive loading.

3 COLLAPSE OF UNIAXIALLY LOADED STIFFENED PANELS

The analyses of stiffened plates have been performed by many researchers over the years. Design methods to determine the ultimate compressive load of the stiffened panels have been presented in a fundamental paper by Faulkner et al. (1973). Guedes Soares and Soreide (1983) studied the behaviour and design of stiffened plates under predominantly compressive loads where they have compared the collapse strength predicted by various methods with available experiments. Several

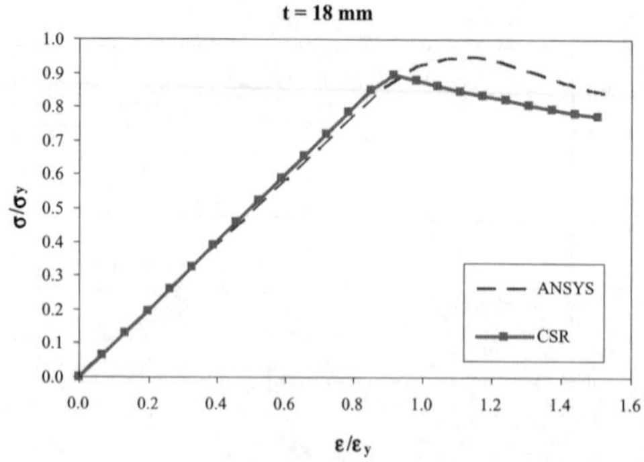

Figure 5. Load end shortening curves for t = 18 mm.

Table 1. Results of ultimate strength for models of different plate thickness.

Plate		t = 12 mm	t = 15 mm	t = 18 mm
σ_U/σ_{yd}	ANSYS	0.73	0.85	0.95
	CSR	0.73	0.82	0.89
Difference	Δ(%)	0.30	3.87	6.14

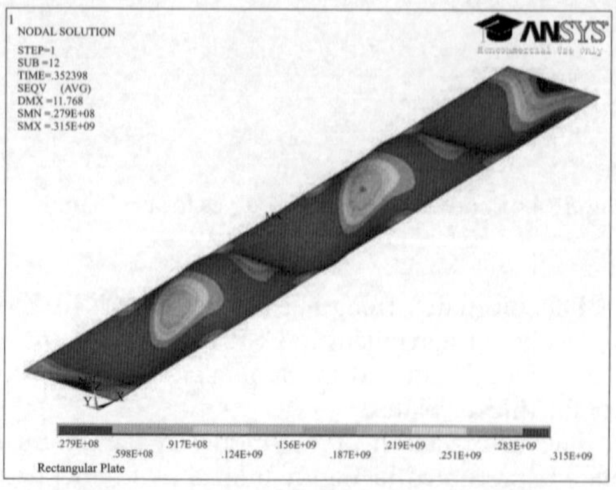

Figure 6. Deformed shape and the Von Mises stress distribution at the ultimate limit state for t = 12 mm.

studies have been dealing with analytical formulations of ultimate strength for stiffened panels trying to embrace important variables (Gordo & Guedes Soares 1993, Guedes Soares & Gordo, 1997). Large part of that knowledge has been incorporated in the newly developed CSR.

The second part of this study deals with the comparison between load–end shortening curves of stiffened plates given by CSR formulations and the results produced by NLFEA.

Two different stiffened panels have been considered. The first one is the panel stiffened by T profile, as given by Paik et al. (2008b). The other one is the panel of the main deck of an existing Aframax oil tanker, stiffened by bulb profile (which is for the need of the modeling converted to L profile).

The extent of the FE model used in the present paper is one half of stiffener spacing on each side of the longitudinal in transverse direction and half of the web frame spacing on each side of web frame in the longitudinal direction. The same assumptions regarding material and residual stress are considered as for the unstiffened plate, while the mesh size is about 40 × 40 mm. Such model extent is suitable for uniaxial load, while for biaxial stresses and lateral pressure model should be extended both in longitudinal and transverse directions (Paik et al. 2008b).

The extent of the FE model and the dimensions of stiffeners are presented in Figure 7 and Table 1.

3.1 *Boundary conditions and initial deflections*

For the present NFEA, the following boundary conditions are applied where T[x, y, z] indicates translational constraints, and R[x, y, z] indicates rotational constraints about x-, y-, and z-coordinates, respectively; a "0" indicates constraint, and a "1" indicates no constraint (Figure 8) (Paik et al. 2008b).

- Symmetric condition at A—A0: R [1, 0, 0] with all plate nodes and stiffener nodes having an equal x—displacement,
- Symmetric condition at B—B0: T [0, 1, 1] and R [1, 0, 0],
- Symmetric condition at A—B and A0—B0: T [1, 0, 1] and R [0, 1, 1],
- At transverse floors: T [1, 1, 0] for plate nodes, and T [1, 0, 1] for stiffener web nodes.

The stiffened panel model is subjected to uniaxial compression. The loading is imposed as a gradually increasing forced displacement at the short edge similarly as for unstiffened plates (Sec. 2.1).

Three types of initial deflections are considered in the study (Paik et al. 2008b):

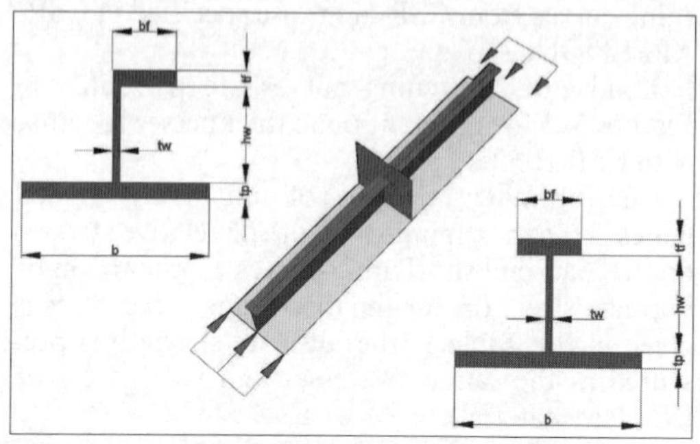

Figure 7. Extent of the model of deck stiffened panel subjected to uniaxial load with main dimensions.

Table 2. Geometric properties of stiffeners (mm).

Profile	a	b	t_p	h_w	t_w	b_f	t_f
T	4300	815	17.8	463.0	8.0	172.0	17.0
L(HP280 × 11)	3840	820	17.5	223.2	11.0	50.8	28.4

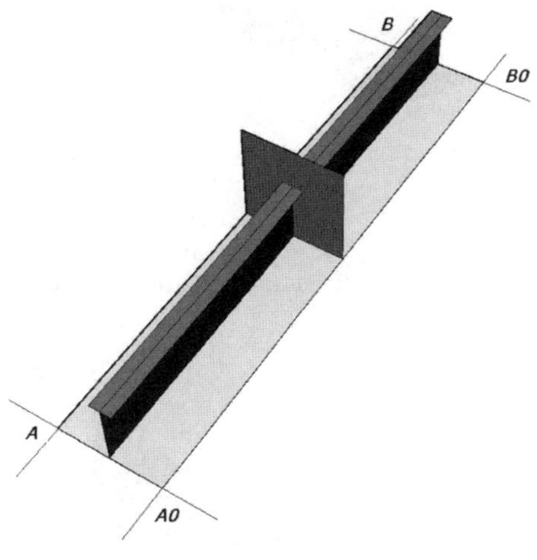

Figure 8. Schematic model for definition of boundary conditions.

- Plate initial deflection with the magnitude of $w_p = b/200$ and shape close to the unstiffened plate buckling mode (m = 5).
- Beam-column type initial deflection with the magnitude of $w_{oc} = a/1000$ and shape as shown in Figure 9.
- Sideways initial deflection of the stiffener web with the longitudinal shape as for beam column type, and vertically half-sine wave with maximum value at stiffener flange, while zero value at intersection of stiffener and the plate. The magnitude of sideways initial amplitude is $w_{os} = a/1000$.

The pattern for plate initial imperfection is assumed as described in Eq. (1) while the pattern of both column–type and sideways initial deflection of the stiffener is supposed to be the buckling mode that results in the minimum buckling strength of the stiffener. Total initial imperfections of the FE model are shown in Figure 10 and 11 for T & L profile respectively.

The model of initial imperfections used in the present study is very similar to the model used by Wang et al. (1996) and Paik et al. (2008b).

3.2 *Results*

Load end shortening curves are presented in Figures 12 & 13 for T and L stiffeners respectively.

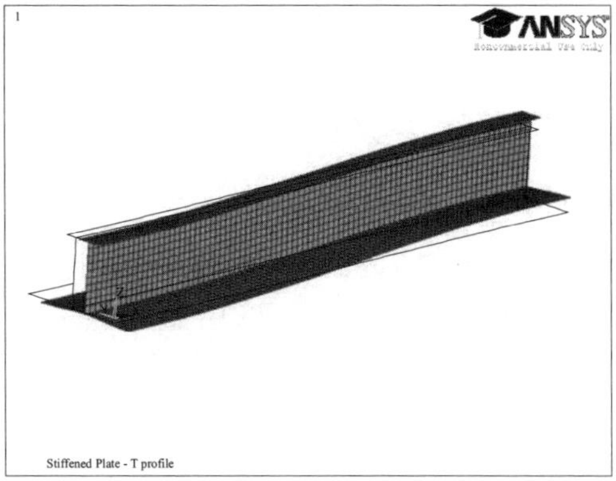

Figure 9. Beam—column initial deflection for T - profile.

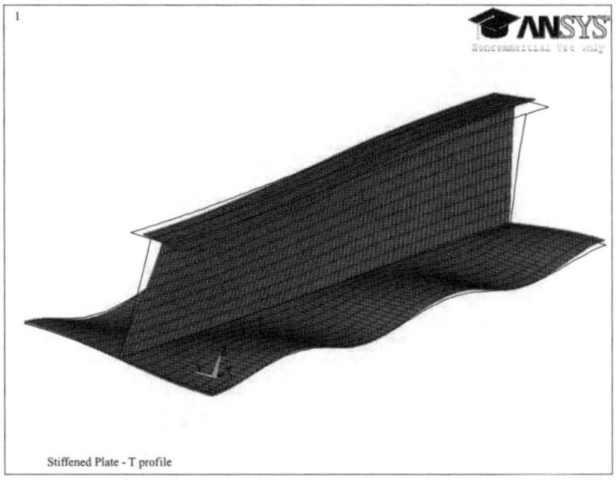

Figure 10. Initial imperfection for T - profile.

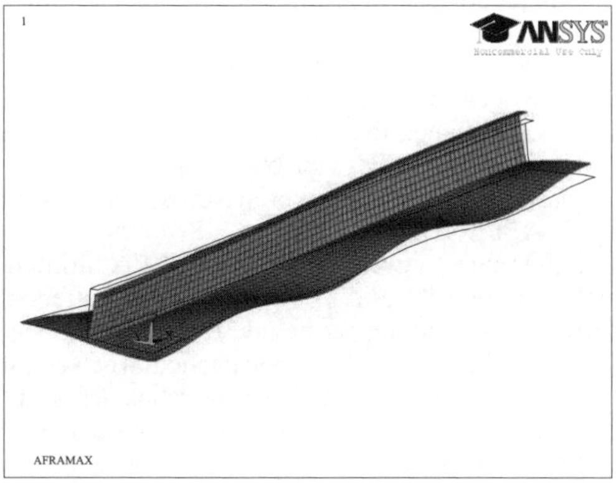

Figure 11. Initial imperfection for L - profile.

In both figures comparison with curves obtained by CSR are also presented (ABS et al. 2006). In addition, results obtained by Paik et al. (2008b) are also included in Figure 11.

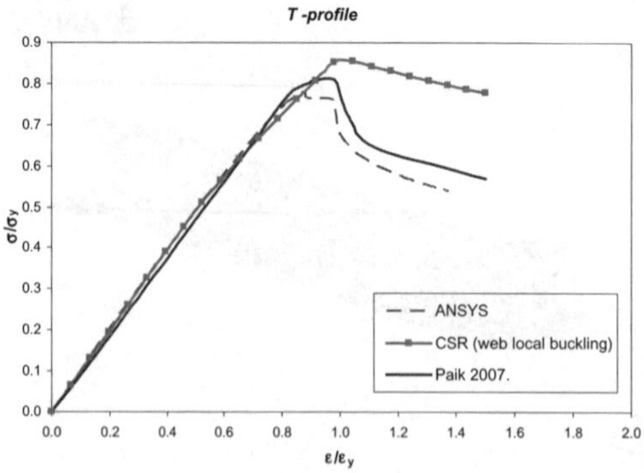

Figure 12. Load end shortening curves for T—profile.

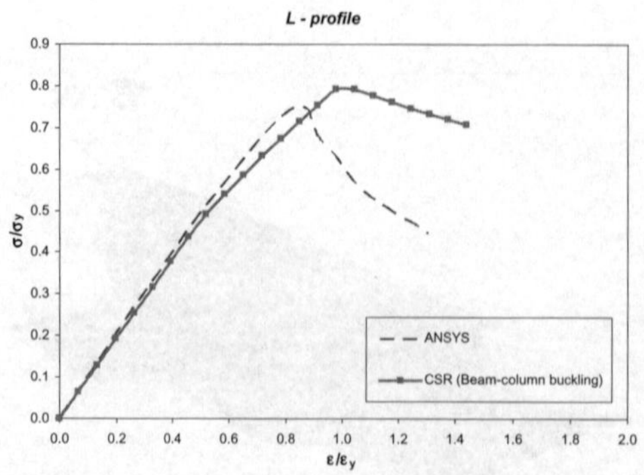

Figure 13. Load end shortening curves for L—profile.

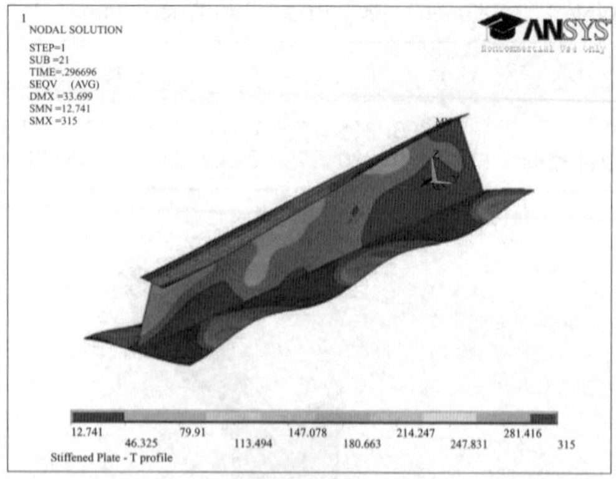

Figure 14. Deformed shape and the Von Mises stress distribution at the ultimate limit state for the T profile.

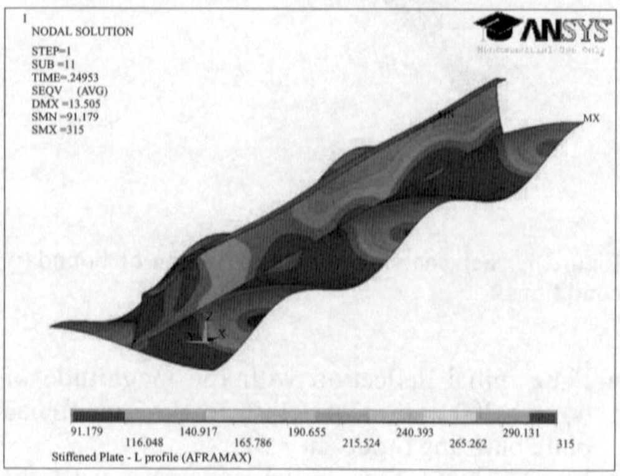

Figure 15. Deformed shape and the Von Mises stress distribution at the ultimate limit state for the L profile.

It should be clarified that CSR curves presented in Figures 11 & 12 are the lowest curves of three different buckling modes: beam column flexural buckling, stiffener torsional buckling and local buckling of the stiffener web. For the panel stiffened by T profile, the most critical CSR failure mode is local buckling of the stiffener web, while for the L stiffener, the most critical failure mode is the beam-column flexural buckling. CSR formulae used in the present paper are those specified in CSR, APPENDIX A, Chapter 2, Section 3.

Differences between CSR and NLFEA ultimate strength are 9.4% and 5.3% for T and L stiffeners respectively. One may conclude that differences in load-end shortening curves are particularly large in post-collapse region. Further research is necessary in order to clarify reasons for those discrepancies.

Provisionally, one may state that the effect of strain hardening could be the reason for discrepancies in the plastic region, as that effect is not considered in the paper. The evidence for that statement is the last ISSC report, where the analysis of the influence of the strain hardening has been performed (Paik et al. 2009). Curves obtained without hardening are quite similar to those obtained herein, while curves obtained with hardening effect are much flatter in post-collapse region, with shapes similar to CSR curves.

The results obtained for the T stiffener agree well with results obtained by Paik et al. (2008b), where difference between two NLFEAs in ultimate strength amounts only 3.5%, while the shapes of load-end shortening curves are similar.

Figures 14 and 15 show the deformed shapes with von Mises stress distribution of the stiffened panels at the ultimate limit state under uniaxial compression. The deformed shapes at the ultimate limit state are slightly different due to the fact that T profile is symmetric regarding flange while L profile is not.

4 CONCLUSION

The present paper is concerned with the ultimate strength of plates and stiffened panels of the main deck of oil tankers, which are predominantly uniaxially loaded.

Load-end shortening curves for unstiffened plates obtained by ANSYS NLFEM agree quite well with the curves proposed by CSR. The differences in predicted collapse stress are up to 6% for thicker plates while slender plates agree even better.

The second part of the study compared various modes of failure of stiffened panels from CSR with NLFEA. Agreement of collapse curves by NLFEA and CSR is less favorable for the stiffened panels than for unstiffened plates. Differences are around 9% for the large T profile while the difference for deck stiffened by bulb profile is around 5%. It seems that inclusion of the strain hardening effect could improve the agreement between load-end shortening curves by NLFEM and CSR.

The conclusions obtained in this study are based on a very limited number of configurations of plates and stiffened panels. Thus, while the results may provide some indication of the performance of the design rules, they cannot be considered as completely general conclusions.

ACKNOWLEDGMENTS

This paper presents material that is in the scope of the course "Ship structural reliability with respect to ultimate strength" within the project "Advanced Ship Design for Pollution Prevention (ASDEPP)" that was financed by the European Commission through the Tempus contract JEP-40037-2005. The work reported was initiated during a visit that the first author made to the Centre for Marine Technology and Engineering at Instituto Superior Técnico, which was funded by the ASDEPP project.

REFERENCES

American Bureau of Shipping, Det Norske Veritas, Lloyd's Register. 2006. *Common Structural Rules for Double Hull Oil Tankers.*

ANSYS. 2009. User's Manual (Version 11). Swanson Analysis System Inc., Houston.

Carlsen, C.A.A Parametric Study of Colapse of Stiffened Plates in Compression. *The Structural Engineer.* 1980; 58B(2): 33–40.

Crisfield, M.A. Full-Range Analysis of Steel Plates and Stiffened Plating under Uniaxial Compression. *Proc. Instn. Civil Engineers*; 1975: 595–624.

DNV. 1992. Buckling Strength Analysis, Classification Note No. 30.1, *Det Norske Veritas, Hovik.*

Faulkner, D. 1975. A review of effective plating for use in the analysis of stiffened plating in bending and compression. *J. Ship Research* 19: 1–17.

Faulkner, D., Adamchak, J.C., Snyder, G.J., and Vetter, M.F., 1973. Synthesis of welded grillages to withstand compression and normal loads. *Computers & Structures* 3(2): 221–246.

Frieze, P.A., Dowling, P.J. & Hobbs, R.H., Ultimate load behaviour of plates in compression. In Steel Plated Structures, Dowling, P.J. Harding, J.E. Frieze, P.A., (Editors).Crosby Lockwood Staples, London, 1977, pp. 24–50.

Gordo, J.M. & Guedes Soares, C. 1993. Approximate Load Shortening Curves for Stiffened Plates under Uniaxial Compression. *Integrity of Offshore Structures—5.* EMAS: 189–211.

Guedes Soares, C. 1988a, Design Equation for the Compressive Strength of Unstiffened Plate Elements with Initial Imperfections. *Journal of Constructional Steel Research* 9: 287–310.

Guedes Soares, C. 1988b, Uncertainty Modelling in Plate Buckling. *Structural Safety.* 5: 17–34.

Guedes Soares, C. 1992. Design Equation for Ship Plate Elements under Uniaxial Compression, *Constructional Steel Research* 22 (1992): 99–114.

Guedes Soares, C., 1997a, Probabilistic Modelling of the Strength of Flat Compression Members, *Probabilistic Methods for Structural Design,* Guedes Soares, C. (ed.), Kluwer Academic Publishers; pp. 113–140.

Guedes Soares, C. 1997b, Quantification of Model Uncertainty in Structural Reliability. *Probabilistic Methods for Structural Design,* Guedes Soares, C. (ed.), Kluwer Academic Publishers; pp. 17–38.

Guedes Soares, C. & Gordo, J.M. 1997, Design Methods for Stiffened Plates Under Predominantly Uniaxial Compression, Marine *Structures* 10: 465–497.

Guedes Soares, C. and Kmiecik, M., 1993, Simulation of the Ultimate Compressive Strength of Unstiffened Rectangular Plates, *Marine Structures.* 6: 553–569.

Guedes Soares, C. and Soreide, T.H. 1983. Behaviour and Design of Stiffened Plates under Predominantly Compressive Loads, *International Shipbuilding Progress,* 30(341), 13–26.

Hussein, A.W. and Guedes Soares C. 2009, Reliability and Residual Strength of Double Hull Tankers Designed According to the new IACS Common Structural Rules, *Ocean Engineering*, 36(17–18): 1446–1459.

Hussein, A.W., Teixeira, A.P., and Guedes Soares, C. 2008, Assessment of the IACS Common Structural Hull Girder Check Applied to Double Hull Tankers. *Maritime Industry, Ocean Engineering and Coastal Resources*, C Guedes Soares, P Kolev (Eds), Taylor & Francis Group; London—UK: pp. 175–184.

International Maritime Organization (IMO). 2006. *Goal-based New Ship Construction Standards*, MSC 81/ INF.6.

Luís, R.M., Witkowska, M., and Guedes Soares, C. 2008. Ultimate Strength of Transverse Plate Assemblies Under Uniaxial Loads. *Journal of Offshore Mechanics and Arctic Engineering* 130(2): 021011-1/021011-7.

Luís, R.M., Guedes Soares, C., and Nikolov, P.I., 2008. Collapse strength of longitudinal plate assemblies with dimple imperfections, *Ships and Offshore Structures*, 3(4): 359–370.

Paik, J.K., Kim, B.J., and Seo, J.K., 2008a. Methods for ultimate limit state assessment of ships and ship-shaped offshore structures: Part I—unstiffened plates, *Ocean Engineering* 35 (2008): 261–270.

Paik, J.K., Kim, B.J., and Seo, J.K., 2008b. Methods for ultimate limit state assessment of ships and ship-shaped offshore structures: Part II—stiffened panels. *Ocean Engineering* 35 (2008): 271–280.

Paik, J.K., Kim, B.J., and Seo, J.K., 2008c. Methods for ultimate limit state assessment of ships and ship-shaped offshore structures: Part III—hull girders. *Ocean Engineering* 35 (2008): 281–286.

Paik et al. 2009. Report of the Committee III.1 Ultimate Strength, *Proceedings 17th Int. Ship and Offshore Structures Congress, Korea.*

Parunov, J., and Guedes Soares, C. 2008. Effects of Common Structural Rules on hull-girder reliability of an Aframax oil tanker. *Reliability Engineering and System Safety.* 93, 9; 1317–1327.

Soreide, T.H.; Bergan, P.G., and Moan, T. Ultimate collapse behaviour of stiffened plates using alternative finite element formulations. In *Steel Plated Structures,* Dowling, P.J. Harding J.E. Frieze P.A., (Editors), Crosby Lockwood Staples, London, 1977: 618–637.

Teixeira, A.P. and Guedes Soares, C. 2001. Strength of Compressed Rectangular Plates Subjected to Lateral Pressure. *Journal of Constructional Steel Research* 57: 491–516.

Qi, E. et al. 2005. Comparative study of ultimate hull girder strength of large double hull tankers, *Marine Structures* 18: 227–249.

Ueda, Y. & Tall, L. 1967. Inelastic buckling of plates with residual stresses. *Publications Int. Assoc. for Bridge and Structural Engineering.*

Ueda, Y. and Yao, T. 1979, Ultimate strength of a Rectangular Plate under Thrust—with Consideration of the Effects of Initial Imperfections due to welding. *Journal of Welding Research Institute.* 8(2): 97–104.

Wang, X., Jiao G., and Moan, T., 1996. Analysis of Oil Production Ships Considering Load Combination, Ultimate Strength and Structural Reliability. *Transactions SNAME,* 96.

Advanced Ship Design for Pollution Prevention – Guedes Soares & Parunov (eds)
© 2010 Taylor & Francis Group, London, ISBN 978-0-415-58477-7

Bending and torsion of stiffeners with L sections under the plate normal pressure

R. Pavazza, F. Vlak & M. Vukasović
Faculty of Electrical Engineering, Mechanical Engineering and Naval Architecture, University of Split, Split, Croatia

ABSTRACT: The behaviour of plate stiffeners with L cross-sections, as deck and bottom longitudinals, bulkhead stiffeners, side frames, etc. subjected to normal plate pressure, is examined. The advanced theory of thin-walled beams with open cross-section on elastic foundation is applied. The stiffeners, as thin-walled beams of open cross-section, are subjected to bending with respect to the enforced centroid axis parallel to the plating and, in addition, to torsion on an elastic foundation with respect to the enforced shear centre in the plating. A uniform normal stress distribution in the cross-section plating is assumed. A linear normal stress distribution along the flange contour is obtained. The comparison of the analytical solutions and the results of the finite element analysis by using shell finite elements are provided.

1 INTRODUCTION

The stiffened panel is the fundamental structural component of marine structures. The stiffeners usually have thin-walled open cross-sections, with the symmetric or asymmetric flange contour. Due to low torsional rigidity of such members, tripping (torsional buckling) occur prior to ordinary beam buckling (Hughes 1983). The tripping effect can be considered by employing thin-walled theories, as well as folded plate theories (Vlasov 1959, Kolbruner & Basler 1969, Gjelsvik 1981). The tripping effect occurs also when the stiffeners are subjected to bending only, i.e. under the normal pressure. Recently, the both loading are examined, the axial load and lateral pressure (Hu et al. 2000).

In this work, the panel with stiffeners with L cross-sections, under plate normal pressure is considered. The stiffeners, as thin-walled beams of open cross-section, are subjected to bending and, in addition, to torsion, due to cross-section asymmetry. The beam bending is considered by using simple bending theories, where an effective breadth of the attached plating is assumed. Torsion will be considered according to the advanced theory of torsion of thin-walled beam on the elastic foundation, by which the cross-section distortion is taken into account (Pavazza 1991, 2007).

The analytical solution will be compared to the finite element method using shell finite elements (ADINA).

2 STIFFENERS WITH L SECTION UNDER PLATE NORMAL PRESSURE

The stiffened panel cross-sections, shown in Figure 1a, due to asymmetry of the cross-sections, under uniform normal plate pressure, deform as shown in Figure 1b, where b_f is the stiffener flange breadth, b_p is the stiffener spacing, h is the stiffener height, t_f is the flange thickness, t_p is plate thickness and t_w is the web thickness.

By the dashed line, displacements of the panel cross-section as rigid contour, as in case of symmetrical flanges, are presented (local deformation of the plates between the stiffeners is ignored); by the full line the cross-section torsion-distortion due to flanges asymmetry is shown (Fig. 1b).

A stiffener with appropriate attached plating, where the cross-section displacements are approximated by a full line as displacements of the contour of a thin-walled beam cross-section, is shown in Figure 2. It is assumed that the cross-section is displaced due to bending (dotted line) and due to torsion (rigid full line) as rigid contour. Thus, the displacement of the stiffener web, in the z-direction, can be expressed as (Pavazza 1991)

$$w_w = w + \delta w \tag{1}$$

where $w = w(x)$ is the displacement of the cross-section in the z-direction due to beam bending with respect to the enforced centroid axis y, parallel to

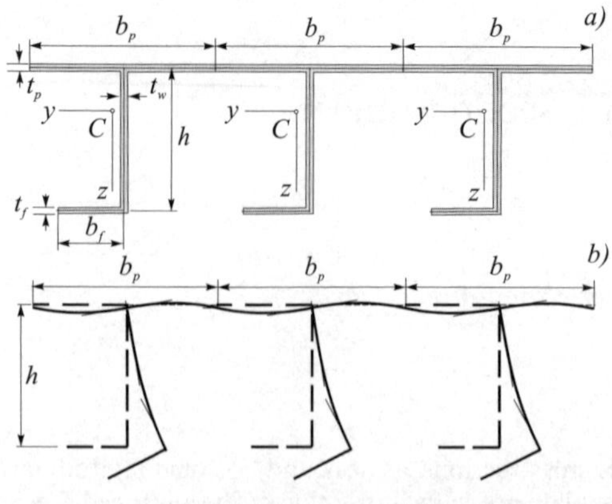

Figure 1. Stiffened plate with L profiles: a) geometrical properties; b) contour under torsion with distortion.

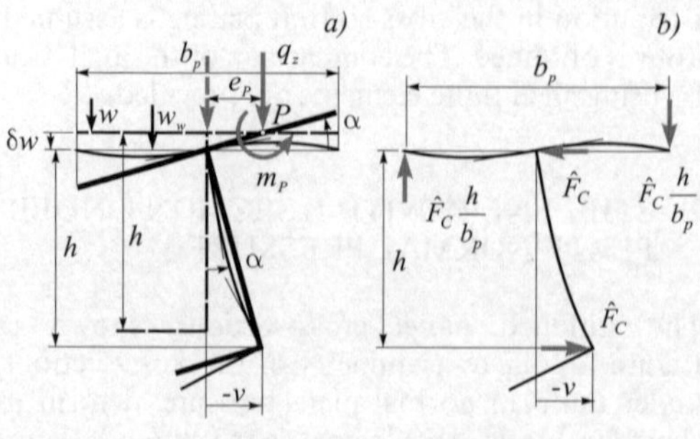

Figure 2. Deformed cross-section contour: a) approximation of deformed contour by rigid contour; b) contour distortion.

the plating, as in the case of symmetrical flanges (Appendix B), i.e. displacement of the cross-section as rigid contour in the z-direction, and $\delta w = \delta w(x)$ is the additional displacement of the web in the z-direction due to beam torsion

$$\delta w = \alpha e_P \tag{2}$$

where $\alpha = \alpha(x)$ is the angle of torsion, i.e. the angle of rotation of the cross-section as rigid contour with respect to the enforced pole P, e_P is the distance of the enforced pole P from the web (Fig. B1).

For the displacements of the flange in the y-direction, $v = v(x)$, one has

$$-v = \alpha h \tag{3}$$

The stiffeners with attached plating can be considered as thin-walled beams with open cross-section subjected to bending with respect to the enforced centroid y-axis, with an effective breadth of the attached plating $b_{pe} \leq b_p$ (B10), due to the non-uniform normal stress distribution in the transverse direction due to the stiffener bending, and in addition to torsion with respect to the enforced pole P, with the attached plating with actual width b_p (Appendix A).

The following differential equations of ordinary theory of thin-walled beams of open cross-section can be employed

$$EI_y \frac{\mathrm{d}^4 w}{\mathrm{d}x^4} = q_z, \quad EI_\omega \frac{\mathrm{d}^4 \alpha}{\mathrm{d}x^4} - GI_t \frac{\mathrm{d}^2 \alpha}{\mathrm{d}x^2} = m_P \tag{4}$$

where $q_z = q_z(x)$ is the force per unit length in the z-direction and $m_P = m_P(x)$ is the moment per unit length with respect to the enforced pole P

$$q_z = pb_p, \; m_P = q_z e_P \tag{5}$$

I_y is the axial moment of inertia of the cross-section area with respect to the enforced axis y (B7a) with the effective breadth b_{pe}, I_ω is the sectorial moment of inertia of the cross-section area with respect to the enforced pole P and I_t is the torsional moment of inertia of the cross-section (B8), both with the actual breadth b_p; E and G are the modulus of elasticity and shear modulus, respectively.

Distortion of the stiffener cross-section can be included by a reactive moment per unit length with respect to the enforced pole P

$$m_{Pc} = k_\alpha \alpha \tag{6}$$

where k_α is the stiffness coefficient, also defined as

$$m_{Pc} = \hat{F}_c h \tag{7}$$

where $\hat{F}_c$ is the reactive force per unit length (Fig. 2b). Thus, for a frame assembled of the portion the stiffener and plating, for the unit length, one has

$$\alpha = \frac{-v}{h} = \frac{\hat{F}_c h^2}{3EI_\xi^{(w)}} \left(1 + \frac{3EI_\xi^{(w)}}{h} C_\beta \right) \tag{8}$$

where

$$C_\beta = \frac{b_p}{12EI_\xi^{(p)}}, \; I_\xi^{(p)} = \frac{t_p^3}{12}, \; I_\xi^{(w)} = \frac{t_w^3}{12}$$

Thus, from (6), (7) and (8), one has

$$k_\alpha = \frac{Et_p^3}{b_p\left[1+4\frac{h}{b_p}\left(\frac{t_p}{t_w}\right)^3\right]} \tag{9}$$

The second equation of (4) may then be rewritten as follows

$$EI_\omega \frac{d^4\alpha}{dx^4} - GI_t \frac{d^2\alpha}{dx^4} = m_P - m_{Pc} \tag{10}$$

i.e., taken into account (6) and (9)

$$EI_\omega \frac{d^4\alpha}{dx^4} - GI_t \frac{d^2\alpha}{dx^2} + k_\alpha \alpha = m_P \tag{11}$$

The normal stress is

$$\sigma_x = \frac{M_y}{I_y} z + \frac{B}{I_\omega} \omega \tag{12}$$

where $M_y = M_y(x)$ is the bending moment and $B = B(x)$ is the bimoment

$$M_y = -EI_y \frac{d^2w}{dx^2}, \quad B = -EI_\omega \frac{d^2\alpha}{dx^2} \tag{13}$$

and z is the rectangular coordinate, i.e. the centroid axis z, and ω is the sectorial coordinate with respect to the enforced pole P (B3).

3 STRESS NON-UNIFORMITY FACTOR

The maximum normal stress occurs at the junction of the stiffener web and flange, which can be expressed, according to (12) and Appendix B, as

$$\sigma_{x1} = \frac{M_y}{I_y} h_C + \frac{B}{I_\omega} h e_P^* \tag{14}$$

So,

$$\sigma_{x,\max} = \sigma_{x1} = \Psi \sigma_x, \quad \Psi \geq 1 \tag{15}$$

where

$$\sigma_x = \frac{M_y}{I_y} h_C \tag{16}$$

as in case of bending a stiffener with symmetric cross-section, and

$$\Psi = 1 + \frac{B}{M_y} \frac{I_y}{I_\omega} \frac{h e_P^*}{h_C} \tag{17}$$

is the factor of the normal stress non-uniformity due to the cross-section asymmetry, where h_c is the distance of the flange from the cross-section centroid. For instance, for the stiffeners with clamped ends at the mid-span where

$$w = 0, \quad \frac{dw}{dx} = 0, \quad \alpha = 0, \quad \frac{d\alpha}{dx} = 0,$$

$$M_y = \frac{q_z l^2}{24}, \quad B = \frac{m_P l^2}{24} \chi_1$$

one has

$$\Psi = 1 + \frac{I_y}{I_\omega} \frac{h e_P^* e_P}{h_C} \chi_1 \tag{18}$$

where χ_1 is the function by which the cross-section torsion with distortion is taking into account (B12, 13).

4 DISPLACEMENT CORRELATION FACTOR

For the horizontal flange displacement it can be written

$$v = \Phi w, \quad \Phi \geq 0 \tag{19}$$

where $w = w(x)$ is the vertical cross-section displacement, according to elementary beam bending theory, and Φ is the displacement correlation factor ($\Phi = 0$ for symmetrical cross-sections).

For the clamped ends and uniform normal plate pressure at the midspan

$$w = \frac{q_z l^4}{384 EI_y}, \quad v = -\alpha h = -\frac{m_P l^4 h}{384 EI_\omega} \varphi_1 \tag{20}$$

where $\varphi_1 = \varphi_1(u,v)$ is the torsion-distortion function (B12, 13). Thus

$$\Phi = -\frac{I_y h e_P}{I_\omega} \varphi_1 \tag{21}$$

5 COMPARISON TO THE FINITE ELEMENT METHOD

The effective flange breadth is calculated for the structure where

$h = 20.3$ cm, $b_p = 73.9$ cm, $b_f = 10.2$ cm,

$t_p = t_w = t_f = 1.3$ cm

for two panel lengths: a) $l = 2.743$ m ($l/b_p = 3.712$) and b) $l = 3.695$ m ($l/b_p = 5$). Then, according to Appendix B,

$A_f = 13.26$ cm^2, $A_w = 26.39$ cm^2, $A'_p = 96.07$ cm^2,

$h'_C = 16.34$ cm, $e_P = 3.22$ cm, $C = -2.632$ cm,

$\overset{*}{e}_P = 3.09$ cm, $I'_y = 6964$ cm^4, $I_\omega = 103338$ cm^4;

a) $b_{pe} = 0.582$ m, $A_p = 75.71$ cm^2,
$h_C = 15.64$ cm, $I_y = 6589$ cm^4;

b) $b_{pe} = 0.644$ m, $A_p = 83.67$ cm^2,
$h_C = 15.94$ cm, $I_y = 6751$ cm^4.

The 3D model of one half of the stiffened panel is analysed with 81378 elements and 245597 nodes for the case a), and 49815 elements and 150502 nodes for the case b), using eight nodded shell elements, utilizing FEM software ADINA. The boundary conditions are applied according to Figure 3.

The applied plate pressure is 50 kN/m^2. Magnified deformed shape of the stiffened panel is shown in the bottom of the Figure 3.

Comparisons of the normal stress distributions along the cross-section contour according the analytic solutions (12) and FEM are presented in Figures 4 and 5 with the same scale for stiffener dimensions (cm) and stresses (MPa).

The results of calculations for the non-uniformity factor Ψ by equation (18), compared to the finite element analysis of the whole panel using equation (15), where the stress $\sigma_x = \sigma_x(x,z)$ is the analytical solution by simple bending theory, are presented in Table 1.

From Table 1, it can be concluded that analytical solution for the maximum stress by simple bending theory underestimates FEM solution (26% for short panel lengths, $l/b_p = 3.712$). On the other hand, for the long panel lengths ($l/b_p = 5$), the difference between these solutions is 8%. Moreover, analytical solutions obtained using the theory of torsion of thin-walled beams on elastic foundation (14) show very good agreement with FEM solutions. In this case, the difference is 3.3% for short lengths and only 0.9% for long panel lengths.

The results for the displacement correlation factor Φ obtained by equation (21), compared to the finite element analysis using equation (19), where the displacement $w = w(x)$ is the analytical solution by simple bending theory, are given in Table 2.

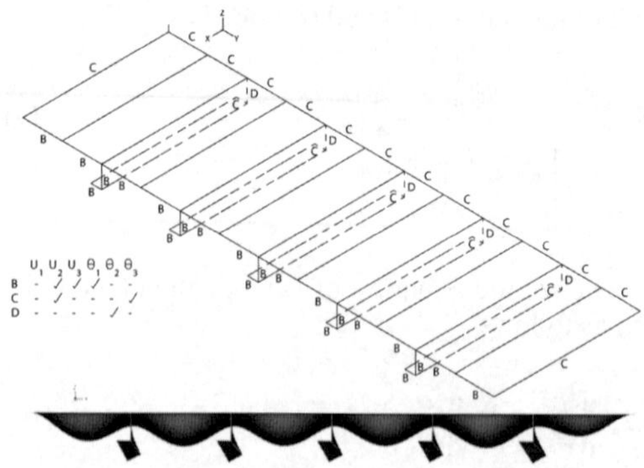

Figure 3. One half of the panel with boundary conditions.

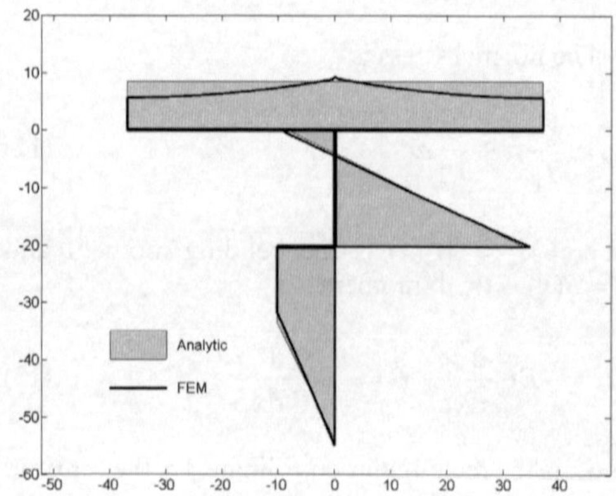

Figure 4. Normal stress distributions along the cross-section contour by analytic solution and FEM for $l/b_p = 3.712$.

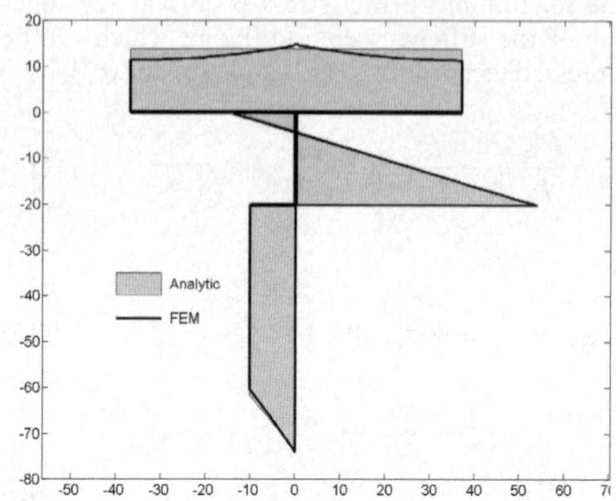

Figure 5. Normal stress distributions along the cross-section contour by analytic solution and FEM for $l/b_p = 5.0$.

Table 1. Normal stress factor Ψ according to the analytic solution (18) and FEM.

	Analytic	FEM	Rel.error
l/b_p	–	–	%
3.712	1.21	1.26	3.3
5.000	1.07	1.08	0.9

Table 2. Displacement factor Φ according to the analytic solution (21) and FEM.

	Analytic	FEM	Rel.error
l/b_p	–	–	%
3.712	1.33	1.42	7.1
5.000	0.58	0.59	2.6

6 CONCLUSION

In stress analysis of the stiffeners of plated structures (as bottom longitudinals, bulkhead stiffeners, side frames, etc.) with L cross-sections, as thin-walled beams of open cross-section subjected to bending and torsion under uniform plate pressure, the cross section distortion must be taken into account. The stiffener bending may be analysed according to the elementary theory of bending (as in case of symmetrical sections). The stiffener torsion may be analysed by using the theory of torsion of thin-walled beams on elastic foundation. By the elastic foundation, the cross-section distortion due to the section asymmetry is taken into account.

The results are proved by the finite element model. The stiffener, as part of the actual ship structures, was modeled by shell finite elements. The comparison of the stiffener maximum normal stresses has shown, for the purpose of preliminary calculations, and an acceptable agreement of the results is obtained, both for short and particularly for long panel lengths.

REFERENCES

ADINA. 1994–2007. Theory and Modeling Guide V.8.4.2, ADINA R&D Inc.

Gjelsvik, A. 1981. *The theory of thin-walled bars*. New York: John Whiley and Sons.

Hu, Y., Chen, B. & Sun, J. 2000. Tripping of thin-walled stiffeners in the axially compressed stiffened panel with lateral pressure. *Thin-Walled Structures* 37:1–26.

Hughes, O.F. 1983. *Ship structural design*. New York: John Wiley and Sons.

Kollbruner, C.F. & Basler, K. 1969. *Torsion in structures*. Berlin: Springer.

Pavazza, R. 1991. Bending and torsion of thin-walled beams on elastic foundation (in Croatian), Thesis, Faculty of Mechanical Engineering and Naval Architecture, University of Zagreb.

Pavazza, R. 2000. An approximate solution for thin rectangular orthotropic/isotropic strips under tension by line loads, *International Journal of Solids and Structures*. 37: 4353–4375.

Pavazza, R., Plazibat, B. & Matoković, A. 2001. On the effective breadth problem of deck plating of ships with longitudinal bulkheads. *International Shipbuilding Progress* 48(1): 51–85.

Pavazza, R. 2007. Introduction to the analysis of thin-walled beams (in Croatian). Zagreb: Kigen.

Pavazza, R., Matoković A. & Vlak, F. 2009. The Effective Breadth Concept for Non-Symmetric Stiffeners. *6th International Congress of Croatian Society of Mechanics; Proc. intern. symp., Dubrovnik, September 30-October 2 2009*. Zagreb: CSM.

Vlasov, V.Z. 1959. *Thin-walled Elastic beams* (in Russian). Moscow: Fizmatgiz.

APPENDIX A

Since the stiffness of the panel in its plane is assumed to be too large, the stiffener bending with respect to the z-axis may be ignored. In that case, the equilibrium of the moments with respect to the z-axis

$$M_z = -\int_A \sigma_x y \mathrm{d}A = 0 \tag{A1}$$

is satisfied if

$$\frac{M_y}{I_y}\int_A zy\mathrm{d}A + \frac{B}{I_\omega}\int_A z\omega\mathrm{d}A = 0$$

i.e.

$$I_{yz} = \int_A zy\mathrm{d}A = 0,\; I_{z\omega} = \int_A z\omega\mathrm{d}A = 0 \tag{A2}$$

i.e. when the centrifugal moments of inertia are equal to zero. This is the case when the cross-section has at least one axis of symmetry, i.e., approximately, when the flange cross-section area is small with respect to the plate cross-section area.

The equilibrium with respect to the x-axis

$$N = \int_A \sigma_x \mathrm{d}A = 0 \tag{A3}$$

is satisfied if

$$\frac{M_y}{I_y}\int_A z\mathrm{d}A + \frac{B}{I_\omega}\int_A \omega\mathrm{d}A = 0 \tag{A4}$$

i.e.

$$S_y = \int_A z\,\mathrm{d}A = 0,\ S_{z\omega} = \int_A \omega\mathrm{d}A = 0 \tag{A5}$$

This is the case when the y-axis is the centroid axis and when ω-coordinate is obtained for the enforced pole P, and corrected to satisfy the sectorial static moment of the cross-section area, given by equation (B3).

APPENDIX B

Centroid coordinate:

$$h_C = h\frac{A_p + \frac{A_w}{2}}{A_w + A_f + A_P},\ A_w = ht_w,\ A_f = b_f t_f,$$

$$A_p = b_{pe}t_p,$$

$$h'_C = h\frac{A'_p + \frac{A_w}{2}}{A_w + A_f + A'_P},\ A'_p = b_p t_p \tag{B1}$$

Distance of the pole P from the web (Pavazza 1991, Pavazza 2007):

$$\Delta_y = \frac{I_{y\omega_1}}{I'_y} = -\frac{hh'_C b_f A_f}{2I'_y} = -e_P,\ I_{y\omega_1} = \int_{A'} z\omega_1\mathrm{d}A \tag{B2}$$

where ω_1 is the sectorial coordinate for the pole P_1.

Corrected sectorial coordinate for the pole P:

$$\omega = \omega_2 + C,\ S_{\omega 2} = \int_{A'} \omega_2\,\mathrm{d}A \tag{B3}$$

where ω_2 is the sectorial coordinate for the pole P.

$$C = -\frac{S_{\omega 2}}{A'} = -\frac{hA_f b_f}{2\left(A_w + A_f + A'_P\right)}\left[\left(2 + \frac{A_w}{A_f}\right)\frac{e_P}{b_f} - 1\right]$$

Corrected distance of the pole P from the web:

$$e^*_P = \left(1 + \frac{C}{he_P}\right)e_P \tag{B4}$$

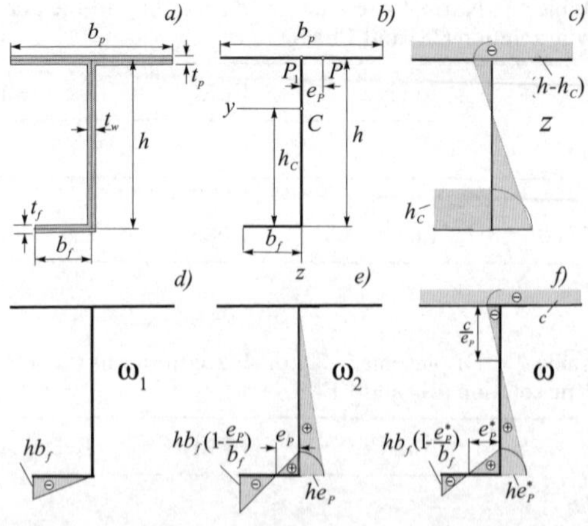

Figure B1. Cross-section sectorial properties: a) cross-section properties; b) contour with enforced centroid axis and poles; c) rectangular z-coordinate; d) sectorial coordinate for the pole P_1; e) sectorial coordinate for the pole P; f) corrected sectorial coordinate ω.

Axial moment of inertia:

$$I_y = \int_A z^2\mathrm{d}A \tag{B5}$$

Sectorial moment of inertia:

$$I_\omega = \int_{A'} \omega^2\mathrm{d}A \tag{B6}$$

Here

$$I_y = h^2\frac{A_f A_p + \frac{A_w}{3}\left(A_f + A_p + \frac{A_w}{4}\right)}{A_w + A_f + A_p},$$

$$I'_y = h^2\frac{A_f A'_p + \frac{A_w}{3}\left(A_f + A'_p + \frac{A_w}{4}\right)}{A_w + A_f + A'_p} \tag{B7}$$

$$I_\omega = \frac{A_f b_f^2 h^2}{3}\left\{\left(1 - \frac{e^*_P}{b_f}\right)^3 + \left(\frac{e^*_P}{b_f}\right)^2\left[\frac{e^*_P}{b_f} + \frac{A_w}{A_f}\left(1 - \frac{|C|}{he_P}\right)\right]\right\}$$
$$+ A'_p C^2\left(1 + \frac{t_w|C|}{3e_P A'_p}\right),$$

$$I_t = \frac{1}{3}\left(b_p t_p^3 + h_w t_w^3 + b_f t_f^3\right) \tag{B8}$$

$$\omega_{f1} = he_P^*,\ \omega_{f2} = -hb_f\left(1 - \frac{e_P^*}{b_f}\right),\ \omega_p = -|C| \quad \text{(B9)}$$

The effective breadth of the attached plating for built in ends approximately is (Pavazza 2000, Pavazza et al. 2001):

$$b_{pe} = \frac{1}{1 + \frac{4}{3}\left(\frac{b_p}{l_0}\right)^2} b_p,\ l_0 \approx 0{,}6l \quad \text{(B10)}$$

Functions χ_1 and φ_1 (Pavazza 1991) are

$$\chi_1 = \frac{6(\mu \text{ch}\mu \sin\nu - \nu \text{sh}\mu \cos\nu)}{u^2(\nu sh2\mu + \mu \sin 2\nu)},$$
$$\varphi_1 = \frac{6}{u^4}\left[1 - \frac{2(\mu \text{ch}\mu \sin\nu + \nu \text{sh}\mu \cos\nu)}{\nu \text{sh}2\mu + \mu \sin 2\nu}\right], \quad \text{(B11)}$$

for $\nu^4 < 16u^4$ with

$$\mu = u\sqrt{2}\cos\frac{\theta}{2},\ \nu = u\sqrt{2}\sin\frac{\theta}{2},\ \theta = Ar\cos\left(\frac{\nu}{2u}\right)^2,$$

and for $\nu^4 > 16u^4$

$$\chi_1 = \frac{6(\omega \text{sh}\lambda - \lambda \text{sh}\omega)}{\lambda\omega(\lambda \text{sh}\lambda \text{ch}\omega - \omega \text{ch}\lambda \text{sh}\omega)},$$
$$\varphi_1 = \frac{24}{\lambda^2\omega^2}\left[1 - \frac{\lambda \text{sh}\lambda - \omega \text{sh}\omega}{\lambda \text{sh}\lambda \text{ch}\omega - \omega \text{sh}\omega \text{ch}\lambda}\right] \quad \text{(B12)}$$

with

$$\lambda = \frac{2}{\sqrt{2}}\sqrt{1 + \sqrt{1-m}},\ \omega = \frac{2}{\sqrt{2}}\sqrt{1 - \sqrt{1-m}},\ m = \left(\frac{2u}{\nu}\right)^4$$

where

$$u = \frac{l}{2}\sqrt[4]{\frac{k_\alpha}{4EI_\omega}},\ \nu = \frac{l}{2}\sqrt{\frac{GI_t}{EI_\omega}} \quad \text{(B13)}$$

Fatigue reliability and rational inspection planning

Advanced Ship Design for Pollution Prevention – Guedes Soares & Parunov (eds)
 ISBN 978-0-415-58477-7

Risk based maintenance of deteriorated ship structures accounting for historical data

Y. Garbatov & C. Guedes Soares
Centre for Marine Technology and Engineering (CENTEC), Technical University of Lisbon, Instituto Superior Técnico, Lisboa, Portugal

ABSTRACT: Maintenance, repair and reliability analyses of tanker and bulk carrier structures subjected to general corrosion are presented. The approach adopts the Weibull model for analysing the failure data. Based on historical data of thickness measurements or corresponding corrosion wastage thickness of structural components in tanker and bulk carriers and the progress of corrosion, critical failure levels are defined. The analysis demonstrates how data can be used to address important issues as the inspection intervals, condition based maintenance action and structural component replacement. It is also demonstrated how to establish practical decisions about when to perform maintenance on structure that will reach a failed state. Different scenarios are analysed and inspection intervals are proposed.

1 INTRODUCTION

Marine structures operate in a complex environment. Water properties such as salinity, temperature, oxygen content, pH level and chemical composition can vary according to location and water depth. Time spent in ballast or cargo, tank washing and inerting (for tankers), corrosion protection effectiveness and component location and orientation have a significant effect on the corrosion. Some types of corrosive attack on metals may be defined as general corrosion, galvanic cells, under deposit corrosion, CO_2 corrosion, top-of-line corrosion, weld attack, erosion corrosion, corrosion fatigue, pitting corrosion, microbiological corrosion and stress corrosion cracking.

Besides the already listed factors some additional ones have been pointed as affecting the corrosion wastage of steel structures such as environment in Melchers, (2003d,e), for corrosion morphological changes in Montero-Ocampo and Veleva, (2002), stress concentrations in Garbatov et al., (2002) and Kobayoshi et al., (1998) and steel surface preparation in Melchers, (2003c). Recently a study of factors governing marine corrosion on the structural steel component level and an identification of the key parameters of corrosion and corrosion fatigue of ballast, oil tanks and cargo holds has been presented by Panayotova et al., (2004a,b).

Two main corrosion mechanisms are present in steel plates, general wastage that results in a generalized decrease of plate thickness and pitting, which consists of much localized corrosion with deep holes appearing in the plate.

When different metals are exposed to a corrosive medium while in metallic contact with each other, an electric cell is formed. The current increases the corrosion rate of the least noble metal, while the nobler one corrodes less and supports most of the cathodic reaction. This is the basis of corrosion protection with sacrificial anodes. Water pollution, in harbours results in the increase of corrosion rate. The water may be more aggressive because of greater concentration of ammonia or sulphides, which is combined with lower oxygen.

The structures are often protected, either with paints or with cathodic systems that deliver a current intensity to the protected metal surface inhibiting the corrosion process. The life of coating typically depends on the coating systems used, details of its application (for example, surface preparation, film thickness, humidity and salt control during application, etc.) of the applied maintenance and other factors.

Corrosion of interior spaces in ship structures has an important role in the long-term structural integrity. Under conditions of high temperature, inappropriate ventilation, high stress concentration, high stress cycling, high rates of corrosion can be achieved in spaces such as ballast tanks and at specific structural details such as horizontal stringers or longitudinal and web frames. Depending on the location of the ship structural elements many results of measurements of corrosion rates may be also found in Purlee, (1965), TSCF, (1992, 1997), Loseth et al., (1994), Yamamoto and Ikagaki, (1998), Paik et al., (1998), Wang et al., (2003a,c), Garbatov et al., (2004) and Garbatov and Guedes Soares, (2008a, 2009a).

Since the of corrosion and wear of structural members are the consequences of complex phenomena governed by many factors, it is necessary to establish corrosion margins and permissible corrosion levels by considering past records. An average annual corrosion rate obtained by dividing the thickness reduction of an aged and worn member by a ship's age at a given time has conventionally been used as the basic criteria, because of easy of assessing and handling, but more rational criteria, assessing it with a probabilistic model are needed.

The second can be based on the results of experiments in specific conditions which suggest laws of growth of corrosion as a function of specific parameters. The corrosion model can be developed by considering all those laws derived from experiments in specific conditions as is being presented by Melchers, (2003a). This approach involves one difficulty in generalizing results from laboratory tests to full-scale conditions. The other difficulty is related to the general lack of data on the environmental conditions which affect corrosion in full-scale.

The third approach, which is the one that is adopted by Guedes Soares and Garbatov, (1998b, 1999a), is to consider that a model should provide the trend that is derived from the dominating mechanism and then it should be fit to the field data. So the model does not represent all corrosion mechanisms as Melchers, (2003b) is pursuing, but only describes a dominant one. The parameters of the dominant mechanism are not derived from experimental work as does Melchers, (2003b), but are fit to full-scale data. By fitting to full-scale data compensates for the potential errors the omission of less important corrosion mechanisms may have. Although the model adopted does not represent the details of all corrosion mechanisms that may develop in time it represents the main trend and by being fit the field data it makes it practical, avoiding any danger of using models that are outside the range of full-scale data.

The large number of parameters that can affect corrosion demonstrates the difficulty of developing a model of corrosion wastage that explicitly considers them. Therefore, the estimation of the corrosion depth and rate needs at the present stage to have an empirical component and to be much based on the historical data collected for a certain ship.

Garbatov and Guedes Soares, (2008b, 2009a), analysed different databases on corrosion wastage to asses the adequacy of a non-linear corrosion wastage model to represent practical situations by adjusting the model parameters to the data of specific situations. An effort is made to establish realistic corrosion rates for ship hull structural components as a function of different locations on the ship hull.

The effect of the different marine atmospheric factors on the behaviour of corrosion in crude oil tanks are analysed by Guedes Soares et al., (2008), in marine atmosphere by Guedes Soares et al., (2009b) and for immersion corrosion by Guedes Soares et al., (2009a), all over the ship's service life. The works consider that the nominal corrosion model is defined based on nominal conditions, and then it is corrected to represent the corrosion behaviour at the actual conditions. The non-linear time variant corrosion model of Guedes Soares and Garbatov, (1998b, 1999a) is used as a nominal corrosion model and the effect of different environmental factors is accounted for. Correction function to each of these factors is developed based on the results published from experimental work of various authors.

Structures degrade with time and need to be inspected and maintained. Decisions about when to perform maintenance on a structure that is subjected to deterioration need information about when the structure will reach a state that requires maintenance, which can be considered a "failure state" according to a maintenance criterion. It is never known exactly when the transition of the structure from a good to a failed state will occur, but it is usually possible to obtain information about the probability of this transition occurring at any particular time. For risk based maintenance decisions, knowledge of probabilistic models related to degradation is needed (Itagaki and Yamamoto, 1977, Fujita et al., 1989, Lotsberg and Kirkemo, 1989, Fujimoto and Swilem, 1992, Madsen, 1997, Faber et al., 1996 and Garbatov and Guedes Soares, 2001, 2009c).

For ship structures the main degradation are fatigue cracks and corrosion (Guedes Soares and Garbatov, 1996a, b). Corrosion and wear of structural members are the result of complex phenomena governed by many factors. The large number of parameters that can affect corrosion demonstrates the difficulty of developing a model of corrosion thickness wastage (sometimes designated as corrosion depth or corrosion wastage) that explicitly considers them. Therefore, the estimation of the corrosion depth needs at the present stage to have an empirical component much based on the historical data collected for a certain ship.

Planning of structural maintenance of ships have been done based on structural reliability approaches (Guedes Soares and Garbatov, 1998a) involving models that represent the time development of corrosion deterioration have been proposed as for example the one of Guedes Soares and Garbatov, (1999a), which was recently calibrated with full-scale data in Garbatov and Guedes Soares, (2007).

This model is able to describe an initial period without corrosion because of the presence of a corrosion protection system, a transition period with a nonlinear increase of corrosion depth up to a steady state of long-term corrosion depth. This model combined with models of probability of detection during inspections is the basis of reliability based maintenance planning of marine structures that have been proposed by Guedes Soares and Garbatov, (1996a).

The model has also been recently applied in Garbatov and Guedes Soares, (2007) to analyse the reliability of a bulk carrier hull subjected to the degrading effect of corrosion. The effect of maintenance actions was modelled as a stochastic process and different repair policies including the ones adapted by the new IACS common structural rules were analysed.

Earlier approaches were based on using structural reliability theory combined with models of corrosion growth with time. The approach in this paper is based on statistical analysis of corrosion depth data leading to probabilistic models of time to failure, which are used as basis for maintenance decisions.

Classical theory of system maintenance describes the failure of components by probabilistic models often of the Weibull family, which represents failure rates in operational phases and in the ageing phases of the life of components as described in various textbooks, such as Moubray, (1997), Rausand, (1998) and Jardine and Tsang, (2005).

Garbatov and Guedes Soares, (2009a, b, c) adopted that type of approaches and demonstrated how they can be applied to structural maintenance of ships that are subjected to corrosion. The approach applied is based on historical data of thickness measurements or corresponding corrosion thickness in ships. Based on the progress of corrosion, critical corrosion thickness levels are defined as "failure", which is modelled by a Weibull distribution. Existing formulations obtained for systems are applied to this case, leading to results that agree with standard practice.

Several sets of corrosion data of structural components of tankers and bulk carriers are analysed here. The analysis demonstrates how this data can be used to address important issues such as inspection intervals, condition based maintenance actions and structural component replacement. An effort is made to establish practical decisions about when to perform maintenance on a structure that will reach a failed (corroded) state. Different scenarios are analysed and optimum interval and age are proposed.

The optimum age and intervals are based on statistical analysis of thickness data using the Weibull model and some assumptions about the inspection and the time required for repair in the case of failure are considered here. The present analysis applies the general framework that was developed for failure of components in a system and adopts it to the corrosion deterioration problem by considering the different corrosion thicknesses as failure criteria.

2 GENERAL CORROSION MODELS

The conventional models for general corrosion wastage presented for example by Guedes Soares, (1988) and Shi, (1993), assumed a constant corrosion rate, leading to a linear relationship between the corrosion thickness and time. Experimental evidence of corrosion, reported by various authors, shows that a non-linear model is more appropriate. Southwell et al., (1979) have observed the wastage thickness increases non-linearly in 2–5 years of exposure, but afterwards it becomes constant. This means that after initial non-linear corrosion, the oxidized material that is produced remains on the surface of the plate and does not allow the continued contact of the plate surface with the corrosive environment stopping corrosion. They proposed a linear and a bilinear model, which is considered appropriate for design purposes. Both models are conservative in the early stages, overestimating the corrosion wastage.

Yamamoto and Ikegaki, (1998) proposed a corrosion model based on a pitting corrosion and plate thickness measurements data. In this model corrosion and wear seen in structural members are assumed to be the consequence of many generated progressive pitting points growing individually. Melchers, (1998) suggested a steady state model for corrosion wastage thickness and he proposed a power approximation for the corrosion wastage. Yamamoto, (1998) has presented results of the analysis of corrosion wastage in different locations of many ships, exhibiting the non-linear dependence of time with the tendency of levelling off. Paik et al., (1998, 2003) suggested that corrosion behaviour could be categorized into three phases. The coating life is assumed to follow the log-normal distribution. The transition time is considered an exponentially distributed random variable.

Guedes Soares and Garbatov, (1998b, 1999b), proposed a model for the non-linear time-dependent function of general corrosion wastage. This time-dependent model separates corrosion degradation into three phases. In the first one there is no corrosion because the protection of the metal surface works properly. The second phase is initiated when the corrosion protection is damaged and corresponds to the start of corrosion, which decreases the thickness of the plate. The third phase corresponds to a stop in the corrosion process and the corrosion rate becomes zero.

Corroded material stays on the plate surface, protecting it from contacting the corrosive environment and the corrosion process stops. Cleaning the surface or any involuntary action that removes that surface material originates the new start of the non-linear corrosion growth process. An investigation of the effect of the different parameters describing above model has been presented by Garbatov et al., (2004) and Garbatov et al., (2007).

A probabilistic model developed by Gardiner and Melchers, (2001), Melchers, (2003a) divides the corrosion process into four stages: initial corrosion; oxygen diffusion controlled by corrosion products and micro organic growth; limitation on food supply for aerobic activity and anaerobic activity. The model consists of phases; each represents a different corrosion controlling process. For more detailed discussion about each phase and the influence of the different environmental factors refer to Melchers, (1999, 2001, 2003b).

Qin and Cui, (2002) proposed a model in which the coating protection system (CPS) such as coating was assumed to deteriorate gradually and the corrosion may start as pitting corrosions before the CPS loses its complete effectiveness. The corrosion rate was defined by equating the volume of pitting corrosion to uniform corrosion.

It has been recognized that corrosion is a complex phenomenon and influenced by many factors. Identifying key issues that can lead to corrosion cannot be achieved through only statistical investigations of corroded ageing ships. There is a need to develop models based on the corrosion mechanisms and to combine them with the corrosion wastage databases to achieve a better understanding and more proper prediction of corrosion in marine structures.

This has been recognized and a new corrosion wastage model was proposed, based on a non-linear time-dependent corrosion model accounting for various immersion environmental factors, including the effects of salinity, temperature, dissolved oxygen, pH and flow velocity including the effect of ships service life in different routes by Guedes Soares et al., (2008, 2009a, b).

While that model will allow changes in the environmental factors to be reflected in the corrosion rates, the historical data available for this paper does not contain that information. Therefore, the model of Guedes Soares and Garbatov, (1998b, 1999b), will be adopted to be adjusted to the full-scale data.

2.1 *Nonlinear corrosion wastage model fit to database of ship hull structures of bulk carriers*

Garbatov et al., (2004) presented the result of a survey of corrosion data on deck and hatch cover plates of three 38,000DW bulk carriers. The corrosion data consists of 2,700 measurements of corrosion thickness collected from four inspections at 16th, 19th, 20th and 24th years of service life of ships. The design thickness of measured deck and hatch cover plates varies between 10 and 30 mm.

The data is fit to the non-linear time domain model by estimating the model parameters. The model is based on the solution of a differential equation of the corrosion wastage (Guedes Soares and Garbatov, 1998b, 1999b):

$$\text{MeanValue}\left[d(t)\right]=\begin{cases} d_\infty\left(1-e^{-\frac{t-\tau_c}{\tau_t}}\right), & t\ge\tau_c \\ 0, & t<\tau_c \end{cases} \quad (1)$$

$$\text{StDeviation}\left[d(t)\right]=\begin{cases} aLog(t-\tau_c-b)-c, & t\ge\tau_c \\ 0, & t<\tau_c \end{cases} \quad (2)$$

where d_∞ is the long-term corrosion wastage, $d(t)$ is the corrosion wastage at time t, τ_c is the time without corrosion which corresponds to the start of failure of the corrosion protection coating (when there is one), τ_t is the transition time duration and a, b and c are coefficients.

The long-term wastage is defined as an extreme value in the service time interval for deck and hatch cover plates. The coating life τ_c, and transition time τ_t, are defined based on the least squares approach and quasi-Newton algorithm to determine the direction to search used at each iteration.

The descriptors of the regression of the corrosion depth as a function of time for the long-term corrosion wastage for hatch cover plates is $d_\infty = 0.99$ mm, for deck plate is $d_\infty = 2.21$ mm and for plates between hatch covers is $d_\infty = 1.98$ mm respectively. The coating life of hatch cover plates is $\tau_c = 15.05$ years, for deck plates is $\tau_c = 13.35$ years and for plates between hatch covers is $\tau_c = 13.56$ years. The transition period of deck plates is $\tau_t = 5.51$ years, for plates between hatch covers is $\tau_t = 4.3$ years and for hatch cover plates $\tau_t = 3.27$ years.

There is some variability of the data around the regressed line, which is expected because of the large variability of factors. However the regression lines show a clear non-linear trend of the type that is reproduced by the non-linear model. To define the probability density function of the corrosion wastage depth, the observed distribution is fit by a theoretical distribution by comparing the frequencies observed in the data to the expected frequencies of the theoretical distribution and for that purpose the Kolmogorov-Smirnov test of fit has been applied. Several distributions were evaluated and it was concluded that corrosion wastage depth is the best fit by the Log-normal distribution.

Fourteen sets of corrosion data of structural components of balk carriers, partially included in the study by Garbatov and Guedes Soares, (2008a) and latter updated in Garbatov and Guedes Soares, (2009a) are analysed here. These sets cover bottom (1), inner bottom (2), below top of bilge—hopper tank-face (3), lower slopping (4), lower wing tank —side shell (5), below top of bilge—hopper tank-web (6), between top of bilge, hopper tank, face (7), between top of bilge, hopper tank, web (8), side shell (9), upper than bottom of top side tank, face (10), upper deck (11), upper slopping (12), upper wing tank side shell (13), upper than bottom of top side tank, web (14) in 8832 measurements. Using regression analysis, the mean value and standard deviation of corrosion wastage as a function of time are fitted.

The probability density function of corrosion wastage as a function of time for bottom plates (1) is shown in Figure 1, the mean value and the standard deviation of corrosion wastage regression equation descriptors for all sets of data are given in Table 1.

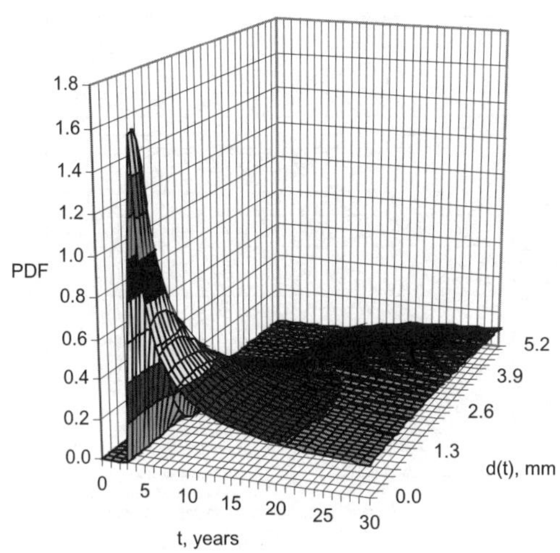

Figure 1. Probability density function of corrosion wastage of bottom plates.

Table 1. Descriptors of Eqns (1) and (2).

	d_∞ mm	τ_t years	τ_c years	a –	b years	c mm
1	1.42	14.14	3.17	7.16	−11.60	7.41
2	3.50	18.29	0.00	1.47	−1.00	−0.23
3	4.68	3.51	9.11	2.08	−1.00	−0.46
4	2.75	19.75	0.00	16.09	−6.18	12.72
5	1.41	14.72	3.35	13.84	−9.16	13.22
6	3.55	3.17	8.73	11.24	−11.65	11.53
7	4.28	8.54	6.85	7.53	7.53	7.53
8	3.35	5.66	7.26	17.93	−7.75	15.70
9	2.25	25.46	0.00	1.91	−1.44	0.01
10	2.83	2.92	8.75	1.60	−1.00	−0.42
11	2.50	19.28	0.00	12.93	−9.83	12.62
12	1.19	15.67	0.69	12.23	12.23	12.23
13	1.58	20.56	0.00	9.88	−13.24	10.89
14	2.60	5.41	7.25	13.41	−16.42	15.72

2.2 *Nonlinear corrosion wastage model fit to database of ship hull structures of tankers*

Two sets of corrosion data, deck plates of ballast and cargo tanks of tankers described by Wang et al., (2003b, c) were analysed in Garbatov et al., (2007) and are summarized here. The first set includes 1,168 measurements of deck plates from ballast tanks with original nominal thicknesses varying from 13.5 to 35 mm on ships with lengths between perpendiculars in the range of 163.5 to 401 m. The second set of data includes 4,665 measurements of deck plates from cargo tank with original nominal thicknesses varying from 12.7 to 35 mm on ships with lengths between perpendiculars in the range of 163.5 to 401 m.

The parameters of the regressed line of corrosion depth as a function of time were determined under the assumption that it is approximated by the exponential function given in Eqn (1). It can be noted the long-term corrosion wastage for deck plates of ballast tanks is $d_{\infty,\text{ ballast}} = 1.85$ mm and $d_{\infty,\text{ cargo}} = 1.91$ mm for cargo tanks respectively. The time without corrosion is $\tau_{c,\text{ ballast}} = 10.54$ years in deck plates of ballast tanks and $\tau_{c,\text{ cargo}} = 11.494$ years for cargo tanks, respectively. Finally, the transition period for deck plates of ballast tanks is $\tau_{t,\text{ ballast}} = 11.14$ years and the one for deck plates of cargo tanks is $\tau_{t,\text{ cargo}} = 11.23$ years.

There is some variability, which is smaller for the deck plates of ballast tanks ($R^2 = 0.9$) than for the deck plates of cargo tanks ($R^2 = 0.85$). The large value of R^2 does not necessarily imply the model will provide accurate predictions of individual plate corrosion thickness as it models the yearly means values and the individual values show a large variability about that value.

Several distributions were evaluated and it was concluded that corrosion wastage depth is the best fit by the Log-normal distribution.

The mean value and the variance of the log-normal distribution for the corrosion wastage of deck plates of ballast tanks are –0.544 and 0.919 and for the corrosion wastage of deck plates of cargo tanks are –0.369 and 1.046 respectively.

3 ANALYSYS OF FAILURE DATA

A statistical parameter frequently used in replacement studies is the hazard rate, $h(t)$. An estimate of the hazard rate of a component at any point in time may be thought of as the ratio of a number of components that failed in an interval of time to the number of components in the original population that were operational at the start of the interval. Thus, the hazard rate of a component at time t is the probability that the component will fail in the next interval of time given that it is good at the start of the interval.

The hazard rate of the Weibull distribution $F(t)$, is:

$$h(t) = \frac{\beta}{\eta}\left(\frac{t-\gamma}{\eta}\right)^{\beta-1} \quad \text{when } t > \gamma, \text{ 0 otherwise.} \tag{3}$$

where β is a shape factor, η is a scale factor and γ is a positioning parameter. $h(t)$ varies with the independent variable t and in particular, when the shape factor is $\beta < 1$, $h(t)$ is a decreasing function of t. When $\beta = 1$, $h(t)$ does not vary with t, $h(t)$ becomes an increasing function of t when $\beta > 1$. When $t-\gamma = \eta$, $F(t)$ is approximately 63.2%, for all values of β. Thus, the scale factor η is also known as the characteristic life of the Weibull distribution.

The probability density function of the Weibull distribution, $f(t)$ is zero for $t < \gamma$. This results in no risk of failure before γ, which is therefore termed the location parameter or the failure-free period of the distribution. In practice, γ may be negative, in which case the component may have undergone a run-in process or it had been in use prior to $t = 0$.

A function complementary to the cumulative distribution function is the reliability function, also known as the survival function. It is determined from the probability that the structure will survive at least to some specified time, t. The reliability function is denoted by $R(t)$ and is defined as:

$$R(t) = \int_t^{\infty} f(t)\,dt = 1 - F(t) \tag{4}$$

Maintenance decision analysis requires the use of the failure time distribution of structural components, which may not be known. There may, however, be a set of observations of failure times available from historical records. One might wish to find the Weibull distribution that fits the observations, and to assess the goodness of the fit.

If data in the form of historical records are not available, a specific test or series of tests needs to be made to obtain a set of observations, that is, a sample data. A sample is characterized by its size and by the method by which it is selected. The purpose of obtaining the sample is to enable inferences to be drawn about properties of the population from which it is drawn.

If the Weibull plot is a curve when the location parameter is considered as zero, then a three-parameter Weibull distribution has to be used to model the data set. The curvature of the plot suggests the location parameter $\gamma > 0$. Obviously, γ must be less than or equal to the shortest failure time, t_i. Finding the correct value of γ will produce a linear plot. The probability density function of the fitted distributions with adjusted location parameter for the ballast tank set of data analysed here is shown in Figure 2.

The failure data can be analysed to estimate measures of reliability such as $R(t)$ from the Weibull plot. For more accuracy, a confidence interval can be determined on the estimation of $R(t)$. Suppose one wants to find a $(1-\alpha)$ confidence interval for $R(t)$, where α is the accepted risk that the interval found does not contain the true $R(t)$. For example, when a 95% confidence interval is established for

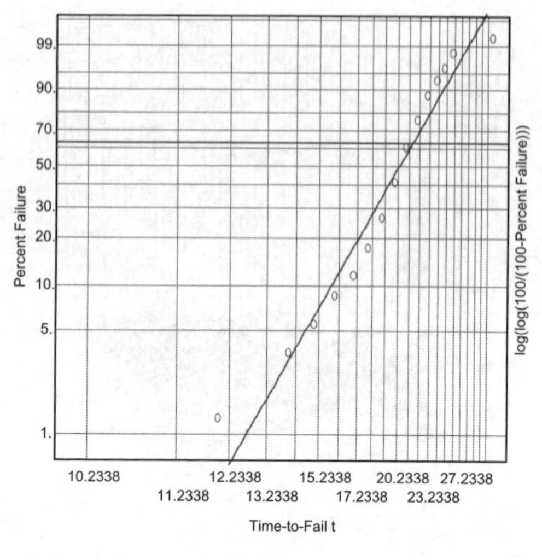

Figure 2. Weibull plot of ballast tank plate's failure data.

$R(t)$, it means that it is 95% confident the real $R(t)$ is contained in the confidence interval.

Reliability of ballast and cargo tanks plates subjected to corrosion with 95% confidence interval is shown in Figure 3.

Figure 2 has shown some underestimating of the failure prediction by the Weibull model. The lack of fitting is explained with the fact that the scatter of the collected data is high. The Weibull probability density function is the best possible fit for the present data. However, in the present figure all the measured corrosion depths are considered as failure, which is not the case that is studied in the followings analysis. A definition for the corrosion depth that provokes failure is given and in all the cases the data related to early corrosion is censored and the Weibull analysis is performed with the remaining data. This procedure, normally avoids the problem registered in Figure 2.

In practice, not every structural component is observed up to failure. When there is only partial information about a components lifetime, i.e. not all the tested components have failed, the information is known as censored or suspended data. Suspended data will not cause complications in the Weibull analysis if all of them are longer than the observed failure times. However, it is necessary to use a special procedure to handle them when some of the failure times are longer than one or more of the suspension times. In the latter case, suspended data are handled by assigning an average order number to each failure time.

The analysis presented here uses data sets of failure times of corroded structural components. Four different levels of censoring related to the failure state of corroded plates are introduced: low corrosion tolerance that is 2% corrosion thickness loss

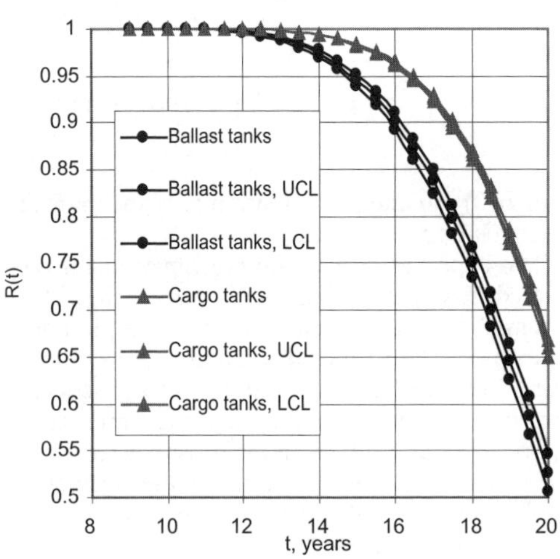

Figure 3. Reliability of ballast and cargo tank plates.

with respect to the initial plate thickness, moderate corrosion tolerance is 4%, high corrosion tolerance is 8% and extreme corrosion tolerance is 12% of the initial plate thickness respectively.

The corrosion thickness levels are set up here as permissible corrosion levels and any time at which the corrosion depth reaches them is classified as failure and others are censored (see Figure 4 and Figure 5).

The completed failure times are described by the Weibull distribution and its statistical descriptors for the different corrosion tolerance levels for corroded deck plates of ballast tank can be seen in

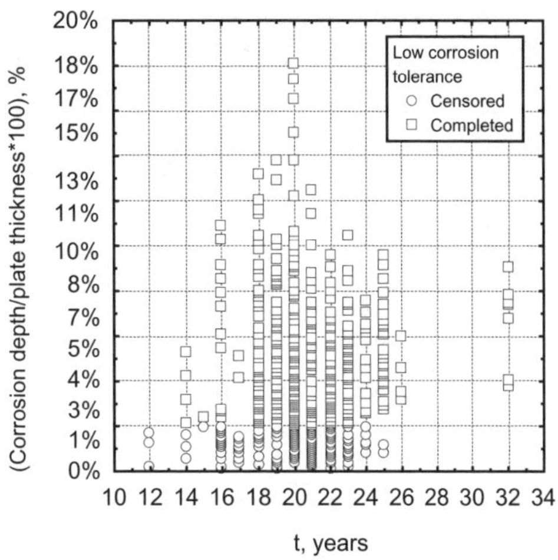

Figure 4. Censored data, 2% corrosion tolerance in ballast tank plates.

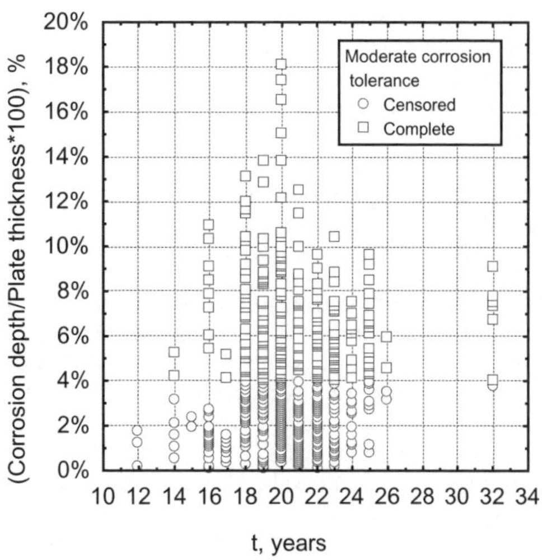

Figure 5. Censored data, 4% corrosion tolerance in ballast tank plates.

Table 2. Weibull distribution function parameters and statistics descriptors, ballast tanks.

	Corrosion tolerance			
Descriptors	Low	Moderate	High	Extreme
Shape factor	3.57	3.42	3.09	3.09
Scale factor	10.96	12.92	21.33	36.36
Location parameter	11.99	11.99	11.99	11.99
Mean value	20.84	20.86	20.04	19.65
St. Deviation	2.71	2.77	2.38	0.79
No of completed	704	413	93	17

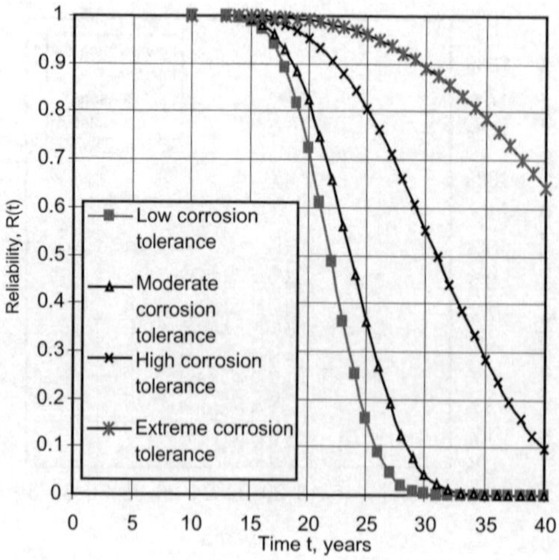

Figure 6. Reliability of ballast tanks plates.

Table 2. The reliability estimates for the different levels of corrosion tolerance limits defined here as low, moderate, high and extreme corrosion tolerance are given in Figure 6.

4 MINIMIZATION OF TOTAL COST

4.1 *Optimal replacement interval*

Structural components are subjected to corrosion and when failure occurs, the plates have to be replaced. Since failure is unexpected then it may be assumed that a failure replacement is more costly than an earlier replacement. To reduce the number of failures, replacements can be scheduled to occur at specified intervals. However, a balance is required between the amount spent on the replacements and their resulting benefits, that is, reduced failure replacements.

It is assumed the problem is dealing with a long period over which the structure is to be in good condition and the intervals between the replacements are short. When this is the case, it is necessary to consider only one cycle of operation and to develop a model for one cycle. If the interval between the replacements is long, it would be necessary to use a discounting approach, and the series of cycles would have to be included in the model to consider the time value of money.

The replacement policy is one where replacements occur at fixed intervals of time; failure replacements occur whenever necessary. The problem is to determine the optimal interval between the planned replacements to minimize the total expected cost of replacing the corroded plates per unit time.

The total cost of a preventive replacement before failure occurred is defined as C_p, while C_f is the total cost of a failure replacement and $f(t)$ is the probability density function of the structural component's failure times. The replacement policy is to perform replacements at constant intervals of time t_p, irrespective of the age of the plate, and failure replacements occur as many times as required in the interval $(0, t_p)$.

To determine the optimal interval between replacements the total expected replacement cost per unit time is minimized. The total expected cost per unit time for replacement at the intervals of a length t_p, denoted $C(t_p)$ equals to the total expected cost in the interval $(0, t_p)$ divided by the interval (Rausand, 1998):

$$C(t_p) = \frac{C_p + C_f H(t_p)}{t_p} \tag{5}$$

where $H(t_p)$ is the expected number of failures in the interval $(0, t_p)$. To determine $H(t)$, the renewal theory approach is to be applied (Lindley, 1976):

$$H(T) = \sum_{i=0}^{T-1}\left[1 + H(T-i-1)\right]\int_{i}^{i+1} f(t)dt, \quad T \geq 1 \tag{6}$$

with $H(0) = 0$ and this equation is termed a recurrence relation.

The expected cost of failure replacement considered here is modelled as $C_f = nC_p$, where C_p is assumed here 10,000 units as an example and n varies as 10, 25, 50 and 300 that results in low, moderate, high and extreme repair cost consequence respectively. The optimal replacement interval for corroded structural components subjected to the replacement strategy is based on the normalized total repair cost with respect to the expected cost of failure replacement.

In this analysis, no-account was taken for the time required to perform replacements since they were considered to be short, compared to the mean time between replacements. When necessary, the replacement durations can be incorporated into the replacement model, as is required when the goal is the minimization of total downtime or, equivalently, the maximization of component availability. However, any cost that is incurred because of the replacement stoppages need to be included as part of the total cost before failure or in the total cost of a failure replacement.

Optimal replacement intervals for the set of corrosion data of deck plates of ballast tanks are given in Table 3 and Figure 7. It can be seen the minimum inspection interval is achieved when there is a combination of low corrosion tolerance and extreme total repair cost consequence, which leads to 2 years optimal replacement interval for ballast tank plates.

The maximum inspection interval for the ballast tank plates is achieved in a combination of low total repair cost consequence and extreme corrosion tolerance, which results in 11 years. Different variations of corrosion tolerance and total repair cost consequence result in different optimal replacements intervals defined as low (white, $8 \leq t_p$ years), moderate (dark, $5 \leq t_p < 8$ years) and high (gray, $2 \leq t_p < 5$ years) frequency repairs (see Table 3 and Figure 7) The solid line in Figure 7 shows the optimal replacement interval.

To recognise how realistic these results might be, it should be referred the present practice adopted by ship Classification Societies is to have main thickness surveys every 5 years and sometimes limited ones at 2.5 years.

4.2 *Optimal replacement age*

This problem is similar to the one presented before, except that instead of making replacements at fixed intervals, with the possibility of performing a replacement shortly after a failure replacement, the time at which the replacement occurs depends on the age of the component. When failures occur, failure replacements are made. When this occurs,

Table 3. Optimal replacement intervals, deck plates in ballast tanks, years.

		Total repair cost consequence			
		Low	Moderate	High	Extreme
Corrosion tolerance	Low	4.0	3.0	3.0	2.0
	Moderate	5.0	4.0	3.0	2.0
	High	8.0	6.0	5.0	3.0
	Extreme	11.0	10.0	8.0	5.0

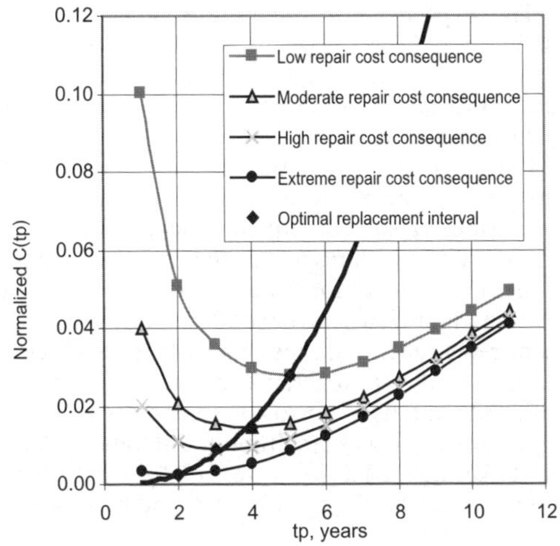

Figure 7. Replacement intervals, deck plates in ballast tanks, moderate corrosion tolerance.

the time clock is reset to zero, and the replacement occurs only when the component has been in use for the specified period.

The problem is to balance the cost of the replacements against their benefits, and this is done by determining the optimal replacement age for the component to minimize the total expected cost of replacement per unit time.

The replacement policy is to perform a replacement when the component has reached a specified age, t_p plus failure replacements when necessary. The objective is to determine the optimal replacement age of the plates to minimize the total expected replacement cost per unit time.

There are two possible cycles of operation: one cycle being determined by the plates reaching their planned replacement age, t_p and the other being determined by the plates ceasing to operate because of a failure occurring before the planned replacement time.

The total expected replacement cost per unit time is defined as $C(t_p)$. The total expected replacement cost per cycle equals the cost of a cycle preventive cost before failure, C_p time the probability of a cycle, $R(t_p)$ plus the cost of a failure cycle, C_f time the probability of a failure cycle, $[1 - R(t_p)]$ divided to the expected length of cycle. The total cost results in (Nowlan and Heap, 1978):

$$C(t_p) = \frac{C_p R(t_p) + C_f \left[1 - R(t_p)\right]}{t_p R(t_p) + \int_{-\infty}^{t_p} t f(t) dt} \tag{7}$$

The expected cost of failure replacement is modelled as $C_f = nC_p$, where C_p is 10,000 units and n varies as 10, 25, 50 and 300 that results in low, moderate, high and extreme repair cost consequence respectively.

Using the set of data analysed here, the optimal replacement age of the deck plates in ballast tanks for various repair consequences as a function of t_p, normalized with respect to the expected cost for failure replacement for ballast and cargo tanks are given in Figure 8, where the solid line shows the optimum replacement age, and Table 4, from which it can be seen the optimal replacement ages for different combinations of corrosion tolerances and total repair consequences resulting in different replacement ages defined as low (grey, $13 \le t_p < 16$ years), moderate (dark, $16 \le t_p < 19$ years) and high (white, $19 \le t_p$ years).

4.3 *Optimal replacement age accounting for the times required for replacements*

The optimal replacement age of the corroded plates is taken as that age which minimizes the total expected cost of replacements per unit time. T_p is the mean time required to make a replacement, T_f is the mean time required to make a failure replacement and $M(t_p)$ is the mean time to failure when replacement occurs at age t_p. The replacement policy is to perform a replacement once the plates have reached a specified age, t_p plus failure replacements when necessary. The objective is to determine the optimal replacement age of the plates to minimize the total expected replacement cost per unit time defined as (Nowlan and Heap, 1978):

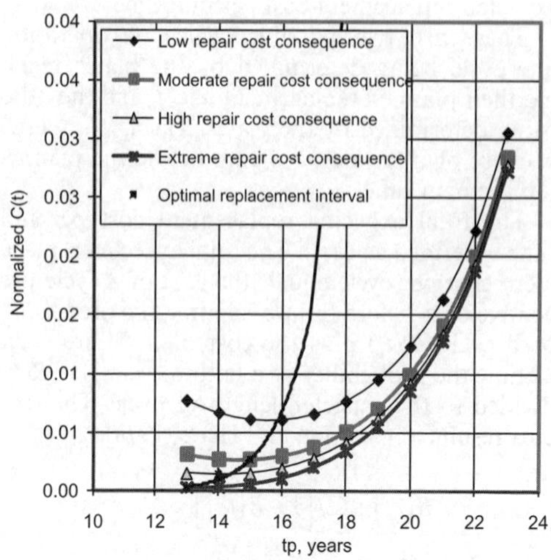

Figure 8. Replacement ages, deck plates in ballast tanks, moderate corrosion tolerance.

Table 4. Optimal replacement ages, deck plates in ballast tanks, years.

		Total repair cost consequence			
		Low	Moderate	High	Extreme
Corrosion tolerance	Low	15.0	14.0	13.0	13.0
	Moderate	16.0	15.0	14.0	13.0
	High	17.0	15.0	14.0	13.0
	Extreme	22.0	19.0	17.0	14.0

$$C(t_p) = \frac{C_p R(t_p) + C_f \left[1 - R(t_p)\right]}{(t_p + T_p) R(t_p) + \left[M(t_p) + T_f\right]\left[1 - R(t_p)\right]} \tag{8}$$

For the data sets of corroded plates, replacement times $T_p = 0.5$ week, $T_f = 1$ week and $C_f = nCp$, where C_p is 10,000 units and n varies as 10, 25, 50 and 300 that results in low, moderate, high and extreme repair cost consequence respectively it is determined the optimal replacement age of the structural components subjected to corrosion.

Various optimal replacement ages related to different levels of corrosion tolerances and total repair cost consequences result in different replacement ages defined as low (grey, $13 \le t_p < 14$ years), moderate (dark, $14 \le t_p < 15$ years) and high (white, $15 \le t_p$ years) replacement age as can be seen in Table 5.

Figure 9 shows replacement ages for deck plates in ballast tanks conditional to moderate corrosion tolerance and the solid line shows the optimum replacement age.

5 MINIMIZATION OF DOWNTIME

Sometimes because of difficulties in costing or the desire to get maximum or utilization of structures, the replacement policy required may be one that minimizes total downtime per unit time or, equivalently, maximizes availability. The problem is to determine the best times at which replacements should occur to minimize total downtime per unit time. The basic conflicts are that as the replacement frequency increases, there is an increase in downtime because of these replacements, but a consequence of this is a reduction of downtime because of failure replacements, and we wish to get the best balance between them has been defined.

Table 5. Optimal replacement ages, deck plates in ballast tanks, years.

		Total repair cost consequence			
		Low	Moderate	High	Extreme
Corrosion tolerance	Low	14.0	14.0	13.0	13.0
	Moderate	15.0	14.0	14.0	13.0
	High	15.0	14.0	14.0	13.0
	Extreme	15.0	14.0	14.0	13.0

5.1 *Optimal replacement interval*

The objective is to determine the optimal replacement interval t_p between replacements to minimize the total downtime per unit time. The total downtime per unit time, for replacement at time t_p, is denoted as $D(t_p)$ and it is defined as the number of failures, $H(t_p)$ in the time interval $(0, t_p)$ time the time required to make a failure replacement, T_f plus the time required to make replacement before failure divided to the interval, $t_p + T_p$, results in:

$$D(t_p) = \frac{H(t_p)T_f + T_p}{t_p + T_p} \tag{9}$$

The replacement interval to minimize total downtime for the sets of data of failure time studied here is modelled as $T_f = nT_p$.

Four different levels of downtime consequence are defined for n equals to 2, 5, 10 and 15 conditioning to $T_p = 2$ weeks as low, moderate, high and extreme downtime consequence respectively. The optimal replacement interval for different corrosion tolerances and downtime consequences result in different optimal replacement intervals defined as low (gray, $4 \le t_p < 7$ years), moderate (dark, $7 \le t_p < 11$ years) and high (white, $11 \le t_p$ years) replacement interval as can be seen in Table 6. Replacement intervals for deck plates in ballast tanks conditional to moderate corrosion tolerance are given in Figure 10, where the solid line shows the optimum replacement interval.

5.2 *Optimal replacement age*

The objective is to determine the optimal age, t_p, at which replacements should occur such the total downtime per unit time is minimized. The total downtime per unit time for replacements once the component becomes of age t_p is $D(t_p)$, which is defined as the total expected downtime in a cycle divided by the expected cycle length (Nowlan and Heap, 1978):

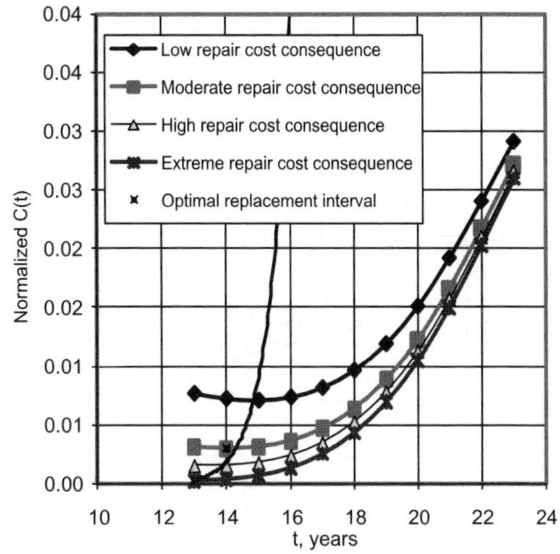

Figure 9. Replacement ages, deck plates in ballast tanks, moderate corrosion tolerance.

Table 6. Optimal replacement intervals, deck plates in ballast tanks, years.

		Downtime consequence			
		Low	Moderate	High	Extreme
Corrosion tolerance	Low	8.0	6.0	4.0	4.0
	Moderate	9.0	6.0	5.0	5.0
	High	11.0	10.0	8.0	7.0
	Extreme	12.0	12.0	12.0	12.0

$$D(t_p) = \frac{T_p R(t_p) + T_f\left[1 - R(t_p)\right]}{(t_p + T_p)R(t_p) + \left[M(t_p) + T_f\right]\left[1 - R(t_p)\right]} \tag{10}$$

The replacement age to minimize the total downtime is modelled as $T_f = nT_p$, where n varies as 2, 5, 10 and 15 and T_p equals to 2 weeks respectively. The results can be seen in Table 7 where different optimal replacement ages are defined as low (gray, $12.2 \le t_p < 12.6$ years), moderate (dark, $12.6 \le t_p < 13$ years) and high (white, $13 \le t_p$ years) replacement age.

6 MAXIMIZE THE AVAILABILITY

The basic purpose behind an inspection is to determine the state of structure. One indicator, such as corrosion deterioration, which is used to describe

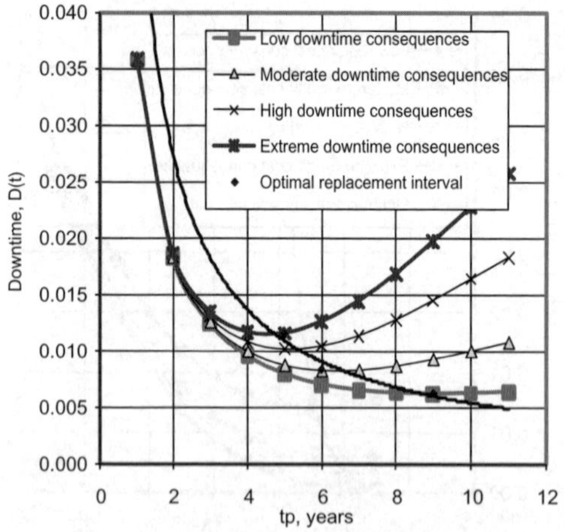

Figure 10. Replacement intervals, deck plates in ballast tanks, moderate corrosion tolerance.

Table 7. Optimal replacement ages, deck plates in ballast tanks, years.

		Downtime consequence			
		Low	Moderate	High	Extreme
Corrosion tolerance	Low	12.2	12.2	12.2	12.2
	Moderate	12.4	12.4	12.4	12.2
	High	12.4	12.4	12.4	12.4
	Extreme	12.8	12.8	12.8	12.8

the state, has to be specified, and the inspection is made to determine the values of this indicator. Then some maintenance action may be taken, depending on the state identified. The decision about when the inspection should take place ought to be influenced by the costs of the inspection and the benefits of the inspection, such as detection and correction of minor defects before major breakdown occurs.

The primary goal addressed here is that of making the structure more reliable through inspection because of establishing the optimal inspection interval for structures, and this interval is called the failure-finding interval.

The time required to conduct an inspection is T_i. It is assumed that after the inspection, if no major faults are found requiring repair or complete component replacement, the component is in the as-new state. This may be because of minor modifications being made during the inspection. T_r is the time required to make a repair or replacement. After the repair or replacement it is assumed the component is in the as-new state.

The objective is to determine the interval t_i, between inspections to maximize availability per unit time. The availability per unit time, denoted by $A(t_i)$, is a function of the inspection interval t_i and it is the expected availability per cycle/expected cycle length.

The uptime in a good cycle equals to t_i, since no failure is detected at the inspection. If a failure is detected, then the uptime of the failed cycle can be taken as the mean time to failure of the component, given that inspection takes place at t_i.

The expected uptime per cycle is calculated as (Lindley, 1976):

$$A(t_i) = \frac{t_i R(t_i) + \int_{-\infty}^{t_i} t f(t)\,dt}{(t_i + T_i) R(t_i) + (t_i + T_i + T_r)\left[1 - R(t_i)\right]} \quad (11)$$

The analysis here uses two data sets of failure times of corroded plates of deck structures of tankers related to ballast and cargo tanks corroded plates and the optimal inspection interval to maximize the availability considering $T_i = 1$ week and $T_r = nT_i$, where n varies as 2, 5, 15 and 30 respectively, which results in low, moderate, high and extreme downtime consequence and can be seen in Figure 11, where the solid line shows the optimum replacement interval, and Table 8. The inspection interval defined here are classified as low (gray, $3 \le t_p < 4.5$ years), moderate (dark, $4.5 \le t_p < 6$ years) and high (white, $6 \le t_p$ years) inspection interval.

The important assumption in the model here is that plates can be assumed to be as good as new

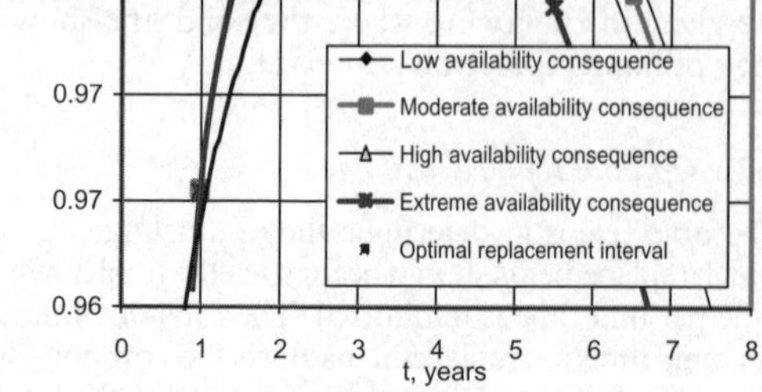

Figure 11. Inspection intervals, deck plates in ballast tanks, moderate corrosion tolerance.

Table 8. Optimal inspection intervals, deck plates in ballast tanks, years.

		Availability consequence			
		Low	Moderate	High	Extreme
Corrosion tolerance	Low	3.5	3.5	3.0	3.0
	Moderate	3.5	3.5	3.5	3.0
	High	5.0	5.0	4.5	4.5
	Extreme	7.5	7.0	7.0	6.5

after inspection if no repair or replacement takes place. In practice, this may be reasonable, and it will be the case if the failure distribution of the component was exponential (since the conditional probability remains constant).

If the as-new assumption is not realistic and the failure distribution has an increasing failure rate, then rather than having inspection at constant intervals, it may be advisable to increase the inspection frequency as the component gets older.

7 COMPARATIVE ANALYSIS

7.1 *Deck plates of ballast and cargo tanks of tankers*

Two sets of corrosion data, deck plates of ballast and cargo tanks of tankers (Garbatov et al., 2007) are analysed here and a comparison between different strategies for inspection, repair and reliability for the deck ballast and cargo tank for plates subjected to high tolerance corrosion and high consequence risk level is presented in Figure 12 to Figure 14.

Both types of deck tank plates behave in the same manner with respect to the optimal repair interval due the total repair cost and downtime consequences.

However, it has to be pointed out that after the 22nd year the cargo tank plates will deteriorate faster as can be seen from Figure 14 and because of that more intensive repair work will be required. Having different assumptions about the cost involved and operational time of inspection and repair will result in differed strategy of maintenance actions.

The analysis just presented is based on failure data collected during the entire life of corroded deck plates of ballast and cargo tanks of tankers. To make use of this analysis the inspection intervals, maintenance and repairs have to be conditioned to a specific ship and its routes and times to

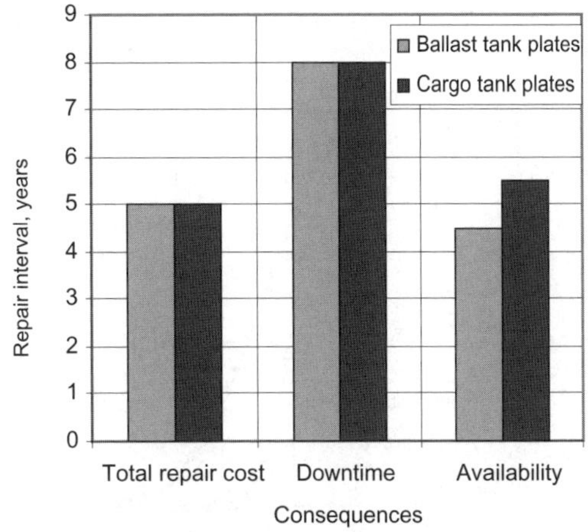

Figure 12. Repair intervals, high corrosion tolerance.

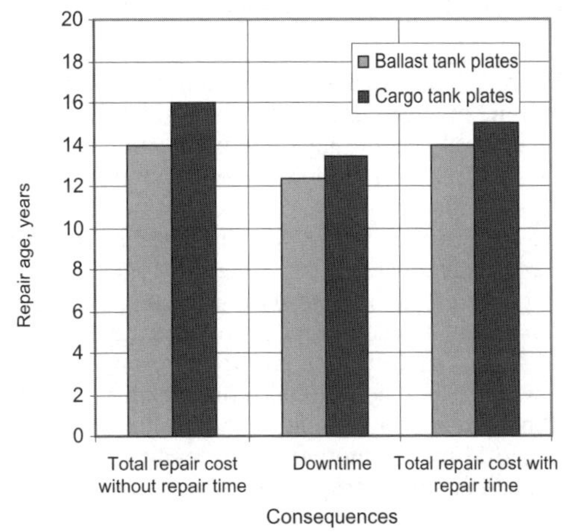

Figure 13. Repair ages, high corrosion tolerance.

arrive to shipyard. Additional constrains have to be applied with respect to the planned inspections for any class notation as defined by Classification Societies.

However there is a variation of the expected cost in the present study. A variation in the cost is not expected to be seen in the term of the preventive cost because it is calculated based on the weight of the structure replaced independently of the corrosion progress. The cost related to failure will have some variation and depends of the severity of the corrosion and the consequences. Such analysis has not been performed in this paper.

7.2 *Structural components of bulk carriers*

Fourteen sets of corrosion data (Garbatov and Guedes Soares, 2008a, 2009a), presented in

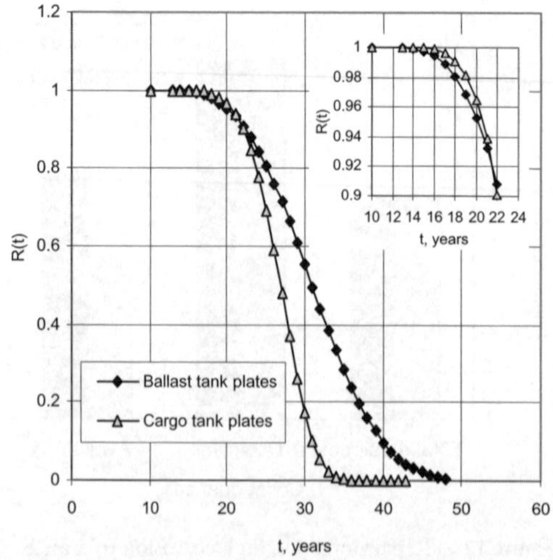

Figure 14. Reliability, high corrosion tolerance.

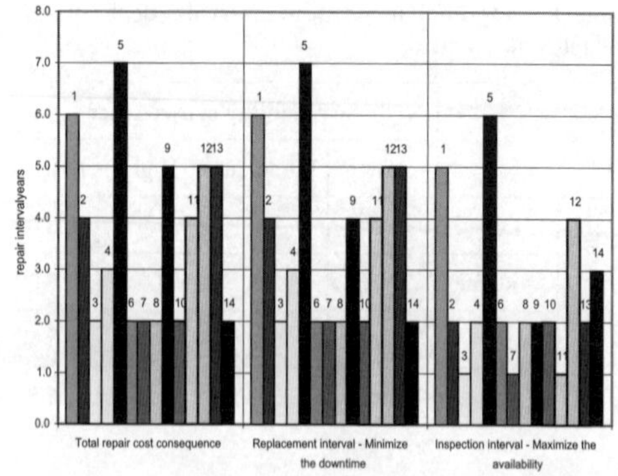

Figure 15. Optimum repair intervals for different criteria.

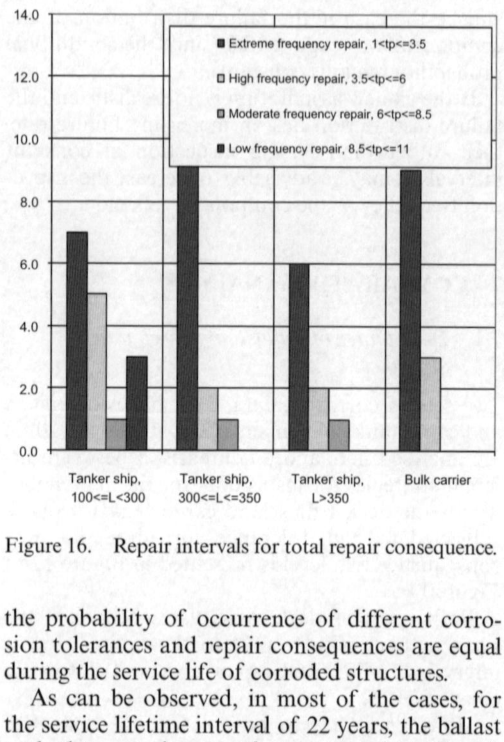

Figure 16. Repair intervals for total repair consequence.

section 2.1 are analysed here and a comparison between different strategies for optimal inspection interval accounting for minimization of total cost, minimization of downtime and maximization of availability of different structures subjected to moderate corrosion and for the resulting moderate cost consequence is presented in Figure 15. As can be observed, ship structural components 3, 6, 7, 8, 10 and 14 require more frequent repair work with respect to the criteria of "total repair cost" and "downtime". When optimal replacement is defined to "maximize the availability" structural components 3, 7 and 11 require more frequent repair work.

The structural components behave in a similar manner with respect to the criteria "minimization of the total repair cost" and "minimization of downtime". However, it has to be pointed out that after some years of service some components will deteriorate faster and because of that more intensive repair work will be required. Furthermore, different assumptions about the costs involved and the operational constraints for the time of inspections will result in different strategies of maintenance.

7.3 *Comparison between deck plates of ballast tanks of tankers and bulk carriers*

Two sets of corrosion data, from deck plates of ballast tanks of tankers and bulk carriers were analysed and a comparison between different strategies for inspection, repair and reliability of plates subjected to various corrosion tolerances and repair consequence levels are presented in Figure 16 to Figure 18. It is considered that the results shown in the above figures have been obtained considering the probability of occurrence of different corrosion tolerances and repair consequences are equal during the service life of corroded structures.

As can be observed, in most of the cases, for the service lifetime interval of 22 years, the ballast tank plates require more frequent repair work with respect to the ability consequences criteria and early repair work because of the downtime and accounting for the total repair cost consequences criteria.

Accounting for that assumption the averaged repair interval for the service life of structures studied for a total cost repair consequence is 4.1 years for tankers of $100 \le L < 300$ m, 2.7 years for tanker ship, $300 \le L <= 350$ m, 3.3 years for tankers with $L > 350$ m and 2.5 years for bulk carriers respectively.

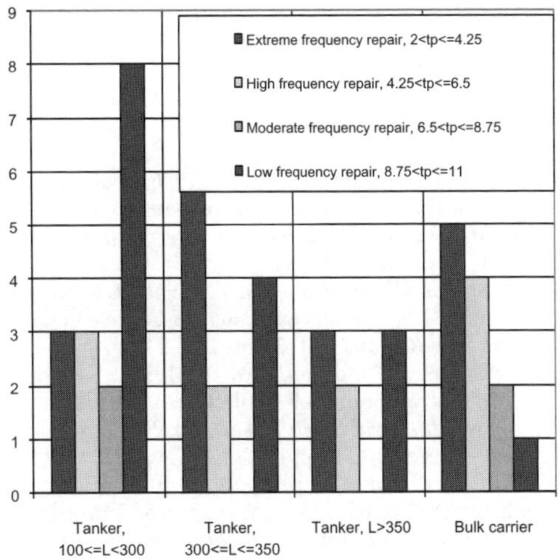

Figure 17. Repair intervals for downtime consequence.

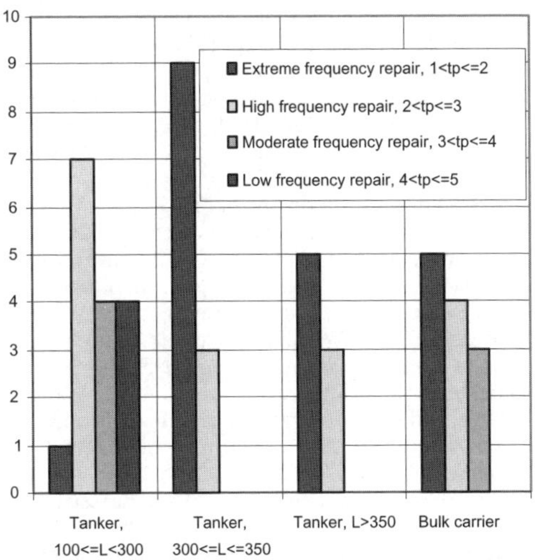

Figure 18. Repair intervals for availability consequence.

The averaged repair interval during the service life for the downtime consequence is 8 years for tankers of $100 \leq L < 300$ m, 5.8 years for tankers of $300 \leq L \leq 350$ m, 6.5 years for tankers of $L > 350$ m and 5.3 years for bulk carriers respectively. The averaged repair interval for the availability consequence is 8 years for tankers of $100 \leq L < 300$ m, 3.5 years for tankers of $300 \leq L \leq 350$ m, 1.8 years for tankers of $L > 350$ m and 2.1 years for bulk carriers respectively.

It can be observed, that in most of the cases the ballast tank plates of bulk carriers require more frequent repair works. If the mean value of the replacement intervals of different length of tankers is averaged then the replacement interval for the total repair cost is 3.3 years, for a downtime consequence is 6.8 years and for the availability consequence is 2.5 years respectively. In this comparison, the bulk carrier structures will take more frequent repairs with respect to cost and downtime consequence and almost the same for availability consequence.

8 CONCLUSIONS

An analysis has been made, which predicts the optimum age and the intervals for plate maintenance accounting for the general corrosion deterioration. The analysis used the general framework that was developed for failure of components of a system and adopted it to the present problem by considering the different corrosion tolerance levels as failure criteria.

It is demonstrated that after fitting a Weibull distribution to corrosion data, different criteria can be adopted and the corresponding optimal repair schedule can be determined. Different criteria for corrosion detection and for the effect of repair can be easily incorporated. Various assumptions were made about different operational times and costs that are not essential to the method but are needed for the example calculation.

It has to be pointed out that collecting more data about the failure time of corroded structures and having more precise information about cost and inspections will enhance the output of the present analysis. Having different assumptions about the cost involved and operational time of inspection and repair will result in different strategies of maintenance actions.

The analysis just presented is based on failure data collected during the entire life of corroded structural components. To make use of this analysis for a specific ship, it would be necessary to conduct a similar analysis for an appropriate data set. Additional constrains have to be applied with respect to the time of planned inspections for any class notation as defined by Classification Societies.

ACKNOWLEDGMENTS

This paper presents part of the teaching material prepared and used in the scope of the course "Fatigue Reliability and Rational Inspection Planning" within the project "Advanced Ship Design for Pollution Prevention (ASDEPP)" that was financed by the European Commission through the Tempus contract JEP-40037–2005.

REFERENCES

Faber, M.H., Kroon, I.B. and Sorensen, J.D., 1996, Sensitivities in Structural Maintenance Planning, *Reliability Engineering and System Safety*, 51 (3), pp. 317–329.

Fujimoto, Y. and Swilem, A.M., 1992, Inspection Strategy for Deterioration Structures Based on Sequential Cost Minimisation Method, *Proceedings of the 11th International Conference on Offshore Mechanics and Arctic Engineering (OMAE'92)*, ASME, pp. 219–226.

Fujita, M., Schall, G. and Rackwitz, R., 1989, Adaptive Reliability Based Inspection Strategies for Structures Subjected to Fatigue, *Proceedings of ICOSSAR*, pp. 1619–1626.

Garbatov, Y. and Guedes Soares, C., 2001, Cost and Reliability based strategies for Maintenance Planning of Floating structures, *Reliability Engineering & System Safety*, 9, pp. 293–301.

Garbatov, Y. and Guedes Soares, C., 2007, Structural Reliability of Ship Hull Subjected to Non-linear Time Dependent Deterioration, Inspection and Repair, *Proceedings of the 10th International Symposium on Practical Design of Ships and other Floating Structures*, Basu, Belenky, Wang & Yu (eds), Houston, USA, pp. 543–553.

Garbatov, Y. and Guedes Soares, C., 2008a, Corrosion Wastage Modelling of Deteriorated Ship Structures, *International Shipbuilding Progress,* 55, pp. 109–125.

Garbatov, Y. and Guedes Soares, C., 2008b, Structural Reliability of Aged Ship Structures, *Condition Assessment of Aged Structures*, Paik & Melchers (eds), Wood head Publishing Limited, pp. 253–313.

Garbatov, Y. and Guedes Soares, C., 2009a, Corrosion Wastage Statistics and Maintenance Planning of Corroded Hull Structures of Bulk Carriers, *Analysis and Design of Marine Structures*, Guedes Soares & Das, (eds), Francis and Taylor, pp. 215–222.

Garbatov, Y. and Guedes Soares, C., 2009b, Maintenance Planning for the Decks of Bulk Carriers and Tankers, *Proceedings of the 10th International Conference on Structural Safety*, Furuta, Frangopol & Shinozuka (eds)., Osaka, Paper ICOSSAR09–0010.

Garbatov, Y. and Guedes Soares, C., 2009c, Optimal Maintenance for Corroded Deck Tanker Ship Structures, *Reliability Engineering and System Safety*, 94 (11), pp. 1806–1817.

Garbatov, Y., Guedes Soares, C. and Wang, G., 2007, Non-linear Time Dependent Corrosion Wastage of Deck Plates of Ballast and Cargo Tanks of Tankers, *Journal of Offshore Mechanics and Arctic Engineering*, 129 (1), pp. 48–55.

Garbatov, Y., Rudan, S. and Guedes Soares, C., 2002, Fatigue Damage of Structural Joints Accounting for Non-linear Corrosion, *Journal of Ship Research*, 46 (4), pp. 289–298.

Garbatov, Y., Vodkadzhiev, I. and Guedes Soares, C., 2004, Corrosion Wastage Assessment of Deck Structures of Bulk Carriers, *Proceedings of the International Conference on Marine Science and Technology*, Union of Scientists of Varna, pp. 24–33.

Gardiner, P. and Melchers, E., 2001, Enclosed Atmospheric Corrosion in Ship Spaces, *British Corrosion Journal*, 36 (4), pp. 272–276.

Guedes Soares, C., 1988, Reliability of Marine Structures, *Reliability Engineering*, Kluwer Academic Publishers, pp. 513–559.

Guedes Soares, C. and Garbatov, Y., 1996a, Fatigue Reliability of the Ship Hull Girder Accounting for Inspection and Repair, *Reliability Engineering and System Safety*, 51 (2), pp. 341–351.

Guedes Soares, C. and Garbatov, Y., 1996b, Reliability of Maintained Ship Hulls Subjected to Corrosion, *Journal of Ship Research*, 40 (3), pp. 235–243.

Guedes Soares, C. and Garbatov, Y., 1998a, Reliability of Maintained Ship Hull Subjected to Corrosion and Fatigue, *Structural Safety*, 20 (3), pp. 201–219.

Guedes Soares, C. and Garbatov, Y., 1998b, Non-linear Time Dependent Model of Corrosion for the Reliability Assessment of Maintained Structural Component, *Safety and Reliability*, Lydersen, Hansen & Sandtorv (eds), Balkema, Rotterdam, pp. 928–936.

Guedes Soares, C. and Garbatov, Y., 1999a, Reliability of Maintained, Corrosion Protected Plates Subjected to Non-Linear Corrosion and Compressive Loads, *Marine Structures*, 12 (6), pp. 425–445.

Guedes Soares, C. and Garbatov, Y., 1999b, Reliability of Plate Element Subjected to Non-linear Corrosion and Biaxial Compressive Loads, *Applications of Statistics and Probability*, Shiraishi, Shinozuka & Wen (eds), A.A. Balkema, pp. 345–355.

Guedes Soares, C., Garbatov, Y. and Zayed, A., 2009a, Effect of Environmental Factors on Steel Plate Corrosion under Marine Immersion Conditions, *Corrosion Engineering Science and Technology*, (in press).

Guedes Soares, C., Garbatov, Y., Zayed, A. and Wang, G., 2008, Corrosion Wastage Model for Ship Crude Oil Tanks, *Corrosion Science,* 50, pp. 3095–3106.

Guedes Soares, C., Garbatov, Y., Zayed, A. and Wang, G., 2009b, Influence of Environmental Factors on Corrosion of Ship Structures in Marine Atmosphere, *Corrosion Science,* 51(9), pp. 2014–2026.

Itagaki, H. and Yamamoto, N., 1977, Bayesian Reliability Analysis and Inspection of Ship Structural Members—An Application to the Fatigue Failures of Hold Frames, *Proceedings of the International Symposium on Practical Design in Shipbuilding (PRADS'77)*, pp. 765–772.

Jardine, A. and Tsang, A., 2005, Maintenance, Replacement and Reliability, Theory and Applications, Taylor & Francis.

Kobayoshi, Y., Tanaka, Y., Goto, H., Matsuoka, K. and Motohashi, Y., 1998, Effects of Stress Concentration Factors on Corrosion Fatigue Strength of a Steel Plate for Ship Structures, *Eng. Mater.*, Trans. Tech, pp. 1037–1042.

Lindley, D.V., 1976, *Introduction to Probability and Statistics from a Bayesian Viewpoint*, Cambridge University Press.

Loseth, R., Sekkesaeter, G. and Valsgard, S., 1994, Economics of High—Tensile Steel in Ship Hulls, *Marine Structures*, 7 (1), pp. 31–50.

Lotsberg, I. and Kirkemo, A., 1989, Systematic Method for Planning In-service Inspection of Steel Offshore Structures, *Proceedings of the 8th International Conference on Offshore Mechanics and Arctic Engineering (OMAE'89)*, ASME, pp. 275–284.

Madsen, H., 1997, Stochastic Modelling of Fatigue Crack Growth and Inspection, *Probabilistic Methods for Structural Design*, Kluwer Academic Publisher, pp. 59–84.

Melchers, R., 1998, Immersion Corrosion of Steels in Marine and Brackish Waters, *Proceeding Corrosion Review*, pp. 31–39.

Melchers, R., 1999, Probabilistic Modelling of Marine Immersion Corrosion of Steels, *Proceedings of the 10th International Congress on Marine Corrosion and Fouling*. University of Melbourne, pp. 212–222.

Melchers, R., 2001, Probabilistic Models of Corrosion for Reliability Assessment and Maintenance Planning, *Proceedings of the 20th International Conference on Offshore Mechanics and Arctic Engineering,* Paper OMAE2001/S & R-2108.

Melchers, R., 2003a, Effect on Marine Immersion Corrosion of Carbon Content of Low Alloy Steels, *Corrosion Science*, 45 (11), pp. 2609–2625.

Melchers, R., 2003b, Mathematical Modelling of the Diffusion Controlled Phase in Marine Immersion Corrosion of Mild Steel, *Corrosion Science*, 45 (5) pp. 923–940.

Melchers, R., 2003c, Modelling of Marine Immersion Corrosion for Mild and Low-alloy Steels—Part 1 Phenomenological Model, *Corrosion*, 4, pp. 319–334.

Melchers, R., 2003d, Probabilistic Models for Corrosion in Structural Reliability Assessment Part 1: Empirical Models, *Journal of Offshore Mechanics and Arctic Engineering*, 125, pp. 264–271.

Melchers, R., 2003e, Probabilistic Models for Corrosion in Structural Reliability Assessment Part 2: Models Based on Mechanics, *Journal of Offshore Mechanics and Arctic Engineering*, 125 (4), pp. 272–280.

Montero-Ocampo, C. and Veleva, L., 2002, Effect of Cold Reduction on Corrosion of Carbon Steel in Aerated 3% Sodium Chloride, *Corrosion*, 58, pp. 601–607.

Moubray, M., 1997, *Reliability Centred Maintenance*, Butterworth Heinemann.

Nowlan, F.S. and Heap, H.F., 1978, Reliability-centred Maintenance, *Technical Report AD/A066-579*.

Paik, J., Lee, J., Hwang, J. and Park, Y., 2003, A Time-Dependent Corrosion Wastage Model for the Structures of Single and Double Hull Tankers and FSOs, *Marine Technology*, 40, pp. 201–217.

Paik, J.K., Kim, S.K., Lee, S. and Park, Y.E., 1998, A Probabilistic Corrosion Rate Estimation Model for Longitudinal Strength Members of Bulk Carriers, *Journal of Ship and Ocean Technology*, 2, pp. 58–70.

Panayotova, M., Garbatov, Y. and Guedes Soares, C., 2004a, Factor Influencing Atmospheric Corrosion and Corrosion in Closed Spaces of Marine Steel Structures, *Proceedings of the International Conference on Marine Science and Technology*, Union of Scientists of Varna, pp. 286–292.

Panayotova, M., Garbatov, Y. and Guedes Soares, C., 2004b, Factor Influencing Corrosion of Steel Structural Elements Immersed in Seawater, *Proceedings of the International Conference on Marine Science and Technology*, Union of Scientists of Varna, pp. 280–286.

Purlee, L., 1965, Economic Analysis of Tank Coating for Tankers in Clean Service, *Material Protection*, pp. 50–58.

Qin, S. and Cui, W., 2002, Effect of Corrosion Models on the Time-Dependent Reliability of Steel Plated Elements, *Marine Structures*, 15, pp. 15–34.

Rausand, M., 1998, Reliability Centred Maintenance, *Reliability Engineering & System Safety*, 60, pp. 121–132.

Shi, W., 1993, In-service Assessment of Ship Structures: Effects of General Corrosion on Ultimate Strength, *Transactions Royal Institution of Naval Architects*, 135, pp. 77–91.

Southwell, C., Bultman, J. and Hummer, J.C., 1979, Estimating of Service Life of Steel in Seawater, *Seawater Corrosion Handbook*, Noyes Data Corporation, pp. 374–387.

TSCF, 1992, *Condition Evaluation and Maintenance of Tanker Structures*, Tanker Structure Cooperative Forum.

TSCF, 1997, *Guidance Manual for Tanker Structures*, Tanker Structure Cooperative Forum.

Wang, G., Spencer, J. and Elsayed, T., 2003a, Estimation of Corrosion Rates of Oil Tankers, *Proceedings of the 22nd International Conference on Offshore Mechanics and Arctic Engineering*, ASME, paper OMAE 2003-37361.

Wang, G., Spencer, J. and Sun, H., 2003b, Assessment of Corrosion Risks to Aging Ships using an Experience Database, *Proceedings of the 22nd International Conference on Offshore Mechanics and Arctic Engineering*, ASME, paper OMAE 2003-37299.

Wang, G., Spencer, J. and Sun, H., 2003c, An Update on Corrosion Additions and Corrosion Wastage Allowances Based on an Experience Data Base of Oil Tankers in Service, *Proceeding of 22nd International Conference on Offshore Mechanics and Arctic Engineering*, ASME, paper OMAE 2003-37362.

Yamamoto, N., 1998, Reliability Based Criteria for Measures to Corrosion, *Proceedings of the 17th International Conference on Offshore Mechanics and Arctic Engineering*, ASME, paper OMAE 1998-1234.

Yamamoto, N. and Ikegaki, K., 1998, A Study on the Degradation of Coating and Corrosion on Ship's Hull Based on the Probabilistic Approach, *Journal of Offshore Mechanics and Arctic Engineering*, 120 (3), pp. 121–128.

Advanced Ship Design for Pollution Prevention – Guedes Soares & Parunov (eds)
© 2010 Taylor & Francis Group, London, ISBN 978-0-415-58477-7

Hull girder fatigue strength of corroding oil tanker

J. Parunov & K. Žiha
Faculty of Mechanical Engineering and Naval Architecture, University of Zagreb, Zagreb, Croatia

P. Mage
Centre for Technology Transfer, Zagreb, University of Zagreb, Croatia

P. Jurišić
Croatian Register of Shipping, Split, Croatia

ABSTRACT: The paper describes the fatigue strength assessment of oil tanker hull girder taking into account degradation of ship hull due to the corrosion. Hull girder degradation is represented as time-varying stress increase caused by time-varying loss of midship section modulus. Nonlinear function representing hull girder section modulus loss is based on recently published corrosion measurements on large number of single-hull oil tankers. Fatigue damage accumulation is calculated for different corrosion severities by integrating time-varying instantaneous fatigue damage. Proposed method represents an improvement of the fatigue assessment procedure compared to Common Structural Rules, as the "rule" method is based on simplified assumption that corrosion wastage is constant in whole lifetime of the vessel. The paper also considers the fatigue yielding due to possible worsening of fatigue resistance due to alternating load cases and load variability in addition to the lifetime shortening due to corrosion.

1 INTRODUCTION

Fatigue and corrosion are recognized as the predominant factors which contribute to the structural failures observed on ships in service. Until recently, the fatigue was considered as a serviceability problem rather than a global hull girder strength problem. However, the latest researches conducted for development of the new Common Structural Rules (CSR) for Double-hull Oil Tankers showed that the majority of observed cracks of ship structures are caused not only by the local dynamic loads but also by the global dynamic hull girder loads such as the wave bending moments (ABS et al. 2006). In other words, fatigue of the hull girder may be a governing strength criterion for oil tankers, in particular if higher tensile steel is implemented.

The aim of the present paper is to perform hull girder fatigue analysis of an oil tanker using two approaches. First approach is the "rule" approach, proposed by CSR, assuming constant loss of hull girder section modulus (HGSM) throughout the whole ship lifetime. Second approach is the improvement of the CSR fatigue analysis procedure by introducing the concept of time-dependant HGSM. Nonlinear function representing HGSM loss is based on recently published corrosion measurements on large number of single hull oil tankers (Wang et al. 2008). The obtained results are compared to those given by CSR approach and corresponding conclusions are drawn. Finally, the effect of ship structural repair on fatigue life is also discussed. The paper demonstrate that newly proposed approach for hull girder fatigue assessment of oil tankers may be used with only slight increase in computational effort.

2 SHIP DESCRIPTION

The ship analyzed in the present study is a double-hull oil tanker with centre line plane bulkhead. The main particulars of the tanker are presented in Table 1. Deck and bottom areas of the ship are produced of higher tensile steel AH32, while the region around neutral axis is made of mild steel ST235. In addition, the whole centre line bulkhead is made of higher tensile steel AH32 due to shear stress requirements.

3 FATIGUE ASSESSMENT OF DECK LONGITUDINALS ACCORDING TO CSR

Deck longitudinals of an oil tanker are predominantly stressed by hull-girder vertical bending moments. Fluctuating stresses that may cause fatigue

Table 1. Main characteristics of oil tanker.

Scantlings		Units, m
Length between perpendiculars	Lpp	235
Moulded breadth	B	41.0
Moulded depth	D	20.0
Scantling draught	T	14.0
Deadweight	DWT	105000 dwt

failure of deck longitudinals are therefore mainly induced by vertical wave bending moments.

The calculation of hull girder stress ranges for fatigue strength assessment of deck longitudinals is based on the fatigue HGSM, calculated by deducting a quarter of the CSR corrosion addition ($-0.25\ t_{corr}$) from the gross thickness of all structural elements comprising the hull girder cross section. Only 25% of CSR corrosion addition is used for HGSM calculation because CSR corrosion additions are extreme values applicable for local strength assessment of individual structural elements. Considering whole ship transverse section, these extreme local corrosion losses are not expected to occur simultaneously in all elements contributing to longitudinal strength. Therefore, CSR proposes to use 25% of extreme corrosion losses in calculating HGSM for fatigue assessment.

The capacity of welded steel joints with respect to fatigue strength is characterized by Wöhler curves (S-N curves) which give the relationship between the stress ranges applied to a given detail and the number of constant amplitude load cycles to failure, with the zero mean stress. The hull detail which is taken into consideration for fatigue assessment of deck structure is the connection of a deck longitudinal and a typical web frame, classed as F-detail (CSR Table C.1.7-Clasification of Structural Details).

The fatigue assessment of the structural details is based on the application of the Palmgren-Miner cumulative damage rule. When the cumulative fatigue damage ratio, DM, is greater than 1, the fatigue capability of the structure is not acceptable. DM is determined as (CSR Appendix C 1.4.1.):

$$DM = \sum_{i=1}^{2} DM_i \tag{1}$$

Where:
DM_i = cumulative fatigue damage ratio for the applicable loading condition,
$i = 1$ for full load condition
$i = 2$ for normal ballast condition.

Assuming that the long term distribution of stress ranges fit a two-parameter Weibull probability distribution, the cumulative fatigue damage DM_i for each of two loading conditions is taken as, (CSR Appendix C, Sec.1.4.1.4):

$$DM_i = \frac{\alpha_i N_L}{K_2} \cdot \frac{S_{Ri}^m}{(\ln N_R)^{\frac{m}{\xi}}} \cdot \mu_i \cdot \Gamma\left(1+\frac{m}{\xi}\right) \tag{2}$$

Where:
N_L = number of cycles for the expected design life. N_L is generally between 0.6×10^8 and 0.8×10^8 cycles for a design life of 25 years.

$$N_L = \frac{f_0 \cdot U}{4 \cdot \log L} \tag{3}$$

$f_0 = 0.85$, factor taking into account non-sailing time for operations such as loading and unloading, repairs, etc.
U = design life (s) = 0.788×10^8 for design life of 25 years,
L = rule length,
m = S-N curves exponent as given in CSR Table C.1.6,
K_2 = S-N curves coefficient as given in CSR Table C.1.6,
α_i = proportion of the ship's life:
$\alpha_1 = 0.5$ for full load condition; $\alpha_2 = 0.5$ for ballast condition,
S_{Ri} = stress range at the representative probability level of 10^{-4} (N/mm^2),
$N_R = 10000$, number of cycles corresponding to the probability level of 10^{-4},
ξ = Weibull shape parameter,
Γ – Gamma function,
μ_i – coefficient taking into account the change in slope of S-N curve

$$\mu_i = 1 - \frac{\gamma\left(1+\frac{m}{\xi}, \upsilon_i\right) - \upsilon_i^{-\frac{\Delta m}{\xi}} \cdot \gamma\left(1+\frac{m+\Delta m}{\xi}, \upsilon_i\right)}{\Gamma\left(1+\frac{m}{\xi}\right)} \tag{4}$$

$$\upsilon_i = \left(\frac{S_q}{S_{Ri}}\right)^{\xi} \cdot \ln N_R \tag{5}$$

S_q = stress range at the intersection of two segments ("knee") of the S-N curves (CSR Table C.1.6).
$\Delta m = 2$—slope change of upper-lower segment of the S-N curve
$\gamma(a,x)$ = incomplete Gamma function, Legendre form
Weibull shape parameter ξ is calculated as:

$$\xi = f_{Weibull} \cdot \left(1.1 - 0.35 \cdot \frac{L-100}{300}\right) \quad (6)$$

The cumulative fatigue damage ratio, DM, is finally converted to a calculated fatigue life:

$$Fatigue\_life = \frac{Design\_life}{DM} \text{ (years)} \quad (7)$$

According to CSR requirements, calculated fatigue life should be higher than design life of 25 years.

Very important step in the fatigue analysis is the correction of fatigue stresses due to the mean stress effect. The stress range may be reduced depending on whether mean stress is tensile or compressive. In the event that mean compressive stress exist and can be quantified, the effect of mean stress may be considered by assuming a stress range equal to the tensile component plus 60% of the compressive component (CSR Appendix C, Sec.1.4.5.10). Mean stresses in main deck longitudinals are compressive in the case of sagging still water bending moments. Generally, for double hull oil tankers, this is the case in full load condition. Therefore, stress range appearing in equation (2) may be considerably reduced in full load condition.

4 TIME-DEPENDANT HGSM

Assumption inherent in the CSR fatigue assessment procedure is that HGSM is reduced because of corrosion already for new (as-built) vessel and that ship sails in such unchanged condition until the end of her lifetime. Such assumption is unrealistic, since corrosion propagation is gradual process that may roughly be divided into three phases:

- the duration of protective coating,
- the transition period until the appearance of visible corrosion,
- the progress of already formed corrosion.

The following equation may be assumed for the HGSM loss at t years old (Wang et al. 2008):

$$R(t) = C(t - t_0)^I \quad (8)$$

where $R(t)$ is the HGSM loss at age t, while t_0 is the year when HGSM starts to deviate from the as-built condition. C and index I are constants that can be determined according to the data set. $R(t)$ represents the ratio of the as-gauged HGSM over the as-built, as presented by the following equation:

$$R(t) = 1 - \text{HGSM(as\_gauged\_at-year\_t)/ HGSM(as\_built)} \quad (9)$$

Data used in the present study are those collected in ABS' Safe hull Condition Assessment Program (CAP) (Wang et al. 2008). HGSM loss is calculated based on the gauging results from all longitudinally-effective structural components on 2195 transverse sections of 211 single-hull oil tankers. Results are presented as average parameters of the Equation (9) for four different levels of the corrosion severity, Table 2 and Figure 1.

Results from the Figure 1 indicate that corrosion loss of HGSM will be less than 10% during whole lifetime of the vessel, in cases of slight and moderate corrosion severities. For severe corrosion rate, HGSM will become less than 90% of its initial value after approximately 23 years while in extremely unfavourable corrosion conditions 10% loss limit will be exceeded after 18 years. This value of 10% of permissible loss of initial HGSM represents important industry standard that must be respected during whole lifetime of the ship.

As the consequence of the reduction of the HGSM, stresses in deck structure induced by hull girder bending moments will increase. Stresses relevant for fatigue assessment of deck longitudinals are those resulting from vertical wave bending moments with exceeding probability of 10^{-4}. Those stresses are presented in Figures 2 & 3 for different levels of corrosion severity and for full load and ballast conditions respectively.

Table 2. Parameters of Equation (2) for different levels of corrosion severity.

Corrosion severity	C	t_0, years	I
Slight	0.62	6.5	0.67
Moderate	0.80	5	0.75
Severe	0.84	3.5	0.83
Extreme	0.80	2	0.91

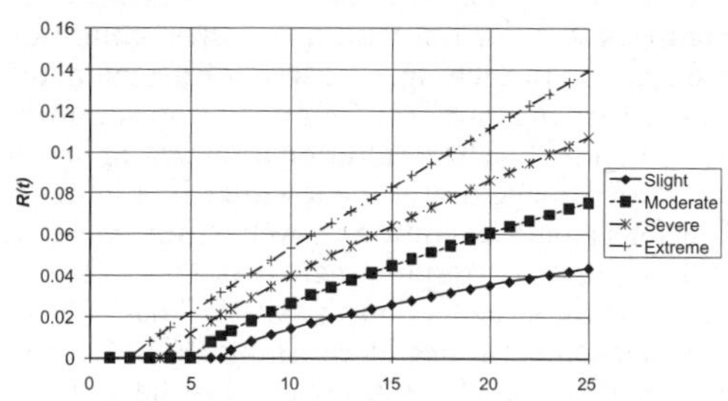

Figure 1. HGSM loss for different levels of corrosion severity.

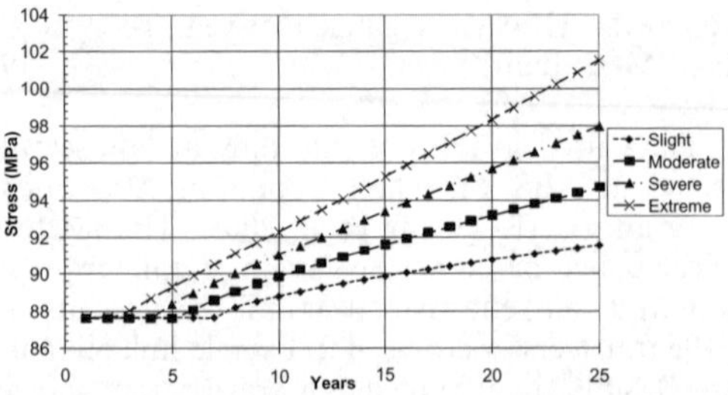

Figure 2. Increase of stress range at probability level 10^{-4} for different levels of corrosion severity (full load condition).

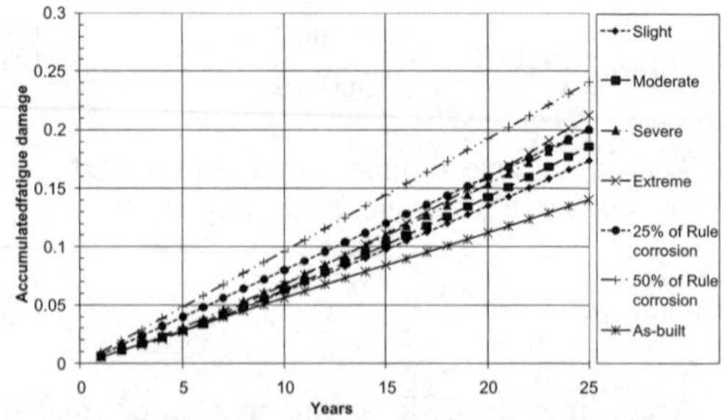

Figure 4. Accumulated damages in full load condition.

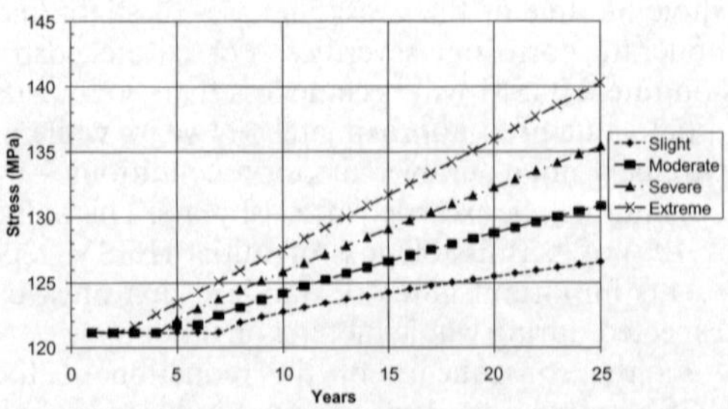

Figure 3. Increase of stress range at probability level 10^{-4} for different levels of corrosion severity (ballast condition).

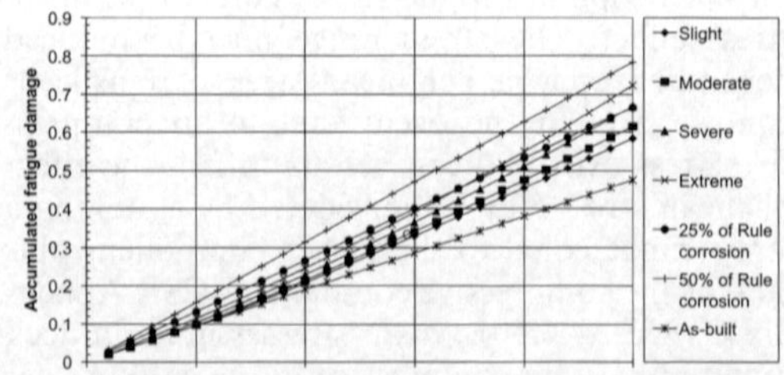

Figure 5. Accumulated damages in ballast condition.

5 RESULTS OF THE ANALYSIS

The main idea of the present paper is to use results presented in Figs. 1–3 in calculation of accumulated fatigue damage instead of assuming constant corrosion loss and constant fatigue stresses during whole ship's lifetime. This is done in a way to assume constant fatigue stresses during time interval of one year, and then to calculate accumulated damage by Equation (2) for one year period. Accumulated damage in 25 years is then obtained by summing contributions from each of previous years.

Results of the analysis are presented in Figures 4 & 5 for full load and ballast condition respectively. In each figure, curves of accumulated damage are presented for four corrosion scenarios from Figure 1. Three additional curves of accumulated damage are also presented in Figures 4 & 5 for as-built condition as well as for constant HGSM losses corresponding to 50% and 25% of CSR corrosion deduction thickness. These corrosion deductions result in constant HGSM losses of abt. 10% and 5% respectively. The latter value i.e. 25% of CSR corrosion deduction is the "rule" case.

Comparison of results on Figures 4 & 5 indicates that accumulated damage in full load condition is much lower compared to ballast as a consequence of the mean stress effect and corresponding reduction of fatigue stresses. One may notice that damage accumulation is linear for cases of constant corrosion losses, since contribution to the accumulated damage is equal in each year. Curves obtained by taking into account time-dependent HGSM loss have concave shape since accumulated damage per year is increasing with ship age as a consequence of increasing fatigue stresses. While this concave shape is moderate for slight and moderate corrosion, the concave shape is quite pronounced for severe and extreme corrosions. The CSR "rule" accumulated damage, calculated for 25% of CSR corrosion deduction thickness, corresponds approximately to the severe corrosion case. All results for accumulated damages are bounded from the lower side by the curve for "as-built" HGSM, while from the upper side by the curve for HGSM loss corresponding to 50% deduction thickness according to CSR. The latter case, i.e. 50% deduction thickness is proposed in CSR for calculation of hull girder properties for strength and buckling assessment.

Total damage in ship's lifetime is obtained by summing accumulated damages in ballast and full load conditions. After that, fatigue life is obtained employing equation (7) and results are presented in Table 3.

Table 3. Accumulated damages and fatigue lives for different levels of corrosion severity.

Corrosion severity	DM ballast	DM full load	DM total	Fatigue life (years)
Slight	0.59	0.17	0.76	33
Moderate	0.61	0.19	0.80	31
Severe	0.67	0.20	0.87	29
Extreme	0.73	0.21	0.94	27
As-built	0.48	0.14	0.62	40
25% CSR ("rule" case)	0.67	0.20	0.87	29
50% CSR	0.79	0.24	1.03	24

6 INFLUENCE OF SHIP REPAIR

As an industry standard, ship in service should maintain their HGSM at least 90% of the required as-built value. Therefore, HGSM loss curve for extreme corrosion presented in Figure 1 is not realistic as HGSM after 18 years becomes less than 90% of its initial as-built value. Shipowner of such ship would be obliged by rules of International Association of Classification Societies (IACS) and by regulations of International Maritime Organization (IMO) to repair ship in order to achieve 90% of as-built HGSM. This effect of ship repair is presented in Figure 6.

Although it is not clearly defined what should be the extent of the repair, it is assumed herein that repair is such to have at 25 years HGSM at approximately 90% of its as-built value. That could be a reasonable assumption in order to avoid new repair of HGSM during ship lifetime.

The accumulated damages with and without repair are presented in Figures 7 & 8 for full and ballast conditions respectively. The effect of the repair is the only slight increase of the fatigue life from 27 to 28 years.

7 FATIGUE YIELD ASSESSMENT

This section considers two components of fatigue yield in addition to the lifetime shortening due to corrosion.

The first aspect of fatigue yield is the fatigue strength worsening between successive load cases.

For practical purposes the S-N fatigue test data for welded joints in shipbuilding according CSR are given by the general fatigue strength and life relation $\Delta\sigma = C \cdot N^{-1/m}$. The fatigue yield approach applied to S-N data considers that the strength worsening $W = \Delta\sigma\,(1-D)^{-1/m}$ reduces the lifetime N_i in proportion to the damage progression $D_{j/i} = n_j / N_i$. Thus the fatigue yield rate is $dD'/dD = W/\Delta\sigma = (1-D)^{-1/m}$. Fatigue yielding based on S-N data represents the fatigue strength worsening due to earlier accumulated damages under alternating load cases with respect to the linear damage progression applied to an intact specimen (Ziha & Blagojevic 2009), as shown:

$$D'(D) = \int_0^D \frac{dD'}{dD} dD = \alpha \cdot \frac{m}{m-1} \cdot \left[1 - (1 - \alpha \cdot D_{j/i})^{\frac{m-1}{m}}\right] \tag{10}$$

where α is the worsening intensity rate.

The second aspect accounts for unpredictable stress amplitude variability during a single load case.

The load variability factor allows the assessment of the fatigue damage fraction $D'_{j/i} = \nu_i \cdot D_{j/i}$ under jth variable loading block for ith stress amplitude

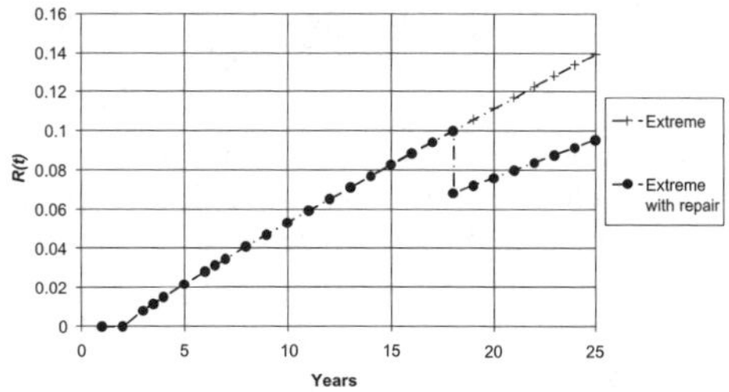

Figure 6. HGSM loss for extreme corrosion without and with effect of ship repair.

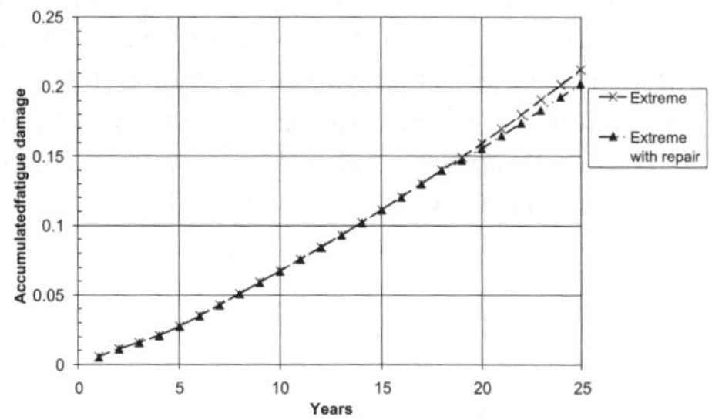

Figure 7. Accumulated damages in full load condition.

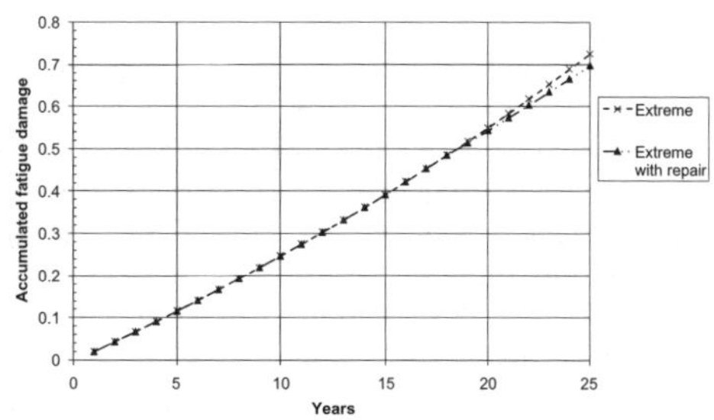

Figure 8. Accumulated damages in ballast condition.

$\Delta\sigma_i$ with respect to constant loads appropriate to linear damage progression for belonging fatigue lifetime N_i. The cumulative damage up to kth block of each variable stress amplitude load $\Delta\sigma_i$ for all i is then $D'(k) = \sum_{j=1}^{K} v_i \cdot D_{j/i}$. The approximation can benefit from piece-wise linearization of load variability factors for the expected range of stress amplitudes, as $D'(K) \approx V \cdot D(K)$, (Ziha & Blagojevic 2009).

The lifetime service of the example ship consists of a large number of alternating block loadings in fully loaded and in ballast conditions exposed to corrosion, Figs. 7 and 8. The additional damages are the consequence of fatigue yield due to worsening of primarily assumed fatigue properties caused by interchanging of intermittent service load conditions and of unpredictable load variability during a single load case, as it is summarized below:

$$D'(D) = \varepsilon_h \cdot \frac{3}{2} \cdot \left[1-(1-D)^{\frac{2}{3}}\right] + \varepsilon_l \cdot \frac{5}{4} \cdot \left[1-(1-D)^{\frac{4}{5}}\right] + V \cdot D \qquad (11)$$

In (11) $\varepsilon_h = 0.5$ and $\varepsilon_l = 0.5$ are the fractions of exposure to load amplitudes above and below the S-N curve knuckle at 10^7 cycles, that is for $m = 3$ and $m = 5$, respectively. The effect of fatigue yield under alternating load sequences shortens the lifetime at 85% of the lifetime calculated under linear damage progression assumption. Thus, the lifetime forecast of 27 shortens at 23 years due to yielding, Fig. 9.

However, an average load variability factor $V = 1.3$ in (11) for full load and ballast condition induces another lifetime reduction for additional 15%, that is altogether 19 year or 0.7 of the forecasted lifetime, Fig. 9. Note how the ship's operational profile is already accounted for in damage progression calculation what for the fatigue yield correction depends only on damage progression upon appropriate S-N data.

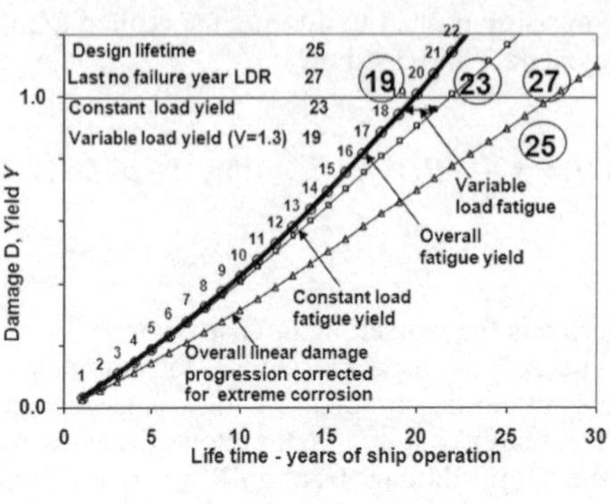

Figure 9. Fatigue yield by years of operation.

8 CONCLUSIONS

The fatigue failure is recognised as one of the governing failure modes in newly developed CSR for Double Hull Oil Tankers. Fatigue is not important only for design of ship structural details, but also may be governing criterion for the required HGSM at midship, i.e. for ship longitudinal strength, affecting thus the overall dimensions of structure subjected to fatigue.

The fatigue failure is accelerated by the corrosion. In first place, corrosion reduces structural scantlings and consequently increases fatiguing stresses. The paper proposes the methodology how to efficiently use recently published data on HGSM loss due to corrosion in the fatigue life assessment. This could lead to more refined and more rational results of the fatigue analysis. The procedure represents an improvement compared to commonly used approach proposed in CSR, as the rules propose constant corrosion loss throughout whole ship's lifetime, being an unrealistic assumption.

The presented analysis could be useful in inspection and maintenance planning, since it enables better prediction of remaining fatigue life of ships 10–15 years old. For those oil tankers, measurements of hull girder section modulus loss could be performed and results fitted to some of curves for different corrosion severities (Fig. 1). Then, the presented fatigue analysis could be performed providing better indication of hull girder fatigue behaviour compared to the rule approach.

The methodology is applicable also in case when ship repair is necessary because of loss of HGSM by more than 10% comparing to as built value. In the present example, ship repair didn't have significant effect on the remaining fatigue life of the main deck longitudinals.

The additional fatigue life shortening due to yielding calculated in the paper closes to the practical observation of 30% fatigue life reduction for a number of uncertain variable stress amplitude load cases.

REFERENCES

American Bureau Of Shipping, Det Norske Veritas, Lloyd's Register, 2006. *Common Structural Rules for Double Hull Oil Tankers.*

Wang, G., Lee, A.-K., Ivanov, Lj., Lynch, T.J., Serratella, C. & Basu, R. 2008. A statistical investigation of time-variant hull girder strength of aging ships and coating life, *Marine Structures* 21, 240–256.

Ziha, K. & Blagojevic, B. 2009, Fatigue Yield of Ship Structures, *Proceedings of the 28th International Conference on Offshore Mechanics and Arctic Engineering, OMAE2009, Honolulu, Hawaii*: 1211–1214.

Advanced Ship Design for Pollution Prevention – Guedes Soares & Parunov (eds)

Fatigue crack propagation in stiffened panels under tension loading

Ž. Božić & J. Parunov
Faculty of Mechanical Engineering and Naval Architecture, University of Zagreb, Zagreb, Croatia

M. Fadljević
Brže Više Bolje d.o.o., Zagreb, Croatia

ABSTRACT: Fatigue crack growth in stiffened panels under cyclic tension loading was studied in this paper. Results of fatigue crack propagation tests with cyclic stress of constant amplitude and frequency on centrally notched plate and stiffened panel specimens were reported. A numerical crack growth simulation procedure based on integration of Paris equation was introduced and fatigue lifetime was simulated for the specimens. Implementing the introduced procedure crack propagation in a main deck longitudinal of a double-hull oil tanker was analyzed. The Mode I, II and III Stress Intensity Factors (SIFs) were determined in a FEA using singular elements. Compared to Mode I SIF values, the Mode II and III values were small and did not influence structural components fatigue lifetime significantly. Secondary bending which occurs due to the cut stiffeners or flanges has been taken into account in the stress intensity factor determination. The study showed a significant influence of secondary bending on crack growth rate in stiffened panels. Also, it was shown that an increase in nominal loading stress range of 10% or decrease of 5%, can shorten or extend fatigue lifetime of the structural component by 30% or 20%, respectively.

1 INTRODUCTION

Fatigue cracking of stiffened panels is an important issue for structural integrity of aged ships. Under a variety of loading and environmental conditions fatigue cracks may initiate at sites of stress concentration due to geometrical discontinuities. For damage tolerance design it is important to determine fatigue crack growth lifetime of structural parts with damage cracks (Broek 1989). Fatigue crack propagation in stiffened panels has been investigated experimentally and by numerical simulations (Božić 1997). The difference in fatigue life of stiffened panels under cyclic tensile loading due to a single and multiple cracks damage has been demonstrated through comprehensive fatigue tests and numerical simulations (Sumi et al. 1996). When cracked stiffened panels undergo tensile loading, so called secondary bending occurs due to the cut stiffeners. The influence of local bending stresses on crack propagation life of stiffened panels was studied by Božić et al. (2009). It was shown in that paper that secondary bending in a stiffened panel may significantly increase local tensile stresses and consequently the crack propagation rate, which reduces components' lifetime.

In this paper a procedure for analysis of crack propagation in stiffened panels was described. The results of fatigue tests carried out on a centrally notched plate specimen and a stiffened panel specimen with a single central crack were presented. A crack growth simulation procedure based on integration of Paris' equation was given. The Paris' constants used in this study were determined using crack growth results of the centrally notched plate specimen. By using the simulation procedure fatigue life was simulated for the specimens. Mode I stress intensity factors (SIF), K_I, were calculated by ANSYS FEM program using shell elements and assuming plane stress conditions (ANSYS User's Manual 2008). FE analysis for the stiffened panel specimen showed high bending stresses in the intact ligament, which should be taken into account in the crack growth simulation. The SIF values calculated in the FE analysis using singular elements were scaled by a factor which depends on the ratio of bending and membrane stress components in the crack tip region.

Crack propagation in L type longitudinal of the main deck of a double-hull oil tanker was analyzed. Due to the flange non-symmetry all three crack tip opening modes: Mode I, II and III occur under tension loading. The K_I values were dominant compared to other two modes. Contribution of K_{II} and K_{III} to the effective SIF, K_{eff}, was small. Consequently, Modes II and III SIF did not influence fatigue lifetime of the analyzed longitudinal stiffener significantly.

2 FATIGUE TESTS

Fatigue tests were carried out on a centrally notched plate specimen (P1) and stiffened panel specimen (SP1). The specimens' geometry is given in Figures 1 and 2. A specimen was fixed by rigid tab plates at the ends, and it was loaded using a hydraulic testing machine. Fatigue test conditions applied in the experiment are listed in Table 1.

The cross sectional area of the intact section, and the average stress range away from the notch, are denoted as, A_0 and $\Delta\sigma_o$, respectively. The force range, and the stress ratio are denoted by $\Delta F = F_{max} - F_{min}$, and $R = F_{min}/F_{max}$, respectively. The average applied stress range was $\Delta\sigma_o = 80$ MPa for both specimens. Initial notch length was 2a = 8 mm. The loading frequency was 5 Hz.

The material used for the specimens is a conventional mild steel for weld construction with the material properties specified as follows: ultimate strength is over 400 MPa, yield strength is over 235 MPa, Young's modulus is 206 GPa, and Poisson's ratio is 0.3. Crack lengths were measured using crack-gauges and optically by a microscope. Experimental crack propagation results for P1 and SP1 specimens are given in Figure 3. Here the crack lengths *a* are to be considered as averaged half crack lengths.

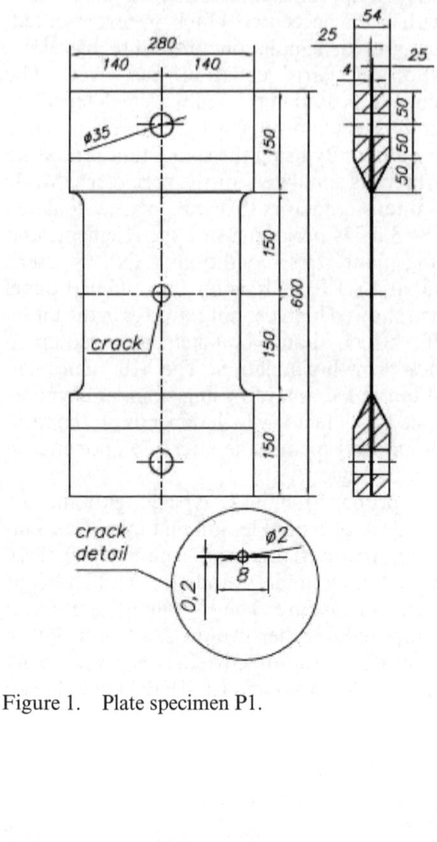

Figure 1. Plate specimen P1.

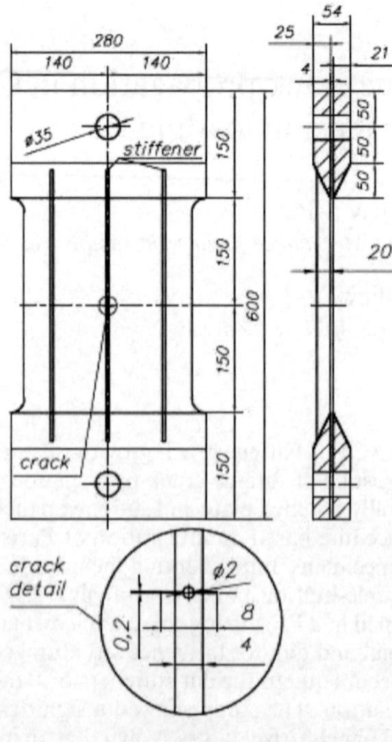

Figure 2. Stiffened panel specimen SP1.

Table 1. Fatigue test conditions.

	A_0 [mm²]	ΔF [N]	$\Delta\sigma_0$ [MPa]	R
P1	960	76800	80	0,0253
SP1	1200	96000	80	0,0204

Using experimental a-N data of P1 specimens the crack growth rate diagram was obtained as given in Figure 4.

In Figure 4 SIF range values ΔK correspond to those given in Section 3.1, Figure 6, for P1 specimen. By means of the interpolated line in the rate diagram the material constants of the Paris' equation are estimated as $C = 0.75 \cdot 10^{-12}$ and m = 3.5. The units for ΔK and $\Delta a/\Delta N$ are $[MPa\sqrt{m}]$ and [*m*], respectively. The m value is within a standard range for such materials, (Broek 1989). It was observed in the experiment that cracks propagated along a straight line without curving, which was to expect since only Mode I loading occur due to double symmetry.

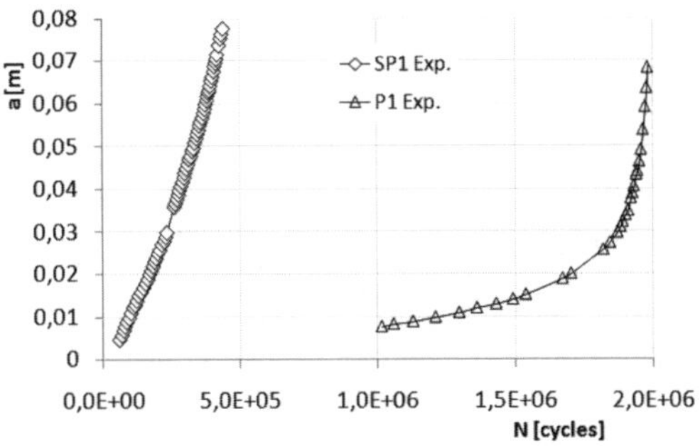

Figure 3. Experimental a-N data for P1 and SP1 specimens.

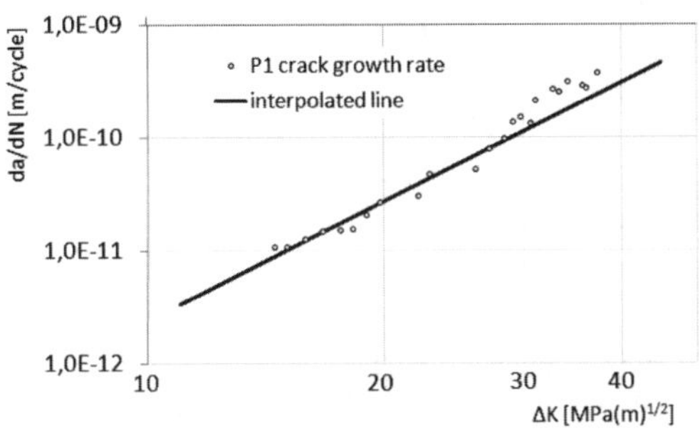

Figure 4. Crack growth rate data for P1 specimen.

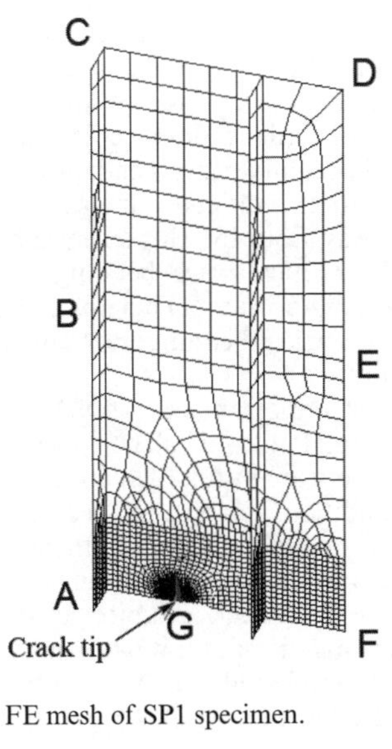

Figure 5. FE mesh of SP1 specimen.

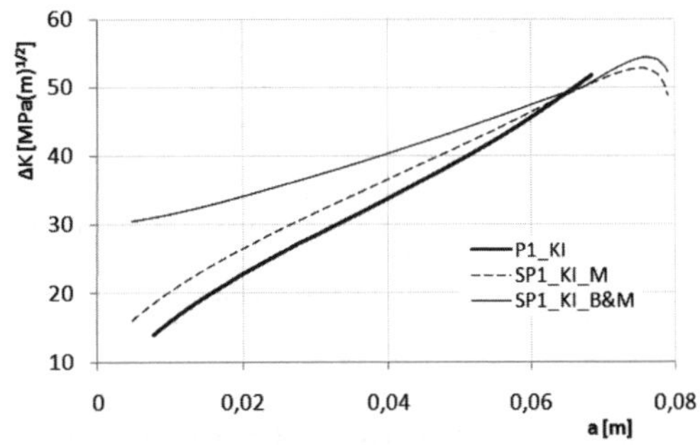

Figure 6. SIFs for P1 and SP1 specimens.

3 CRACK GROWTH SIMULATION

3.1 *Fatigue test specimens*

The crack propagation simulation procedure is based on numerical integration of the Paris' equation (1), where the stress intensity factors were calculated by ANSYS FEM program, (ANSYS 2008).

$$\frac{da}{dN} = C(\Delta K)^m \tag{1}$$

In the FE analysis eight node quadratic isoperimetric shell elements assuming plane stress conditions were used. The region surrounding the crack tip was meshed with singular elements, having midside nodes adjacent to the crack tip placed at the quarter points.

In the FE modeling due to the symmetry of specimen geometry and loading conditions it is sufficient to model one quarter of the specimens. FE mesh of the SP1 specimen is given in Figure 5. Loading conditions are given in Table 1. The applied boundary conditions are as given in Table 2. It should be clarified that cracked surface is represented by straight line AG, where G is a crack tip. Therefore, AG part of the model boundary is free to deform. Rectangular surface between corners B, C, D and E represents thick plate on Figure 2. That part of the model can displace in axial direction only because of the settings of the testing equipment.

With shell elements employed in the analysis the SIF values were calculated from FE results for nodal displacements of singular elements in a standard post processing procedure, (ANSYS 2008). The standard procedure for calculation of SIF by shell elements do not take into account local bending in the vicinity of a crack tip. Under tension loading in SP1 specimen significant bending occurs due to the change of cross sectional area

Table 2. Boundary conditions for SP1 specimen.

Plane	Boundary conditions
ABC	symmetry
CD	axial load applied
BCDE	displacements in axial plane allowed
GF	symmetry
AG	free (cracked surface)

geometry characteristics in the cracked surface. Due to cut (cracked) area the second central main axis is shifted towards the plate.

In order to determine the SIF values associated with the plate surface on which positive bending occurs a simple linear extrapolation procedure was employed. A correction factor was introduced, $cfb = 1 + \sigma_{y\ Bend}/\sigma_{y\ Memb}$, where $\sigma_{y\ Bend}$ and $\sigma_{y\ Memb}$ are positive bending stress component and membrane stress component in front of a crack tip in direction perpendicular to the crack face, respectively. SIF values associated with the plate surface on which positive bending occur, $\Delta K_I(_{B\&M})$, are calculated as: $\Delta K_I(_{B\&M}) = cfb \cdot \Delta K_I(_M)$, where $\Delta K_I(_M)$ are SIFs calculated by ANSYS in a postprocessing procedure using shell elements as described above. Figure 6 shows the SIF values for specimens P1 and SP1.

SP1 has higher SIF values than P1, which is due to stiffener influence and larger cut area. $\Delta K_I(_{B\&M})$ values which take account of bending stresses are higher compared to $\Delta K_I(_M)$, especially for the shorter a values where secondary bending is emphasized. It is to observe that ΔK_I values decrease as the crack tip approaches to the intact stiffener. Simulated crack growth lives for the specimens are given in Figures 7 and 8. For P1 specimen a good agreement of numerical simulation and experimental results is observed. Regarding SP1 specimen, the case based on $\Delta K_I(_M)$ values which do not consider local bending stresses, (M), gives fatigue life which is significantly longer compared to experimental results. The analyzed case with bending included (B&M) gives shorter fatigue life than one obtained in the experiment. Residual welding stresses, which were not taken into account in the simulation, could in this case decrease crack growth rate as the crack tip traverses the plate and approaches to the intact stiffener. This could be a possible reason for discrepancy between simulated B&M fatigue life and the experimental one.

In the numerical model presented herein it is assumed that a double symmetry exists and correspondingly only Mode I SIFs occur. This assumption is based on the experiment, where cracks propagated along a straight line until specimens failed. Crack curving was not observed.

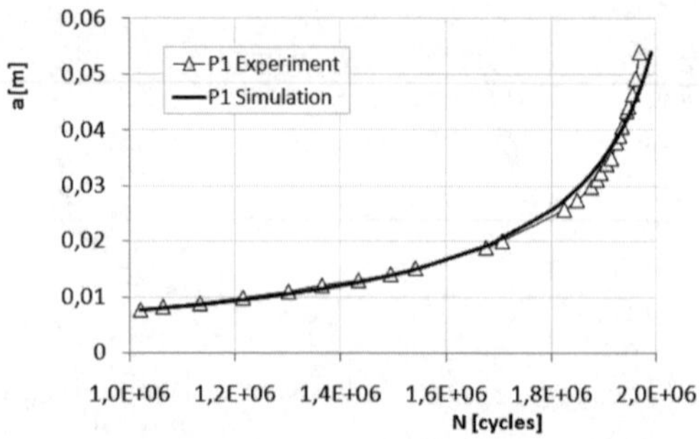

Figure 7. Simulated fatigue lifetime for P1 specimen in comparison with experimental result.

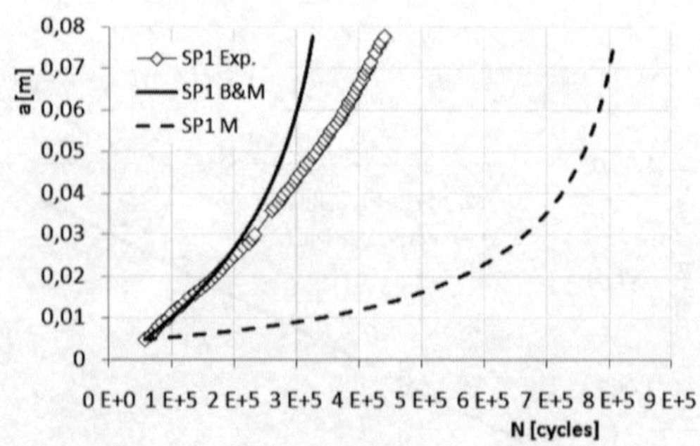

Figure 8. Simulated fatigue lifetime for SP1 specimen in comparison with experimental result.

3.2 *Fatigue stresses of main deck of oil tanker*

Main deck of a double-hull oil tanker is loaded by two types of global loads: still water bending moments and wave-induced bending moments. Still water loads are slowly-varying loads that typically do not induce fatigue damage, while fluctuating wave-induced loads are considered as the main generators of fatigue damage in ship structure. Although horizontal wave bending moments may contribute to the fatigue damage of main deck elements away from the ship's centre line, stresses induced by vertical wave bending moments are much more important. Therefore, only stresses induced by vertical bending moments are considered herein.

Two most important operational modes of oil tankers are full load and ballast. Normally, double-hull oil tankers are in full load condition sailing in sagging condition, i.e. the main deck is subjected to the compressive still water loads. In ballast, double-hull tankers are usually sailing in hogging condition, i.e. the main deck is subjected to the tension loads. The analysis of still water bending moments of modern double-hull tankers of different sizes is provided by Parunov et al. (2009). It is generally accepted by classification societies that oil tankers spend approximately same amount of time in ballast and full load conditions (ABS et al. 2006).

Amplitudes of fluctuating, wave-induced stresses, are of the random nature. The stress range is not constant, but randomly distributed in long-term according to the Weibull probability distribution law. Normally, extreme stresses induced by global wave bending moments are larger than stresses induced by still-water bending moments. This has a consequence that main deck may be loaded by tension stresses in sagging and also by compression stresses in hogging. However, such extreme wave-induced stresses occur seldom, with very low probability, and they are practically insignificant for fatigue life of the main deck structure considering the S-N approach to the fatigue life analysis. Much more important for fatigue life are lower stress ranges, occurring fairly often and having relatively high probability of occurrence. It is well known that such stresses have significant impact on fatigue life of ship structures. Based on these considerations, one may conclude that the main deck structure in ballast is predominately loaded in tension and that assumptions used in the present paper are primarily applicable for ballast loading condition.

Considering the Linear Elastic Fracture Mechanics (LEFM) based crack propagation assessment a random fluctuation of loading stress amplitudes may has significant influence on crack growth rate and corresponding fatigue life. For example, so-called retardation is a well-known effect where a larger crack tip plasticity zone is generated due to an overload cycle, which temporarily slows down crack propagation. Also, loading block sequencing may have significant influence on fatigue crack propagation rate.

As the stress intensity factor range ΔK depends on the applied stresses, and since the methodology employed in the paper requires deterministic stress range to be used, it is of interest to demonstrate how these deterministic stress range may be approximated considering the SN approach.

Stochastic stress ranges $\Delta\sigma$ in long term time period may be represented by Weibull probability distribution:

$$F_w(\Delta\sigma) = 1 - e^{\left(-\left(\frac{\Delta\sigma}{q}\right)^k\right)} \quad (2)$$

where q and h are Weibull scale and shape parameter respectively. Values of q and h for given vessel depend on the wave environment where ship sails. For an example double-hull oil tanker sailing in North Atlantic, h is close to unity, while $q = 14$ MPa (Jurišić et al. 2008).

Accumulated damage, using Weibull long term stress range distribution and commonly used S-N approach, reads:

$$D = \frac{\upsilon_0 T_d}{K_P} q^m \Gamma\left(1 + \frac{m}{h}\right) \quad (3)$$

where υ_0 represent average frequency of wave load cycles, while T_d is design lifetime of the ship. K_P and m are S-N curve coefficient and exponent respectively while Γ represents Gamma function. By analogy with deterministic approach, one finds out that

$$q^m \Gamma\left(1 + \frac{m}{h}\right) \quad (4)$$

represents an equivalent to $\Delta\sigma^m$ in pure deterministic approach. By taking into account that m is usually equal to 3, then $\Gamma = 6$, while an equivalent deterministic stress range $\Delta\sigma$ reads 25 MPa.

It should be clearly stated that the deterministic equivalent to the stochactic stress ranges is only an approximation used in the present paper and that further research is necessary to demonstrate its real applicability and accuracy in crack propagation problems. It was necessary in the present paper to determine at least approximately equivalent deterministic stress range $\Delta\sigma$ and corresponding SIF range in a cracked component in order to check whether the values approach to a critical K_{Ic} value.

Because of uncertainty in material constants C and m, and due to the reasons mentioned above the estimated fatigue crack propagation lifetimes are often given as normalized values rather then as absolute number of cycles.

3.3 *Crack propagation in a deck longitudinal stiffener*

In deck structure of a ship, cracks may initiate by stress concentration due to geometrical discontinuity. Scantlings of the analyzed ship deck structure are shown in Figure 9. Web frame spacing a reads 4.3 m, while stiffener spacing b reads 815 mm. Deck longitudinal is L profile with web 337 × 8 mm and flange 96 × 13 mm. A typical location where fatigue crack may initiate is the welded connection of the stiffener of the transversal deck girder and the face plate (flange) of a deck longitudinal. The initiated crack can further propagate through the stiffener and penetrate into the deck plate. FEM mesh of the model is illustrated in Figure 10. In the finite element modeling, the length of the model extends three bays of transverse frames. The width of the model is nine stiffener spacing, while the symmetry

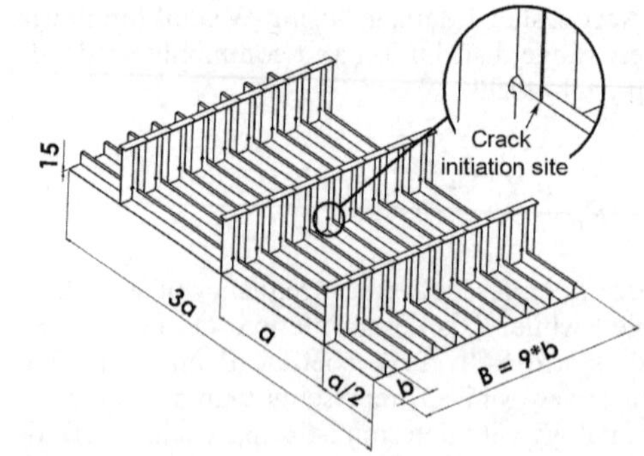

Figure 9. Deck ship structure part.

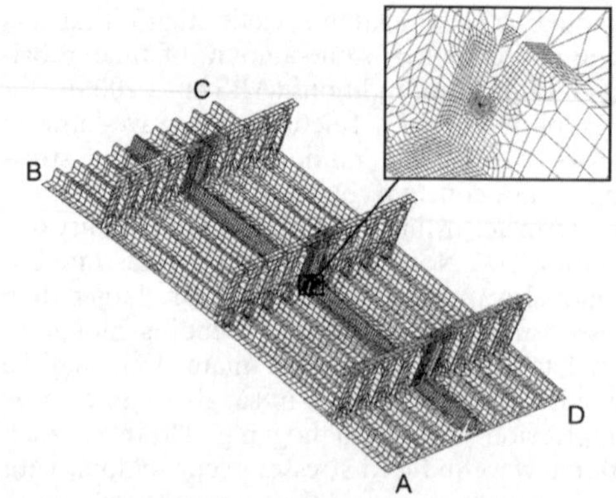

Figure 10. FE mesh of deck ship structure part.

boundary conditions are imposed on both sides. In FE modeling several intact stiffeners should be included in order to take into account the influence of the load carrying capacity of the surrounding structure. It was found out using FEM analysis (Božić 1997) that two and half bays is enough for the half model to take into account the load carrying capacity of the surrounding structure.

Applied boundary conditions are summarized in the Table 3. Assuming the stress range of 25 Mpa, as determined in Section 3.2, the Mode I, II and III stress intensity factors, K_I, K_{II} and K_{III} have been calculated for several crack lengths within the range, 30–320 mm. It is assumed herein that crack propagates along a straight vertical line towards the deck plate.

Mode I SIF values $\Delta K_I(_M)$ are calculated by ANSYS in a post-processing procedure using singular elements as described in Section 3.1. The SIF values associated with the plate surface on which positive bending occur, $\Delta K_I(_{B\&M})$, are calculated by implementing the *cfb* factor and according to the procedure described in Section 3.1. $\Delta K_I(_{B\&M})$ values are higher compared to $\Delta K_I(_M)$ values, particularly for shorter crack lengths where local bending is emphasized. Mode II and III SIFs are much less compared to the Mode I values. When Mode I and III values occur in loading simultaneously then effective SIF value, K_{eff}, should be considered in a fatigue crack propagation analysis. The K_{eff} value is calculated according to Equation 5, (Broek 1989).

$$K_{eff} = \sqrt{K_I^{\,2} + \left(0.8 \cdot K_{III}\right)^2} \tag{5}$$

$\Delta K_I(_{B\&M})_{eff}$ and $\Delta K_I(_M)_{eff}$ are effective SIFs calculated according to Equation 5 assuming $\Delta K_I(_{B\&M})$ and $\Delta K_I(_M)$ as Mode I values. Since contribution of K_{III} values in Equation (5) is small the $\Delta K_I(_{B\&M})_{eff}$ and $\Delta K_I(_M)_{eff}$ values are only slightly higher than $\Delta K_I(_{B\&M})$ and $\Delta K_I(_M)$ values, respectively, as shown in Figure 11. Correspondingly, for the considered structural detail the influence of Mode III SIFs on fatigue life is negligible and it is sufficient to consider only Mode I in fatigue life analysis. In order to simplify the analysis it is assumed in this study that ΔK_{II} values do not cause significant crack curving, since Mode II SIF values are small compared to Mode I SIF values.

Influence of bending stresses on fatigue life is demonstrated in Figure 12. The figure shows normalized fatigue lifetime associated with $\Delta K_I(_{B\&M})$ and $\Delta K_I(_M)$ SIF ranges, where the normalization is performed according to Equation 6.

One can see from the diagram that simulated fatigue life which considers local bending, B&M, is only 50% of that without including bending, (M). It is to observe that bending stresses significantly increase crack growth rate for shorter crack lengths.

$$\frac{N_{(B\&M)}}{N_{(M)}} = \frac{\displaystyle\int_{a_0}^{a} \frac{da}{\left(\Delta K_{I(B\&M)}\right)^m}}{\displaystyle\int_{a_0}^{a} \frac{da}{\left(\Delta K_{I(M)}\right)^m}} \tag{6}$$

Therefore, it is important to consider local bending in fatigue life analysis of ship structural components. In further analyses only Mode I SIF taking account of bending stresses (B&M) will be considered in fatigue lifetime assessment.

Furthermore, the influence of nominal tension stress on fatigue lifetime of the deck longitudinal is considered. For that purpose three load cases were studied: a nominal tension stress range increased for 10% and decreased for 5%. SIF values for the three studied case are given in Figure 13. Fatigue lifetime results $N_{B\&M}$, $N_{B\&M+10\%}$ and $N_{B\&M-5\%}$,

Table 3. Boundary conditions.

Vertical plane	Boundary conditions
AD	axial and transversal displacement allowed
AB	symmetry
BC	symmetry
CD	symmetry

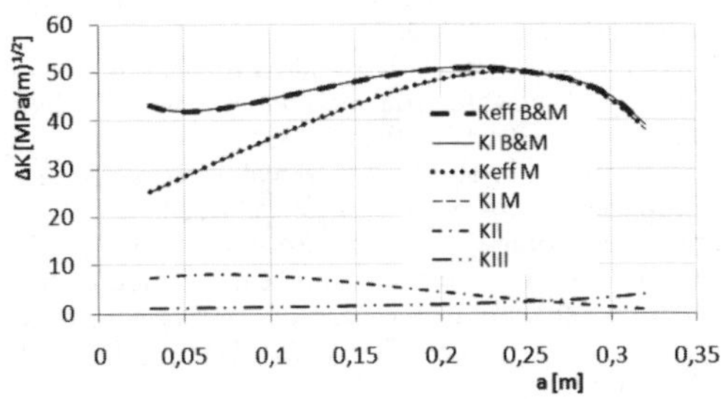

Figure 11. SIF values in the deck longitudinal.

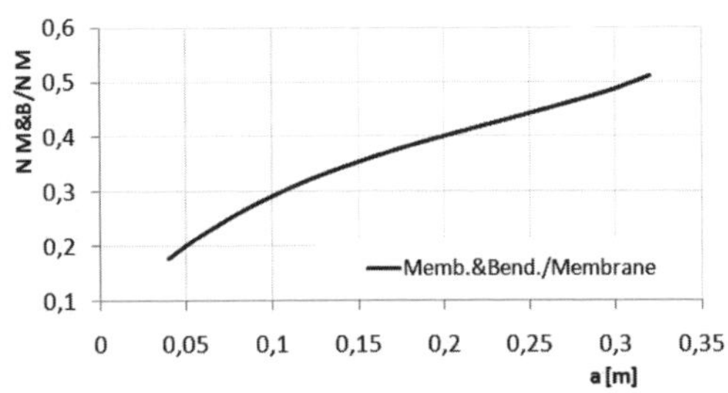

Figure 12. Normalized fatigue lifetime considering B&M and M cases.

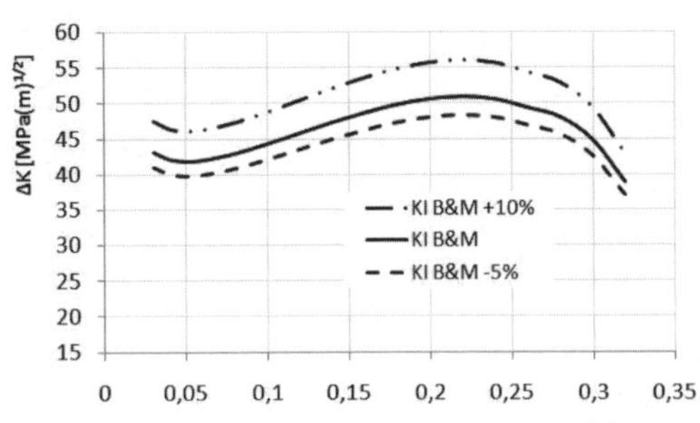

Figure 13. SIF values in the deck longitudinal considering 10% load increase and 5% load decrease.

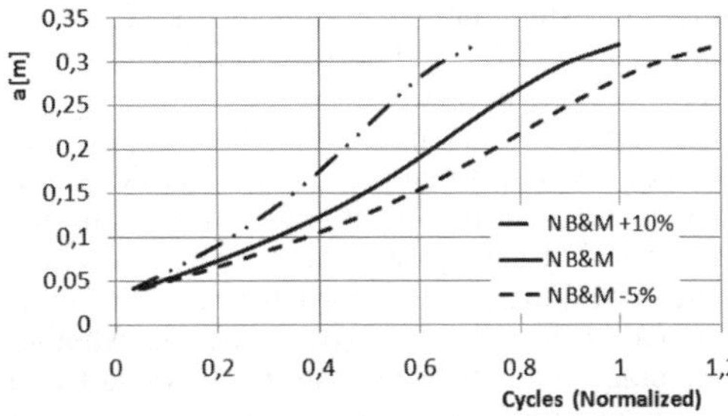

Figure 14. Normalized fatigue lifetime.

normalized with respect to $N_{B\&M}$ are shown in Figure 14. Normalization is performed in the following way:

$N_{Nnormalized}(a) = N(a)/N_{B\&M}$, where N(a) is a calculated fatigue life for the considered case and $N_{B\&M}$ is total lifetime of the B&M case associated with the crack length 320 mm.

From the Figure 14 one can see that a 10% increase in nominal tension stress decreases fatigue life for 30%, and that a 5% decrease in nominal tension stress extends fatigue life for 20%.

10% increase of stresses may typically occur as a consequence of the hull-girder section modulus degradation due to corrosion. 5% increase is used herein to represent the influence of the hull-girder section modulus increase because of the new Common Structural Rules for design and construction of double-hull tankers (ABS et al. 2006).

4 CONCLUSION

FE analysis showed that in cracked stiffened panel specimen under tensile loading high local bending stresses occur, which significantly influence SIF values and crack growth rate. Crack propagation simulations based on modified SIFs, which take account of bending, proved to be conservative, giving shorter crack propagation life, compared with the experimental results. The reason for it could be the welding residual stresses which were not taken into account in the simulation.

Considering crack propagation in a deck longitudinal with an L-type flange of an actual ship the Modes I, II and III SIFs were determined in a FEA using singular elements. Mode II and III SIF values are much less then Mode I values and do not influence structural components fatigue lifetime significantly.

Local bending increases SIF values in stiffened panels, particularly for shorter crack lengths, significantly reduces fatigue lifetime of the component. Also, it was shown that an increase in nominal loading stress range of 10% or decrease of 5%, can shorten or extend fatigue lifetime of the structural component by 30% or 20%, respectively.

The presented methodology of crack propagation analysis may have important application in condition assessment of aged oil tankers and rational inspection and maintenance planning. Determined crack growth rate and a-N curve shape in conjunction with regularly spaced inspections can help ship surveyor to decide whether some crack requires immediate repair or not and how long it could take that crack propagates from certain small size to the unacceptably large size (e.g. through the whole stiffener web).

REFERENCES

American Bureau of Shipping, Det Norske Veritas, Lloyd's Register, 2006. *Common Structural Rules for Double Hull Oil Tankers.*

Bozic, Z. 1997. *Fatigue and Fracture of Multiple Site Cracks in Stiffened Panels.* Ph.D. Dissertation: Department of Naval Architecture and Ocean Engineering, Yokohama National University, Japan.

Bozic, Z., Wolf, H., Semenski, D., Bitunjac, V. 2009. Fatigue of Stiffened Panels with Multiple Interacting Cracks—an Experimental and Numerical Simulation Analysis, *12th International Conference on Fracture, Ottawa, Canada.*

Broek, D. 1989. *The Practical Use of Fracture Mechanics.* Dordrecht, The Netherlands: Kluwer Academic Publishers.

Dexter, R.J., Pilarski, P.J., Mahmoud, H.N. 2003. Analysis of crack propagation in welded stiffened panels. *International Journal of Fatigue*, 25 (9–11):1169–1174.

Jeom Kee Paik, Bong Ju Kim, Jung Kwan Seo 2008. Methods for ultimate limit state assessment of ships and ship-shaped offshore structures: Part II stiffened panels, *Ocean Engineering* 35:271–280.

Jurišić, P., Parunov, J., Senjanović, I. 2007. Assessment of Aframax Tanker Hull-Girder Fatigue Strength According to New Common Structural Rules. *Brodogradnja* 58(3); 262–267.

Mahmoud, H.N., Dexter, R.J. 2005. Propagation rate of large cracks in stiffened panels under tension loading. *Marine Structures* 18 (3): 265–288.

Paris, P., Erdogan, F. 1963. A critical analysis of crack propagation laws. *J Basic Eng* 85:528–534.

Parunov, J., Ćorak, M., Guedes Soares, C. 2009. Statistics of still water bending moments on double hull tankers, *Analysis and Design of Marine Structures, Eds. C.Guedes Soares & P.K Das,* London:Taylor & Francis Group.

Sumi, Y., Bozic, Z., Iyama, H., Kawamura, Y. 1996. Multiple Fatigue Cracks Propagating in a Stiffened Panel. *Journal of the Society of Naval Architects of Japan*: 407–412.

Swanson Analysis System, Inc. 2008. *ANSYS User's Manual Revision 10.0.*

Advanced Ship Design for Pollution Prevention – Guedes Soares & Parunov (eds)
© 2010 Taylor & Francis Group, London, ISBN 978-0-415-58477-7

High cycle and low cycle fatigue in ship structures

S. Rudan
Faculty of Mechanical Engineering and Naval Architecture, University of Zagreb, Zagreb, Croatia

ABSTRACT: Ship structures are particularly prone to fatigue of material. As they sail wave loads and other cycling loads are acting on ship structural details and fatigue damage is being accumulated. Two main types of fatigue may be distinguished: high cycle fatigue, with moderate loads in big number of cycles, and low cycle fatigue, with high loads in medium number of cycles. Both can cause fatigue failure, but a combination of the two even more. However, the problem of combined HCF and LCF is yet to be understood thoroughly. This paper gives an introduction in the state-of-the-art HCF and simplified LCF calculations. It points out to several models proposed for HCF and LCF combination into resulting fatigue. Procedures are illustrated through one numerical example and results are discussed.

1 INTRODUCTION

Fatigue of materials in ships, and wide range of other types of structures, is a long known problem. It occurs due to inevitable microscopic damage accumulation in material due to cyclic loading, such as e.g. wave loads, so that after certain amount of time structural failure occurs. This may lead to problems that vary in scale from unwanted and costly repairs to a large crack propagation that endangers entire structure. Despite of significant scientific effort involved in understanding fatigue phenomena fatigue failure is still a significant source of problems.

Oceangoing ships are particularly prone to fatigue due to several reasons occurring simultaneously: they operate in the corrosive environment for 20 years or more, they are exposed to various loads and at times sail in harsh environment, ship structure contains very big number of common or special types of structural details, the length of the welds is measured in kilometers etc.

Two basic types of material fatigue may be distinguished: first, high cycle fatigue (HCF) that occurs due to a large number of medium strength loads and second, low cycle fatigue (LCF) that occurs due to a small number of high strength loads.

In shipbuilding industry HCF is commonly investigated using statistical and probabilistic methods in combination with hydrodynamic and structural analysis, either by simplified method or by direct calculations. Classification societies Rules are followed whenever possible. Fracture analysis of crack growth is usually not of main concern as cracks appear on various locations (structural details), so statistic and probabilistic approach better fits the purpose. In addition, the cracks in ship structures may sometimes grow for meters in length before becoming a threat to the ship integrity. On the other hand LCF is not as extensively investigated and the reasons for that will be explained. LCF tends to shorten, sometimes significantly, the life of structural details that include notches, imperfections, micro cracks and any other possible source of highly stressed local regions. Finally, combination of HCF and LCF in ship structures is only recently being systematically examined.

This paper addresses some of the procedures available for estimating HCF, LCF and their combined effect on fatigue life of ship structural details. HCF analysis is presented through state-of-the-art spectral fatigue analysis, which is a procedure for direct fatigue damage calculation. LCF analysis is done using simple closed form equations due to lack of consistent and more sophisticated procedures. Several proposed models on how to combine HCF and LCF is presented.

The application of described procedures is presented on fatigue analysis of particular problem commonly found in LPG ships. An imperfection in Y-joint of shells and longitudinal bulkhead in bilobe tanks leads to high stress concentration that significantly shortens fatigue life. The problem is not negligible and must be controlled.

2 HIGH CYCLE FATIGUE

High cycle fatigue (HCF) of structural details occurs due to a high number of loading cycles of moderate amplitude, typically from 10^4 to 10^8 or more cycles. During their lifetime ships are exposed to a number of loads that vary in time, such as wave loads, vibrations, thermal stresses etc. Among these, wave loads are considered to be the main cause of high cycle fatigue for structural details directly exposed to wave loads (e.g. side shell longitudinals), but also for a number of other structural details that "feel" the wave induced hull structure deformation.

Structural details life-span is commonly estimated by fatigue damage summation based on S-N (Wöhler) curves. S-N curves are constructed using the results of a number of cyclic load tests until the structural detail failure. Two main groups of standard S-N curves are UK DEn (HSE) (Bureau Veritas 1997) and IIW S-N curves (IIW 1995), covering a range of structural details in classes and categories.

Although these S-N curves are being recognized as the industry standard, they are not always easily applicable to problems in shipbuilding industry as they are not able to take into account a variety of structural details found in ship structures and also they do not address the corrosion environment issues directly. Due to that certain Classification societies define universal S-N curves that are applicable to any type of structural details, both in aggressive and non-aggressive environment (DNV 2003). For example, DNV defines four universal S-N curves: S-N I for welded joints in air or with cathode protection against corrosion, S-N II for welded joints in corrosive environment and, in the similar manner, S-N III and S-N IV curves for base material.

DNV universal S-N curves are based upon UK DEn S-N curves of class F and F2 but modified in a way that resulting lifetime is equal when using F and F2 S-N curves together with adequate stress concentration factors, and universal DNV S-N curves together with stress concentration factors evaluated from FEM analyses. Once the proper S-N curve is selected, fatigue damage may be evaluated using Palmgren-Miner summation:

$$D = \sum_{i=1}^{k} \frac{n_i}{N_i} \tag{1}$$

where: k is a number of reference stress ranges, n_i is a number of stress cycles at $\Delta\sigma_i$ and N_i is a number of cycles at $\Delta\sigma_i$ until failure. If the structural load is of constant amplitude, the number of cycles till failure may be determined directly from S-N curve at a single point ($\Delta\sigma$, N) or using analytical expression for S-N curve:

$$N = \bar{a}\Delta\sigma^{-m} \tag{2}$$

where $\bar{a}$ and m are S-N curve parameters, Figure 1.

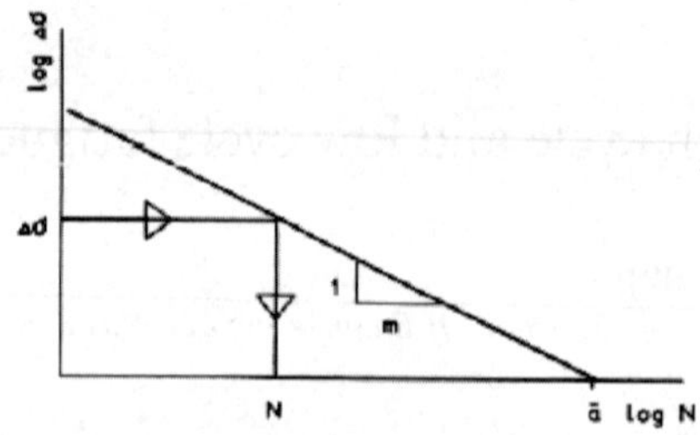

Figure 1. One slope S-N curve.

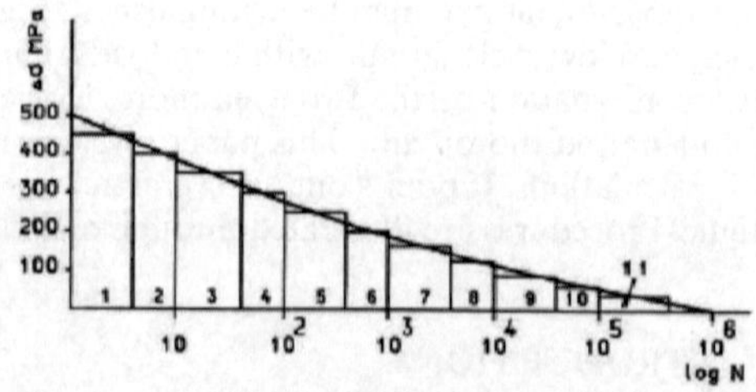

Figure 2. Stress blocks discretization of S-N curve.

If the amplitude varies, a long-term distribution of stress ranges may be divided approximately in blocks (intervals) of constant stress range, Figure 2. Then particular damage is calculated for a number of cycles for each block and total damage according to Palmgren-Miner summation rule.

Due to complexity of ocean waves the stress range cycles on oceangoing vessels are considered to be distributed by the probability function $f(\Delta\sigma)$. Since a structural detail is exposed to n_0 stress cycles during its lifetime it follows that a number of cycles for each stress range interval [$\Delta\sigma$, $\Delta\sigma + d\Delta\sigma$] is equal to $n_0\ f(\Delta\sigma)d\Delta\sigma$. Then Palmgren-Miner summation takes the form:

$$D = \int_0^\infty \frac{n_0 f(\Delta\sigma)}{N(\Delta\sigma)} d\Delta\sigma \tag{3}$$

where $N(\Delta\sigma)$ is a number of cycles till failure at constant stress range $\Delta\sigma$. If stress distribution probability function is a two-parameter Weibull function:

$$f(\Delta\sigma) = \frac{h}{q}\left(\frac{\Delta\sigma}{q}\right)^{h-1} \exp\left[-\frac{\Delta\sigma}{q}\right]^h \tag{4}$$

where q is stress range and h is shape parameter, a resulting expression for high cycle fatigue damage summation then takes the form (DNV 2003):

$$D_{HCF} = \frac{T_d}{T_0 \cdot \bar{a}} p_n q_n^m \Gamma\left(1 + \frac{m}{h_n}\right) \tag{5}$$

where D_{HCF} is high-cycle fatigue damage, $\bar{a}$ and m are S-N curve parameters, T_d is design life of ship in seconds, T_0 is long-term average response zero-crossing period, p_n is fraction of design life in loading condition n, and q_n and h_n are Weibull stress range scale and shape parameters, respectively, for each loading condition n.

High-cycle fatigue analysis aims to take into account most effects that contribute to fatigue damage accumulation in structural details on the ocean-going vessels such as sailing route, loading condition, sea state etc. It normally consists of four steps:

1. hydrodynamic analysis, where wave loads are calculated for different combinations of wave length, heading angle, ship speed and ship loading conditions;
2. structural analysis, where ship structure is subjected to hydrodynamic loads;
3. statistic and stochastic analysis, where stress transfer functions are combined with particular sea-state occurrence probability and so the long-term stress range distribution is evaluated;
4. fatigue damage calculation.

3 LOW CYCLE FATIGUE

Highly stressed structural details are prone to low cycle fatigue (LCF) when fatigue of material occurs after 10 to 10^4 cycles. Although Classification societies verify structural details against fatigue for the entire life of ship in service, often fatigue damage occurs much earlier, after several years of sailing or even before. This then leads to costly ship repairs and should therefore be avoided. LCF will occur at locations where micro fractures exist, at the weld root notches or any other location where local stress is very high and higher than yield stress at times. LCF in ship structures is only recently being investigated by for example (Urm et al. 2004) and (Heo et al. 2004). In (Urm et al. 2004) authors list the main reasons for the lack of comprehensive understanding of LCF in ship structures:

- cyclic curves σ-ε for base and weld materials in shipbuilding are not available in literature,
- ε-N and S-N curves do not exist for $N < 10^4$ cycles,
- practical and simple LCF calculation procedure for ship structures is not defined.

Currently, LCF calculation methods may fit in one of two groups: methods using range of local deformations and method of pseudo-elastic stresses. Of the two, only former method will be mentioned here. Method of local deformations aims to determine deformation and stresses at highly stressed micro-locations as a function of global deformation and stresses of structural details. Figure 3 sketches the relation between elastic global behavior of structural detail and elastic-plastic local behavior of crack tip at micro-location.

Tests with controlled deformation are the main source of information about behavior of materials with cracks and micro-cracks. Global elastic behavior affects local elastic-plastic behavior in a way that local stress-deformation curve is formed in a shape of hysteresis. If the material behaves in a stable manner, after a number of cycles the shape of hysteresis doesn't undergo any further changes and then all the points on hysteresis satisfy Ramberg-Osgood equation:

$$\varepsilon_{loc} = \frac{\sigma_{loc}}{E'} + \left(\frac{\sigma_{loc}}{K'}\right)^{1/n'} \tag{6}$$

where σ_{loc} and ε_{loc} are local stress and strain, respectively, E' is stable cyclic Young's modulus of elasticity, K' is strength coefficient and n' is strain hardening exponent. By replacing σ and ε with $\Delta\sigma_{loc}/2$ and $\Delta\varepsilon_{loc}/2$, local deformation range may be expressed as:

$$\frac{\Delta\varepsilon_{loc}}{2} = \left(\frac{\Delta\sigma_{loc}}{2E'}\right) + \left(\frac{\Delta\sigma_{loc}}{2K'}\right)^{1/n'}. \tag{7}$$

Neuber rule relates these local stresses and deformation with the global ones, enabling simple empirically based LCF calculation. To achieve that, two different stress concentration factors are to be

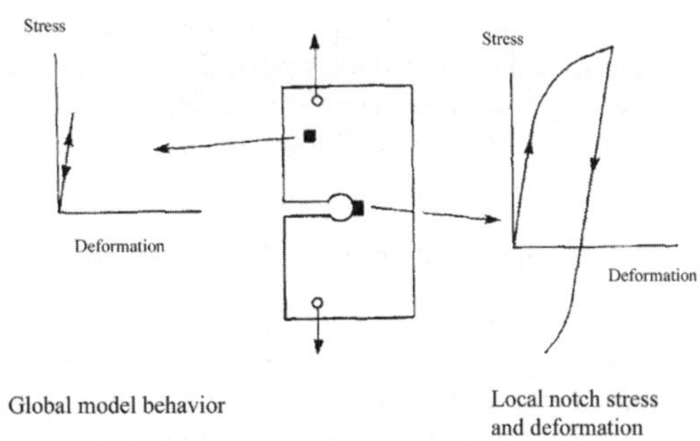

Figure 3. Global to local (notch) stress and deformation relationship.

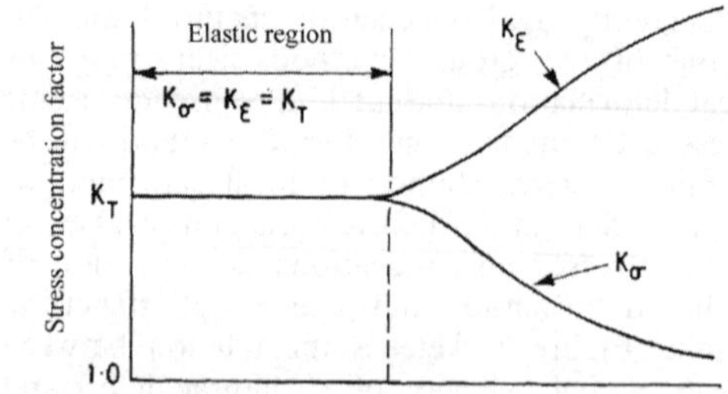

Figure 4. Stress concentration factors in elastic and elastic-plastic region.

defined: K_σ and K_ε, being stress concentration factor and strain concentration factor, respectively:

$$K_\sigma = \frac{\text{maximum local stress}}{\text{global stress}} \qquad K_\varepsilon = \frac{\text{maximum local strain}}{\text{global strain}} \tag{8}$$

Neuber showed that product of the two is constant i.e. that $K_\sigma \cdot K_\varepsilon = K_T^2 = const.$, Figure 4.

By defining $K_\sigma = \sigma_{loc}/\sigma$ and $K_\varepsilon = \varepsilon_{loc}/\varepsilon$, where σ and ε are global stress and deformation respectively, according to Neuber rule it follows:

$$K_\sigma \cdot K_\varepsilon = \frac{\sigma_{loc} \cdot \varepsilon_{loc}}{\sigma \cdot \varepsilon} = K_T^2 \tag{9}$$

In the elastic region $\varepsilon = \sigma/E$ and introducing this into equation above:

$$\sigma_{loc} \cdot \varepsilon_{loc} = \frac{\left(K_T^2 \cdot \sigma^2\right)}{E}, \text{ and} \tag{10}$$

$$\Delta\sigma_{loc} \cdot \Delta\varepsilon_{loc} = \frac{\left(K_T^2 \cdot \Delta\sigma^2\right)}{E} \tag{11}$$

Simultaneously solving equations above by iteration method, it becomes possible to determine the range of local deformations in dependence on given global stress range (Sleczka 2004).

4 CUMULATIVE HCF AND LCF LIFE

A combination of HCF and LCF is recognized in studies of jet engine turbine blades: after an airplane lifts off, which rises mean stress value, a high cycle stress oscillations occur. Landing of the airplane decreases stresses to zero and then entire cycle is repeated next time. A graphical representation of this "start-stop" model is presented in Figure 5,

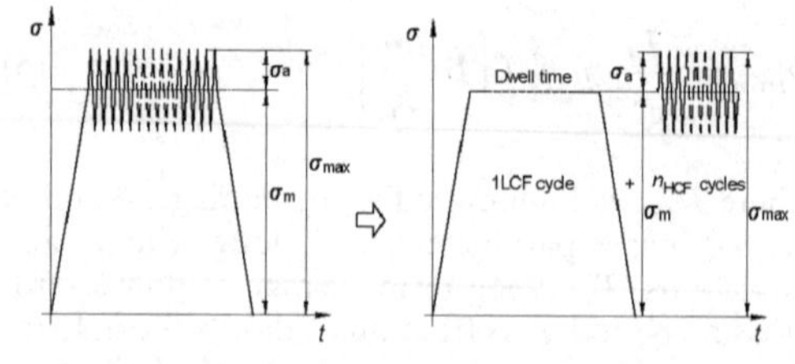

Figure 5. "Start-stop" model of combined HCF and LCF (Jelaska et al. 2003).

where a single flight stress history is split into one LCF cycle and a number of HCF cycles. Single and multiple overloads in "start-stop" model were analyzed by (Byrne et al. 2003), while the effect of LCF on the acceleration of HCF crack growth is presented by (Russ 2005).

Once the HCF and LCF lifes of a structural detail are known they need to be combined so that total fatigue life may be determined. A straightforward combination, linear life model for titanium alloy turbine blades, is presented in (Hou et al. 2009):

$$\frac{1}{t_{TC}} = \frac{1}{t_{HCF}} + \frac{1}{t_{LCF}} \tag{12}$$

A more general approach is presented by (Jelaska et al. 2003) where S-N curves are used to construct Goodman curves of constant fatigue life. Then, using Smith diagrams, which are again constant fatigue life curves but for different r values, a closed form equations for combined HCF and LCF values are established. Yet, they do not take into account stress concentration, residual stresses, imperfections in welded joints and are therefore of limited use.

Shipbuilding community has yet to invest in research of cumulative effect of HCF and LCF in ship structures. This conclusion is supported by the fact that relevant research is nearly non existing in the open literature and, at the same time, it is not recognized by the Classification societies Rules. (Hansen & Thayambali 1995) presented fatigue damage estimation due to a combination of wave load induced HCF and whipping and slamming induced LCF. (ISSC 2000) lists a number of papers analyzing "storm model" assumption—that a ship periodically experience extra high stresses during sailing in storms. A sequence of storms in time is supposed to affect the damage accumulation. (Urm et al. 2004) made a systematic research of HCF and LCF in ship structural details due to wave loads, within DNV research department, and estimated cumulative fatigue life to be:

$$D = D_{HCF}\left(1-\frac{\nu_{LCF}}{\nu_0}\right) + \nu_{LCF}\left\{\left(\frac{D_{LCF}}{\nu_{LCF}}\right)^{1/m} + \left(\frac{D_{HCF}}{\nu_0}\right)^{1/m}\right\}^m$$

where:

D_{LCF} – low cycle fatigue damage

$$D_{LCF} = \sum_{i=1}^{k} \frac{n_i}{N_i}$$

D_{HCF} – high cycle fatigue damage

$$D_{HCF} = \frac{\nu_0 T_d}{\bar{a}} \sum_{n=1}^{N_{load}} p_n q_n^m \Gamma\left(1+\frac{m}{h_n}\right)$$

ν_{LCF} – mean zero-crossing frequency for LCF
ν_0 – mean zero-crossing frequency for HCF

$$\nu_0 = \frac{1}{4Log_{10}L}$$

L – ship length.

5 NUMERICAL EXAMPLE

Numerical example illustrates HCF, LCF and combined HCF and LCF calculation procedure for structural detail on 6500 cbm oceangoing LPG vessel, having one 4500 m^3 bilobe tank and one 2000 m^3 cylindrical tank. It transports different types of liquefied petroleum gases, from ethylene to vinyl-chloride monomers cooled at −104°C and pressurized at 4.5 bars. Figure 6 presents general drawing and Figure 7 cross section of LPG vessel in concern.

Table 1 lists the ship main particulars and Table 2 cylindrical and bilobe tank particulars, respectively.

Critical structural detail is Y-joint of shells and longitudinal bulkhead in bilobe tank. Due to difficulties in manufacturing process a misalignment of shells occur during welding and an eccentricity is introduced into what should be a perfect Y-joint, Figure 8. Tanks are designed and behave as membrane structures and can stand the bending stress due to Y-joint misalignment only to a certain extent.

Eleven Y-joint models are generated employing the volume finite elements, each having different eccentricity of Y-joint ranging from 0 mm to 20 mm in step of 2 mm. One of this models, i.e. the one with e = 14 mm is presented in Figure 8. Particular attention is paid to weld modelling so that exact hot-spot stress value may be determined. Notch stress values are then determined according to Rules (DNV 2003).

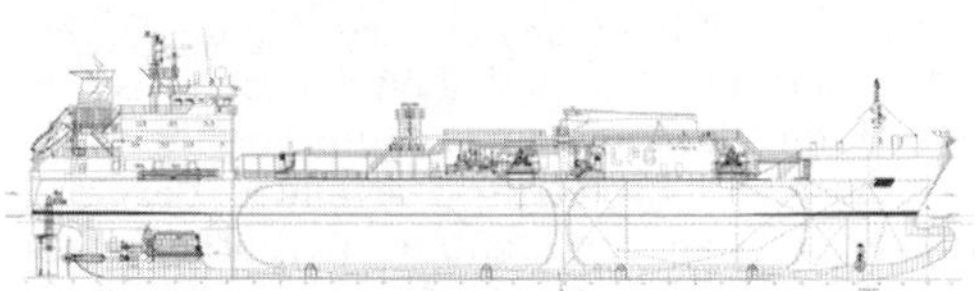

Figure 6. LPG Carrier 6500 cbm.

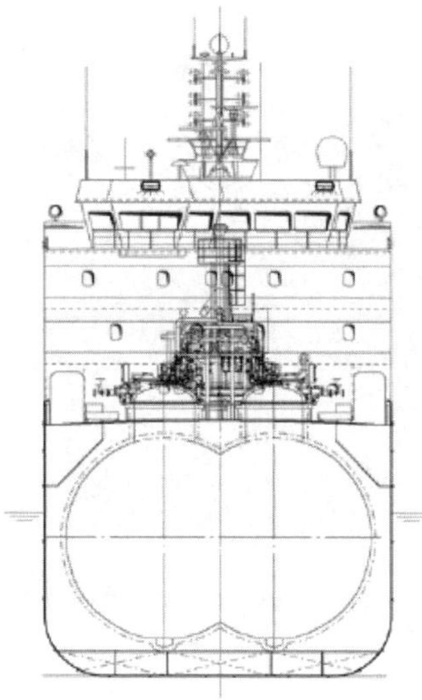

Figure 7. LPG Carrier 6500 cbm—cross section.

Table 1. Ship main particulars.

Length over all	114.89 m
Length between perpendiculars	109.211 m
Breadth	16.80 m
Draught	7.60 m
Draught—ethylene load	6.60 m
Speed—ethylene load	16 kn

Table 2. Cylindrical and bilobe tank dimensions.

Cylinder tank length	29.29 m
Cylinder tank radius	4.75 m
Bilobe tank length	40 m
Bilobe tank breadth	14.8 m
Bilobe tank radius	4.75 m

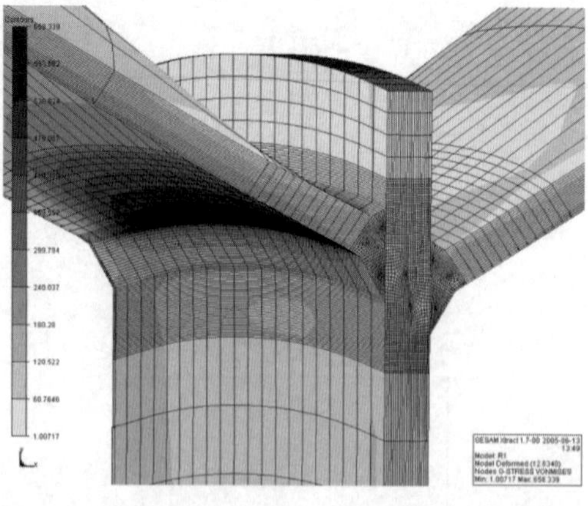

Figure 8. Solid element FEM model of Y-joint, eccentricity e = 14 mm.

Table 3. Notch stress concentration factors as a function of Y-joint eccentricity.

	Eccentricity	σ_{vMt}
Model 0	0 mm	213
Model 1	2 mm	338
Model 2	4 mm	384
Model 3	6 mm	431
Model 4	8 mm	510
Model 5	10 mm	567
Model 6	12 mm	584
Model 7	14 mm	679
Model 8	16 mm	715
Model 9	18 mm	819
Model 10	20 mm	845

Notch stress concentration factors, for different values of eccentricites in imperfect Y-joint, are presented in Table 3. The details of calculation procedure are given in (Rudan and Senjanović 2005). Note that yield stress for tank material is 390 N/mm^2 and tensile strenth 540 N/mm^2. Therefore, even minor eccentricites leads to unacceptable high stress values and eccentricities exceeding 10 mm may be considered critical and need to be controled.

6 HCF FATIGUE ANALYSIS

Full spectral analysis is performed for the 6500 cbm Gas Tanker and for the Y-joint of the aft bilobe tank, for various amounts of joint imperfection (eccentricity from 0 to 20 mm).

Hydrodynamic analysis is performed using SESAM software package and strip method. A ship is divided into 56 strips, Figure 9, and subjected to a number of analyses: two loading conditions (full and ballast), 30 different wave lengths (ranging from 0.105 to 2.6 ship lengths), 13 different heading angles (from 0 to 180 degrees in increments of 15 degrees) and for one ship speed. Thus, in addition to hydrostatic analysis, a total of 780 different load cases are analyzed in both real and imaginary domain for each loading condition.

Structural analysis is performed using SESAM as well. Coarse mesh finite element model of the entire LPG ship, including cylindrical and bilobe tanks, is generated. A superelement modeling technique is used with direct mesh refinement in the Y-joint area on the bilobe tank. Figure 10 presents finite element mesh of the ship and tanks (hull mesh is hidden in the midship).

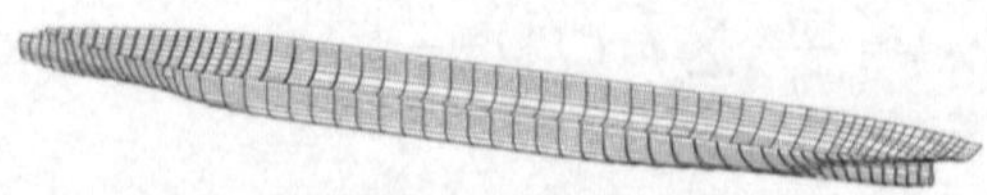

Figure 9. Waveship model of LPG ship.

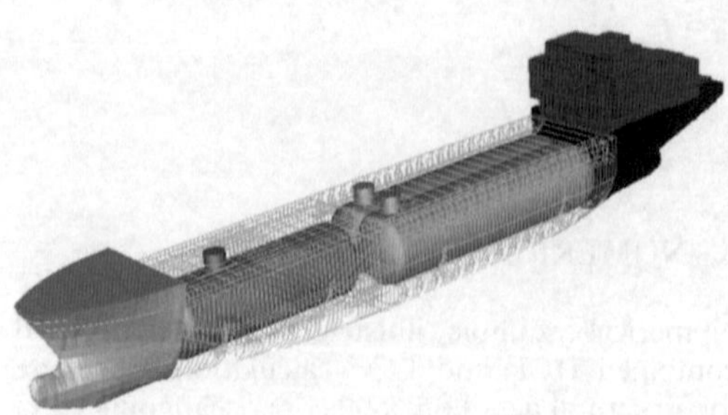

Figure 10. Finite element model of LPG—part of hull hidden.

Figure 11 presents finite element models of the cylindrical and bilobe tanks in a cut-through view so that structural elements may be distinguished i.e. stiffening rings, vacuum rings etc.

The entire model consists of some 170,000 shell finite elements and some 140,000 nodes. The model is subjected to each of 780 load cases which consist of: wave pressure at the wet hull surface, inertial forces and internal tank pressure. As a result of structural analysis a set of notch stress transfer functions for Y-joint are obtained, Figure 12.

The statistic and probability analysis is performed for both loading conditions, the entire range of the Y-joint eccentricities and for the North Atlantic sea-state condition. As a result long-term stress distribution is obtained. Table 4 presents the Weibull distribution parameters for different values of the Y-joint eccentricity.

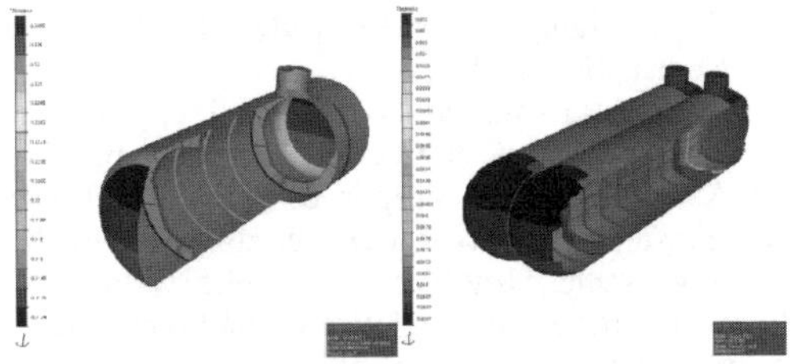

Figure 11. FEM models of cylindrical (left) and bilobe tank (right).

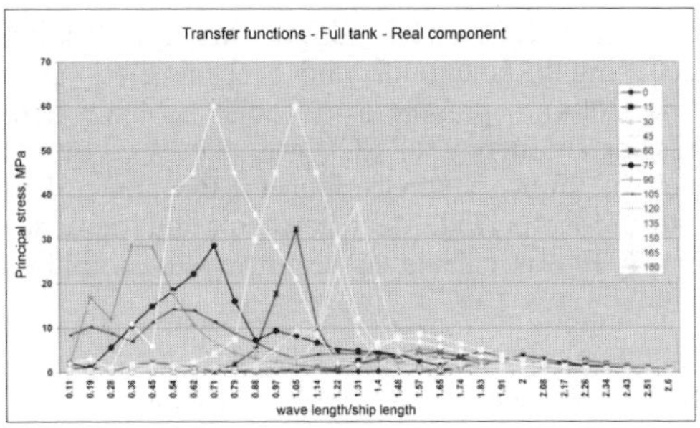

Figure 12. Notch stress transfer functions for Y-joint detail (typical).

Table 4. Weibull parameters for ballast (LCO3) and full (LC05) loading condition.

$\frac{K_{scf,e}}{K_{scf,0}}$	LC03 q	LC03 h	LC03 T_0 [s]	LC05 q	LC05 h	LC05 T_0 [s]
1.000	3.354	0.7893	8.667	8.473	0.798	8.744
1.271	4.2632574	0.7893	8.667	10.77	0.798	8.745
1.442	4.8372336	0.7893	8.667	12.22	0.798	8.744
1.617	5.4230851	0.7893	8.667	13.70	0.798	8.745
1.947	6.5314529	0.7893	8.667	16.50	0.798	8.745
2.131	7.148972	0.7893	8.667	18.06	0.798	8.744
2.192	7.3508533	0.7893	8.667	18.57	0.798	8.745
2.562	8.5938086	0.7893	8.667	21.71	0.798	8.745
2.697	9.0450726	0.7893	8.667	22.85	0.798	8.744
3.109	10.426574	0.7893	8.667	26.34	0.798	8.745
3.208	10.759084	0.7893	8.667	27.18	0.798	8.745

Fatigue damage is calculated according to the DNV Rules:

$$D_{HCF} = \frac{T_d}{T_0 \cdot \bar{a}} p_n q_n^m \Gamma\left(1 + \frac{m}{h_n}\right) \tag{13}$$

Fatigue damage results are presented in Table 5. The analysis revealed that full loading condition is the main cause of fatigue damage and that is in line with the simplified fatigue analysis (Senjanović et al. 2007). It is also clear from Table 5 that eccentricity of 12 mm ($D_{HCF} < 1$) may be tolerated when considering high cycle fatigue damage analysis only, assuming a life time of ship in service of 25 years.

Table 5. HCF fatigue damage as a function of Y-joint eccentricity.

e	LC03	LC05	D_{HCF}	Fatigue life [years]
0 mm	0.005	0.064	0.069	362.617
2 mm	0.010	0.132	0.142	176.587
4 mm	0.014	0.192	0.207	120.878
6 mm	0.020	0.271	0.291	85.792
8 mm	0.035	0.474	0.509	49.108
10 mm	0.046	0.621	0.668	37.446
12 mm	0.050	0.675	0.726	34.449
14 mm	0.081	1.079	1.160	21.559
16 mm	0.094	1.258	1.352	18.489
18 mm	0.144	1.927	2.071	12.071
20 mm	0.158	2.117	2.276	10.986

7 LOW CYCLE FATIGUE ANALYSIS

LPG ship in service operates worldwide, usually on a fixed sailing route under a long-term contract. The tank is loaded with the liquefied gas at the terminal of departure and then emptied at the terminal of arrival, so that pressure in the tank oscillates between approximately 0 bars (vacuum can happen during unloading) and design pressure of 5 bars in case of a bilobe tank on LPG carrier. Assuming that no damage occurs during voyage (between terminals), oscillation of the global stresses in the tank will be as presented in Figure 13 (Røren et al. 1975).

However, local stresses will not follow the stable pattern from Figure 13. Roots of the notches and any other highly stressed local areas will inevitably be subjected to the low cycle fatigue damage. Since in elastic region $\sigma = E \cdot \varepsilon$, the Neuber formula for the stress and strain range becomes:

$$\Delta\sigma_{loc}\Delta\varepsilon_{loc} = \frac{\left(K_T^2 \Delta\sigma^2\right)}{E} \tag{14}$$

For a given nominal stress range $\Delta\sigma$, the intersection of Ramberg-Osgood and Neuber equations creates a hysteresis loop. A set of eleven hysteresis loops, corresponding to each one of the Y-joint finite element models, is created (Rudan & Senjanović 2005), Figure 14.

Fatigue life may be calculated according to the Morrow equation, including the correction for the mean stress effects (Xu 1997), that relates fatigue life N_f to the maximum local strain amplitude $\Delta\varepsilon_{loc}$:

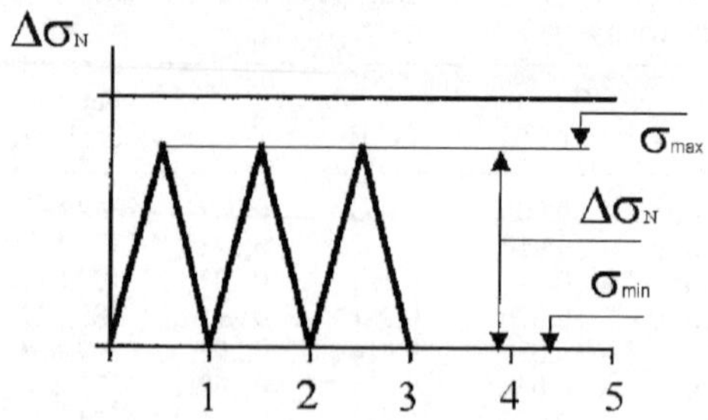

Figure 13. LPG nominal stresses due to cargo loading and unloading.

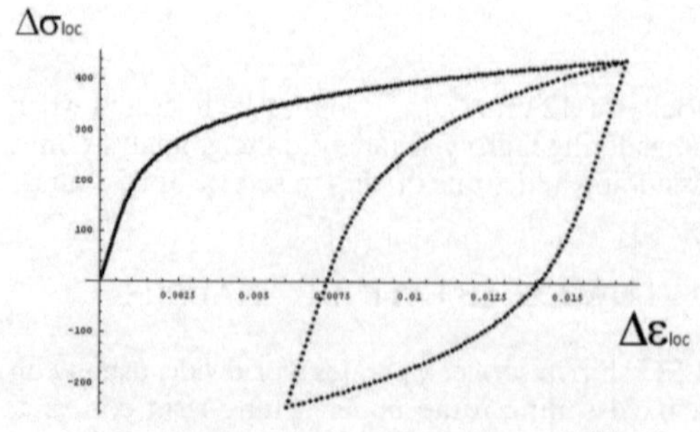

Figure 14. Hysteresis loop for local noth behavior (typical).

$$\frac{\Delta\varepsilon_{loc}}{2} = \frac{\left(\sigma'_f - \sigma_{mean}\right)}{E}\left(2N_f\right)^b + \varepsilon'_f\left(2N_f\right)^c \quad (15)$$

where σ'_f is fatigue strength coefficient, σ_{mean} is the mean stress, b is fatigue strength exponent, ε'_f is fatigue ductility coefficient, c is fatigue ductility exponent and E is Young's modulus of elasticity.

Since obtaining the exact material properties was not possible, a general purpose steel SAE1020 properties were chosen. The fatigue properties of SAE1020 are described by the following parameters:

$E = 2.05$ GPa, $K' = 941$ MPa, $n' = 0.18$,
$\sigma'_f = 815$ MPa, $\varepsilon'_f = 0.25$, $b = -0.114$, $c = -0.53$.

Fatigue life is estimated as a parameter of the Y-joint eccentricity e and with the assumption of 5 days load cycle for typical LPG carrier. Nominal stress in the Y-joint is chosen to be $\sigma_N = \sigma_{vMt} = 183.34$ MPa as this is theoretical stress in the Y-joint excluding the effects of weld geometry, notch presence and eccentricity (i.e. $e = 0$ mm). Hot-spot extrapolation of principal stresses and notch principal stresses are obtained according to DNV fatigue assessment Rules (DNV 2003). Table 6 lists the results.

The LPG tank material is 5% nickel steel, a high-strength material with excellent cryogenic properties. Assuming fatigue properties of SAE1020 steel all the results will be on the conservative side by the unknown scale. However, results clearly indicate the strong influence of relatively small eccentricity to the estimated lifetime of Y-joint.

8 COMBINED HCF AND LCF

Combined HCF+LCF fatigue life for Y-joint, according to linear life and according to (Urm et al. 2004) is presented in Table 7. Mean zero-crossing frequency for LCF was calculated on the basis of 25 years service life of LPG and 5 days load-unload period. Mean zero-crossing frequency for HCF was calculated according to simplified

Table 6. Estimated LCF life-time of Y-joint.

Eccen-tricity e	Hot-spot principal stress [N/mm²]	Notch stress conc. factor	Notch Strain range [mm]	D_{LCF}	Estimated life [years]
0 mm	307	2.51	0.002309	0.492	50.8
2 mm	389.5	3.19	0.00512	2.101	11.9
4 mm	442	3.62	0.00633	3.472	7.2
6 mm	496	4.06	0.007718	5.434	4.6
8 mm	597.5	4.89	0.010254	10.869	2.3
10 mm	654	5.35	0.012287	14.705	1.7
12 mm	672.5	5.50	0.012906	16.667	1.5
14 mm	786	6.43	0.016685	27.778	0.9
16 mm	828	6.77	0.018181	35.714	0.7
18 mm	953	7.80	0.022851	50	0.5
20 mm	984.5	8.05	0.024119	62.5	0.4

Table 7. Combined HCF and LCF fatigue damage.

e [mm]	HCF [years]	LCF [years]	Total life *Linear life model* [years]	Total life Urm et al. (2004) [years]
0	362.617	50.8	44.558	42.974
2	176.587	11.9	11.149	10.816
4	120.878	7.2	6.795	6.599
6	85.792	4.6	4.366	4.243
8	49.108	2.3	2.197	2.138
10	37.446	1.7	1.626	1.583
12	34.449	1.5	1.437	1.399
14	21.559	0.9	0.864	0.841
16	18.489	0.7	0.674	0.657
18	12.071	0.5	0.480	0.468
20	10.986	0.4	0.386	0.376

expression in (Urm et al. 2004). Full loading condition (LC05) is taken into account.

Table 7 presents decrease of total fatigue life due to combination of HCF and LCF effects. Due to large influence of LCF, total life is close to LCF life estimation. This is partly due to inadequate (unavailable) material properties used for LCF calculation. It is easy to conclude that having better information about specific material characteristics, calculations would lead to more precise fatigue life estimation. This however does not exclude further discussion about both models proposed.

9 CONCLUSION

Fatigue assessment of ship structural details is a complex task. A ship is a large structure, yet fatigue cracks initiate in microscopic notches or very local highly stressed areas. Service lives of oceangoing vessels are 20 to 25 years and they sail in corrosive and sometimes very harsh environments. Fatigue of ship structural details may be divided into HCF and LCF component. Both components are specific since they originate from different loads, include different number of load cycles and stress ranges and are evaluated either statistically (HCF) or in particular way (LCF). State-of-the-art HCF calculation procedure and simplified LCF calculation procedure are presented and illustrated through numerical example: highly stressed imperfect Y-joint of bilobe tank shells and longitudinal bulkhead in LPG cargo carrier.

It may be concluded that:

- Spectral-analysis of HCF is state-of-the-art direct fatigue calculation procedure that takes into account a number of parameters that affect structural detail fatigue life: ship loading condition, sailing route etc. through direct hydrodynamic, structural and probabilistic analysis.
- LCF may be conveniently determined by simplified methods; however, advanced methods are needed to obtain more precise results.
- HCF is recognized in Classification societies Rules, while LCF is not and is commonly analyzed either by closed-form equations or directly upon specific request (research),
- Cumulative effect of HCF and LCF interaction is only recently being investigated, mostly by DNV and probably other in-house research, common procedure is not available,
- LCF calculation requires S-N curves in low cycle and high stress range and additional material properties, which are not readily available for standard ship materials (steel), let alone specific materials as cryogenic 5% Nickel steel in LPG tanks.

Future research of fatigue in ship structures should take into account problems mentioned in this paper and try to overcome the problem of unknown material properties.

REFERENCES

Bureau Veritas, 1997. *Fatigue Strength of Welded Ship Structures.*

International Institute of Welding, 1995, *Recommendations on Fatigue of Welded Components*, IIW Document, XIII-1539-95.

Byrne J., Hall R.F., & Powel B.E. 2003. Influence of LCF overloads on combined HCF/LCF crack growth, *International Journal of Fatigue*, 25 (2003) 827–834.

Det Norske Veritas, 2003. *Fatigue Assesment of Ship Structures.*

Technical Committee II.1, 2000. Quasi-Static Response. *Proceedings of the 14th International Ship and Offshore Structures Congress*, Vol. 1:133–195, Nagasaki.

Jelaska D., Glodez S. & Podrug S. 2003. Closed form expression for fatigue life prediction of combined HCF/LCF loading, *Facta Universitatis*, Vol. 3, No. 13: 635–646.

Urm, H.S., Yoo, I.S., Heo, J.H., Kim, S.C. & Lotsberg, I. 2004. Low Cycle Fatigue Strength Assesment for Ship Structures, *9th PRADS Symposium*, 2004, Germany.

Heo J.H., Kang J.K., Kim Y., Yoo I.S., Kim K.S. & Urm H.S. 2004. A Study on the Design Guidance for Low Cycle Fatigue in Ship Structure, *9th PRADS Symposium*, Germany.

Hou N.X., Wen Z.X., Yu Q.M. & Yue Z.F. 2009. Application of a combined high and low cycle fatigue life model on life prediction of SC blade, *International Journal of Fatigue, Vol. 31, Issue 4*: 616–619.

Sleczka L. 2004. Low cycle fatigue strength assesment of butt and fillet weld connection, *Journal of Constructional Steel Research* 60: 701–712.

Hansen P.F. & Thayamballi A.K. 1995. Fatigue Damage Considering Whipping Arising from Slamming, *OMAE'95*, New York: 155–163.

Rudan S. & Senjanović I. 2005. Fatigue Strength Assesment of a Weld Connection Misalignment in LPG Bilobe Cargo Tanks, *IMAM 2005*, Lisbon, Portugal.

Røren E.M.Q., Vedeler B., Flatseth J.H. & Johannessen Th., 1975. Design of ships carrying LPG and LNG, Det Norske Veritas.

Russ S.M. 2005. Effect of LCF on HCF crack growth of Ti-17, *International Journal of Fatigue, Vol. 27, Issues 10–12*: 1628–1636.

Senjanović I., Rudan S. and Slapničar V. 2006. Design and construction of bilobe cargo tanks, *International Conference: Design, Construction & Operation of Natural Gas Carriers & Offshore Systems*, ICSOT 2006.

Xu T. 1997. Fatigue of Ship Structural Details—Technical Development and Problems, *Journal of Ship Research, Vol. 41*: 318–331.

Advanced Ship Design for Pollution Prevention – Guedes Soares & Parunov (eds)
© 2010 Taylor & Francis Group, London, ISBN 978-0-415-58477-7

Design optimization and product reliability—responsibility of designers and producers

V. Grubišić
Technical Consulting, Reinheim, Germany

N. Vulić
Croatian Register of Shipping, Split, Croatia

ABSTRACT: Industrial components must comply with the requirements for reliability and safety throughout their entire life cycle. Anybody taking part in their design, production and/or operation may be responsible for a possible failure. The types of responsibility are design responsibility, production responsibility and exploitation (operation) responsibility. This paper presents two actual cases of failure: the case of production responsibility where the bearing girders of Diesel engines in several large tankers cracked, as well as the case of design responsibility where a failure of a redesigned wheel on a high speed train failed caused a disaster. The authors want to point out that, in order to fulfill the reliability and safety requirements for a component, proper care is to be taken about the following influences of utmost importance: the actual operating loads and usage conditions, the design shape and proper dimensioning, the material, as well as the production process.

1 INTRODUCTION

Transport means of today (e.g. vehicles, trains, ships, aircrafts, etc.) are very sophisticated and complex industrial products. Their reliability and safety is highly dependent upon their components. The basic influences during the development of an industrial product component are material (necessary for its production), energy (necessary for production and return of the component into the environment), time (needed for the development and production) and living space (environment with the ecological requirements). In accordance with the presently valid definition an industrial product is considered improper when, during its intended use, the product does not ensure required action on the basis of all the conditions and knowledge at the time when it was brought into use.

It is of utmost importance that proper care is taken about the following influences: the actual operating loads, the design shape and proper dimensioning, the material, as well as about the production process of the component. If these influences are somehow neglected in any sense, failures with high costs and/or severe consequences may occur. The paper presents the two actual cases.

The first case is the failure of Diesel bearing girders in several large tankers where cracks were found. Forensic analysis showed that the operating loads were properly determined, the design and dimensioning was proper, and, finally that the material was properly selected and certified. However, the production process proved to be the actual cause for the failures. This situation caused a consequence related with the operation of a functional (*class*) component and was resolved between the engine licensor, the ship builder and ship owner, based upon the fact that it did not cause either loss of lives or environment pollution.

The second case presented is a much more severe one: the infamous Eschede train accident. Wheel designers replaced monoblock wheels with the newly designed resilient multipiece wheels used to improve the driving capabilities of the high-speed train. After a certain period in operation the cracks initiated in the wheel rims. The cracks were not discovered at usual inspections and the reshaping of the wheel rim (necessary for assure driving behavior of the train) and they propagated during another period of time, finally becoming critical and causing the wheel rim to break. This finally caused the train to slip out of the rails, hitting a viaduct on its way and causing loss of more than 100 lives and severe injuries of many passengers. Later on it has been found out that the improper design was the actual cause: though the design would be correct for a new wheel, it is unfortunately improper for a used wheel after a certain period of time in operation.

The authors conclude that, in general, improper approach to design, manufacture or maintenance of the components can generate in operation the cracks that can have extremely severe consequences. Any person taking part in any of these three activities shall bear in mind whether the component in question is a safety one (where fractures are not allowed at all) or a functional one (where fractures are to be avoided). From the point of view of costs it is very important whether a failure occurs during development, manufacturing, assembly or usage phase of the component. In general, in each of these subsequent phases actual costs needed to deal with the failures are increased by a factor of ten. Finally, in order to fulfill the reliability and safety requirements for a component, proper care shall always be taken about the actual operating loads, the design shape and proper dimensioning, the material, as well as the production process.

Table 1. Classification of components on the basis of the requirements for their reliability.

	Primary	Primary	Secondary
Components	A	B	C
Fractures	not allowed	to be avoided	
Influence	fractures lead to accident and danger for user and environment	fractures interrupt the function	fractures influence neither safety nor function
Examples	wheel	connecting rod	various accessories
	hub	crankshaft	
	spindle	starter	
	brake	cooler	
	suspension arm	gears	

2 CLASSIFICATION OF COMPONENTS AND TYPES OF RESPONSIBILITY

Concerning their reliability requirements, components may be classified into primary and secondary ones. Primary components can be of type A (safety components) and of type B (functional components). Table 1 presents a typical subdivision among these types of components with respect to their reliability. Note: that the examples presented in Table 1 are related to motor vehicle industry.

In marine and offshore business primary type A components are the so-called safety components (related either to the safety of human lives or to environment protection). They are to follow the requirements of various international conventions: SOLAS, MARPOL, Load Line, Colreg, Torremolinos, etc. Primary type B components are to follow the class requirements, i.e. the requirements of the classification society, commonly based upon IACS requirements and recommendations. Generally, there are not any requirements the secondary components should follow other than those imposed by the future ship owner. Typical example for a primary component type A is a pressure vessel safety valve or an oily water separator. For a primary type B it would be a windlass or a mooring winch and for a secondary component a lathe in the ship's workshop.

There are situations where a primary component can under certain circumstances belong to type A and under different conditions to type B. Ship engine, under normal weather conditions and calm seas, is certainly only a functional (class) component of type B. However, at heavy seas, survival of the ship can depend upon its proper functioning and then it becomes a safety, type A primary component. Engine designers, manufacturers and ship owners shall be always aware of this.

The presented component classification has a consequence in the following types of responsibilities in a case of failure:

a. the design or development responsibility, resulting in a design failure;
b. the manufacturing (production) responsibility, resulting in a manufacturing failure;
c. operational (exploitation) responsibility, resulting in an assembly or a maintenance failure.

There are actual practical cases where a failure and its less or more severe consequences require an investigation to find out who was actually responsible. The two selected typical cases will be presented further on.

3 CASE I: DAMAGES ON BEARING GIRDERS IN LARGE TANKERS PROPULSION ENGINES (GRUBIŠIĆ ET AL. 2008)

After operational usage from the minimum time of 11 000 hours, up to 30 000 hours, on 22 bearing girders of 11 large tankers, cracks were detected in similar zones. 52 engines of the same type with the total of 466 bearing girders were in service at that time, meaning that on 21 per cent of ships the girders were cracked, whereby 4.7 per cent of the total number of girders cracked. Three different positions of the cracks were found. However, the cracks were not always located in the position of highest stress on the girder.

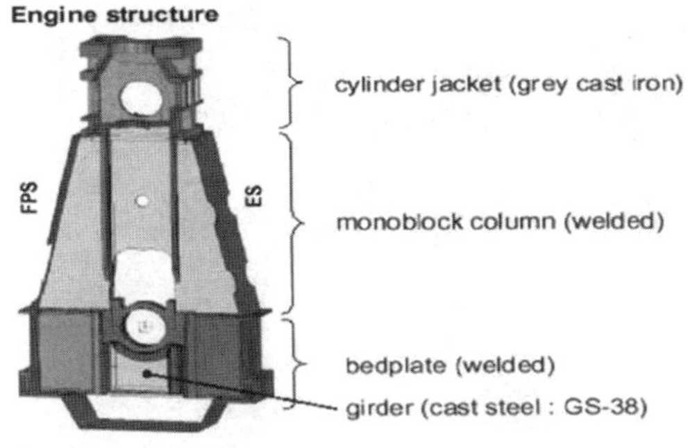

Figure 1. A typical part of the Diesel engine structure showing the actual position of bearing girders.

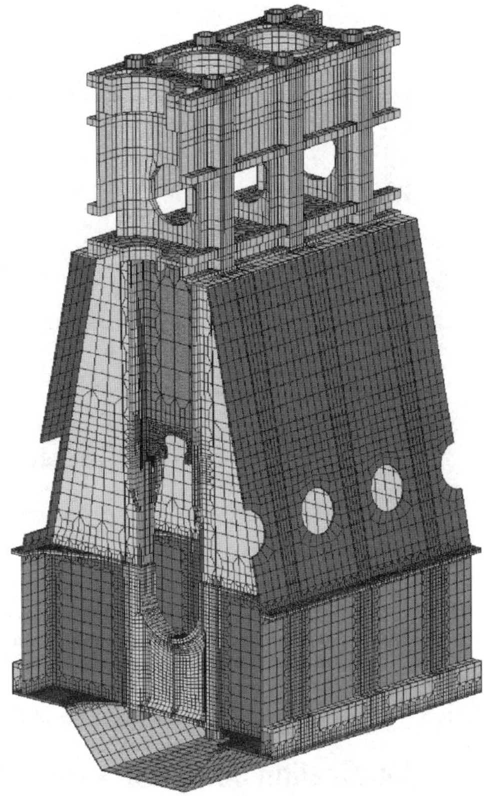

Figure 2. Finite element model of the Diesel engine part.

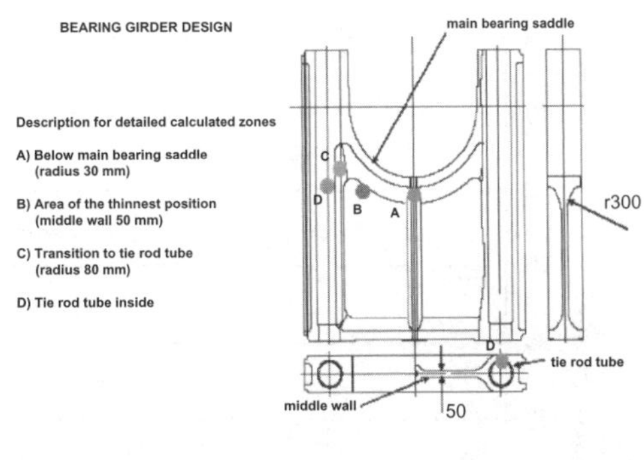

Figure 3. Zones of strength validation based on FEM analysis.

Design validation started with determination of operating loads originating from cylinder pressures and inertial forces of the reciprocating mechanisms in each of engine cylinders. The calculation model took into account all important parameters including the influence of oil film forming in each of the journal bearings. A typical FEM model is presented in Figure 2 and it takes one whole engine cylinder and the two halves of the neighboring cylinders.

The zones of strength validation based on FEM analysis are presented in Figure 3 and the fatigue assessment results in Figure 4.

The material of the bearing girders was cast steel of the GS-38 type. It has been found that the material itself was supplied with all the necessary certificates, proving its chemical composition and mechanical properties.

As presented in Figure 4, the calculated stresses in all areas are below the allowable values.

Verification of the FEM calculation model was performed by actual measurements. Stresses were measured on the two engine girders in several typical positions. The results of this verification are presented in Figure 5 showing a very good agreement between the calculated and measured results.

The agreement between the calculated and the measured stresses verifies the design, proving the calculation presumptions and the results. It is very important to verify the calculation model, best by measurements and experimental tests, because, in case of failure this is certainly a proof that the responsibility does not lie on the designers. That was the actual outcome in this very case.

However, the cracks began to appear in exploitation after the number of operation hours stated at the beginning of this section. For the reasons just described, the responsibility for the cracks was to be looked for at some other party, rather than the designers of the engine bearing girders. To look for operational responsibility in this case does not have any sense at all, owing to the fact that the engine bearing girders are not foreseen to be exposed to any maintenance operations-they are simply girders. Operators of the engines could not have made any kind of misuse of their engines that could affect girders and initiate cracks in them. This reveals that there is obviously a problem to be looked for in the manufacturing process of the

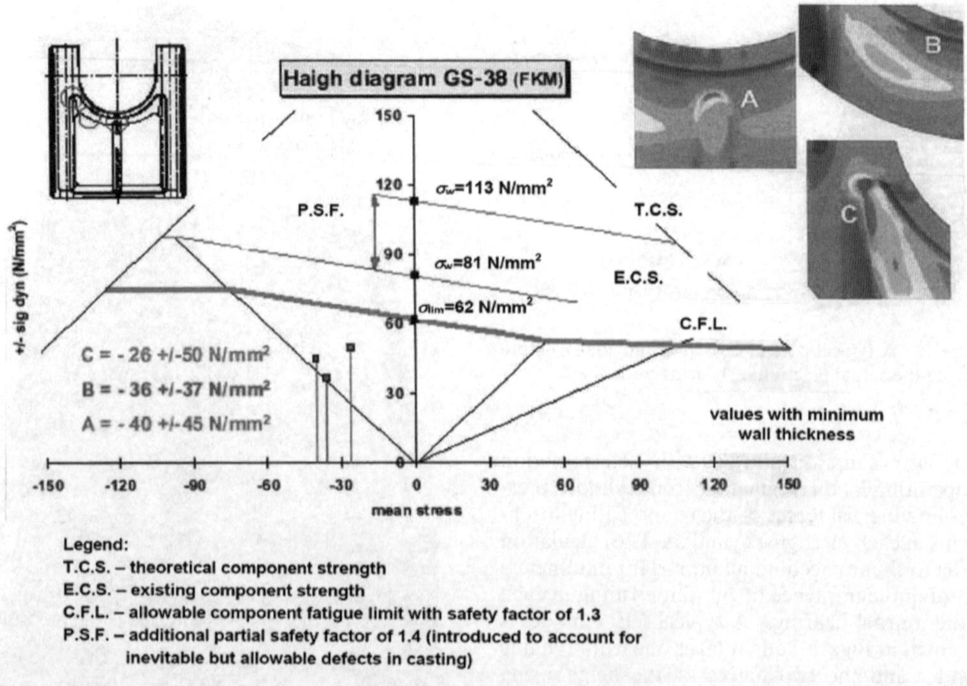

Figure 4. Fatigue assessment results on the basis of the FKM-Guidance data.

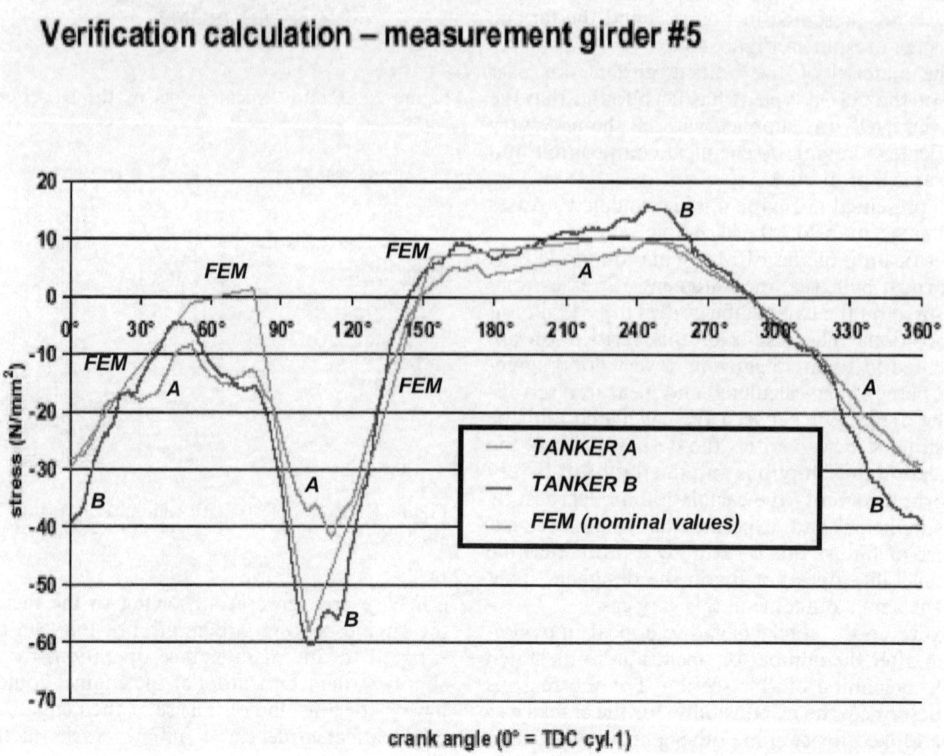

Figure 5. Results of design verification by practical measurement on the two bearing girders of actual ship engines.

bearing girders. Types of damage on the bearing girders are presented in Figure 6.

The thickness in the midsection of the girders was reduced from previously 60 mm in similar engine types to 50 mm in this one. This reduction proved to be correct from the design point of view, as the stresses remained within the allowable limits. However, care had to be taken in the manufacturing process of the girders.

The process of casting of these components results in casting defects, that are easily found by inspection after the casting has been completed and the component has cooled down. These defects are normally to be expected and are solved by repair welding. All the cracks found have initiated in the zones of repair welding as presented in Figure 7.

The fact that the cracks were related to the repair welding shed some more light to the whole situation. The attention was drawn to the procedure and quality assurance of the repair welding performed. Final outcome proved manufacturers lack of attention with respect to the performance of this procedure. The procedure was correctly prescribed, but incorrectly followed by the manufacturers and their quality control. The analysis of the implementation of the repair welding quality assurance procedure proved the responsibility of the manufacturers of the bearing girders.

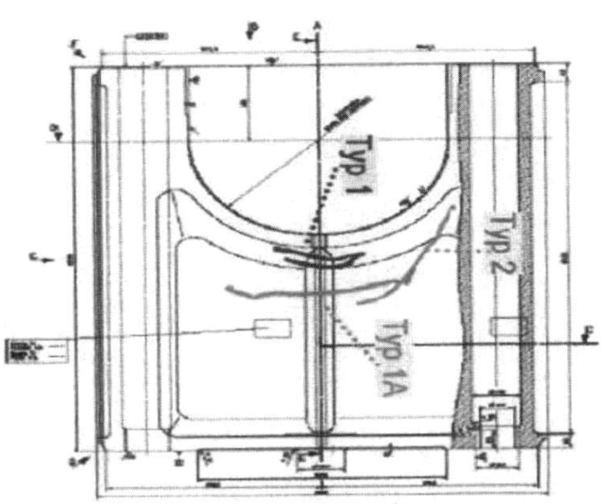

Figure 6. Types of damage on the bearing girders in service.

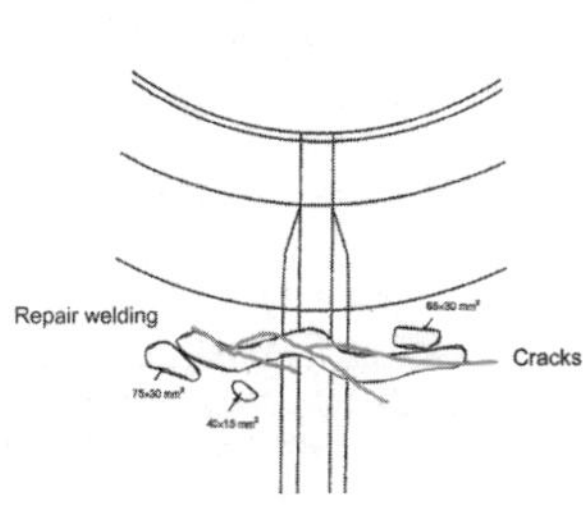

Figure 7. Zones of repair welding.

4 CASE II: HIGH SPEED TRAIN ACCIDENT CAUSED BY A WHEEL FRACTURE (FISCHER & GRUBIŠIĆ 2007)

This case describes another situation where the final responsibility lied on another party rather than the manufacturers. This was the accident of the high speed German train ICE 1 (No. 884), of which the bogie is shown in Figure 8.

In order to improve comfort of the passengers the new concept of wheel design was introduced. Previously, the wheels were designed as monoblock, stiff wheels. Wheels were then redesigned, thus obtaining the new concept, introducing resilient part (rubber elements) between the outer wheel rim and the wheel disc to improve the dynamic behavior of the bogie. Monoblock (solid) and the resilient wheel design are presented in Figure 9.

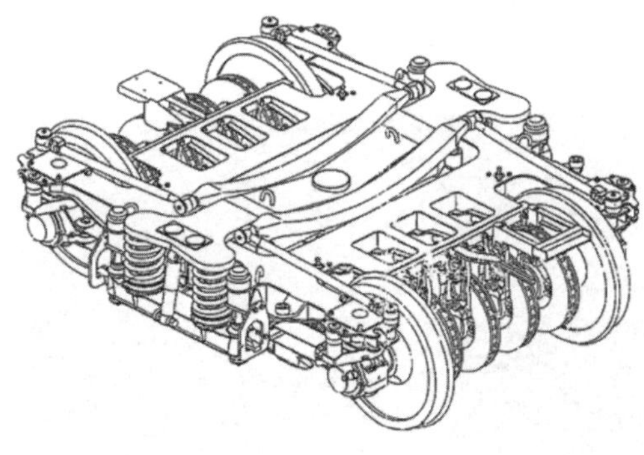

Figure 8. Bogie of the train ICE 1 (static axle load 16.2 t; static wheel load 80 kN).

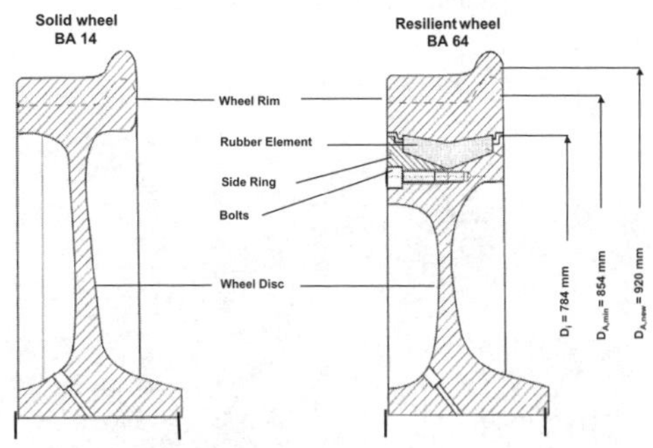

Figure 9. Design of old (solid, monoblock) and new (resilient, multipiece) wheels.

Figure 11 shows the photo of the actual finding at the site of the accident: the broken wheel rim. It is evident that the fatigue crack initiated from the lower part of the inner side of the rim, then spread upwards and finally caused the sudden break of the rest of the rim, with all the chaotic consequences shown in Figure 10.

The investigation at the site of the accident revealed the actual cause of the accident: one of resilient wheel rims broke, causing the train composition to slip out of the railroad, hit the viaduct supports located sideways and finally causing the disaster. The actual photo of the accident site is shown in Figure 10.

Forensic analysis of this problem showed that this was the situation of design responsibility. Rims of wheels actually wear in operation; Figure 12 shows the diameter of the new wheel (920 mm) and the usage limit of 854 mm, which was allowed based on experience with monoblock, solid wheels. In the case of the resilient, multipiece wheel the critical usage limit was 880 mm at which the initial cracks can start at operational usage. The outer diameter of the broken wheel rim measured 862 mm.

After the accident in Eschede the resilient wheels were replaced by solid wheels.

Figure 10. Train accident near Eschede.

Figure 11. Fracture surface of the rim (1 889 000 km).

Further investigation, with the details beyond the scope of this paper, resulted in the conclusion that can be drawn from the Figure 13.

The safety ratio is defined as the ratio of mean values of fatigue strength and the mean value of operational stress. If the designers think this way, they would not expect a problem after a long period in operation, though the safety ratio is reduced by a factor of two. The problem lies in the fact that both the fatigue strength and the operational stress cannot be represented only by their mean values. They have their distribution around the mean value.

As the operational time passes and the wheel rims achieve their wear and the number of fatigue cycles the operational stresses increase, whereas the fatigue strength decreases. The upper tail of stress distribution overlaps the lower tail of fatigue strength, forming the area of damage (Figure 13). This is the area where the stresses exceed the strength. Designers should be aware of this situation and its consequences.

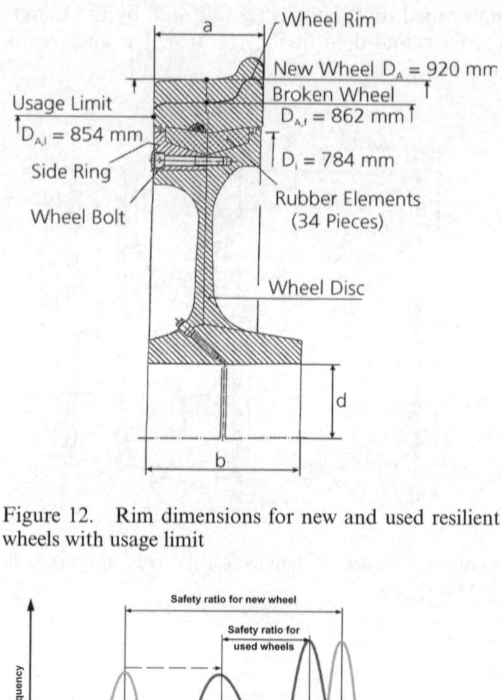

Figure 12. Rim dimensions for new and used resilient wheels with usage limit

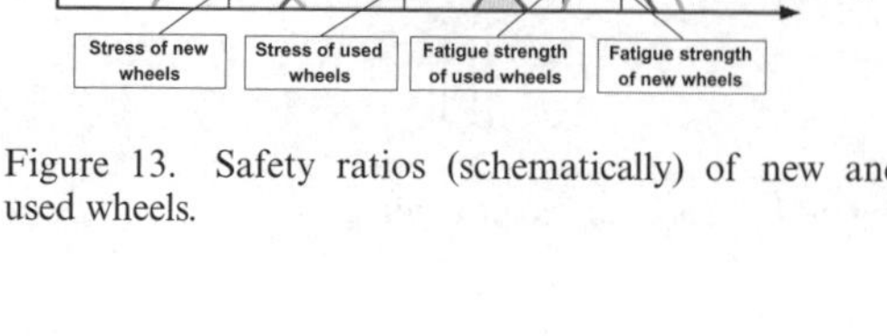

Figure 13. Safety ratios (schematically) of new and used wheels.

5 CONCLUSION

Machinery components essential for the reliable and safe functioning of surface, water and air borne means of transportation, as well as for many other types of machines, are designed, produced and operated every day. There exist very sophisticated means of design, based upon numerical solid modeling, as well as several types of numerical calculations (finite elements, boundary elements, etc.). There are also smart means of verification of numerical models by measurements. The experimental approval is the state-of-the-art in the vehicle industry (Grubišić 1986).

Nowadays, it is very common that the designer believes into own results of numerical calculations without any doubt and considers verification by measurement unnecessary. Designers shall always ask themselves how to prove to some third party that they determined operational loading and boundary conditions in their calculations correctly.

Every designer shall think about the possible consequences of misunderstanding their basic presumptions. The same is valid for the manufacturers and operators.

If a component failure happens somebody is responsible. Designers should be able to prove that their design was not the cause, showing that they took into account all the conditions and available state-of-the-art when the product was put in operation. Manufacturers should be able to prove that they strictly followed the design specifications, material verification, work instructions and the prescribed quality control procedures. Operators should prove the correct operation and maintenance in accordance with the operational and maintenance instructions, without any misuse.

The final message of the authors to designers, manufacturers and operators is that they shall always bear in mind the types of components (primary safety, primary functional or secondary), as well as the types of responsibility: design, manufacturing and operational. This is not only related to machinery components, but a general case valid for all the technical branches.

REFERENCES

Fischer, G. & Grubišić, V. 2007. Praxisrelevante Bewertung des Radreifenbruchs von ICE 884 in Eschede (Usage related evaluation of wheel fracture from ICE 884 at Eschede). *Materialwissenschaft und Werkstofftechnik*, 38 (10): 789–801.

Grubišić, V. 1986. Criteria and methodology for lightweight design of vehicle components. *Report No. TB-176. Darmstadt: Fraunhofer Institute for Structural Durability (LBF).*

Grubišić, V., Vulić, N. & Sönnichsen, S. 2008. Structural durability validation of bearing girders in marine Diesel engines. *Engineering Failure Analysis* 15 (4): 247–260.

Advanced Ship Design for Pollution Prevention – Guedes Soares & Parunov (eds)
© 2010 Taylor & Francis Group, London, ISBN 978-0-415-58477-7

Maritime component reliability assessment and maintenance using bayesian framework and generic data

J. Barle & D. Ban
Faculty of Electrical Engineering, Mechanical Engineering and Naval Architecture, University of Split, Split, Croatia

M. Ladan
Brodosplit-Shipyard Ltd., Split, Croatia

ABSTRACT: The use of reliability models is often limited by a lack of satisfactory data for the estimates of system parameters. This paper discusses the problem of life cycle management with insufficient failure record of the equipment. This deficiency is addressed by using expert opinion and Bayesian modeling. The framework for Bayesian updating of the reliability models is presented and the different aspects of updating process are discussed. The use of expert opinion, generic data and the types of uncertainties associated is presented. Since the most of the generic data provides only constant failure rate, updating process for the exponential sampling distribution is described in detail.

1 INTRODUCTION

Through the use of reliability models, reliability and maintenance engineers use equipment life data to estimate its probability to perform the required functions and to support activities like: planning of the spare parts, inventory and maintenance activities, evaluating the personnel or choosing manufacturer or supplier, Murthy & Rausand & Osteras (2008) and Percy (2008). In growing complexity of systems and the need to achieve more with leaner budget, making the right decisions immediately after the problem arises is crucial.

Some facts should be mentioned. Failures encountered during the operation of the equipment are well founded on the fact that the data undoubtedly reflect all influential factors on its reliability and as a consequence the engineering reliability is failure oriented. The main goal of engineering maintenance, on the other hand, is to reduce the number of failures. Therefore, even if we are dealing with a component that is in service for some time, the use of reliability models is often limited by a lack of satisfactory data upon which it can be based. This stands for the so-called 'frequentists' approach to estimation of unknown model parameters where null hypotheses are possible if a sufficient data are observed. To solve this problem we may build reliability model on the data gathered with similar components. Considering the fact that operating conditions are not necessarily the same, one may say that we are often dealing with one-of-a-kind situations and the reliability estimates founded on previous experience alone are not always applicable.

Above mentioned clearly states that neither failure record or previous experience alone can solve reliability and maintenance problems and it makes a great deal of practical sense to use all the information available.

It should be pointed out that the non-safety equipment is expected to fail and the failure record will increase with time so the concept of learning from the data is applicable. Safety components that are impractical to be monitored and due to respective failure modes are expected to fail are usually in parallel or redundant configuration. Therefore the well-prepared maintenance actions significantly contribute to reliability of the system.

The Bayesian framework for model development offers a possibility of taking the previous knowledge and failure data into account. Therefore the model development is a learning process and knowledge is continually updated as more information becomes available. Such analyses are most credibly performed when subject matter experts are involved to play a key role.

Almost all commercially available reliability databases provide only constant failure rates, see Rausand & Høyland (2004) and OREDA (2002). This may cause some problems. First of all, the exponential model is not a reasonable choice for equipment that could endanger people or environment. Secondly, many failure modes listed are prone to some kind of deterioration process, so the exponential model is not expected. A further fact should be mentioned in order to support the use of this paper. Although the non-aging property would seem to limit the usefulness of the constant failure rate (exponential model), it has continued to play a critical role in reliability calculations. Many probabilistic models, like reliability block diagrams or Markov models, are founded on exponential probabilistic model of failure and can be used without too much difficulty whenever the respective failure rates are given.

In this paper, first we present the interpretation of the constant failure rate in reliability modeling. Next we provide a brief overview of the key concepts of Bayesian modeling relevant to later discussion. Since the validity of the results depends on the validity of the assumptions required by the model, the derivation of exponential model is presented. Finally, through some examples we will explain how to incorporate prior knowledge such as generic data and expert opinion into the estimation process.

2 CONSTANT FAILURE RATE IN RELIABILITY MODELING

The example of generic data base is OREDA (2002). OREDA (Offshore Reliability Data) is the most known database whose main objectives are: "*collection and analysis of maintenance and operational data, establishment of a high quality reliability database, and exchange of reliability, availability, maintenance and safety technology among the participating companies*". Failure events are gathered from two or more installations, and reflect a weighted average of the experience. All the failure rates presented are based on the assumption that the failure rate is constant and independent of time. The data compiled in OREDA are directly relevant for offshore conditions. However, in some cases, considering the maritime operating conditions and particular failure mechanism we might justifiably expect similar frequencies.

For example the failures of valves are presented from population of 1170 items installed on 40 different offshore platforms. The accumulated (calendar) time in service is $36.67 \cdot 10^6$ hours, and accumulated operational time is $31.62 \cdot 10^6$ hours. During that period the 1017 failures were recorded. Failure rate is given as mean and its value is 31.05 per 10^6 and standard deviation is 18.89 per 10^6 hours for calendar time. The lower and upper bounds with 90% confidence intervals are respectively $\lambda_L = 7.74$ and $\lambda_U = 67{,}14$ failures per 10^6 hours.

From the foregoing discussion, the analyst must have an opinion about parameters that must be expressed through a probability distribution. It is desirable that the distribution parameters have proper operational interpretation. First question immediately rises: why the constant failure rate is given even for the failures of mechanical components that are clearly caused by fatigue or wearout. Therefore the exponential approximation of increasing failure rate should be considered. The data available for the analysis is usually the number n of failures during a observation time in service $\Delta t = t_2 - t_1$. The failure rate estimated by n/t will thus be an Average Failure Rate (AFR)

$$AFR(t_1, t_2) = \frac{1}{t_1 - t_2} \int_{t_1}^{t_2} \lambda(u)\mathrm{d}u = \frac{\ln R(t_1) - \ln R(t_2)}{t_2 - t_1} \tag{1}$$

where the $R(t)$ is general reliability function and $\lambda(t)$ is the respective failure rate function. For the Weibull model, the failure rate and reliability functions are

$$\lambda(t) = \frac{\beta}{\theta}\left(\frac{t}{\theta}\right)^{\beta-1} \tag{2}$$

$$R(t) = \exp\left[-\left(t/\theta\right)^{\beta}\right] \tag{3}$$

The Weibull probability density is

$$f(t) = \frac{\beta}{\theta}\left(\frac{t}{\theta}\right)^{\beta-1} \exp\left[-\left(t/\theta\right)^{\beta}\right] \tag{4}$$

where β is the shape parameter and θ is the scale parameter. Last three equations are different representations of the same model. If the 'real' life distribution is Weibull distribution with an increasing failure rate function, $\beta > 1$, and we use a constant failure rate estimate, we overestimate the failure rate in the initial phase and underestimate the failure rate in the last part of the observation or prediction interval. This is illustrated in Figure.1.

This general conclusion rejects all information regarding most of the mechanical failures with increasing failure rate, expressed in the form of exponential model. To deal with the above shortcoming we should consider the exponential model when 'real' Weibull model estimates β is in the range $1 \div 2$.

Generally, in that range of β, it is unlikely that the any kind of periodic maintenance is practically feasible. In such cases, even if the failure rate is increasing, reliability model can be reasonably well

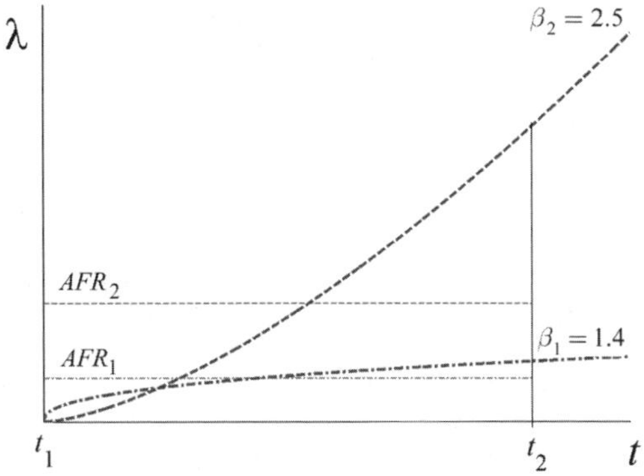

Figure 1. *AFR* of the Weibull models with β parameter set to 1.4 and 2.5.

developed on exponential basis. Validity of this approximation rises when observation window Δt decreases.

In the case of the Weibull distribution, the shape parameter β is usually related to particular failure mode and the quality measures of the component in hand and therefore has no operational meaning. Therefore the parameter β can be assumed as failure specific, and constant across the range of stress levels. This cannot be assumed for the scale parameter θ because it generally changes with the stress applied. Consequently θ can be assumed as usage or operational specific parameter.

3 BAYESIAN INFERENCE

This section describes the concept of subjective or personal probability. In Bayesian reliability analysis, the statistical model consists of two parts: the likelihood function, $f(x|\varphi)$ and the prior distribution, $\pi_0(\varphi)$. The distribution that represents our knowledge about these parameters is the prior distribution, $\pi_0(\varphi)$. The likelihood function is typically constructed from the sampling distribution of the data and is considered as fixed. The sampling distribution contains the vector of unknown parameters φ. Once we acquire the failure data, we regard the sampling distribution as a function of the unknown parameters. In Bayesian analysis, the likelihood function and the prior distribution are the basis for parameter estimation and inference. Mathematically we can combine prior knowledge with current data through equation:

$$\pi_1(\varphi | x) = \frac{f(x,\varphi)}{f(x)} = \frac{f(x|\varphi)\,\pi_0(\varphi)}{\int_\Omega f(x|\varphi)\,\pi_0(\varphi)\,\mathrm{d}\varphi} \tag{5}$$

Now, the failure model parameters are random variables Φ in space Ω, it values φ (for example Weibull parameters β and θ), are random variables which behave according to distribution $\pi_0(\varphi)$. Hence, before any failures being observed, a priori estimate about the expected value of φ from Φ is $E[\pi_0(\varphi)]$. Furthermore

$$f(x,\varphi) = f(x|\varphi)\,\pi_0(\varphi) \tag{6}$$

constitutes the joint density and expression in denominator of Equation 5

$$f(x) = \int_\Omega f(x|\varphi)\,\pi_0(\varphi)\,\mathrm{d}\varphi \tag{7}$$

is called the marginal density and can be interpreted as normalizing constant, i.e. a constant whose role is to ensure that posterior $\pi_1(\varphi | x)$ is a proper density function. Hence, posterior distribution is always proportional to joint density:

$$\pi_1(\varphi | x) \propto f(x|\vartheta)\,\pi_0(\varphi) \tag{8}$$

This process must be repeated as another failure is acquired and the posterior distribution of Φ becomes the prior. The available failure data is denoted in chronological order by $\boldsymbol{T} = \{T_1, T_2, ..., T_n\}$. After the n-th failure is observed, the prior distribution of Φ, $\pi_{n-1}(\varphi)$, is updated to the posterior distribution of Φ, $\pi_n(\varphi | \boldsymbol{T})$.

From computational viewpoint regarding the foregoing scheme the first problem arises in solving the integral in denominator of Equation 5. It is clear that for the different choices of the prior distribution $\pi_0(\varphi)$ the joint density, Equation 6, may take one algebraic form or another. For certain choices of the prior, the posterior has the same algebraic form as the prior. Such a choice is a conjugate prior. Conjugate priors are shown in Table 1.

There is no one 'correct' way of inputting prior information and different approaches can give different results. From practical viewpoint, prior distribution should reflect the best available knowledge or information about unknown parameters and should not be specified simply for computational convenience. If the conjugate prior distribution that provides an adequate representation of information cannot be found, numerical technique, such as MCMC should be used, Hamada et al. (2008) and Singpurwalla (2006).

The first order, or so called 'classical', approach to inference on future failures replaces the unknown parameters with respective mean value approximation $E[\varphi]$

Table 1. Conjugate priors.

Sampling distribution	Conjugate prior
Binomial (π)	Beta
Exponential (λ)	Gamma
Gamma (κ)	Gamma
Multinomial (π)	Dirichlet
Multivariate Normal (μ, Σ)	Normal Inverse Wishart
Negative Binomial (π)	Beta
Normal (μ, σ^2 known)	Normal
Normal (σ^2, μ known)	Inverse Gamma
Normal (μ, σ^2)	Normal Inverse Gamma
Pareto (β)	Gamma
Poisson (λ)	Gamma
Uniform (0, β)	Pareto

$$f(x \mid \varphi) \approx f(x \mid \hat{\varphi}) \tag{9}$$

which generally yields good approximation in presence of vast amount of data. Within the Bayesian framework it is correct to predict the future failures with prior-predictive distribution

$$f(x) = \int_{\Omega} f(x \mid \varphi)\, \pi_0(\varphi)\, \mathrm{d}\varphi \tag{10}$$

or with posterior-predictive distribution

$$f(x \mid \boldsymbol{T}) = \int_{\Omega} f(x \mid \varphi)\, \pi_1(\varphi \mid \boldsymbol{T})\, \mathrm{d}\varphi \tag{11}$$

Since only difference between the two prediction models is in chronological order, the distribution that contains more data (the latest) is used for prediction.

Typically physical and statistical arguments regarding the prior distribution model should be given by the informed opinion of the analyst and/or any chosen subject matter specialists. Arguments must be supported on: physics of failure theory and computational analysis, prototype testing, generic reliability data and past experience with similar devices.

The foregoing Bayesian updating framework is a general one. Now, its application will be illustrated by considering revision of an initial estimate of the failure rate in the gamma-exponential conjugate model.

The exponential distribution in terms of the failures is

$$f(x) = \Lambda\, exp(-\Lambda x) \tag{12}$$

with Λ, $x > 0$. The constant failure rate Λ should be treated as a random variable, and it is necessary to specify the form of the distribution that it follows. The model assumed is again exponential

$$\pi_0(\lambda) = \nu\, exp(-\nu\lambda) \tag{13}$$

with ν, $\lambda > 0$. In Equation 13, ν is the parameter of this distribution. The dimension of ν is lifetime, being the inverse of those of λ. If λ designates an estimate of Λ then Equation 12 becomes

$$f(x \mid \lambda) = \lambda\, exp(-\lambda x) \tag{14}$$

with λ, $x > 0$. And after the occurrence of the first failure T_1, Equation 5 results in:

$$\pi_1(\lambda \mid T_1) = \frac{f(T_1 \mid \lambda)\, \pi_0(\lambda)}{\int_0^{\infty} f(T_1 \mid \lambda)\, \pi_0(\lambda)\, \mathrm{d}\lambda} \tag{15}$$

In this equation, λ is the unknown parameter of interest distributed with posterior $\pi_1(\lambda \mid T_1)$ and $\pi_0(\lambda)$, given in Equation. 13, is the prior distribution of λ. Subsequently $f(T_1 \mid \lambda)$, given in Equation 14, is the likelihood function that updates a prior distribution. Equation 13 is computationally convenient but seriously limited in expressing prior knowledge because it has only one parameter. Within the Bayesian framework we are free to choose other forms of priors. Equation 13, can be assumed in gamma form

$$\pi_0(\lambda) = \frac{\kappa_0^{\alpha_0}}{(\alpha_0 - 1)!}\, \lambda^{\alpha_0 - 1} \exp(-\kappa_0 \lambda) \tag{16}$$

with κ, α, $\lambda > 0$. Expected value and variance of the gamma distribution are

$$\hat{\lambda}_0 = E\left[f(\lambda_0)\right] = \frac{\alpha_0}{\kappa_0};\quad Var(\lambda_0) = \frac{\alpha_0}{\kappa_0^2} \tag{17}$$

Gamma prior, Equation 16, is equal to exponential prior, Equation 13, when $\alpha_0 = 1$ and $\kappa_0 = \nu$. Gamma distribution can generate a wide variety of shapes for the prior distribution by simply modifying the numerical values assigned to parameters α_0 and κ_0 without impacting mathematical simplicity.

The posterior mean (expected value) is the most frequently used Bayesian parameter estimator. The posterior mode (maximum belief) and median (central value) are less commonly used alternative estimators. Prior to first failure all inference should be founded on the α_0, κ_0. The occurrence of the first failure T_1 gives the evidence that must be considered in order to validate the assumptions stated by the prior distribution. The marginal density becomes

$$f(T_1)=\int_0^\infty \left(\lambda e^{-\lambda T_1}\right)\left(\frac{\kappa_0^{\alpha_0}}{(\alpha_0-1)!}\lambda^{\alpha_0-1}\exp(-\lambda\kappa_0)\right)d\lambda = \frac{\alpha_0\kappa_0^{\alpha_0}}{(T_1+\kappa_0)^{\alpha_0+1}} \tag{18}$$

and by applying Equation 15 the posterior is

$$\pi_1(\lambda\mid T_1)=\frac{\left(\lambda e^{-\lambda T_1}\right)\left(\frac{\kappa_0^{\alpha_0}}{(\alpha_0-1)!}\lambda^{\alpha_0-1}e^{-\lambda\kappa_0}\right)}{\frac{\alpha_0\kappa_0^{\alpha_0}}{(\kappa_0+T_1)^{\alpha_0+1}}} = \frac{(\kappa_0+T_1)^{\alpha_0+1}}{\alpha_0!}\lambda^{\alpha_0}e^{-(\kappa_0+T_1)\lambda} \tag{19}$$

That is according to Equation 16 also the gamma distribution with parameters $\alpha_1 = \alpha_0 + 1$, $\kappa_1 = \kappa_0 + T_1$ and expected value

$$\hat{\lambda}_1 = E\left[f(\lambda\mid T_1)\right]=\frac{\alpha_0+1}{\kappa_0+T_1} \tag{20}$$

After the second failure T_2 the calculation is repeated starting from Equation 18. If the parameters α_0 and κ_0 are replaced with α_1 and κ_1, and time to failure T_1 with T_2 we obtain:

$$f(\lambda\mid T_1,T_2)=\frac{(\kappa_1+T_2)^{\alpha_1+1}}{\alpha_1!}\lambda^{\alpha_1}e^{-(\kappa_1+T_2)\lambda} \tag{21}$$

which is gamma distributions with parameters $\alpha_2 = \alpha_0 + 2$ and $\kappa_2 = \kappa_0 + T_1 + T_2$. Consequently, for the failures $\boldsymbol{T} = \{T_1, T_2, ..., T_n\}$ we get gamma distribution with parameters

$$\alpha_n = \alpha_0 + n,\ \kappa_n = \kappa_0 + \sum_{i=1}^{n} T_i \tag{22}$$

This form is applicable only if the parameters are integers. Further generalization can be presented in the general form of gamma distribution

$$f_{\Lambda\mid T}(\lambda\mid T)=\frac{\kappa_n^{\alpha_n}}{\Gamma(\alpha_n)}\lambda\alpha_n - 1_{\exp(-\kappa_n\lambda)} \tag{23}$$

and any real and positive values can be used for expressing the prior distribution. The expected value of the failure rate is

$$\hat{\lambda}_n = E\left[f(\lambda\mid \boldsymbol{T})\right]=\frac{\alpha_0+n}{\kappa_0+\sum_{i=1}^{n}T_i} \tag{24}$$

and variance.

$$Var(\lambda_n)=\frac{\alpha_0+n}{\left(\kappa_0+\sum_{i=1}^{n}T_i\right)^2} \tag{25}$$

Some comments are in order now. From Equation 24 and 25, it is clear that the weight given to the prior decreases as the sample size increases. In other words: evidence from the data has higher weight than the prior information. By employing Equations 24 and 25 with previously presented example of generic data regarding the valve one may calculate parameters α_0 and κ_0.

Considering the maintenance reality much of the failure data will be subjected to censoring. That is dealt with ease by employing the likelihood principle inherent to foregoing scheme

$$f(\boldsymbol{T}\mid\varphi)\propto\prod_{i=1}^{n}p_i(T_i\mid\varphi) \tag{26}$$

where

$$p_i(T_i\mid\varphi)=\begin{cases} f(T_i\mid\varphi); & T_i \text{ observed}\\ R(T_i\mid\varphi); & T_i \text{ right censored}\\ F(T_i\mid\varphi); & T_i \text{ left censored}\end{cases} \tag{27}$$

with $F = 1-R$. Posterior-predictive distribution of the gamma-exponential model is

$$f(x\mid\boldsymbol{T})=\int_\Omega \lambda\exp(-\lambda x)\frac{\kappa_n^{\alpha_n}}{\Gamma(\alpha_n)}\lambda^{\alpha_{n-1}}\exp(-\lambda\kappa_n)\,d\lambda = \frac{\alpha_n\kappa_n^{\alpha_n}}{(x+\kappa_n)^{\alpha_n+1}} \tag{28}$$

4 ILLUSTRATIVE EXAMPLES

The key to the Bayesian modeling, is the specification of a prior probability distribution on φ, before the data analysis. After describing how to calculate Bayesian model, we are presenting some examples in order to illustrate some of the main aspects related to prior specification. The data chosen to demonstrate different aspects of updating in chronological order is: 16, 4, 55, 10, 8, 7, 19, 22, 13 and 11. We assume that failures occur independently according to the exponential distribution. As stated before we are about to estimate the constant failure rate after each of the failures. Therefore we initially carry out a qualitative check to see how the all of data fits to exponential distribution, Figure 2. The graph appears to indicate a trend in the range 0 ÷ 20 but the lack of more data and the absence of additional information clouds this conclusion.

It is well known that the maximum likelihood estimate is strongly biased for the small data sets, but just, for comparison purposes the same data were used in maximum likelihood estimate (MLE) of the reliability model, Figure 3.

The failure rate $\lambda = 0.055$ will be used as the reference value in the future discussion. This should not be considered as the correct reliability model, but just as the result that will be reached MLE in case all of the data is available.

The gamma distribution is flexible and therefore capable to express a prior knowledge. By varying the parameters α_0 and κ_0 of gamma prior distribution Equation 16, we can find a distribution that approximately represents our prior belief about λ, and a standard Bayesian analysis can be carried out. When there is very little prior knowledge about the model parameters an non-informative or diffuse prior distribution is preferred. In this case, we would typically specify a prior distribution that

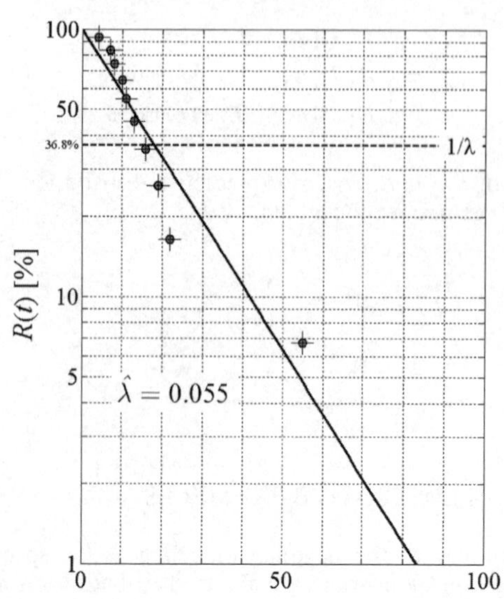

Figure 2. Exponential probability plot of complete data set.

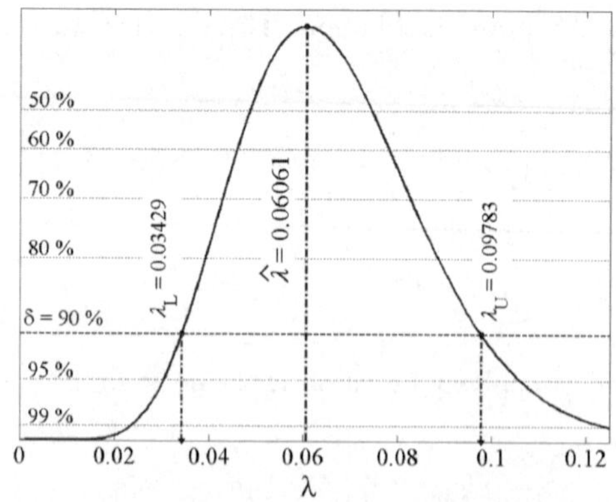

Figure 3. MLE estimate for the complete data set with exponential model (90% confidence interval).

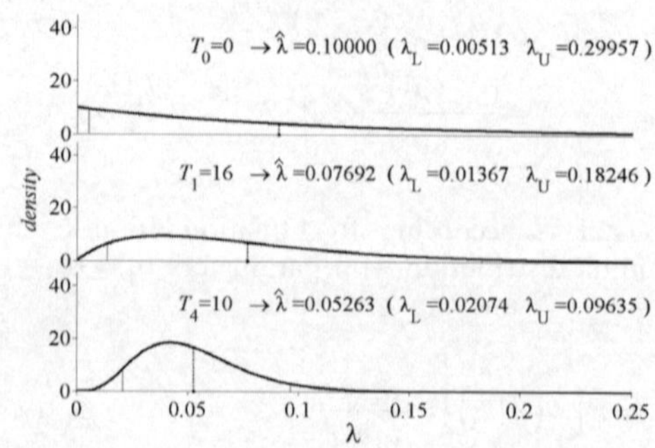

Figure 4. Bayesian updating process with diffuse prior distribution.

is at least approximately uniform over the range of indifference, Figure 4.

The first distribution ($T_0 = 0$) shows our prior belief about λ, and since the $\alpha_0 = 1$ the prior is exponential. Note that the mass of this prior distribution is spread over a wide range. The lower and upper credibility bounds of our estimate are denoted by λ_L and λ_U respectively. In the two subsequent graphs are the posterior distributions. In the middle is a posterior distribution after the first failure $T_1 = 16$ and in the bottom is a posterior distribution after the fourth failure $T_4 = 10$. It is evident that following the information gained from the data, updating process relatively quickly adjust our estimate of the first stage prior. Also, with more data the estimate bounds narrows. In Bayesian updating scheme we are generally interested on estimates after the first few failures. After the 10th failure, Figure 5 the Bayesian results resemble to MLE shown on Figure 3. This illustrates a very general property of Bayesian statistical procedures. In plain terms, the data easily swamp the information in the prior and diffuse priors allow us to compare the results with the MLE. In large samples,

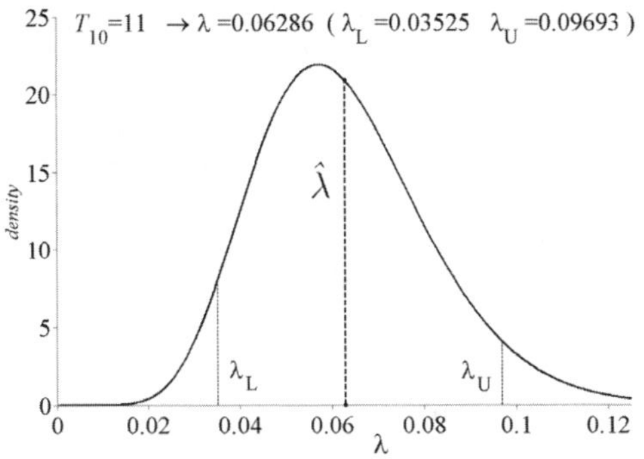

Figure 5. Bayesian updating process with diffuse prior distribution (90% credibility interval).

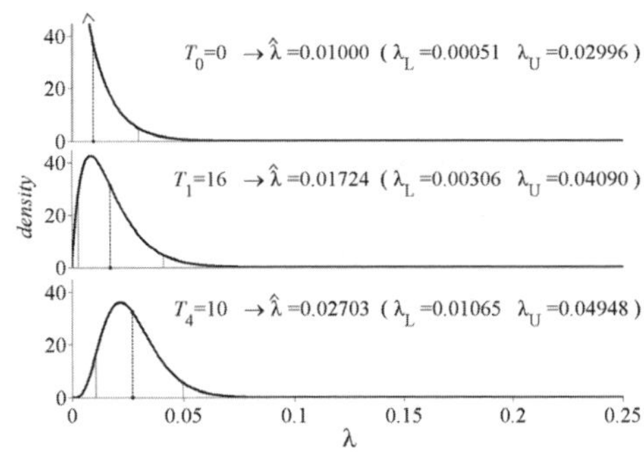

Figure 6. Bayesian updating process with an 'optimistic' prior distribution.

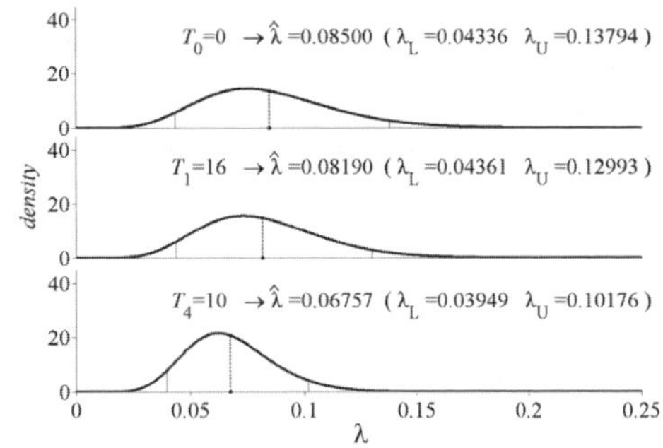

Figure 7. Bayesian updating process with an 'broad' gamma prior distribution.

they give answers that are similar to the answers provided by MLE.

Bayesian interval on Figure 5 agrees with the MLE confidence interval on Figure.3, though their probabilistic interpretations are different. The MLE is a frequency statement about the likelihood that numbers calculated from a sample capture the true parameter and provides us a confidence interval for an unknown parameter. On the Bayesian side the parameter estimates, along with credibility intervals, are calculated directly from the posterior distribution. Credibility intervals are legitimate probability statements about the unknown parameters, since these parameters now are considered random, not fixed. The credibility level for Bayesian model is also set to 90%.

Although in large samples broad prior distribution give answers that are very similar to the answers provided by classical statistics, in small samples results depend on the chosen prior distribution. If we choose prior distributions that assign non-negligible mass to the region surrounding the 'true' value of a parameter, then the posterior distribution will slowly converge, Figure 6. Such a scenario is likely to occur if low failure rate, given by a manufacturer is adopted for analysis. Because of the attitude of people involved in estimate we call this prior to be 'optimistic'.

This demonstrates how the Bayesian solution depends on the prior adopted. It can be viewed as either advantage or disadvantage, depending on how you regard the prior density; as representative one to real prior information, and the aim of the investigation.

The informative prior distributions assign most of the prior weight around the 'true' value of the parameter estimated. They are appropriate only if our belief is founded on a quality information about the parameters. Therefore the prior model must be strongly supported by experts. If not, our updating process will converge more slowly even than the MLE and the evidence provided by early failures will not provide any significant impact on posterior distribution. This is shown on Figures 7 and 8. On the Figure 7 we use a wide, but still informative prior, and the 'true' estimate is well covered. The expected value of our estimate quickly converges to the estimate shown in Figure 2. In this case the prior is gamma with $\alpha_0 = 8.5$ and $\kappa_0 = 100$.

The results for gamma prior with $\alpha_0 = 12.5$ and $\kappa_0 = 350$ are shown on Figure 8.

In this case the mean of the prior is closer to the 'true' value than in Figure 7, but the 'true' value is less covered. This result suggests that small intervals should be adopted only when there is plenty of relevant information, and larger intervals when there is a lack of information. Note that the standard deviation in OREDA (2002) is relatively high comparing to mean values of failure rate.

The main valid assumption if a successful reliability model is to be established is the specification of the sampling distribution. These arguments should be founded from the physics-of-failure on the device in question. Weibull distribution is frequently used in reliability analysis to fit failure data, because it is capable to handle decreasing, constant and increasing failure rates.

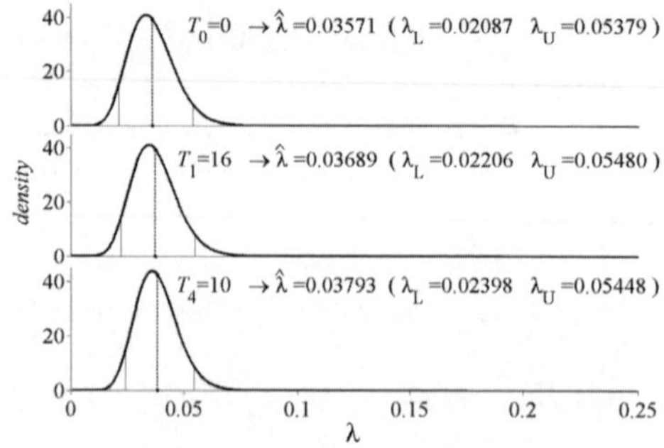

Figure 8. Bayesian updating process with a 'narrow' gamma prior distribution.

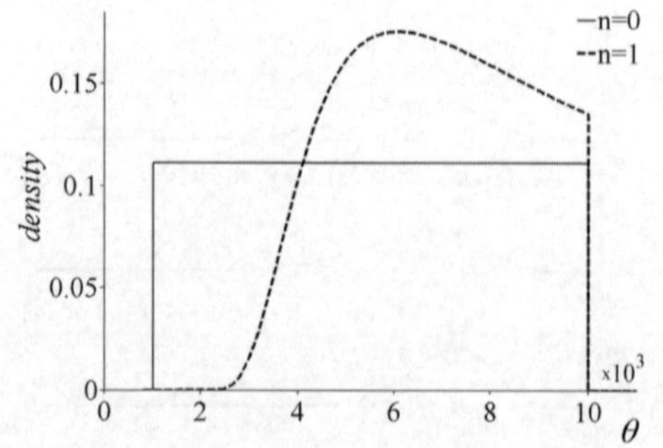

Figure 9. Prior and posterior distributions for Weibull example.

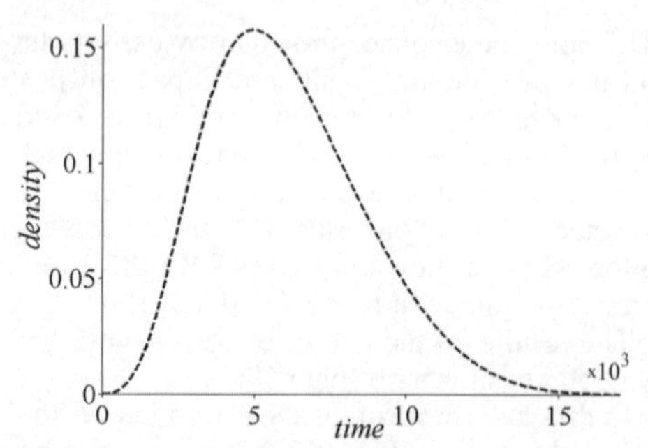

Figure 10. Posterior predictive distribution for Weibull example after the first failure.

Also the Weibull distribution is very flexible and it is not uncommon to fit a small set of failure data equally well as the 'real' failure distribution. Since the Weibull distribution is not a member of the exponential family Bayesian models are not amenable to simple analyses with conjugate priors. Weibull distribution has two parameters that must be considered.

We may suppose that components in this multi-parameter problem are independent so that their joint prior density is the product of corresponding univariate marginal priors. From the previous discussion it seems reasonable to simplify the analysis by assuming constant parameter $\beta = 3.5$. We assumed the uniform prior on θ over the range $1 \div 10$, and through the numerical methods the posterior is calculated for one failure $T_1 = 4.25 \cdot 10^3$, Figure 9.

The respective posterior predictive distribution is calculated numerically and is presented in Figure 10.

5 CONCLUSIONS

When dealing with small data sets Bayesian inference represents excellent alternative to frequentist approach. It is founded on the previous knowledge and on the concept of subjective probability that is non existent in the 'objective' concepts of frequentists approach.

Since it is unlikely in most applications that data will ever exist to objectively validate the reliability model our estimate must relay on the other sources of information. Since the most of the generic data provides only constant failure rate, updating process for the exponential sampling distribution is described in detail. In this paper the probabilistic concept for such analysis is discussed and, considering the closed form easily applicable solution is given.

The accompanying discussion describes the special care that should be given to the prior model selection. Narrow intervals should be adopted only if there is plenty of relevant information and larger intervals when there is a lack of information. In other words the analyst must subjectively find balance between optimistic and indifferent attitudes. The first results in narrow priors, and it is unlikely that data will lead the model to better estimates. The second attitude produces broad priors, the results are similar to classic inference and the advantage of Bayesian method is not fully utilized.

REFERENCES

Det Norske Veritas 2002. *OREDA—Offshore Reliability Data Handbook (4th ed)*, Høvik, DNV.

Hamada, M.S., Wilson, A.G., Reese, C. & Shane, M.H.F. 2008. *Bayesian Reliability*, New Jersey, Springer.

Murthy, D.N.P., Rausand, M. & Osteras, T. 2008. *Product Reliability: Specification and Performance*, New Jersey, Springer.

Percy, D.F. 2008. Maintenance Based on Limited Data. In K.A.H. Kobbacy & D.N.P. Murthy (eds), *Complex System Maintenance Handbook*: 133–154. New Jersey, Springer.

Rausand, M. & Høyland, A. 2004. *System Reliability Theory; Models, Statistical Methods and Applications (2nd ed)*, Chichester, Wiley.

Singpurwalla, N. 2006. *Reliability and Risk: A Bayesian Perspective*, Chichester, Wiley.

Collision and grounding as criteria in design of ship structures

Advanced Ship Design for Pollution Prevention – Guedes Soares & Parunov (eds)

Ship collision as criteria in ship design

P. Varsta, S. Ehlers, K. Tabri & A. Klanac
Marine Technology, Helsinki University of Technology, Finland

ABSTRACT: This paper presents a brief insight into the ASDEPP course 'Collision and Grounding as Criteria in ship design'. However, in this paper the focus is placed on the collision aspect alone. The general aim is to raise the awareness of the methods and procedures to be used to assess the behaviour of ship structures subjected to a collision. After a short background chapter, the following chapter describes how to consider and simulate the internal mechanics of ship collisions. Furthermore, the dynamics involved in ship collisions are presented, and at last a coupled numerical approach to solve, both the internal mechanics and outer dynamics simultaneously. Each chapter is followed by discussing a short example.

1 INTRODUCTION

A ship collision accident can result in severe environmental damage and loss of life. See for example Cahill (2002); Buzek and Holdert (1990). Looking at the IMO data (IMO 1999 to 2003) shows that the share of collision accidents is about 20% of all serious and very serious ship accidents.

A possible way to mitigate consequences of these adverse events is to design the structure in such way to withstand these accidental loads besides the operational loads. In this sense, a ship collision is effectively treated as ship design criteria.

In order to apply a structure tolerant to collision we need to analyse first the:

a. Structural behaviour in collision, or the so-called *internal mechanics*, and the
b. Ship motions during contact, or the *external dynamics*.

It is a common approach to decouple the analyses of structural deformations from the analysis of ship motions. There, the structural response is evaluated in the so-called displacement controlled analysis—the struck ship is kept fixed and the striking ship collides with it at a constant velocity along a prescribed path. Simulations itself are carried out in a quasi-static fashion; see ISSC (2006). By this means the deformations of the ship side structure are alone in contributing to the crashworthiness. This allows the structural collision simulation to be uncoupled from the outer dynamics, as this approach results in the maximum energy to be absorbed by a specific structure.

This internal mechanics analysis provides the contact force as a function of the prescribed penetration path. The actual extent of the penetration is obtained by comparing the area under the force-penetration curve to the deformation energy evaluated with some calculation model for external dynamics based on the conservation of momentum in collision (Pedersen and Zhang, 1998). Under the prescribed displacement the collision process does not describe any actual collision scenario, but gives an assessment of the crashworthiness of the side structure.

The accuracy of the displacement controlled approach depends on the level of precision on predicting the penetration path. This can be done rather precisely for symmetric collisions, where the striking ship collides under a right angle at the amidships of the struck ship and only few motion components are excited. Statistical studies have, however, indicated that the majority of collisions are non-symmetric in one or another way. Often the collision angle deviates from 90 deg or the contact point is not at the amidships. As in non-symmetric ship collisions more motion components are excited, the penetration path cannot be predefined with reasonable precision, but it should be evaluated in parallel with the ship motions. Numerical simulations are to be carried out in a coupled manner, where the ship motions and the structural deformations are evaluated in the same calculation run. This allows evaluating the contact force and deformation energy based on the actual penetration and not based on the predictions made beforehand.

2 INTERNAL MECHANICS

The finite element method is commonly used to carry out the collision simulations, as this numerical method is flexible and widely applicable for complex structures. These simulations contain highly non-linear structural deformations, including rupture. Therefore, the finite element analyses of ship collision incidents require the input of the true strain and stress relation until failure. In other words, the material relation and a failure criterion to determine the failure strain are needed.

Finite element-based analysis of ship collision simulations has been performed in many commercial codes, such as LS-DYNA, ABAQUS, and MSC/DYTRAN, by (Kitamura 1996; Sano *et al.* 1996 and Kuroiwa 1996). The material's true strain and stress relation is commonly selected in the form of a power law. The power law parameters can be obtained from standard tensile experiments for a chosen finite element length in an iterative procedure. In this paper, the true stress-strain curve is obtained as presented by Peschmann (2001) and by reverse engineering resulting in the following power law form

$$\sigma = K \cdot \varepsilon^n \qquad (1)$$

where the strength coefficient K is equal to 730 and the strain hardening index n is equal to 0.2, see also Figure 1. The experimental yield stress is found to be 284 MPa. The Young's Modulus is equal to 206 GPa, Poisson ratio equals 0.3 and the steel density is 7850 kg/m³. The simulation terminates ones the plate ruptures, namely when the critical through thickness strain according to Zhang *et al.* (2004) is reached

$$\varepsilon_f(l_e) = \varepsilon_g + \varepsilon_e \cdot \frac{t}{l_e} \qquad (2)$$

where ε_g is the uniform strain and ε_e is the necking strain, t is the plate thickness and l_e is the individual element length. It is commonly recommended that the ratio l_e/t is not less than 5 for shell elements. The values of the uniform and necking strain are achieved from thickness measurements related to the calculated stress states given by Zhang *et al.* (2004) are 0.056 for the uniform strain and 0.54 for the necking strain in the case of shell elements.

Implementation of those values is performed for material model 123 of LS-DYNA. Material 123 is a modified piecewise linear plasticity model using the stress versus strain curve defined in Figure 1. This material model allows fracture based on plastic thinning. The pre-calculated critical strain values are input for the material definition. The critical strain values are a function of the actual shell thickness and the prescribed element size. For further details on the crashworthiness simulation procedure see Ehlers *et al.* (2007) and Ehlers *et al.* (2008).

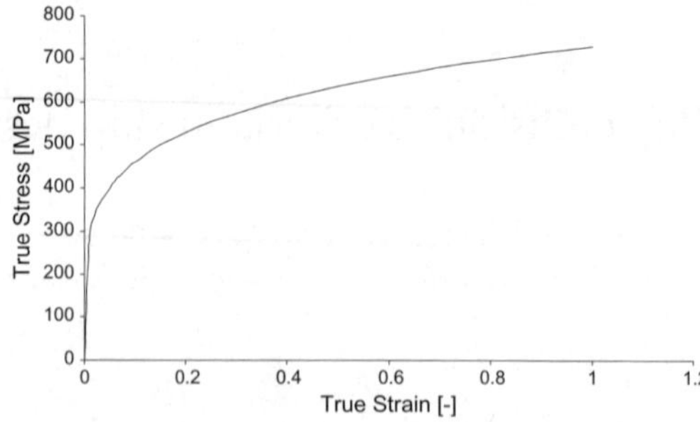

Figure 1. True stress versus strain relation.

2.1 *Finite element simulation*

The solver LS-DYNA is used for the collision simulations presented in this paper. Furthermore, the ANSYS parametric design language served to build the finite element model of the ship cross-sections; see Ehlers *et al.* 2008. The example model is built between two transverse bulkheads and the translational degrees of freedom are restricted at the bulkhead locations. The remaining edges are free. The structure is modelled using four noded, quadrilateral Belytschko-Lin-Tsay shell elements with 5 integration points through their thickness. Standard LS-DYNA hourglass and timestep control is a good choice for the simulations; see Hallquist (2007). The collision simulations are displacement-controlled. The rigid bulbous bow is moved into the ship side structure at a constant velocity of 10 m/s. This velocity is reasonably low so as not to cause inertia effects resulting from the ships' masses; see Konter *et al.* (2004). The ship motions are not yet considered in this analysis stage, as the maximum crashworthiness is of interest. The automatic single surface contact of LS-DYNA—see Hallquist (2007)—is used to treat the contact occurring during the simulation with a static friction coefficient of 0.3. The reaction forces between the striking bow and the side structure are obtained by a contact force transducer penalty card; see Hallquist (2007) and Ehlers *et al.* (2008). Integrating the contact force over the penetration depth leads to the energy absorbed by the structure, see for example Figure 2 and 3.

2.2 *The parametric finite element model*

This chapter presents the basis of a parametric finite element model using the ANSYS parametric design language. It includes a changing number of stiffeners and stiffener types and plate thickness. In this parametric modelling procedure, all parameters of a finite element model can be treated as variables. In other words, thicknesses, stiffener dimensions and stiffener spacing can be varied.

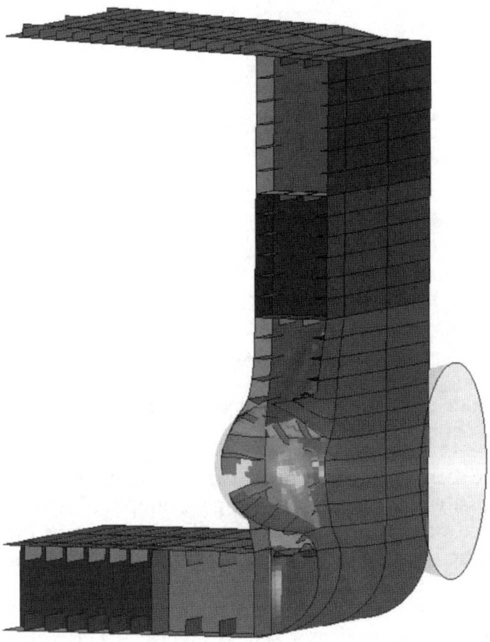

Figure 2. Deformed tanker side structure.

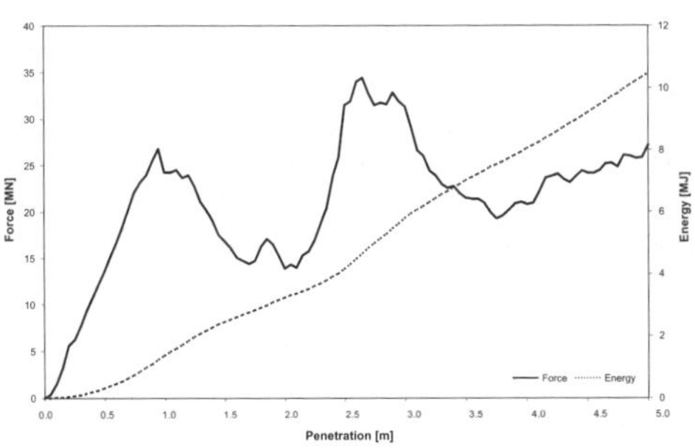

Figure 3. Force (bold) and Energy (dashed) versus penetration curve.

The parametric finite element model is built based on the assumption that the structural cross section can be split into a discrete number of strakes, i.e. plate with stiffening, bounded by 'hard-points', i.e. connections of strong structural elements. As an example a ship cross-section is given in Figure 4 adopting the strake definition principle. Figure 5 presents a single strake between hard points. In this paper, strakes are assumed to be longitudinally stiffened between webframes, the hard-points are marked as *P1* and *P2*, the thickness of the plate is *t* and the frame spacing *s*. Other strake variables are: the number of stiffeners and their type, and the web-frame spacing *S*.

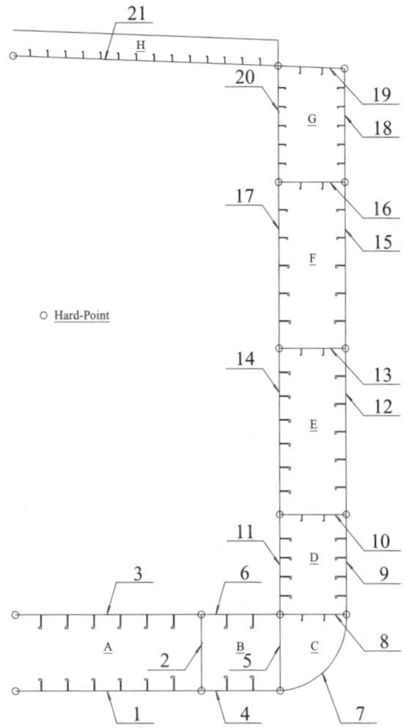

Figure 4. Example ship cross section split into 21 strakes.

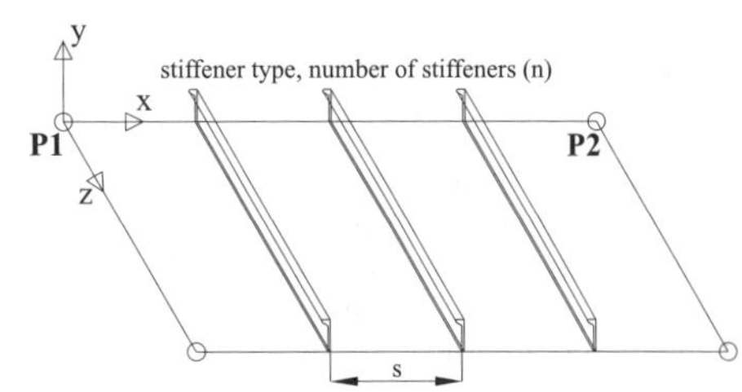

Figure 5. Strake variables.

The parametric finite element model adopts the steps given in the flowchart presented in Figure 6. The detailed ANSYS parametric design language code is given in Ehlers *et al.* 2008.

2.3 *A crashworthy stiffened plate*

To demonstrate how to combine the presented FE modelling procedure with optimization, a single stiffened plate is designed to maximize the ratio between the capacity to absorb energy in collision with a rigid spherical indenter and the mass of the structure, see Figure 7. The geometry of the stiffened plate is of similar type as shown in Figure 5. The plate and stiffener thicknesses range from 1 to 10 mm. The stiffeners are flat bars, with an overall varying height between 50 to 200 mm. The number of stiffeners ranges from 2 to 19. The variables are

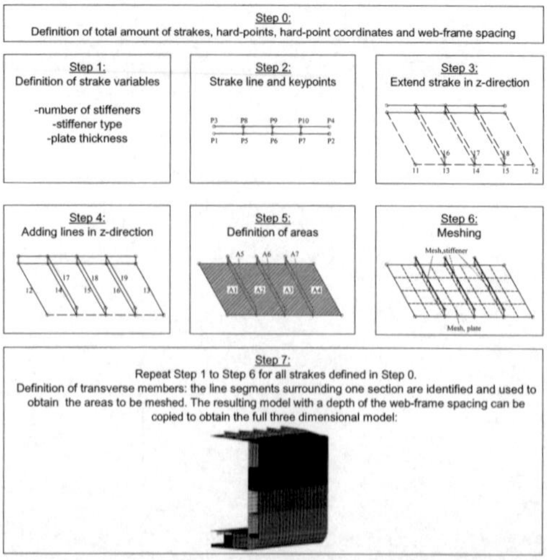

Figure 6. Steps of the parametric finite element model generation.

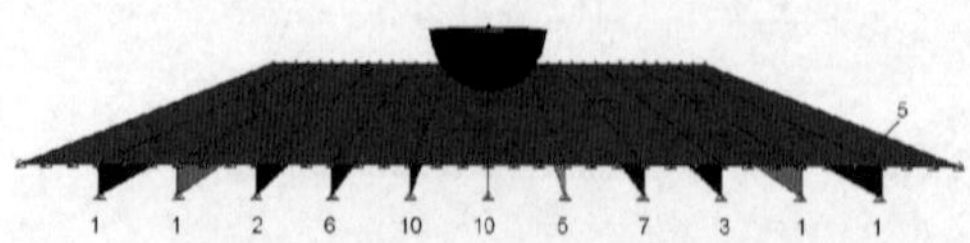

Figure 7. Finite element model of the optimal stiffened plate with indicated plate thicknesses in mm.

changed in discrete steps of 1 for the stiffeners, or 1 mm for the thicknesses. The overall dimensions of the plate are 3 m × 4.5 m, having the short edges fixed and long edges freed. The displacement-controlled rigid sphere indents the plate in the centre.

The crashworthiness optimization is unconstrained, and it is performed with Particle Swarm Optimization (PSO) algorithm (Eberhart and Kennedy 1995). MATLAB is utilized as a shell program managing the process and communication between LS-DYNA and PSO. The particle swarm size is set to 20 with 40 calculation rounds. Inertia in PSO is at start 1.4, the dynamic inertia reduction factor is 0.8 and the number of rounds to improve the solutions before the inertia is reduced is 4. Optimization is initiated with a random population. Each structural alternative is indented with the rigid sphere until the plate ruptures. The resulting energy versus penetration curve is stored for comparison. Furthermore, the maximum energy value divided by the specific mass of the structural alternatives, namely the Energy-per-Mass ratio (E/M), serves as the optimisation objective.

After 40 calculation rounds PSO converges to a solution, see Figure 8. All together 800 alternatives are generated. Their energy vs. penetration plots

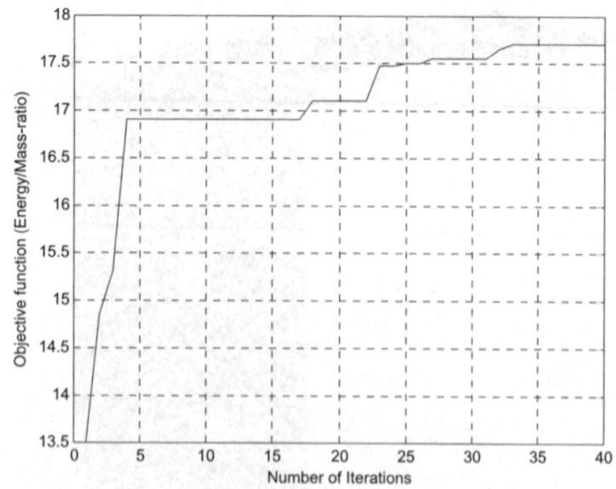

Figure 8. Convergence of objective function.

are seen in Figure 9. From the same figure it is also possible to see that this simple example of a stiffened plate can produce a significant scatter in the capacity to absorb energy until the plate ruptures. The alternative with the highest energy capacity per mass ratio has 11stiffeners with the stiffener and plate thicknesses as given in Figure 7. This panel weighs 695 kg (51.5 kg/m^2) and absorbs 11.9 kJ until rupture. To indicate significance of these values, the panel with the minimum plate thickness (1 mm) and 11stiffeners weighs 144 kg (10.6 kg/m^2) and absorbs 0.9 kJ and the panel with the maximum plate thickness (10 mm) and 11stiffeners weighs 1448 kg (107.2 kg/m^2) and absorbs 15.4 kJ. However, in reality the striking location is unknown, and therefore the structure should be optimised for various striking locations, see Ehlers (2009).

2.4 *Example 2: Multi-objective optimization of a crashworthy tanker (based on the study of Klanac et al. 2009)*

Here we extend the previous example onto more complex problem, involving constrained multi-objective optimization of a crashworthy tanker. Optimization of crashworthiness of the structure is now decoupled from the mass optimisation of the structure, so the result of optimization involves a set of competitive alternatives with minimum of mass for a given crashworthiness level.

Actually, three objectives are considered for optimization: a) maximize the crashworthiness, i.e. hull capacity to absorb collision energy before breaching (abbreviated onwards as ENERGY), b) minimize the hull mass (abbreviated as HULL) and c) minimize the mass of stainless steel *Duplex* (abbreviated as DUPLEX).

Observed tanker is 180 m long, 32 m wide and 18 m deep, with a draught of 11.5 m. It has a typical product/chemical carrier construction as seen in

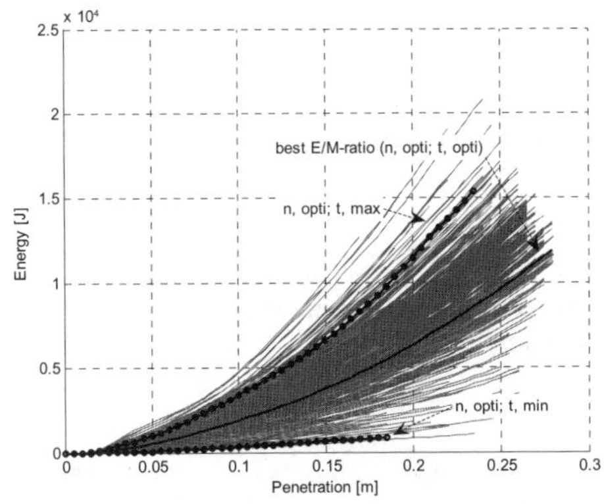

Figure 9. Energy versus penetration curves.

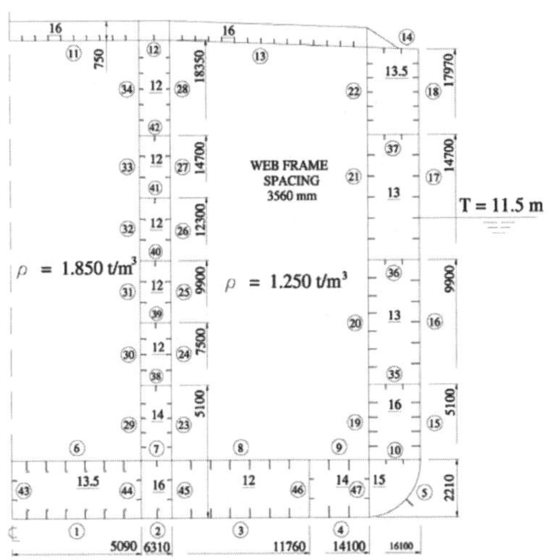

Figure 10. Half of the main frame section of a considered tanker with dimensions in mm, indicated strake numbers (in circles) and scantlings of the transverse structures (underlined).

Figure 10, where the tanks are plated with stainless steel—*Duplex*. The collision is assumed to occur in the side shell of the cargo area amid two transverse bulkheads and at the draught of approx. 4 m, as seen in Figure 11.

Tanker's longitudinally-stiffened cargo space structure is split with transverse bulkheads placed every 17.8 m. The cargo tanks' plating is assumed to be built from *Duplex* steel to resist the transported chemicals, while the remaining structure is assumed to be of high tensile steel. *Duplex* steel yield strength is 440 MPa and that of high tensile steel is 355 MPa.

Tanker's structure is symmetric about the centre line. It is optimized through 47 longitudinal strakes of one half of the midship section. No fore- and

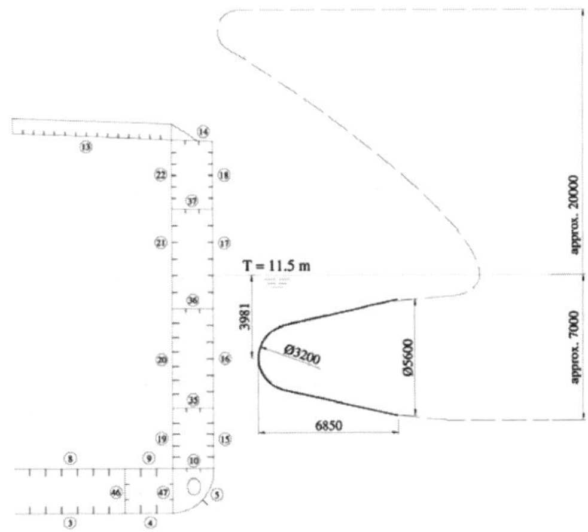

Figure 11. Collision scenario (in actual scale) with indicated rigid bulb shape (bolded lines) used in FEM simulations.

aft- ship structure is considered, neither in analysis of hull response to normal service loads and collision loads.

The assumed section affected by collision (named onwards as 'the crash section') includes the outer structure, starting 6310 mm from the centre line, i.e. the outer cargo tank, part of the double bottom and the whole double side as seen also in Figure 11. Longitudinally, the section is bounded by two transverse bulkheads.

Each strake is generally described with five parameters: plate thickness, stiffener size, number of stiffeners, stiffener type and panel's material type. The former two parameters are varied in optimization, while the latter three are fixed. Their adopted values are seen also in Figure 10. In total, the optimization is conducted with 94 variables.

Variables are considered discretely. The plate thicknesses are assumed to be available for every whole millimetre, from 5 to 24 mm outside the crash section, and in the crash section between 4 and 42 mm. The stiffener sizes are taken from the standard tables of profiles, e.g. Ruuki (2008), for the whole available range of HP profiles and flat bars. Flat bars are applied exclusively in the crash section only, due to their improved collision properties over HP profiles; see Alsos and Amdahl (2009). In other ship sections, HP profiles are used normally. Optimization of the transverse structures, i.e. the structure of the web frames, is not conducted, while its assumed scantlings are given in Figure 10.

During the optimization the response under the hull girder loads is calculated applying the Coupled Beam (CB) method (Naar *et al.* 2004). Local response of strake structures under hydrostatic loading is calculated with the uniformly loaded simple beam model, and it is added to the response of the structure to the hull girder loading. The structure of each strake is checked for eight standard failure criteria concerning: plate yield and buckling, stiffener yield, lateral and torsional buckling, stiffener's web and flange (where appropriate) buckling for each loading condition, see DNV (2005) for more on their definition. Additional crossing over criterion Hughes *et al.* (2004) is used to ensure controlled panel collapse due to extensive in-plane loading. Altogether 376 failure criteria are calculated for each design alternative and for each loading condition.

The FE model for the analysis of crashworthiness, adopting the approach presented in the previous example, is constrained in the planes of transverse bulkheads to translate in all three directions. The size of elements for plating is established on the basis of stiffener spacing, being on average 0.8×0.9 m. To model the stiffeners, one element is used per stiffener height. The structure is modelled using strictly the four noded, quadrilateral *Belytschko-Lin-Tsay* shell elements (Flanagan and Belytschko 1981) with 5 integration points through their thickness.

Klanac *et al.* (2009) perform the optimization of this problem using a *Genetic Algorithm* (GA) code VOP in a so-called 'two-stage optimization procedure', where masses are minimized first without computing crashworthiness. Once the minima of masses are attained, crashworthiness maximization is added to the optimization, alongside that of mass minimization. This approach saves computational time since optimization-difficult minimization of masses is performed without evaluation of time-expensive crashworthiness. On the other hand, attained designs with minimal mass hold commonalities with designs with maximal crashworthiness, so they can be effectively utilized to attain crashworthiness maxima more quickly.

The attained Pareto frontier is fairly evenly developed between the objectives' extremes, allowing us to make relevant observations as seen in Figure 12. From Figures 12a and 12b we can now notice that the significant increase in crashworthiness, from the lightest alternative with 10 MJ to the alternative with 90 MJ, is achieved practically without any increase in DUPLEX. Seemingly, only the outer shell is modified, but also mildly as seen also in Figure 14. The increase in HULL, for a 'tenfold' increase in ENERGY, is only 25%. Furthermore, the amount of expensive *Duplex* steel can be kept at minimum for a fairly crashworthy design. This reduces the chance of unacceptable rise in costs for the increase in ship safety. Figure 13 depicts the difference in scantlings for the two distinctive alternatives, 1 and 32, respectively being the HULL minimal and ENERGY maximal designs.

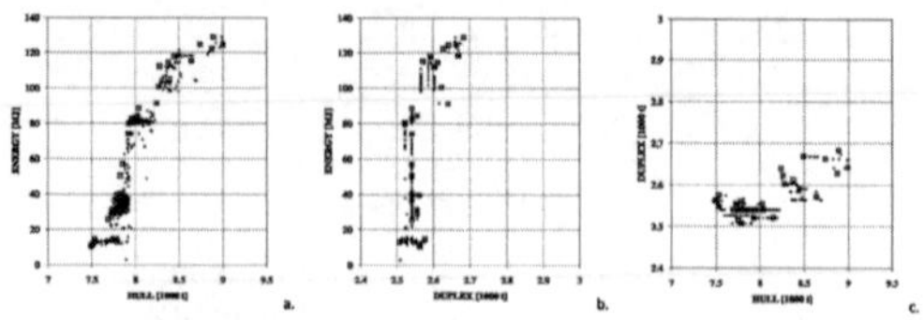

Figure 12. Pareto optima (o) spanned between the three objectives, and a representative set of alternatives selected for further analysis (μ).

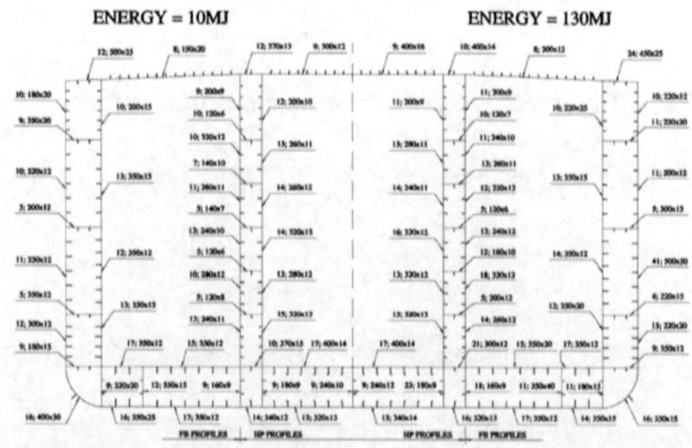

Figure 13. Scantlings of the min HULL (10 MJ) and the max ENERGY (130 MJ) design alternatives (first number, before semicolon, represents the strake thickness, and the second represents the profile size).

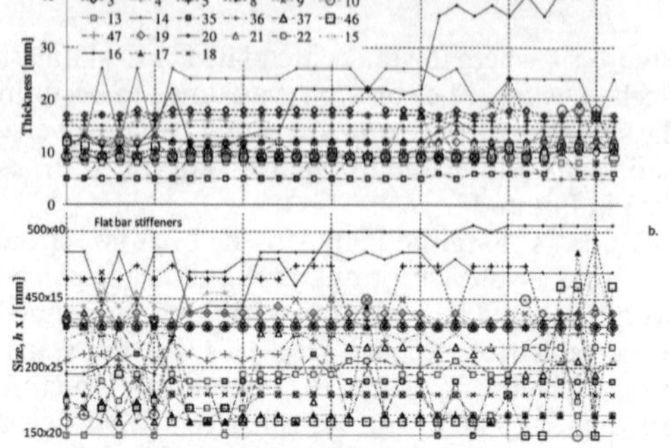

Figure 14. 32 sorted Pareto optima according to the rising ENERGY: a) correspondence in rise between ENERGY, HULL and DUPLEX with b) plate thickness and c) stiffener size.

If we observe the rise in ENERGY in more detail, from the lightest to the heaviest design, we can easily notice where and when material is added to increase the crashworthiness of the structure. As seen from Figures 14b and c, material addition is most intensive for the strake 16 in the outer shell, exactly where we expect the collision to occur, and in the strake 14. For the higher levels of ENERGY, above 100 MJ, material is slightly added to the inner shell, in strake 20, being directly behind the strake 16. Furthermore, if we compare Figures 12 and 14, we can also see that for the fine mesh calculations, ENERGY is strongly linked to the strake 16, where the major increases in ENERGY are analogous to the increases in the plating thickness and stiffener size of the strake 16; take also a special notice of the changes in ENERGY and plate thickness of the strake 16 between the design alternatives 21 and 22. On the other hand, the neighbouring side shell strakes, as well as the horizontal strakes of deck and double bottom, show no significant consistent changes, as indicated in Figure 14. This leads to the conclusion that the increase in crashworthiness has been attained predominantly through the local stiffening, at the place of presumed collision.

This finding is physically justified with the observation of McDermott *et al.* (1972) that most of the collision energy is absorbed by the membrane tension of the structural elements. So, if the structure is globally stiff, the initial collision deformations arising locally in the structure at the striking location will be constrained to dissipate and membrane tension will not develop sufficiently. This input of energy will raise plastic strains locally and eventually initiate fracture in the vicinity of the contact. If on the contrary a structure at the striking location is supported flexibly, but is locally stiff, the deformations will spread more easily, initiating membrane tension and dissipating the strains widely over the structure. For this reason their intensity will be low and fracture will be postponed. This can be clearly seen in Figure 15 that depicts the calculated collision deformations at the moment of inner hull rupture for a low crashworthy (design alternative 1) and a highly crashworthy (design alternative 32) hull structure. The evidence of this reasoning can be found to exist in some other reported studies, including the results of the full scale collision tests (Pedersen 2003), or in the numerical simulations analyzing these tests (Redhe *et al.* 2002).

If we assume that serious and very serious collisions predominantly involve contact with bulbous bow, than we can suppose that an efficient mean to elevate crashworthiness is through the local stiffening that extends below ship's waterline. This proposal is symbolically depicted in Figure 16, and it is similar to the better known 'ice belt', or more appropriately, to the 'torpedo belt' of the early 20th century naval ships. Obviously, if the collision would occur above or below such a locally stiffened location, a significant difference in crashworthiness can be expected. To understand the extent of this difference, we perform additional collision simulations on the studied tanker assuming the same characteristics of the striking bulbous bow. The fine mesh model is applied for the considered 32 Pareto optimal alternatives, and collision simulations are performed accordingly with striking positions on strakes 15 and 17, i.e. below and above the original struck strake 16. The results of these collision simulations are presented in Figure 17, which shows that all simulated collisions

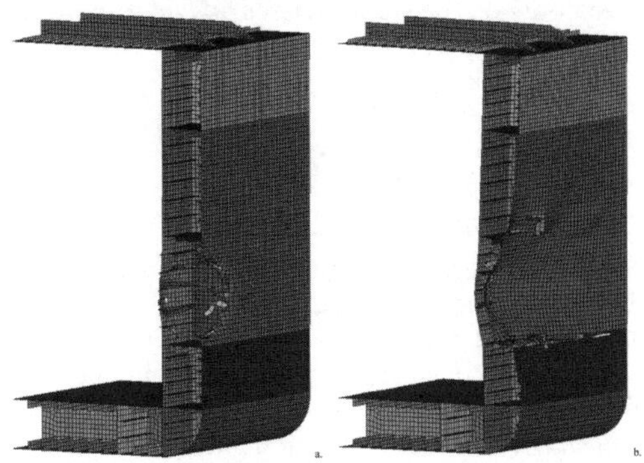

Figure 15. Computed collision deformations, applying the fine mesh FE model, for the a) min HULL/low crashworthy alternative 1, and b) max ENERGY/high crashworthy alternative 32, at the point of inner hull rupture.

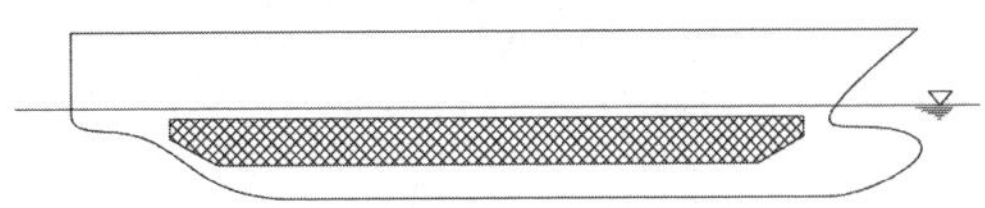

Figure 16. Collision belt of increased side shell stiffness depicted with the hatched surface.

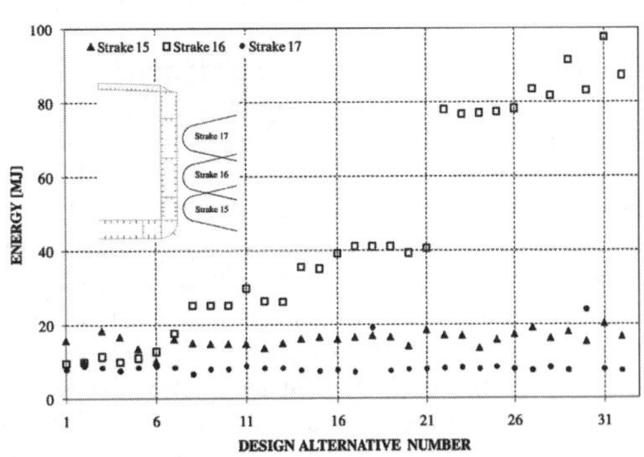

Figure 17. Results of the collision simulation for different striking locations.

on strakes 15 and 17 result in relatively similar values of ENERGY. Moreover, these ENERGY values do not reduce below the levels attained for the least crashworthy alternatives.

3 OUTER DYNAMICS

One of the first calculation models to describe the external dynamics of ship collisions was proposed by Minorsky (1959). This single-degree-of-freedom (dof) model is based on the conservation of linear momentum and there is no coupling with the inner mechanics. With this classical model the loss of kinetic energy E_D in collision is evaluated based only on the ships' masses m^A, m^B and on the collision velocity v_0

$$E_D = \frac{\left(a_2^B + m^B\right)}{\left(a_1^A + m^A\right) + \left(a_2^B + m^B\right)} \frac{\left(a_1^A + m^A\right)}{2} v_0^2 \qquad (3)$$

where superscripts A and B denote, respectively, the striking and the struck ship, a_2 is hydrodynamic added mass in sway and a_1 is hydrodynamic added mass in surge. This loss in energy is absorbed by structural deformations. The interaction between the ship and the surrounding water was through a constant added mass. The model allows fast estimation of the energy available for structural deformations without providing exact ship motions. Woisin (1988) extended the collision model to consider three dof—surge, sway and yaw. Later, in 1998, Pedersen and Zhang included the effects of sliding and elastic rebounding. These two-dimensional models in the plane of the water surface are capable of analysing non-symmetric collisions and evaluating the loss of kinetic energy in a collision absorbed by structural deformations. However, there only the inertial forces are taken into account and thus, the effects of several hydromechanical force components and the contact force are neglected.

For precise inclusion of all the relevant forces and to consider the coupling between the ship motions and the contact force, a precise description of the whole collision process, together with the full time histories of the motions and forces involved, is to be obtained. These are provided in coupled time-domain simulations, where a system of equations of ship motion is solved using a numerical integration procedure. Motion-dependent forces, such as the radiation forces arising from the interaction with the surrounding water, can be included in the analysis. As a result of the complexity of solving the equations, these simulation models are often reduced to include the motions in the horizontal plane only (Petersen, 1982; Brown, 2002). Tabri *et al.* (2009b) presented time-domain simulation model, where all the motion components were included. This model was validated experimentally with model-scale collision experiments (Tabri *et al.* 2009b).

Table 1. Deformation energy obtained by different methods.

Test	Experiment (total/plastic)	Decoupled model (Zhang, 1999)	Coupled model (Tabri *et al.* 2009b) (total/plastic)
301	4.20/4.14 [J]	4.30/- [J]	4.62/4.21 [J]
313	3.14/3.09 [J]	3.7/- [J]	3.45/3.14 [J]

To present the differences between the coupled and decoupled approaches, we predict the outcomes of two non-symmetric model-scale experiments that were described in Tabri *et al.* (2009b). The deformation energy as evaluated experimentally and calculated with the two approaches is presented in Table 1. This energy consists of plastic and elastic parts, the latter is returned to the kinetic energy of the colliding ships through the elastic springback of the deformed structures. With typical ship structures the elastic energy is an order of magnitude smaller compared to the plastic energy. Here, the sum of the elastic and plastic energies is referred to as total deformation energy. For the decoupled model only the total deformation energy is presented, while for the other two methods the plastic deformation energy is also presented. In the decoupled model the decomposition of the total energy into its plastic and elastic components requires knowledge of the ships' velocities immediately after the contact, which cannot be precisely defined on the basis of the decoupled approach alone.

Computational models tend to overestimate the total deformation energy by approximately 10%. In the decoupled analysis the total deformation energy is the only outcome and the penetration is assumed to follow the direction of the initial velocity of the striking ship (Zhang, 1999). To solve the final value of the penetration corresponding to the deformation energy obtained, the contact force model from Tabri *et al.* (2009) is exploited. The penetration paths are depicted in Figure 18, where x^B denotes longitudinal and y^B transversal extent of penetration with respect to the struck ship. The decoupled analysis resulted in slightly deeper, but shorter penetration in longitudinal direction compared to the coupled time domain simulations. The penetration paths differ as in time-domain simulations the penetration is evaluated considering the motions of both ships during the collision. In the decoupled analysis such information is not

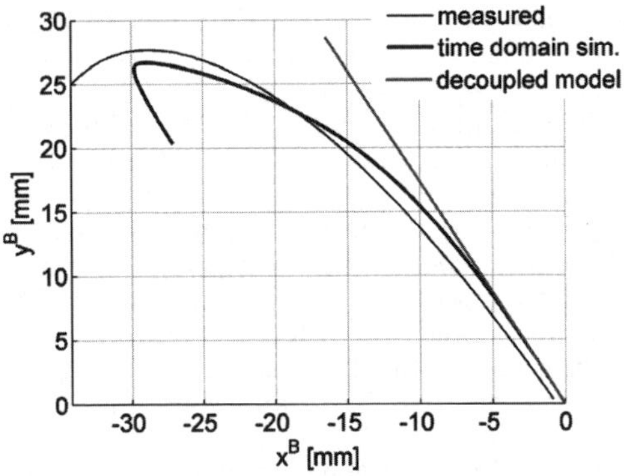

a) 301 (β = 120 deg, L_c = 0.16 L^B = 0.37m)

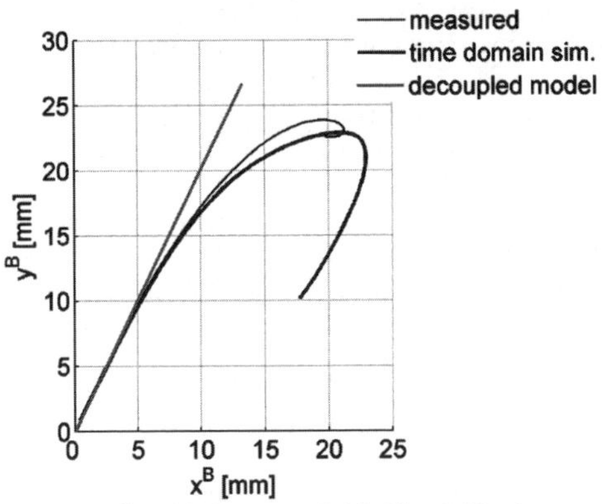

b) 313 (β = 60 deg, L_c = 0.13 L^B = 0.29m)

Figure 18. Penetration paths of the bulb in the struck ship, L^B-length of the struck ship model, β-collision angle (see Appendix B in Tabri *et al.* 2009 for test matrix).

available and it is assumed that the penetration follows the direction of the initial velocity between the ships Zhang (1999).

4 COUPLED MECHANICS AND DYNAMICS IN FE-SIMULATIONS

Several studies have revealed that, in addition to the contact force, the inertial forces and the corresponding energies are the main components in collision (Tabri *et al.* 2009a). Linear effects associated with inertial forces can conveniently be included in FE simulations as they are proportional to the acceleration. The frictional force is proportional to the square of ships velocity and the hydrodynamic damping force requires the evaluation of convolution integral (Cummins 1962). Their inclusion in the simulations is not straight forward and as their share in the energy balance is relatively low, about 10% of the total available energy, they are not yet included in the coupled FE analysis procedure proposed by Pill and Tabri (2009).

The main challenge to include the inertia effects is to provide a proper description of the ship masses, inertias and added masses. These properties should be modelled with as few elements as possible in order to add little to the simulation time. Mass m and inertia about the vertical axis I_{zz} are the main properties of the ship when evaluating the planar motions under the external forces.

In a FE model, the masses and inertias of the colliding ships are modelled by using a small number of mass points. The principle of the modelling is presented in Figure 19. The striking ship consists of a modelled bow region and three mass points m_j. Correspondingly, the struck ship consists also of three mass points and a part of the side structure. If the mass of the modelled structure is m_{STR}, the total mass of the ship is obtained as

$$m = \sum_{j=1}^{3} m_j + m_{STR} \tag{4}$$

If one of the mass points is located at the COG of the ships, the distance $k_{zz,j}$ of the other two mass points from COG is evaluated from

$$I_{zz} + \mu_{66} = \sum_{j=1,3} m_j \cdot \left(k^*_{zz,j}\right)^2 + m_{STR} \cdot k^2_{STR} \tag{5}$$

where μ_{66} is the yaw added mass, $k^*_{zz,j}$ is the radius of gyration of the j-th mass element, which also takes into account the yaw added mass; and k_{STR} is the radius of gyration of the modelled structure about the centre of gravity (COG) of the ship. Therefore, the yaw added mass μ_{66} is included by calculating the suitable value for k^*_{ZZj}. Using three mass points enables to model the mass and inertia of the ship with respect to the vertical axis and to control the initial location of the COG. The mass points are constrained to move together with the boundary nodes of the modelled parts of the ship.

As yaw is the only rotational motion component, its added mass is conveniently included by proper positioning of the mass points in the longitudinal direction of the ship. The added mass components associated with translational motions should be included in a certain directions only. The surge added masses of the striking and struck ships are marked as $\mu_{11}{}^A$ and $\mu_{11}{}^B$ in Figure 19. These added masses are positioned in the centres of gravity of the ships. The approach causes the surge added mass of to be included also in the sway direction. In the case of the striking ship the sway added mass is not properly modelled as her motions are predominantly in surge direction. Neglecting the

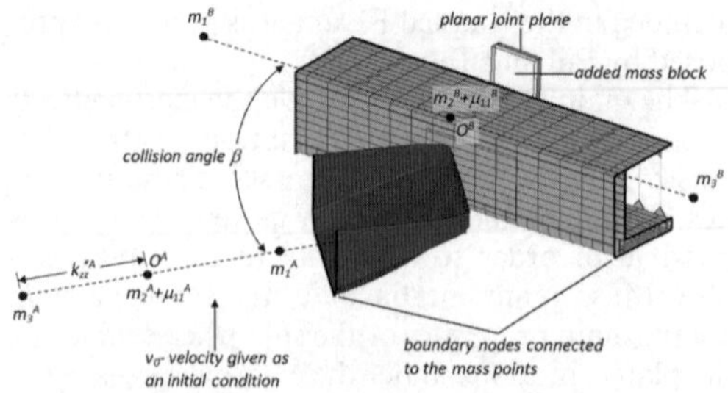

Figure 19. Calculation setup for dynamic collision simulations.

sway added mass of the striking ship results in only minor error in the simulation results.

In the case of the struck ship the sway added mass cannot be neglected and it is modelled as a single block of additional mass that is located on the side opposite to the side that is hit, see Figure 19. The mass of the sway added mass block is calculated by subtracting the surge added mass, already added to the node at the COG, from the sway added mass. The sway added mass of the struck ship is connected to a rigid support plate on the ship using a planar joint, which restricts relative movement in one direction and allows the joined entities to move in the other directions with respect to each other. Therefore it is possible to take added masses into account in one direction only. In the case of surge motions, the sway added mass block remains in its initial position and is not included in the mass of the struck ship.

The side structure included in the FE model is connected through its boundary nodes to the mass points. Thus, any motion of the structure is transmitted to the mass nodes and the whole ship mass is included. Desired collision angle β and eccentricity are achieved by rotating the striking ship to the required angle and moving it to the proper location. The striking ship is given the initial velocity v_0.

The advantage of the dynamic simulations is that they simulate an actual collision event and no prescriptions other than the initial conditions are required. The drawback is that the mass-scaling would lead to larger errors compared to displacement controlled simulations and is therefore not suggested. As both ships are moving, the whole contact process lasts longer as it takes longer time for the striking ship to penetrate deep enough into the struck ship to cause breaching of the inner hull. Therefore, coupled outer dynamics and internal mechanics simulations require significantly longer computational time, than uncoupled simulations.

4.1 *Dynamic simulations of LNG-tug collision*

To demonstrate the dynamic FE collision simulations four different collisions between a LNG tanker and a harbour tug are simualted. In all the simulated collisons, the contact point is at amidships and the collision angle is 60 deg according to the definition in Figure 20. The collision at amidships presents the most critical scenario as the largest amount of the initial kinetic energy of the striking ship is transmitted into the structrual deformation energy. Main parameters defining the collision scenario are given in Table 2. The vertical position is depicted in Figure 21. It is assumed that all the deformations are limited to the struck ship and the striking ship is treated as rigid.

In first three collision scenarios it is assumed that the tug collides with the struck ship at its maximum speed of 13 kn. Three different plating thicknesses at the contact region are used—17 mm, 25 mm and 35 mm. In fourth simulated scenario the speed of the striking ship is reduced to 5 kn and the plating thickness of the struck ship is 17 mm.

The deformations at the end of the collision are presented in Figure 22 to 25. The simulated scenarios revealed that in collisions at the maximum speed of the striking ship, the outer hull of the struck ship will be penetrated even when its plating thickness is as high as 35 mm. The size of the damage opening increases significantly as the thickness becomes lower. However, with the location of the contact location at the amidships, the most unfavourable scenario is simulated. The energy to be absorbed by structural deformations becomes less when the contact point is at some distance from the amidships. Reduction of the speed of the striking ship to 5 knots reduces the amount of energy available for structural deformations and the outer plating of the struck ship remains intact even in the case of 15 mm plating.

Time-histories of the deformation energy in collision are presented in Figure 26 to 29. Part of the energy is absorbed by structural deformation and part by friction between the ships, the sum of these two energies is referred to as total deformation

Table 2. Collision parameters.

Collision velocity	$v_0 = 13$ kn or 5 kn
Collision angle (Figure 12)	$\beta = 60$ deg
Mass of the striking ship	
Structural mass	$m^A = 930$ [ton]
Hydrodynamic added mass in surge	$m_1{}^A = 46.5$ [ton] (5% of m^A)
Mass of the struck ship	
Structural mass	$m^B = 179\,211$ [ton]
Hydrodynamic added mass in surge	$m_1{}^B = 8960$ [ton] (5% of m^B)
Hydrodynamic added mass in sway	$m_2{}^B = 35842$ [ton] (20% of m^B)

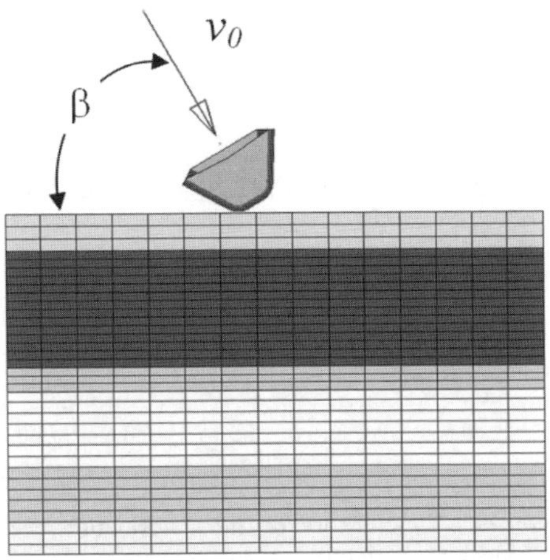

Figure 20. Collision angle and velocity.

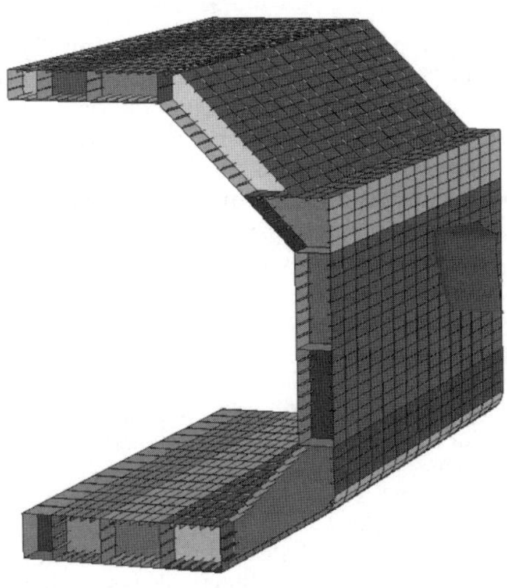

Figure 21. Position of the contact point.

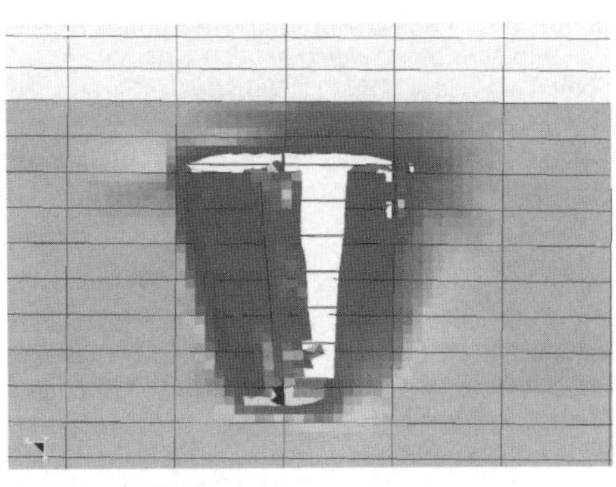

Figure 22. Deformations at the end of collision ($v_0 = 13$ [kn], $t = 17$ [mm]).

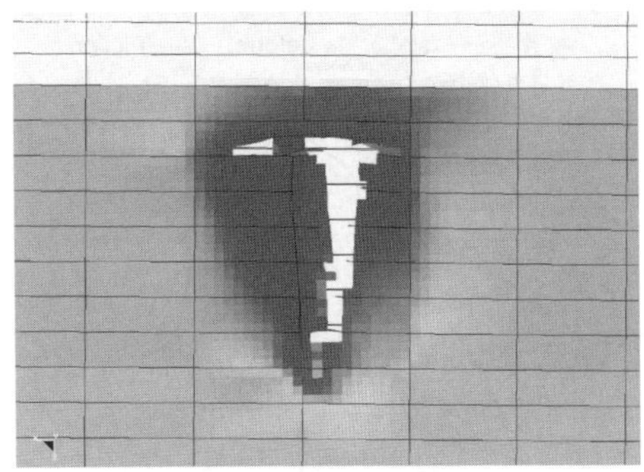

Figure 23. Deformations at the end of collision ($v_0 = 13$ [kn], $t = 25$ [mm]).

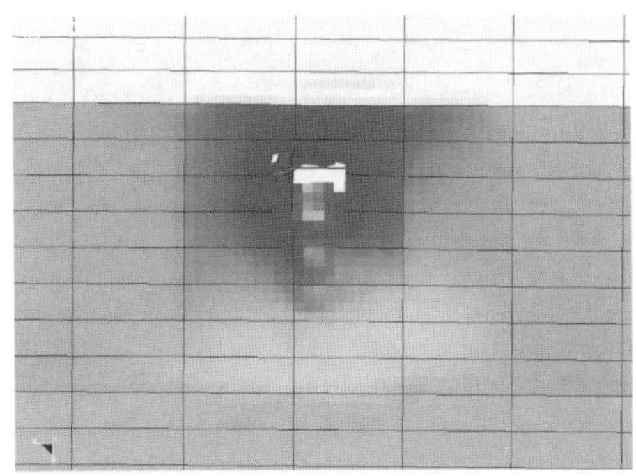

Figure 24. Deformations at the end of collision ($v_0 = 13$ [kn], $t = 35$ [mm]).

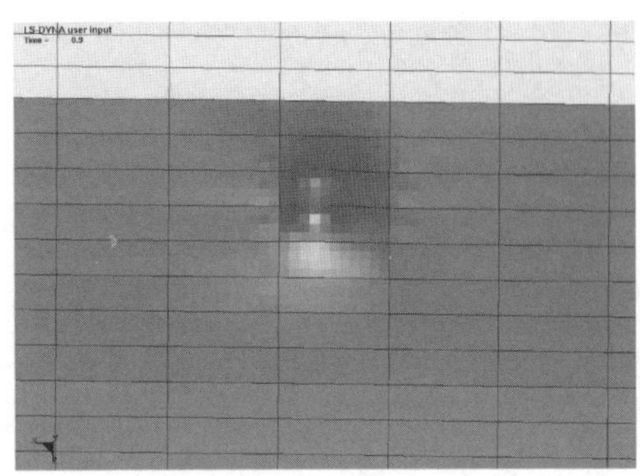

Figure 25. Deformations at the end of collision ($v_0 = 5$ [kn], $t = 17$ [mm]).

energy. For the first three collision scenarios the total deformation energy is similar, about 21 MJ, but the split between the deformation and friction energy varies because of different contact kinematics and the shape of the contact surface.

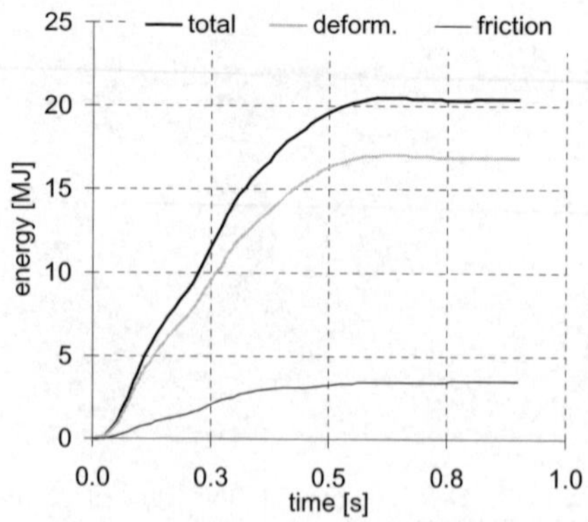

Figure 26. Time-history of the deformation energy ($v_0 = 13$ [kn], $t = 17$ [mm]).

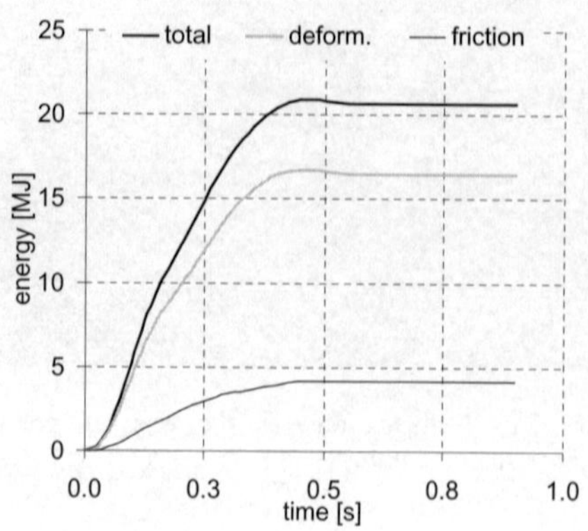

Figure 27. Time-history of the deformation energy ($v_0 = 13$ [kn], $t = 25$ [mm]).

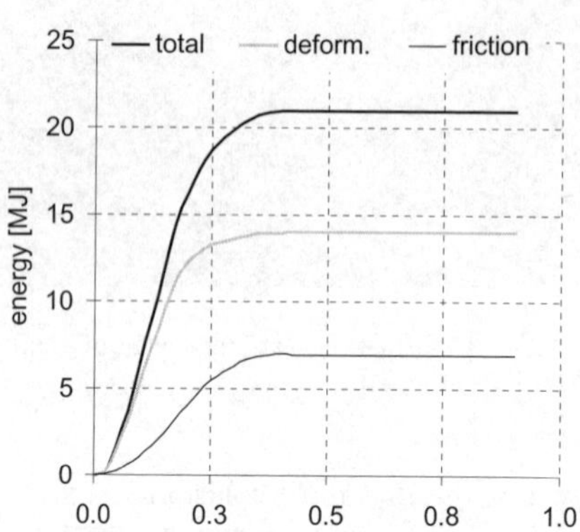

Figure 28. Time-history of the deformation energy ($v_0 = 13$ [kn], $t = 35$ [mm]).

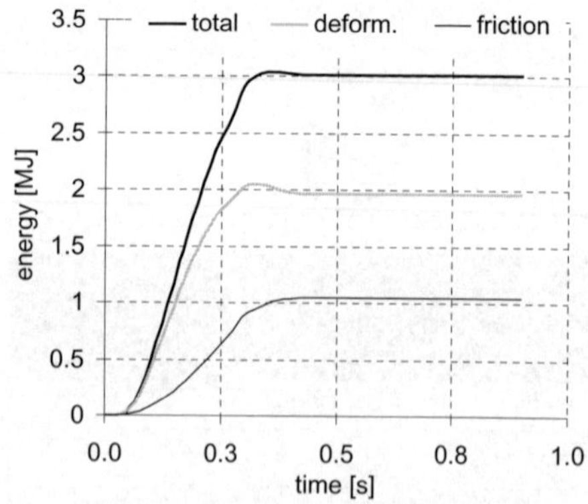

Figure 29. Time-history of the deformation energy ($v_0 = 5$ [m/s], $t = 17$ [mm]).

5 SUMMARY

This paper presents and summarizes briefly the inner mechanics and outer dynamics of ship collisions to be considered if accidental loads are of interest besides the operational loads as criteria in ship design. It outlines further important aspects to be considered and presents a brief insight into the sensitivity of those aspect on the design based on selected examples.

REFERENCES

Alsos HS, Amdahl J. 2009. On the resistance to penetration of stiffened plates, Part I—Experiments. *Int J Impact Eng*, 36(6):799–807.

Brown AJ. 2002. Collision scenarios and probabilistic collision damage; *Marine Structures*, 15:335–364.

Buzek FJ, Holdert HMC. 1990. *Collision cases judgments and diagrams*. Second Edition, Lloyd's of London Press, England.

Cahill RA. 2002. *Collisions and their causes*. Third edition, The Nautical Institute, England.

Cummins WE. 1962. The Impulse Response Function and Ship Motions; *Schifftechnik*, 9:101–109.

Det Norske Veritas (DNV) 2000. *Rules for ships*. Pt.3 Ch.1 Sec.14 Buckling control. Hovik.

Eberhart RC, Kennedy J. 1995. A new optimizer using particle swarm theory. In *Proceedings of the Sixth International Symposium on Micromachine and Human Science*, Nagoya, Japan; 39–43.

Ehlers S, Tabri K, Schillo N, Ranta J. 2007. Implementation of a novel ship side structure into a tanker and a ropax vessel for increased crashworthiness. *6th European LS-DYNA Users' Conference*, Gothenburg, Sweden, 4.111–4.120.

Ehlers S, Broekhuijsen J, Alsos HS, Biehl F, Tabri K. 2008. Simulating the collision response of ship side structures: A failure criteria benchmark study. *Int Ship Progress*; 55:127–144.

Ehlers, S, Klanac, A, Kõrgesaar, M. 2008. A design procedure for structures against impact loading. *STG Yearbook*, in press.
Ehlers, S. 2009. A procedure to optimise ship side structures for crashworthiness. *Journal of Engineering for the Maritime Environment*. doi: 10.1243/14750902JEME179.
Flanagan DP, Belytschko TA. 1981. Uniform Hexahedron and Quadrilateral with Orthogonal Hourglass Control. *Int J Numer Meths Eng* 17(5):679–706.
Hallquist, JO. 2007. *LS-DYNA. Keyword User's Manual*, Version 971. Livermore: Livermore Software Technology Corporation.
Hughes OF, Gosh B, Chen Y. 2004. Improved prediction of simultaneous local and overall buckling of stiffened panels. *Thin-Walled Structures* 42:827–856.
International Maritime Organisation. 1999. *Casualty statistics and investigations: Very serious and serious casualties for the year 1999*. FSI.3/Circ.2, 2001; Available at: http://www.imo.org/includes/blastDataOnly.asp/data_id%3D5397/2.pdf, accessed on: 4 Aug 2008.
International Maritime Organisation. 2000. *Casualty statistics and investigations: Very serious and serious casualties for the year 2000*. FSI.3/Circ.3, 2002; Available at: http://www.imo.org/includes/blastDataOnly.asp/data_ id%3D5118/3.pdf, accessed on: 4 Aug 2008.
International Maritime Organisation. 2001. *Casualty statistics and investigations: Very serious and serious casualties for the year 2001*. FSI.3/Circ.4, 2004; Available at: http://www.imo.org/includes/blastDataOnly.asp/data_id%3D8934/4.pdf, accessed on: 4 Aug 2008.
International Maritime Organisation. 2002. *Casualty statistics and investigations: Very serious and serious casualties for the year 2002*. FSI.3/Circ.5, 2005; Available at: http://www.imo.org/includes/blastDataOnly.asp/data_id%3D11539/5.pdf, accessed on: 4 Aug 2008.
International Maritime Organisation. 2008. *Casualty statistics and investigations: Very serious and serious casualties for the year 2003*. FSI.3/Circ.6, 2005; Available at: http://www.imo.org/includes/blastDataOnly.asp/data_id%3D11540/6.pdf, accessed on: 4 Aug 2008.
International ship and offshore structures congress (ISSC). 2006. Collision and Grounding, Committee V.1, Vol 2.
Kennedy J, Eberhart RC. 1995. Particle swarm optimization. In *Proceedings of IEEE International Conference on Neural Networks*, Piscataway, NJ. 1942–1948.
Kitamura O. 1996. Comparative Study on Collision Resistance of Side Structure; *International Conference on Design and Methodologies for Collision and Grounding Protection of Ships*, San Francisco, California, USA, August 22–23, Also in Marine Technology, Vol. 34, No. 4, 293–308.
Klanac A, Ehlers S, Jelovica J. 2009. Optimization of crashworthy marine structures. *Marine Structures* 1–21.
Konter, A, Broekhuijsen, J, Vredeveldt, A. 2004. A quantitative assessment of the factors contributing to the accuracy of ship collision predictions with the finite element method. In *Proc Third Int Conf Collision and Grounding of Ships*. Japan. p. 17–26.
Kuroiwa T. 1996. Numerical Simulation of Actual Collision & Grounding Accidents; *International Conference on Design and Methodologies for Collision and Grounding Protection of Ships*, San Francisco, California, USA, August 22–23.
McDermontt JF, Kline RG, Jones Jr EL, Maniar NM, Chiang WP. 1974. Tanker structural analysis for minor collisions. *SNAME Trans* 82:382–414.
Minorsky VU. 1959. An analysis of ship collision with reference to protection of nuclear power plants; *Ship Research* 3:1–4.
Naar H, Varsta P, Kujala P. 2004. A theory of coupled beams for strength assessment of passenger ships. *Marine Structures* 17(8):590–611.
Pedersen CBW. 2003. Topology optimization for crashworthiness of frame structures. *Int J Crashworthiness* 8(1):29–39.
Pedersen PT, Zhang S. 1998. On impact mechanics in ship collisions; *Marine Structures*, 11:429–449
Pedersen P, Zhang S. 2000. Absorbed Energy in Ship Collisions and Grounding; *Journal of Ship Research*, 44:2.
Peschmann J. 2001. *Energy absorption computations of ship steel structures under collision and grounding* (translated from German). Doctoral Dissertation. Technical University of Hamburg.
Petersen MJ. 1982. Dynamics of ship collisions, *Ocean Engineering* 9:295–329
Pill I, Tabri K. 2009. Finite element simulations of ship collisions: A coupled approach to external dynamics and inner mechanics. In *Proceedings of MARSTRUCT 2009, The 2nd International Conference on Marine Structures*, Lisbon, Portugal, 16–18 March 2009, pp. 103–109.
Redhe M, Forsberg J, Jansson T, Marklund PO, Nilsson L. 2002. Using the response surface methodology and the D-optimality criterion in crashworthiness related problems: an analysis of the surface approximation error versus the number of function evaluations. *Structural and multidisciplinary optimization* 24(3), 185–194.
Ruukki 2008. *Hot rolled shipbuilding profiles*. Available at: http://www.ruukki.com/www/materials.nsf/materials/6EBCE827F1708096C2257242004C0D65/$File/Shipbuilding_profiles_EN_0806.pdf?openElement, accessed on: 4 Aug 2008.
Sano A, Muragishi O, Yoshikawa T. 1996. Strength Analysis of a New Double Hull Structure for VLCC in Collision; *International Conference on Design and Methodologies for Collision and Grounding Protection of Ships*, San Francisco, California, USA, August 22–23.
Tabri K, Broekhuijsen J, Matusiak J, Varsta P. 2009a. Analytical modelling of ship collision based on full-scale experiments; *Marine Structures*, 22:42–61.
Tabri K, Määttänen J, Ranta J. 2008. Model-scale experiments of symmetric ship collisions. Marine Science and Technology, 13:71–84.
Tabri K, Varsta P, Matusiak J. 2009b. Numerical and experimental motion simulations of non-symmetric ship collisions; *Marine Science and Technology*, in press.
Woisin G. 1988. Instantaneous loss of energy with unsymmetrical ship collisions; *Schiff & Hafen*, 40:50–55
Zhang S. 1999. *The mechanics of ship collisions*. Ph.D. thesis, Technical University of Denmark. p. 262.
Zhang L, Egge ED, Bruhms H. 2004. Approval procedure concept for alternative arrangements. In *Proceedings of the Third International Conference on Collision and Grounding of Ships*, Japan, pp. 87–97.

Advanced Ship Design for Pollution Prevention – Guedes Soares & Parunov (eds)
 ISBN 978-0-415-58477-7

Autonomous guidance and navigation based on the COLREGs rules and regulations of collision avoidance

L.P. Perera
Centre for Marine Technology and Engineering (CENTEC), Technical University of Lisbon, Instituto Superior Técnico, Lisboa, Portugal

J.P. Carvalho
INESC-ID, Technical University of Lisbon, Instituto Superior Técnico, Lisboa, Portugal

C. Guedes Soares
Centre for Marine Technology and Engineering (CENTEC), Technical University of Lisbon, Instituto Superior Técnico, Lisboa, Portugal

ABSTRACT: Autonomous Guidance and Navigation (AGN) with intelligent decision making capabilities will be an important part of future ocean navigation. This paper is focused on an overview of the AGN systems with respect to the collision avoidance in ocean navigation. In addition, a case study of a fuzzy logic based decision making process accordance with the International Maritime Organization (IMO) Convention on the International Regulations for Preventing Collisions at Sea (COLREGs) will be illustrated.

1 INTRODUCTION

1.1 *Autonomous guidance and navigation*

"Automatic steering is a most valuable invention if properly used. It can lead to disaster when it is left to look after itself while vigilance is relaxed. It is on men that safety at sea depends and they cannot make a greater mistake than to suppose ..."

The statement was made by the Justice Cairns with respect to a collision that had occurred due to the fault in the automatic pilot system when the British cargo ship "Trentbank" was overtaking the Portuguese tanker "Fogo" in the Mediterranean in September 1964 (Cockcroft and Lameijer, 2001). The verdict not just shows the importance of the AGN systems in ocean navigations but also of the human vigilance of its capabilities and requirements for further developments.

The automatic pilot systems are primary level developments units of the Autonomous Guidance and Navigation (AGN) Systems and their applications have been in the dreams of ship designers in several decades. The development of computer technology, satellite communication systems, and electronic devices, including high-tech sensors and actuators, have turned these dreams into a possible reality when designing the next generation ocean AGN systems.

The initial step of the AGN system, which made the foundation for applied control engineering, was formulated around 1860 to 1930 with the invention of the first automatic ship steering mechanism by Sperry (1922). Sperry's work was based on replication of the actions of an experienced helmsman that was formulated as a single input single output system (SISO).

Similarly, the research work done by Minorski (1922) is also regarded as the key contribution to the development of AGN Systems. His initial work was based on the theoretical analysis of automated ship steering system with respect to a second order ship dynamic model. The experiments were carried out by Minorski on New Mexico in cooperation with the US Navy in 1923 and reported in 1930 (Bennett 1984).

Hence both works done by Sperry and by Minorski are considered as the replication of non-linear behaviour of an experienced helmsman (Roberts et al., 2003) in ocean navigation. From Sperry's time to present, much research has been done and considerable amount of literature has been published on AGN systems with respect to the areas of marine vessel dynamics, navigation path generations, and controller applications, environmental disturbance rejections and collision avoidance conditions.

The functionalities of the Multipurpose Guidance, Navigation and Control (GNC) systems are summarized by Fossen (1999) on a paper that focuses not only on course-keeping and course-changing maneuvers (Conventional auto pilot system) but also integration of digital data (Digital charts and weather data), dynamic position and automated docking systems.

Recent developments of design, analysis and control of AGN systems are also summarized by Ohtsu (1999), and several ocean applications of AGN systems have been further studied theoretically as well as experimentally by (Moreira et al., 2007; Healey and Lienard, 1993; Do and Pan, 2006). This area is bound to become increasingly important in the future of ocean navigation due to the associated cost reduction and maritime safety.

The block diagram of main functionalities of a AGN system integrated with collision avoidance facilities is presented in Fig. 1. The AGN system contains four units of Collisions Avoidance Conditions (CAC), Control System (CS), Filters and Sensors. The sensor unit consists of sensors that could measure course and speed characteristics of the vessel navigation. The filter unit is formulated for filter-out noisy signals. The output of the Filter unit will be feedback into the CAC unit as well as the CS unit. The sensor signals will be used by the CS unit to control overall vessel navigation of the vessel and used by the CAC unit to make decisions on collisions avoidance process.

1.2 *Collision avoidance*

Having an intelligent decision making process is an important part of the future AGN systems in ocean navigation. However, conventional ocean navigational systems consist of human guidance, and as a result, 75–96% of marine accidents and causalities are caused by some types of human errors (Rothblum et al., 2002; Antão and Guedes Soares, 2008). Since most of the wrong judgments and wrong operations of humans at the ocean ended in human casualties and environmental disasters, limiting human subjective factors in ocean navigation and replacing them by an intelligent Decision Making (DM) system for navigation and collision avoidance could reduce maritime accidents and

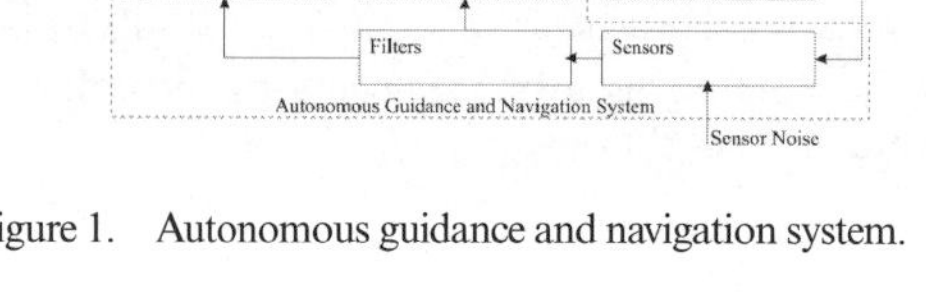

Figure 1. Autonomous guidance and navigation system.

its respective causalities. However development of collision avoidance capabilities into the next generation AGN systems in ocean navigation is still in the hands of future researchers and this part of the intelligent AGN systems has been characterized as e-Navigation (eNAV, 2008).

The terminology used in recent literature regarding the collision avoidance conditions designates the vessel with the AGN system as the "Own vessel", and the vessel that needs to be avoided as the "Target vessel". With respect to the Convention on the International Regulations for Preventing Collisions at Sea (COLREGs) rules and regulations, the vessel coming from the starboard side has higher priority for the navigation and is called the "Stand on" vessel and the vessel coming from the port side has lower priority and is called the "Give way" vessel. These definitions have been considered during the formulation of collision situations in this work.

The decision making process and strategies in interaction situations in ocean navigation, including collision avoidance situations, are presented by Chauvin and Lardjane (2008). The analysis of quantitative data describing the manoeuvres undertaken by ferries and cargo-ships and behaviour of the "Give way" and the "Stand on" vessels with respect to verbal reports recorded on board a car-ferry in the Dover Strait are also presented in the same work.

The detection of the Target vessel position and velocity are two important factors assessing the collision risk in ocean navigation as illustrated in recent literature. Sato and Ishii (1998) proposed combining radar and infrared imaging to detect the Target vessel conditions as part of a collision avoidance system. The collision risk has been presented with respect to the course of the Target vessel and image processing based course measurement method proposed in the same work.

The vessel domain could be defined as the area bounded for dynamics of the marine vessel and the size and the shape of the vessel domain are other important factors assessing the collision risk in ocean navigation. Lisowski et al. (2000) used neural-classifiers to support the navigator in the process of determining the vessel's domain, defining that the area around the vessel should be free from stationary or moving obstacles. On a similar approach, Pietrzykowski and Uriasz (2009) proposed the notion of vessel domain in a collision situation as depending on parameters like vessel size, course and heading angle of the encountered vessels. Fuzzy logic based domain determination system has been further considered in the same work.

Kwik (1989) presented the calculations of two-ship collision encounter based on the kinematics and dynamics of the marine vessels. The analysis of collision avoidance situations is illustrated regarding the vessel velocity, turning rate and direction, and desired passing distance in the same work. Yavin et al. (1995) considered the collision avoidance conditions of a ship moving from one point to another in a narrow zig-zag channel and a computational open loop command strategy for the rudder control system associated with the numerical differential equation solver is proposed. However, the dynamic solutions based on differential equations could face implementation difficulties in real-time environment.

The design of a safe ship trajectory is an important part of the collision avoidance process and has usually been simulated by mathematical models based on manoeuvring theory (Sutulo et al., 2002). An alternative approach based on neural networks has also been proposed by Moreira and Guedes Soares (2003). Optimization of a safe ship trajectory in collision situations by an evolutionary algorithm is presented by Smierzchalski and Michalewicz (2000), where comparison of computational time for trajectory generation with respect to other manoeuvring algorithms, and static and dynamic constrains for the optimization process of the safe trajectories are also illustrated. However, the optimization algorithms always find the solution for the safe trajectory based on assumptions; hence the optimum solutions may not be realistic and may not have intelligent features. As an example, it is observed that some of the optimization algorithms always find the safest path behind the Target vessel and that may lead to a conflict situation with the COLREGs rules and regulations.

1.3 *The COLREGs*

It is a fact that the COLREGs rules and regulations regarding collision situations in ocean navigation have been ignored in most of the optimization algorithms. The negligence of the IMO rules may lead to conflicts during ocean navigation. As for the reported data of the maritime accidents, 56% of the major maritime collisions include violation of the COLREGS rule and regulations (Statheros et al., 2008). Therefore the methods proposed by the literature ignoring the COLREGs rules and regulations should not be implemented in ocean navigation.

On the other hand, there are some practical issues on implementation of the COLREGs rules and regulations during ocean navigation. Consider the crossing situations where the Own vessel is in "Give way" situations in Figures 2, 3, 4, and 5 and in "Stand on" situations in Figures 6, 7, 8 and 9, there are velocity constrains in implementing COLREGs rules and regulations of the "Give way" and the "Stand on" vessels collision situations when the

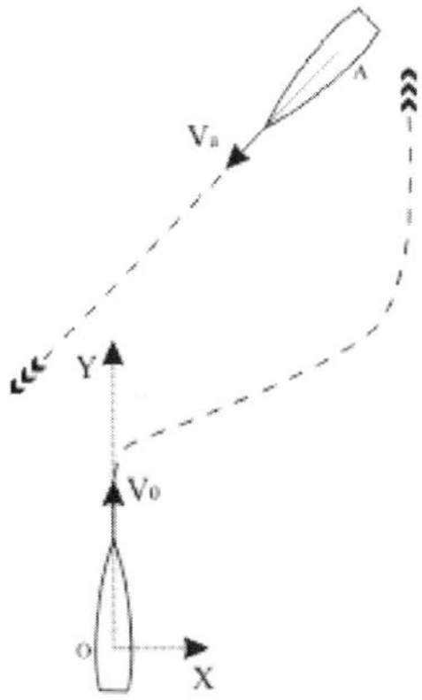

Figure 2. Crossing.

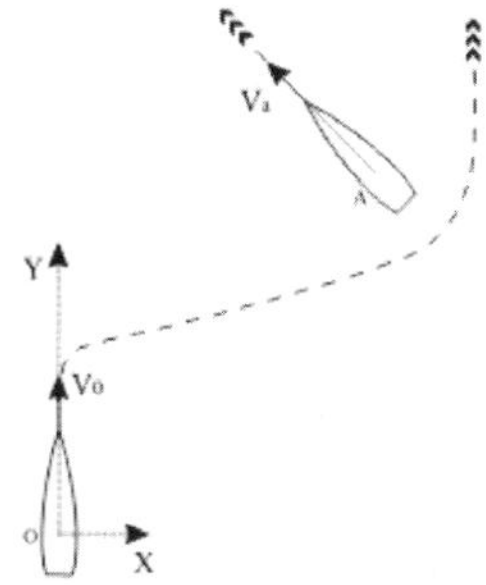

Figure 3. Crossing.

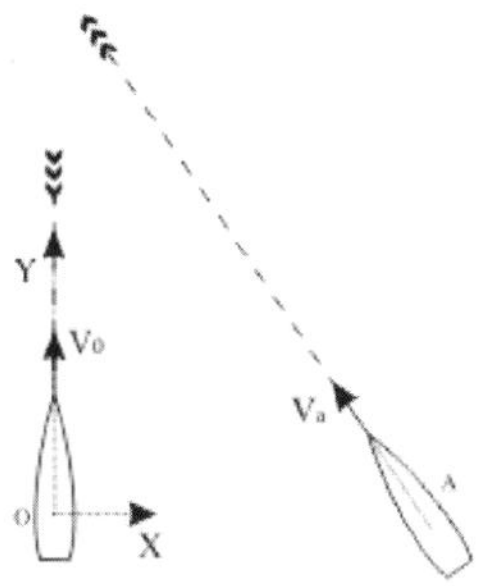

Figure 4. Parallel-crossing.

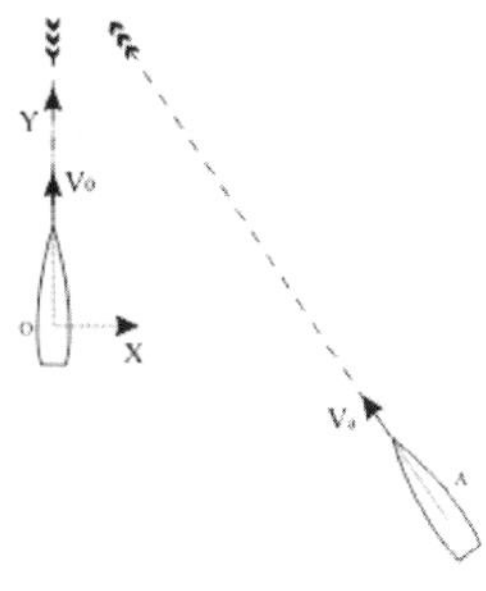

Figure 5. Crossing.

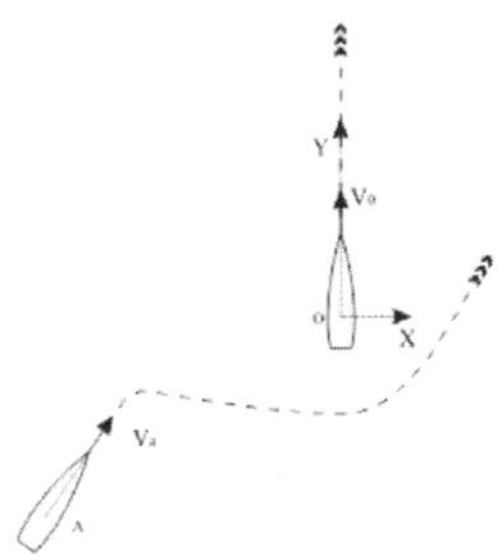

Figure 6. Crossing.

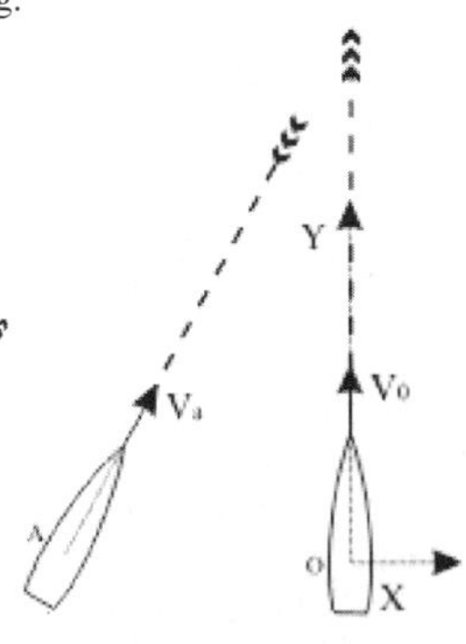

Figure 7. Parallel-crossing.

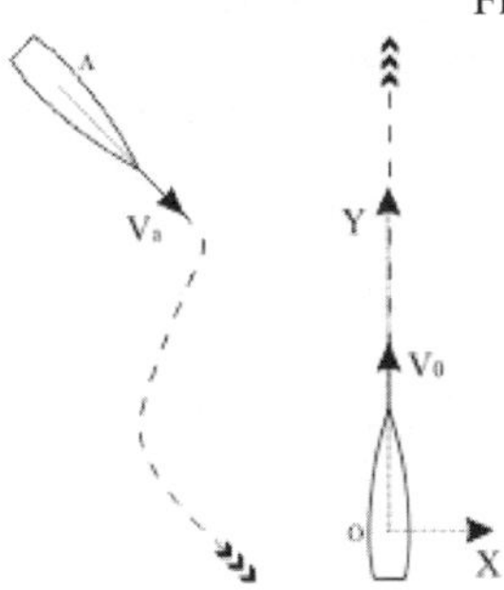

Figure 8. Crossing.

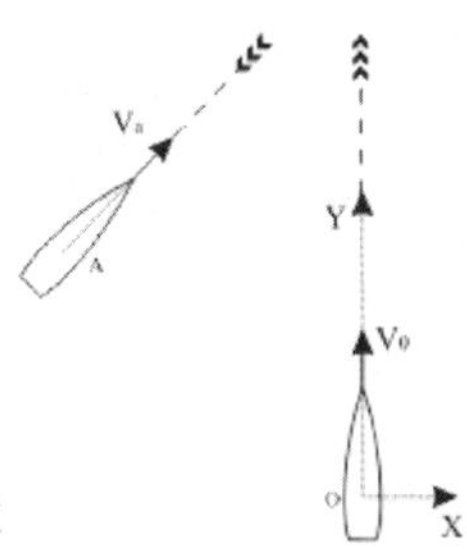

Figure 9. Crossing.

Target vessel has very low or very high speed compared to the Own vessel.

In the collision avoidance approach of repulsive force based optimization algorithms proposed by Xue et al. (2009), the Own vessel is kept away from the obstacles by a repulsive force field. However this approach may lead to conflict situations when the moving obstacles present a very low speed or very high speed when compared to the Own vessel speed. In addition, complex orientations of obstacles may lead to unavoidable collision situations. On the other hand, repulsive force based optimization algorithms are enforced to find the global safe trajectory for Own vessel navigation, and this might not be a good solution for the localized trajectory search. In addition the concepts of the "Give way" and the "Stand on" vessels that are derived on COLREGs rules and regulations during the repulsive force based optimization process are not taken into consideration and therefore may not be honoured.

The intelligent control strategies implemented on collision avoidance systems could be categorized as Automata, Hybrid systems, Petri nets, Neural networks, Evolutionary algorithms and Fuzzy logic. These techniques are popular among the machine learning researchers due to their intelligent learning capabilities. The soft-computing based Artificial Intelligence (AI) techniques, evolutionary algorithms, fuzzy logic, expert systems and neural networks and combination of them (hybrid expert system), for collision avoidance in ocean navigation are summarized by Statheros et al. (2008).

Ito et al. (1999) used genetic algorithms to search for safe trajectories on collision situations in ocean navigation. The approach is implemented in the training vessel of "Shioji-maru" integrating Automatic Radar Plotting Aids (ARPA) and Differential Global Position System (DGPS). ARPA system data, which could be formulated as a stochastic predictor, is designed such that the probability density map of the existence of obstacles is derived from the Markov process model before collision situations as presented by Zeng et al. (2001) in the same experimental setup. Further, Hong et al. (1999) have presented the collision free trajectory navigation based on a recursive algorithm that is formulated by analytical geometry and convex set theory. Similarly, Cheng et al. (2006) have presented trajectory optimization for ship collision avoidance based on genetic algorithms.

Liu and Liu (2006) used Case Based Reasoning (CBR) to illustrate the learning of collision avoidance in ocean navigation by previous recorded data of collision situations. In addition, a collision risk evaluation system based on a data fusion method is considered and fuzzy membership functions for evaluating the degree of risk are also proposed. Further intelligent anti-collision algorithm for different collision conditions has been designed and tested on the computer based simulation platform by Yang et al. (2007) Zhuo and Hearn (2008) presented a study of collision avoidance situation using a self learning neuro-fuzzy network based on an off-line training scheme and the study is based on two vessel collision situation. Sugeno type Fuzzy Inference System (FIS) was proposed for the decision making process of the collision avoidance.

1.4 *Fuzzy-logic based systems*

Fuzzy-logic based systems, which are formulated for human type thinking, facilitate a human friendly environment during the decision making process. Hence several decision making systems in research as well as commercial applications have been presented before (Hardy, 1995). The conjunction of human behavior and decision making process was formulated by various fuzzy functions in Rommelfanger, (1998) and Ozen et al. (2004). A fuzzy logic approach for the collision avoidance conditions with integration of the virtual force field has been proposed by Lee et al. (2004). However the simulations results are limited to the two vessel collision avoidance situations. The behaviour based controls formulated with interval programming for collision avoidance of ocean navigation are proposed by Benjamin et al. (2006). Further, the collision avoidance behaviour is illustrated accordance with the Coast Guard Collision Regulations (COLREGS-USA).

Benjamin and Curcio (2004) present the decision making process of ocean navigation based on the interval programming model for multi-objective decision making algorithms. The computational algorithm based on If-Then logic is defined and tested under simulator conditions by Smeaton and Coenen (1990) regarding different collision situations. Further, this study has been focused on the rule-based manoeuvring advice system for collision avoidance of ocean navigation.

Even though many techniques have been proposed for avoidance of collision situations, those techniques usually ignore the law of the sea as formulated by the International Maritime Organization (IMO) in 1972. These rules and regulations are expressed on the Convention on the International Regulations for Preventing Collisions at Sea (COLREGs)(IMO (1972)). The present convention was designed to update and replace the Collision Regulations of 1960 which were adopted at the same time as the International Convention for Safety of Life at Sea (SOLAS) Convention (Cockcroft and Lameijer (2001)).

The detailed descriptions of collision avoidance rules of the COLREGs, how the regulations should be interpreted and how to avoid collision, have been presented by Cockcroft and Lameijer (2001). Further, the complexity of autonomous navigation not only in the sea but also in the ground has been discussed by Benjamin and Curcio (2004). The legal framework, rules and regulations, is discussed and the importance of collision avoidance within a set of given rules and regulations is highlighted in the same work.

2 COLREGs RULES AND REGULATIONS

The COLREGs (IMO (1972)) includes 38 rules that have been divided into Part A (General), Part B (Steering and Sailing), Part C (Lights and Shapes), Part D (Sound and Light signals), and Part E (Exemptions). There are also four Annexes containing technical requirements concerning lights and shapes and their positioning, sound signalling appliances, additional signals for fishing vessels when operating in close proximity, and international distress signals.

Three distinct situations involving risk of collision in ocean navigation have been recognized in recent literature (Smeaton and Coenen, 1990), Overtaking (see Figures 11 and 12); Head-on (see Figure 10) and Crossing (see Figures 2 to 9) and the rules and regulations with respect to these collision conditions have been highlighted by the COLREGs.

The maritime collision could be defined as "a brief dynamic event consisting of the close proximity of two or more stationary or moving ocean obstacles or vessels". Hence, maintenance

Figure 10. Head-on.

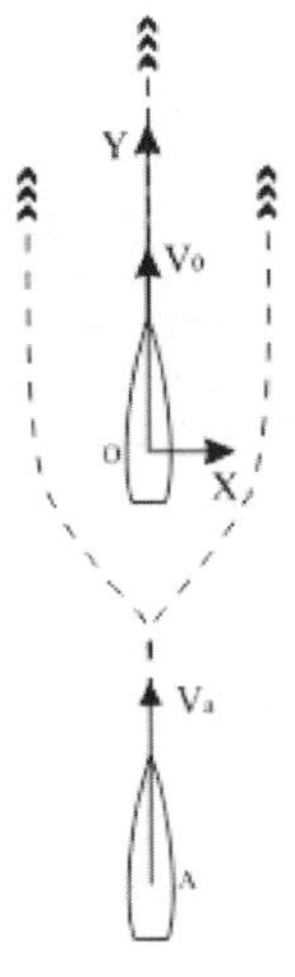

Figure 11. Overtake.

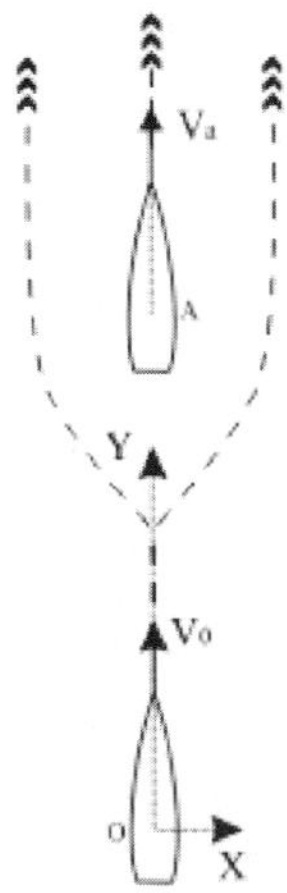

Figure 12. Overtake.

of safe distance among vessels and other obstacles are important factors of the maritime safety as illustrated by the COLREGs.

The safe distance maintenance by both vessel in a overtake situation (see Figures 3 and 8) has been illustrated by the COLREGs rule 13(a) (IMO, 1972):

"..., any vessel overtaking any other shall keep out of the way of the vessel being overtaken".

Hence both vessels have the responsibility to maintain safe distances during the Overtake encounter. Further, the course change of the head-on situation (see Figure 2) has been specified by the COLREGs rule 13(a) (IMO, 1972):

"When two power-driven vessels are meeting on reciprocal or nearly reciprocal course so as to involve risk of collision each shall alter her course to starboard so that each shall pass on the port side of the other."

As required by the COLREGs, the vessels should take early action to avoid situations of crossing ahead with the risk of collision in starboard to starboard and must be passing by port to port. The crossing situations further formulated by the COLREGs and according to rule 15 (IMO, 1972):

"When a two power-driven vessel are crossing so as to involve risk of collision, the vessel which has the other her own starboard side shall keep out of the wan and shall, if the circumstances of the case admit, avoid crossing ahead of the other vessel"

As specified by the COLREGs, the vessel coming from the starboard side has higher priority for the navigation and the vessel coming from the port side has lower priority as mentioned before. This concept is previously defined as the "Give way and the Stand on vessels". However, considering the collision conditions, where the "Give way" vessel did not take any appropriate actions to avoid collisions, as illustrated by the COLREGs rule 17(b) (IMO, 1972), the "Stand on" vessel has been forced to take appropriate actions to avoid collision situation,

"When, from any cause, the vessel required to keep her course and speed finds herself so close that collision cannot be avoided by the action of the "Give way" vessel alone, she shall take such action as will best aid to avoid collision"

However, the decision making process of the Own vessel in this critical collision situation should be carefully formulated, because the collision avoidance in this situation alternatively depends on the "Stand on" vessel manoeuvrability characteristics. Further, this situation might lead to a "Crash stopping" maneuver of the "Stand on" vessel due the lack of distance for speed reductions. As recommended by the COLREGs rule 6 (IMO, 1972) with respect to the reduction of vessel speed for the safe distance:

"Every vessel shall at all times proceed at a safe speed so that she can take proper and effective action to avoid collision and be stopped within a distance appropriate to the prevailing circumstances and conditions. In determining a safe speed the following factors shall be among those taken into account ..."

Hence special factors should be considered for integration of course and speed controls due to the fact that the Own vessel may not response to the required changes of course or speed. Further information with respect to the "Stopping distance" and "Turning circles" should be considered for formations of the decision making process. Vessel course changes and/or speed changes in ocean navigation must be formulated in order to avoid collision situations. The specific controllability of either course or speed change has been highlighted in the COLREGs rule 8(b) (IMO, 1972):

"Any alteration of course and/or speed to avoid collision shall, if the circumstances of the case admit, be large enough to be readily apparent to another vessel observing visually or by radar; a succession of small alterations of course and/or speed should be avoided"

Hence integrated controls of course as well as speed changes should be implemented during ocean navigation to avoid collision situations. Similarly special measures should be considered for integration of course and speed controls due to the fact that the Own vessel may not response to the required changes of course or speed.

3 COLLISION AVOIDANCE METHODOLOGY

3.1 *Identification of obstacles*

The stationary and moving obstacles in ocean navigation can be identified by several instruments and systems like Eye/camera, radar/Automatic Radar Plotting Aid (ARPA), and Automatic Identification System (AIS). ARPA provides accurate information of range and bearing of nearby obstacles and AIS is capable of giving all the information on vessel structural data, position, course, and speed. The AIS simulator and marine traffic simulator have been implemented on several experimental platforms for design of safe ship trajectories (Hasegawa, 2009).

3.2 *Collection of navigational information*

The navigational information of other vessels can be categorized into static, dynamic and voyage related information (Imazu, 2006). Static information is composed of Maritime Mobile Service Identity (MMSI), Call Sign and Name, IMO number, length and beam, type of ship and location and position of communication antenna. Dynamic information can be divided into vessel position, position time stamp, course over ground, speed over ground, heading, navigational status and rate of turn. Finally, voyage related information can be expanded into vessel draft, cargo type, destination and route plan. Collection of navigational information is an important part of the decision making process of the collision avoidance in ocean navigation and can be achieved by collaboration with the Automatic Identification System (AIS).

3.3 *Analysis of navigational information*

The collected obstacles and other vessels information should be considered for further analysis of navigational information. However continuous careful observation, collection and analysis of navigational information should be done in small regular intervals to obtain early warning for collision situations. Further in ocean navigation, complex collision situation with the combination of above situation could be occurred and identification of each situation with respect to the each collision conditions will helpfully for the overall decisions on ocean navigation.

3.4 *Assessments of the collision risk*

The analysis of navigational information will help to assess the collision risk. The assessment of collision risk should be continuous and done in real-time by the navigational system in order to guarantee the Own vessel safety. As illustrated in the literature, the mathematical analysis of collision risk detection can be divided into two categories (Imazu, 2006): Closest Point Approach Method (CPA–2D method) and Predicted Area of Danger Method (PAD–3D method). CPA method consists of calculation of the shortest distance from the Own vessel to the Target vessel and the assessment of the collision risk, which could be predicted with respect to the Own vessel domain. However, this method is not sufficient to evaluate the collision risk since it doesn't take into consideration the target vessel size, course and speed. The extensive study of CPA method with respect to a two vessel collision situation has been presented by Kwik (1989).

The PAD method consists of modelling the Own vessel possible trajectories as an inverted cone and the Target vessel trajectory as an inverted cylinder, being the region of both object intersection categorized into the Predicted Area of Danger. The Target vessel size, course and speed could be integrated into the geometry of the objects of navigational trajectories.

3.5 *Decisions on navigation*

The decisions of collision avoidance in ocean navigation are based on the speed and course of each vessel, distance between two vessels, distance of the Closest Point of Approach (RDCPA), time to DCPA, neighbouring vessels and other environmental conditions. The decision space of collision avoidance can be categorized into three stages for each vessel in open ocean environment:

- When both vessels are at non collision risk range, both vessels have the options to take appropriate actions to avoid a collision situation;
- When both vessels are at collision risk range, the "Give way" vessel should take appropriate actions to achieve safe passing distance in accordance with the COLREGs rules and regulations and the "Stand on" vessel should keep the course and speed;
- When both vessels are at critical collision risk range, and the "Give way" vessel does not take appropriate actions to achieve safe passing distance in accordance with the COLREGs rules, then "Stand on" vessel should take appropriate actions to avoid the collision situation.

3.6 *Implementation of decisions on navigation*

As the final step, the decisions on vessel navigation should be formulated with respect to the collision risk assessments. The actions that are taken by the Own vessel are proportional to the Target vessel behaviour as well as the COLREGs rules and regulations. The expected Own and Target vessel actions of collision avoidance could be formulated into two categories: Course change and speed change. However the initiation for actions should be formulated with respect to the Target vessel range and the rate of change of range. Further if the actions taken by the Target vessel are not clear or there is doubt about actions, sound signals should be used as recommended by the COLREGs.

In a collision situation, when the Target vessel is ahead or fine on the bow of the Own vessel and the Target vessel overtaking the Own vessel from astern or fine on the quarter without the safe distance, alteration of course is more effective than speed alteration. However in a collision situation, when the Target vessel is approaching from abeam or near the abeam of the Own vessel, alteration of speed is more effective than course alteration but course alteration could be achieved the same (Cockcroft and Lameijer, 2001). On conventional navigational systems, power driven vessels usually prefer course changes over speed changes due to the difficulties and delays in controllability of engines from the bridge unless the engines are on stand-by mode. However these problems could be overcome with the AGN systems with the integration of speed and course control systems.

4 CASE STUDY: FUZZY LOGIC BASED DECISION MAKING PROCESS

This section focuses on a fuzzy logic based Decision Making (DM) system to be implemented on vessel navigation to improve safety of the vessel by avoiding the collision situations and is an illustration of the study of Perera et al. (2009). The experienced helmsman actions in ocean navigation can be simulated by a fuzzy logic based Decision Making (DM) process, being this one of the main advantages in this proposal.

4.1 *Formation of collision conditions*

Figure 13 presents two vessels in a collision situation in ocean navigation that is similar to a Radar plot in the Own vessel. The Own vessel is initially located at the point O (x_o, y_o), and the Target vessel is located at the point A (x_a, y_a). The Own and Target vessels velocities and course, are represented by V_o, V_a and ψ_o, ψ_a. The relative speed ($V_{a,o}$) and course ($\psi_{a,o}$) of the Target vessel with respect to the Own vessel can be estimated using the range and bearing values in a given time interval. The relative trajectory of the target vessel has been estimated with the derivation of relative speed,

$V_{a,o}$ and relative course, $\psi_{a,o}$. All angles have been measured regarding the positive Y-axis. Further, it is assumed that both vessels are power driven vessels regarding the IMO categorization.

Further, in Figure 13, the Own vessel ocean domain is divided into three circular sections with radius R_{vd}, R_b and R_a. The radius R_a represents the approximate distance to the Target vessel identification and this distance could be defined as the distance where the Own vessel is in a "Give way" situation and should take appropriate actions to avoid collision. The distance R_b represents the approximate distance where the Own vessel is in "Stand on" situation, but should take actions to avoid collisions due to absence of the appropriate actions from the Target vessel. Further the distance R_{vd} represents the vessel domain. The approximate distances considered for this study are $R_{vd} \approx 1$ NM, $R_b \approx 6$ NM and $R_a \approx 10$ NM.

The Own vessel collision regions are divided into eight regions from I, to VIII. It is assumed that the Target vessel should be located within these eight regions and the collision avoidance decisions are formulated in accordance to each region. As represented in target vessel position II in Figure 13, the Target vessel positions have been divided into eight divisions of vessel orientations regarding the relative course (from II–a to II–h).

4.2 *Fuzzification and defuzzification*

The overall design process of the fuzzy logic based DM system could be categorized into the following six steps.

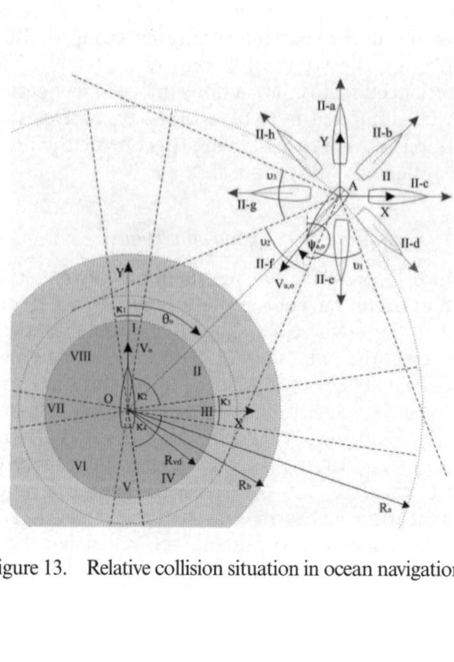

Figure 13. Relative collision situation in ocean navigation.

- Identification of the input Fuzzy Membership Functions (FMFs).
- Identification of the output FMFs.
- Creation of FMF for each inputs and outputs.
- Construction of If-Then fuzzy rules to operate overall system.
- Strength assignment of the fuzzy rules to execute the actions.
- Combination of the rules and defuzzification of the output.

4.2.1 *Fuzzy sets and membership functions*

Fuzzy sets are defined by Fuzzy Membership Functions (FMF), which can be described as mappings from one given universe of discourse to a unit interval. However it is conceptually and formally different from the fundamental concept of the probability (Pedrycz and Gomide, 2007). A fuzzy variable is usually defined by special FMF called Linguistic Terms that are used in fuzzy rule based inference. The Linguistic terms for the inputs (Collision Distance, Collision Region, Relative Collision Angle and Relative Speed Ratio) and outputs (Speed and Course change of the Own vessel) were defined for this analysis.

Mamdani type "IF <Antecedent i> is <Linguistic Term n> and/or <Antecedent j> is <Linguistic Term m> and/or .... THEN <Consequent> is <Linguistic Term p>" rule based fuzzy system and inference via Min-Max norm was used during this study. On Mamdani fuzzy inference systems, the fuzzy sets resulting from the consequent part of each designed fuzzy rule are combined through an aggregation operator (Ibrahim, 2003) according to the activation level of the antecedents. The Min-Max norm is the aggregation operation considered in this study. In this norm the Min (minimum) operator is considered for intersection and Max (maximum) operator is considered for the union of two fuzzy sets. Finally the defuzzification was made using the center of gravity method. In this method, one calculates the centroid of the resulting FMF and uses its abscissa as the final result of the inference.

4.2.2 *Fuzzy Inference System (FIS)*

The block diagram for Fuzzy Inference System (FIS) with integration of navigational instruments is presented in Figure 14. The initial step of the fuzzy inference system consists of data collection of the target vessel position, speed and course. As the next step, the relative trajectory, relative speed and relative course of the Target vessel are estimated. Then, the data is fuzzified with respect to the input FMF of Collision Distance, Collision Region, Relative Speed Ratio and Relative Collision Angle.

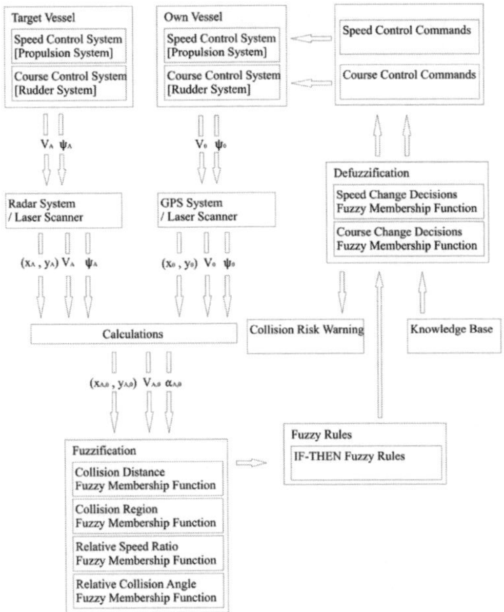

Figure 14. Block diagram for Fuzzy Inference System.

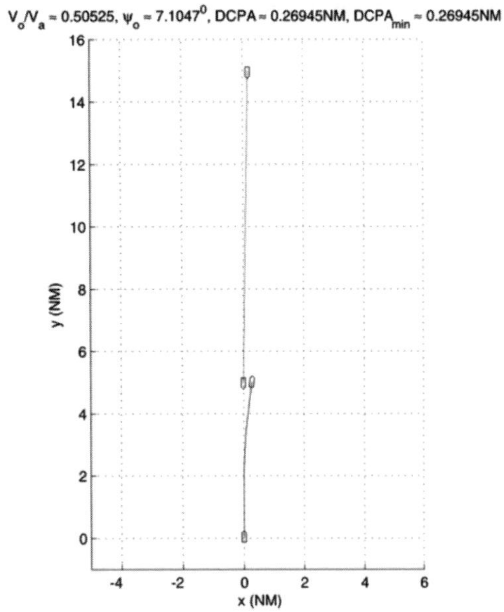

Figure 15. Heading situation.

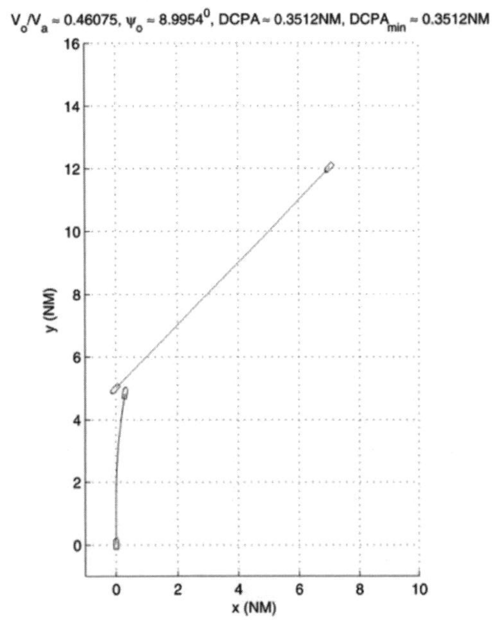

Figure 16. Crossing situation.

The If-Then fuzzy rules are developed in accordance with the COLREGs rules and regulations and using navigational knowledge. The outputs of the rule based system are the Collision Risk Warning and the Fuzzy Decisions. Finally the fuzzy decisions are defuzzified by output FMF of Course Change and Speed Change to obtain the control actions that will be executed in the Own vessel navigation. The control actions are further expanded as Course control commands and Speed control commands that can be implemented on Rudder and Propulsion control systems as presented in the Figure.

4.2.3 *Computational implementation*

The fuzzy logic based DM system has been implemented on the software platform of MATLAB. MATLAB has support for the fuzzy logic schemes Mamdani and Sugeno Types (Sivanandam et al. (2007)). However this work has been implemented on the Mamdani based Fuzzy Inference System (FIS). The Mamdani type fuzzy logic scheme consists of utilizing membership functions for both inputs and outputs. As previously mentioned, If-Then rules are formed by applying fuzzy operations into the Mamdani type membership functions for given inputs and outputs.

The MATLAB simulations for two vessels collision situations with respect to the different speeds and course conditions in the Cartesian coordinate space have been presented in Figures from 15 to 22. These figures contain the start and end positions of the Own and Target vessels with respect to navigational trajectories. The initial vessel speed condition is $V_o/V_a = 0.5$ and initial course of the own vessel is $\psi_o = 0^0$. The star position of the Own ves-

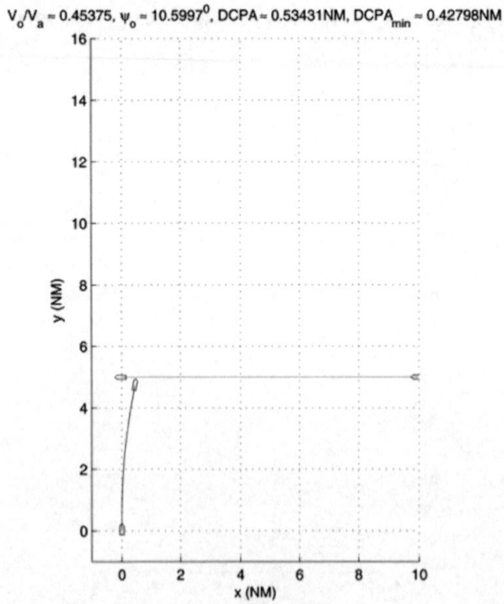

Figure 17. Crossing situation.

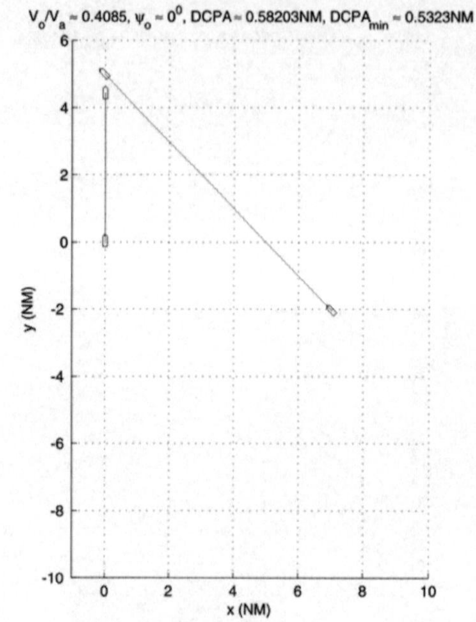

Figure 18. Crossing situation.

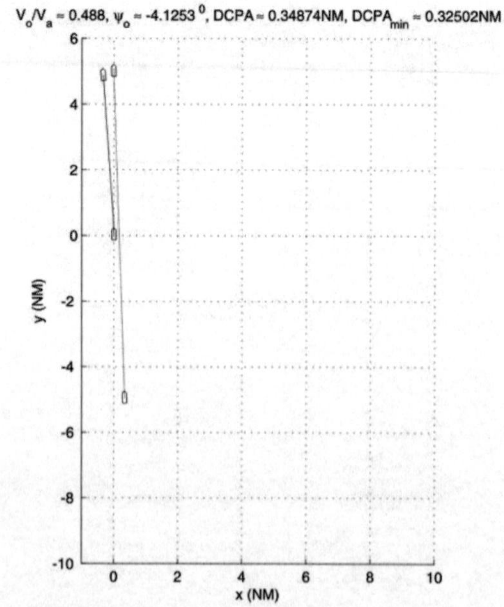

Figure 19. Overtake situation.

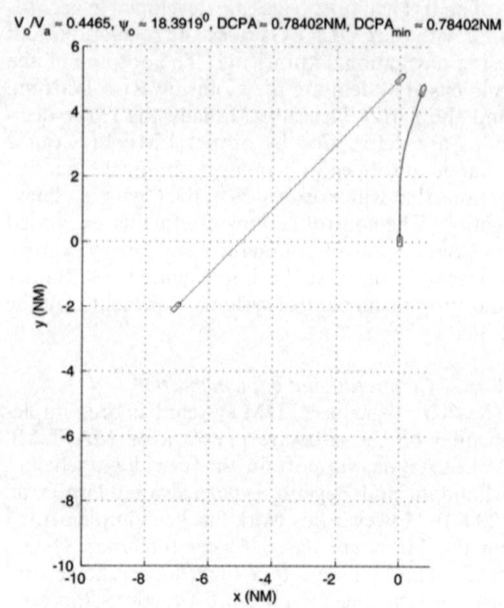

Figure 20. Crossing situation.

sel (0, 0) and the collision point for both vessels (0, 5) has been considered for all simulations.

As noted from the simulations fuzzy logic based DM systems has made proper trajectory for all the collision conditions.

5 CONCLUSION

An overall discussion about inclusion of collision avoidance in AGN systems has been presented. In the case study, the decision making process of

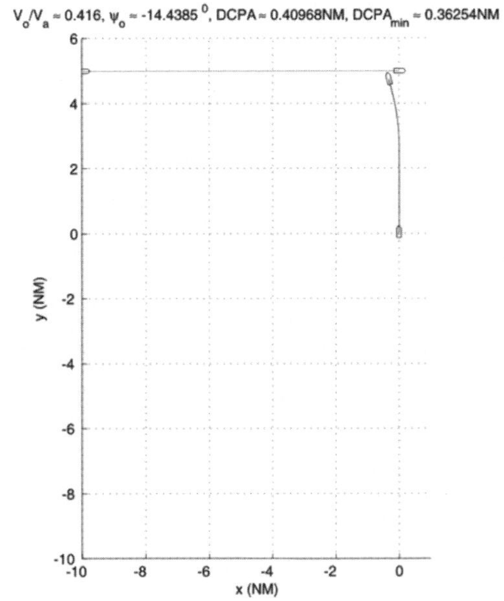

Figure 21. Crossing situation.

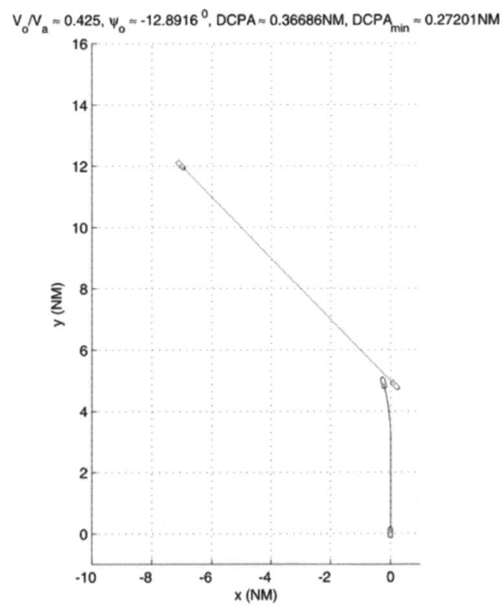

Figure 22. Crossing situation.

AGN system that was based on the fuzzy logic and human expert knowledge in ocean navigation is introduced. As observed, the DM system was able to overcome collision conditions by fuzzy logic based decision making process. Furthermore the DM system has taken proper manoeuvres to avoid close-quarter situation during the ocean navigation where the collision risk is high. Therefore the DM system would not have to take quick decisions based on inadequate information and time as human decisions.

ACKNOWLEDGMENT

The research work of the first author has been supported by the Doctoral Fellowship of the Portuguese Foundation for Science and Technology (Fundação para a Ciência e a Tecnologia) under the contract SFRH/BD/46270/2008. Furthermore, this work contributes to the project "Methodology for ships manoeuvrability tests with self-propelled models", which is being funded by the Portuguese Foundation for Science and Technology (Fundação para a Ciência e Tecnologia) under the contract PTDC/TRA/74332/2006.

REFERENCES

Antão, P. & Guedes Soares, C., 2008. Causal factors in accidents of high speed craft and conventional ocean going vessels. Reliability Engineering and System Safety 93, 1292–1304.

Benjamin, M.R. & Curcio, J.A., 2004. COLREGs—based navigation of autonomous marine vehicles. IEEE/OES Autonomous Underwater Vehicles, 32–39.

Benjamin, M.R., Curcio, J.A. & Newman, P.M., 2006. Navigation of unmanned marine vehicles in accordance with the rules of the road. Proceedings of the 2006 IEEE International Conference on Robotics and Automation, 3581–3587.

Bennett, S.,1984. Nicholas minorsky and the automatic steering of ships. Control Systems Magazine, IEEE 4 (4), 10–15.

Chauvin, C. & Lardjane, S., 2008. Decision making and strategies in an interaction situation: Collision avoidance at sea. Transportation Research Part F (11), 259–262.

Cheng, X.D., Liu, Z.Y. & Zhang, X.T., 2006. Trajectory optimization for ship collision avoidance system using genetic algorithm. Proceedings OCEANS—Asia Pacific, Singapore, 385–392.

Cockcroft, A.N. & Lameijer, J.N.F., 2001. A Guide to The Collision Avoidance Rules. Elsevier Butterworth-Heinemann, Burlington, MA. USA.

Do, K.D. & Pan, J., 2006. Robust path-following of underactuated ships: Theory and experiments on a model ship. Ocean Engineering 33 (3), 1354–1372.

eNAV, eNavigation 2008. URL http://www.enavigation.org/.

Fossen, T.I. (Ed.), 1999. Recent developments in Ship Control Systems Design. World Superyacht Review. Sterling Publication Limited, London.

Hardy, T.L., 1995. Multi-objective decision-making under uncertainty: Fuzzy logic method. NASA Technical Memorandum—Computing in Aerospace 10 Meeting (106796), 1–6.

Hasegawa, K., 2009. Advanced marine traffic automation and management system for congested waterways and coastal areas. Proceedings of International Conference in Ocean Engineering (ICOE), Chenai, 1–10.
Healey, A.J. & Lienard, D., 1993. Multivariable sliding mode control for autonomous diving and steering of unmanned underwater vehicle. IEEE Journal of Oceanic Engineering 18 (3), 327–339.
Hong, X. & Harris, C.J.,Wilson, P.A., 1999. Autonomous ship collision free trajectory navigation and control algorithms. Proceedings of 7th IEEE International Conference on Emerging Technologies and Factory Automation, Barcelona, Spain 923–929.
Ibrahim A.M., 2003, Fuzzy Logic for Embedded Systems Applications, Elsevier Science (USA).
Imazu, H., 2006. Advanced topics for marine technology and logistics. Lecture Notes on Ship collision and integrated information system, Tokyo University of Marine Science and Technology, Tokyo, Japan.
IMO, 1972. Convention on the international regulations for preventing collisions at sea (COLREGs). URL http://www.imo.org
Ito, M., Zhang & F., Yoshida, N., 1999. Collision avoidance control of ship with genetic algorithm. Proceedings of the 1999 IEEE International Conference on Control Applications, 1791–1796.
Kwik, K.H., 1989. Calculations of ship collision avoidance manoeuvres: A simplified approach. Ocean Engineering 16 (5/6), 475–491.
Lee, S., Kwon & K., Joh, J., 2004. A fuzzy logic for autonomous navigation of marine vehicles satisfying colreg guidelines. International Journal of Control Automation and Systems 2 (2), 171–181.
Lisowski, J., Rak, A. & Czechowicz, W., 2000. Neural network classifier for ship domain assessment. Mathematics and Computers in Simulations 51, 399–406.
Liu, Y. & Liu, H., 2006. Case learning base on evaluation system for vessel collision avoidance. Proceedings of the Fifth International Conference on Machine Learning and Cybernetics, Dalian, China, 2064–2069.
Minorski, N., 1922. Directional stability of automatic steered bodies. Journal of American Society of Naval Engineers 34 (2), 280–309.
Moreira, L., Fossen, T.I. & Guedes Soares, C., 2007. Path following control system for a tanker ship model. Ocean Engineering 34, 2074–2085.
Moreira, L. & Guedes Soares, C., 2003. Dynamic model of maneuverability using recursive neural networks. Ocean Engineering 30 (13), 1669–1697.
Ohtsu, K., 1999. Recent development on analysis and control of ship's motions. Proceedings of the 1999 IEEE International Conference on Control Applications, 1096–1103.
Ozen, T., Garibaldi, J.M. & Musikasuwan, S., 2004. Modeling the variation in human decision making. Proceedings IEEE Annual Meeting of the Fuzzy Information Processing—NAFIPS 04 2, 617–622.
Pedrycz, W. & Gomide, E., 2007. Fuzzy Systems Engineering—Toward Human Centric Computing. A John Wiley & Sons, INC, Hoboken, New Jersey.
Perera, L.P., Carvalho, J.P. & Guedes Soares, C., 2009. Decision making system for the collision avoidance of marine vessel navigation based on COLREGs rules and regulations. Proceedings 13th Congress of International Maritime Association of Mediterranean, Istambul, 1121–1128.
Pietrzykowski, Z. & Uriasz, J., 2009. The ship domain—a criterion of navigational safety assessment in an open sea area. The Journal of Navigation 63, 93–108.
Roberts, G., Sutton, R., Zirilli, A. & Tiano, A., 2003. Intelligent ship autopilots—a historical perspective. Mechatronics 13 (1), 1091–1103.
Rommelfanger, H.J., 1998. Multicriteria decision making using fuzzy logic. Proceedings Conference of the North American Fuzzy Information Processing Society—NAFIPS 1998, 360–364.
Rothblum, A.M.,Wheal, D.,Withington, S., Shappell, S.A., Wiegmann, D.A., Boehm,W. & Chaderjian, M., 2002. Key to successful incident inquiry. Proceedings 2nd International Workshop on Human Factors in Offshore Operations (HFW), Houston, Texas, 1–6.
Sato, Y., Ishii, H., 1998. Study of a collision-avoidance system for ships. Control Engineering Practice 6, 1141–1149.
Sivanandam, S.N., Sumathi, S. & Deepa, S. N., 2007. Introduction to Fuzzy Logic using MATLAB. Springer Berlin Heidelberg, New York.
Smeaton, G. P. & Coenen, F. P., 1990. Developing an intelligent marine navigation system. Computing & Control Engineering Journal, 95–103.
Smierzchalski, R. & Michalewicz, Z., 2000. Modeling of ship trajectory in collision situations by an evolutionary algorithm. IEEE Transactions on Evolutionary Computation 4 (3), 227–241.
Sperry, E., 1922. Automatic steering. Transactions: Society of Naval Architects and Marine Engineers, 53–61.
Statheros, T., Howells, G. & McDonald-Maier, K., 2008. Autonomous ship collision avoidance navigation concepts, technologies and techniques. The Journal of Navigation, 61, 129–142.
Sutulo, S., Moreira, L. & Guedes Soares, C., 2002. Mathematical models of ship path prediction in manoeuvring simulation systems. Ocean Engineering 29, 1–19.
Xue, Y., Lee, B.S. & Han D., 2009, Automatic collision avoidance of ships, Proceedings of the Institution of Mechanical Engineers, Part M: Journal of Engineering for the Maritime Environment, 223 (1), 33–46.
Yang, S., Li, L., Suo, Y. & Chen, G., 2007. Study on construction of simulation platform for vessel automatic anti-collision and its test methods. Proceedings IEEE International Conference on Automation and Logistics, 2414–2419.
Yavin, Y., Frangos, C., Zilman, G. & Miloh, T., 1995. Computation of feasible command strategies for the navigation of a ship in a narrow zigzag channel. Computers Math. Applic. 30 (10), 79–101.
Zeng, X., Ito, M. & Shimizu, E., 2001. Building an automatic control system of maneuvering ship in collision situation with genetic algorithms. Proceedings of the American Control Conference, Arlington, VA, USA, 2852–2853.
Zhuo, Y. & Hearn, G. E., 2008. A ship based intelligent anti-collision decision-making support system utilizing trial manoeuvers. Proceedings Chinese Control and Decision Conference—CCDC, Yantai, China, 3982–3987.

Advanced Ship Design for Pollution Prevention – Guedes Soares & Parunov (eds)
© 2010 Taylor & Francis Group, London, ISBN 978-0-415-58477-7

Oil spill incidents in Portuguese waters

J.A.V. Gouveia & C. Guedes Soares
Centre for Marine Technology and Engineering (CENTEC), Technical University of Lisbon, Instituto Superior Técnico, Lisboa, Portugal

ABSTRACT: Shipping is probably one of the biggest economic activities in the world. The very low prices in the maritime transport and the world's dependency of oil trade, which has faced a large increase in the last decades, are seen as a strong sight of the nonstop trend in this context. On the other hand, among other environmental threats to the marine environment, major discharges of hazardous substances, including oil spills, are those that can produce bigger impact on public and cause the greatest concerns to the coastal states' authorities. Some of the world's most important oil trade routes cross Portuguese jurisdictional waters. It is now possible to count some hundreds of vessels—sometimes around one thousand—sailing off Portuguese coast every moment. This paper aims to present the Portuguese experience in terms of incidents involving oil spills through the statistical analysis of available data that have been collected by the Maritime Authority for almost four decades.

1 INTRODUCTION

The transport of goods by sea and the number of vessels that carry them have recognized a large increase over the years. Oil and other hydrocarbons, is by its volume, one of the most important goods transported by sea.

Being undeniable its value to modern society, there has also been situations in which it has been a source of concern, as it is the case of major accidents at sea that result in large amount of oil spilled into the sea. Since 1960 there have been efforts to find new ways to minimize this risk and therefore the possible consequences of such accidents. According to the International Tanker Owners Federation Pollution (ITOPF), the trend of recent decades on this kind of accidents is to decrease in number and volume of product spilled. However, accidents at sea continue to occur with severe damages to the marine environment and the coastal resources.

Technically, in order to avoid these situations, and to minimize their consequences, major efforts have been made to strengthen the quality of ships and provide them with systems that can contribute to a safer navigation. At the same time it is required that the coastal states be prepared for a sustained, rapid and effective response.

As widely reported, the foreign trade of Portugal is made in significant part by sea so ports are the main platforms of entering and exiting of essential goods to the life of the Portuguese people.

The Portuguese mainland proximity of some major shipping routes and the lack of maritime surveillance, which was in part overcome with the VTS and AIS—operating since early 2008—raises additional concerns for Portuguese authorities.

The Oil Pollution Response Service (OPRS) of the Portuguese Maritime Authority Directorate General (MADG) is the body that keeps the records of all spill incidents that are detected in Portuguese Waters. Their data base has presently about 2000 records, many of them from hazardous substances.

This paper will analyze the recorded information in a MADG's oil spills database and will assess the possibility of more effective measures being implemented to prevent such incidents and to manage them when they occurr.

2 FRAMEWORK TO COMBAT POLLUTION AT SEA IN PORTUGAL

2.1 *A brief reference to international regulations*

The matter related with the protection of the oceans and the sea pollution is referred in a great number of international instruments, especially the conventions of the International Maritime Organization (IMO) and European regulations. In terms of prevention, among others, reference should be made to the UN Convention on the Law of the Sea (UNCLOS, 1982), which devotes the entire Part XII to the aspects relating to the protection and preservation of the marine environment; the SOLAS Convention, 1974; the MARPOL Convention 73/78; the London Dumping Convention (LDC, 1972); and OSPAR Convention, 1985.

On the control of pollution, reference is made to the OPRC Convention, 1990, which obliges member countries to set up devices to respond to such incidents, in particular through the development of contingency plans to be used in cases of marine pollution. The OPRC-HNS, 2000 provides for the inclusion in those plans of measures to tackle incidents involving spills of hazardous substances other than oil.

Also important are some other tools related to the mechanisms of compensation for pollution damage, which aim to create mechanisms for compensation to those who may suffer losses coming from spills of hazardous substances at sea, which are the CLC Convention 1969 and FUND 71/92. Portugal has ratified all of these instruments with the exception of OPRC 1990.

The legal framework in the EU context is a very wide picture of documents trying to cover all aspects on this issue. The growing role of European Maritime Safety Agency (EMSA), should be highlighted which since its creation in 2001, in the aftermath of the Erika I package, is tasked to develop procedures for standardization of all matters relating to maritime safety and to the protection and preservation of the marine environment in the EU maritime areas. It also provides technical support to Member States to implement European legislation within this domain.

2.2 *National legal framework and responsibilities*

The national responsibility for protecting the marine environment and, specifically, to combat marine pollution, is included as part of the tasks of the Maritime Authority System (MAS), created by Decree nº. 43/2002 of 2 March, which, among its tasks (Article 6) are the "safety of navigation and control," the "preservation and protection of the marine environment" and "preventing and combating pollution" of the sea.

The coordination of operational activities to combat marine pollution in the terms defined in the Clean Sea plan (CSP) is responsibility of the captain of the port in accordance with the provisions in article 13 of Decree-Law No. 44/2002 of 2 March. This document addresses the responsibilities of the port authorities in this regard, as laid down in Decree-Law No. 46/2002 also March 2 in their areas of jurisdiction, and in respect of 4 and 3 degrees of intervention by the PML. Specifically, within the MAS, is the National Maritime Authority (AMN) and its sphere of responsibility that the current law assigns the responsibility to intervene in the functional area as under the scheme already approved by the Decree-Law No. 44/2002.

The OPRS is within this framework, a technical service nationwide to support the Maritime Authority Director-General decision in this matter. This service was established in the Portuguese Navy on 29th January 1973. Since then, MADG, greatly due to the implementation of its local authorities, which are located to cover the whole country, has tried to record the largest possible number of elements of pollution incidents at sea in Portugal, and commenced this task in 1971.

Portugal has not ratified the OPRC Convention in 1990. However, following the crash of Aragon in January 1990, which spilled about 25,000 tons of oil in waters of Madeira, and especially affected the coast of Porto Santo, the Portuguese government has attached to the Ministers Council Resolution nº. 25/93 of 15 April, the Portuguese Oil Pollution Contingency Plan, so called *Clean Sea Plan* (CSP).

3 OIL SPILL DATA

The oil spills data base is an Access database where it is possible to include information such as the date of the incident, the area of jurisdiction of each maritime department where the incident occurred, the geographic position, type of substance spilled, the origin or source of pollution, the estimated amount of oil spilled, the cause (probable or confirmed), the consequences that resulted from it, the actions that were taken, as well as the source of information, measures were implemented and/or fines applied to the polluters. Although these database fields are appropriate for the purposes of this study, it was possible to detect some problems with the fulfilment and management of the information available that made this job not so easy. The information required to populate the database is sent in almost all cases by the local maritime authority through a standardized format.

Due to uncertainties and gaps in data, it was necessary to make some assumptions in order to be able to take advantage of the maximum number of records. In some cases some information was also inferred in order to exploit the existence of records and to emphasize the number of occurrences rather than ignore those who did not possess all the fields of data. After preliminary analysis, the number of validated records was 1943, serving as a working basis for this study.

3.1 *Maritime surveillance and detection of marine pollution incidents*

In Portugal maritime surveillance for detection of marine pollution incidents has been carried out mainly by Navy ships and aircraft of the Air Force (FAP) but all navigating ships share the responsibility to report the incidents seen at sea. The same is valid to the civil aviation, which should also

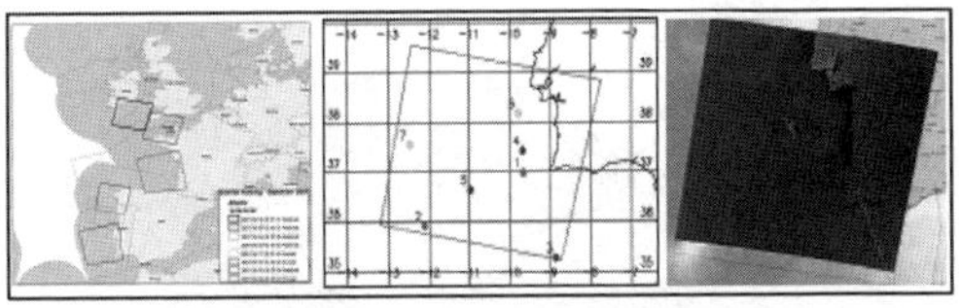

Figure 1. Process of getting *cleanseanet* images: selecting the desired images; image sent to SCPMH identifying possible spills; image SAR (synthetic aperture radar) covering the area of interest.

contribute to this knowledge because of its privileged position to do so. However, there are no systematic monitoring programs, or ways to detect and identify vessels in the vicinity where spills are detected.

Since EMSA started operating the *cleanseanet* project, Portugal has begun to receive sattelite images from this system in August 2007, although that system became operational since April of that year. The satellite images are received by EMSA that sends them to both the MADG and the MRCC Lisbon with circles of 3 colours (red, green and blue) corresponding to the 3 levels of confidence high, medium or low, respectively. In the first case, as a subsequent procedure, a FAP aircraft or Navy ship is sent to confirm the spill based upon the drift of the spill is calculated in Hydrographical Institute. Sampling of the product is made whenever possible.

3.2 *Reporting the incidents to the maritime authority*

The data is reported to MADG in order to inform the OPRS as soon as possible of what happened and to give all the information that exists so far. Later, having more relevant information, a full report of the incident (final POLREP) must also be submitted to that Service within 24 hours. Supplementary information is also sent to OPRS so that the process is continuously updated and the Service can, when appropriate, give technical support to help resolve the incident. Depending on the size and danger of the spill, there are other entities that are also informed to involve them in the respective scope of responsibilities.

4 STATISTICAL ANALYSIS OF THE EXISTING DATA

4.1 *When oil spills occurred*

In a first analysis, the average number of incidents is 52.5 per year, being however, a value that certainly will not match the total spills effectively occurred due to lack of capacity for their detection.

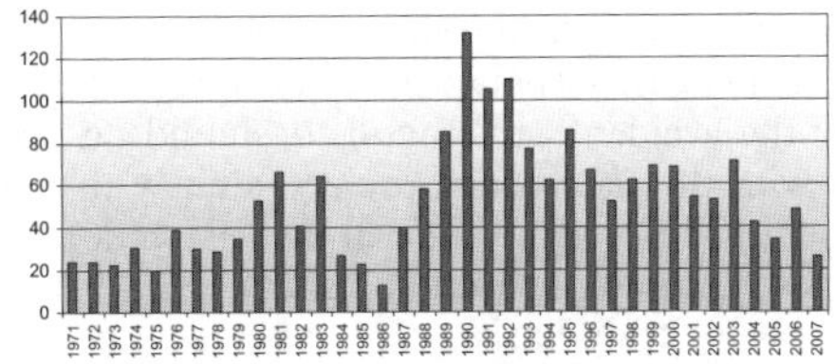

Figure 2. Number of incidents per year.

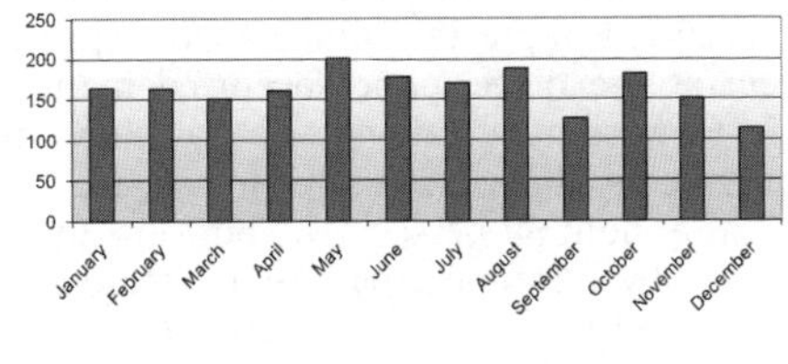

Figure 3. Number of incidents per month.

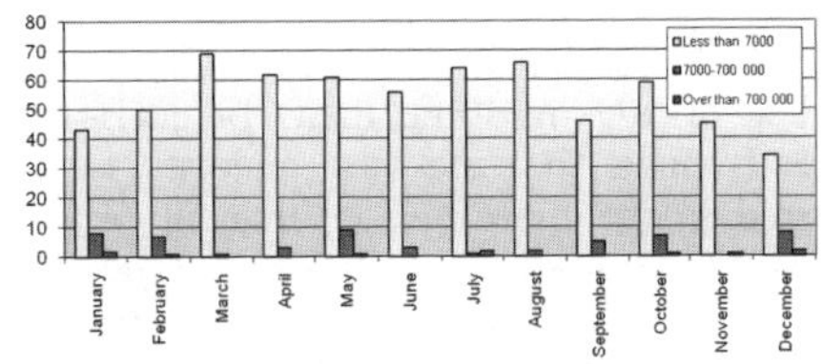

Figure 4. Number of incidents per month and per volume.

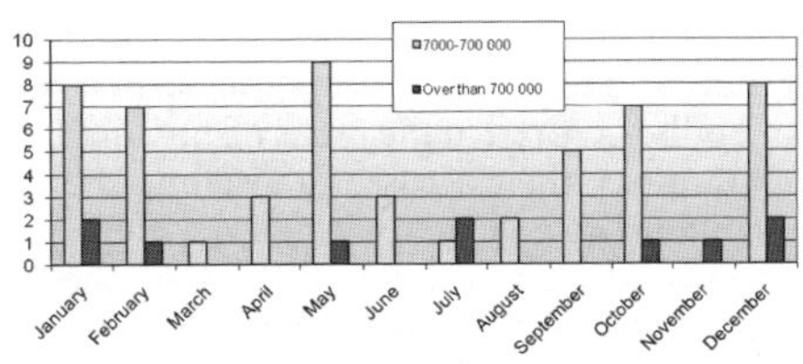

Figure 5. Number of incidents per month considering only the major ones.

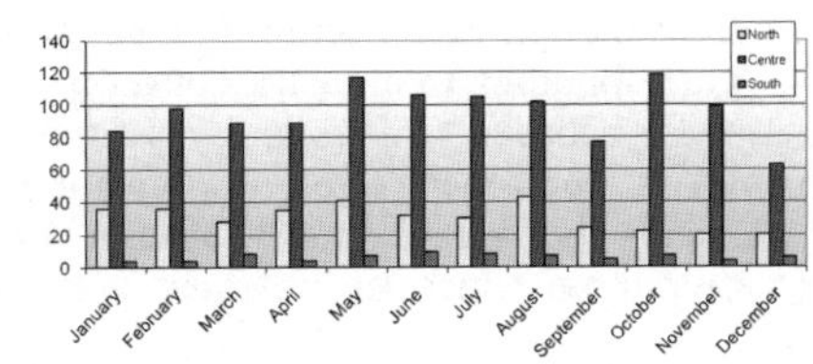

Figure 6. Number of incidents per month and per maritime department in mainland.

It is believed that, given the role of the *cleanseanet* project so far, figure 2 represents only a minority of those who actually take place in Portuguese maritime spaces.

Figure 2 shows an increase from 1971 to 1990, after which there was a gradual decline. The largest number was in 1990 and subsequent years probably due to increased vigilance by the FAP, especially after the accident of Aragon. It should be noted that with the *cleanseanet* project is likely that this number will rise again since in the vast majority of images captured are found oil spills.

On a monthly basis, there appears to be fairly distributed number of incidents throughout the year although when crossing this information with the volume spilled, one can note that small spills occurred in greater numbers from spring to autumn as the biggest ones occurred mainly in winter. However, the statistics of May shows that such accidents can occur at any time of year.

Yet in a monthly perspective, but now making an analysis by maritime department in mainland, it was not found any pattern of events, there are large variations throughout the year instead.

4.2 *Location of incidents*

In view of the maritime areas of jurisdiction, figure 7 shows that the vast majority of spills occurred in the maritime department of centre (59% of the records). The following are the departments of north with 19%, and then the south and the Azores with 4%. Madeira was limited to 3%. The remaining 11% are recorded as having occurred in places further away from the coast so that it was given the designation EEZ.

Figures 8 and 9 show a spatial distribution of incidents. As shown in Figure 8, the largest quantity of spills occurred in the areas of jurisdiction of the main ports, their accesses and along the lanes off the mainland coast.

For spaces under the jurisdiction of the local maritime authority was in Lisbon that there was the greatest number of incidents.

4.3 *Source of oil spills*

The origin of the spill is in Figure 10, highlighting the 14% corresponding to industries onshore or in ports, the 8% between tankers and 7% fitting to fishing vessels. The other vessels account for 34% of cases, being the largest share of events immediately after those which do not have a known source.

By volume of oil spilled, it appears that the incidents with more than 700000 litres spilled are due to other ships and tankers. The medium size spills include, in addition, industries onshore, fishing vessels, general cargo ships and warships.

4.4 *Possible causes*

In this chapter, the variables were grouped together to identify the type of incident that may

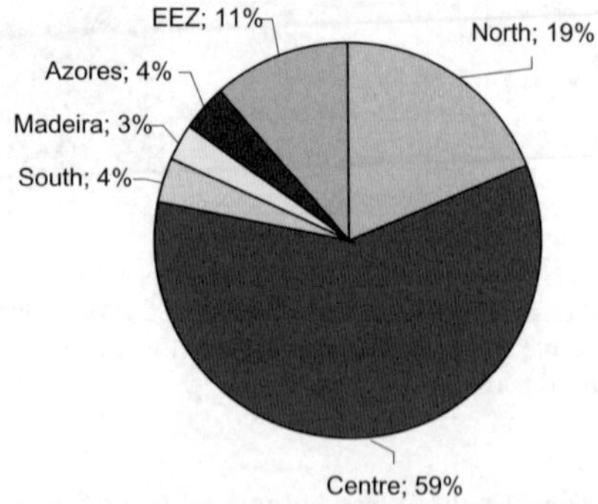

Figure 7. Percentage of incidents by maritime department.

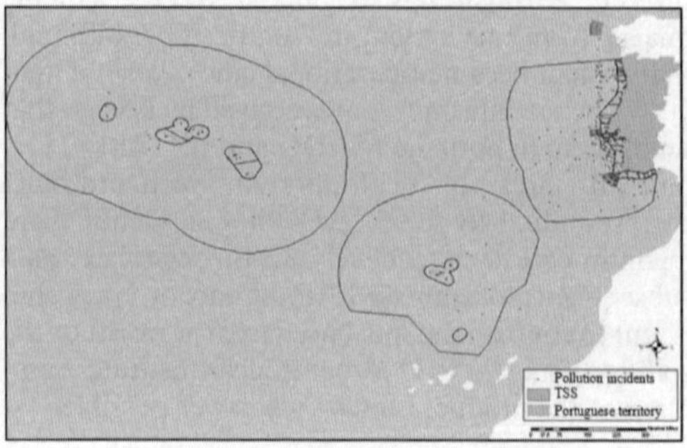

Figure 8. Representation of all incidents between 1971–2007.

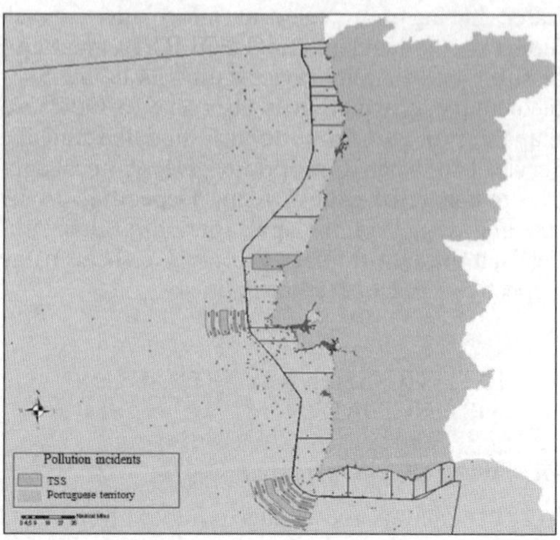

Figure 9. Representation of all incidents 1971–2007 in mainland.

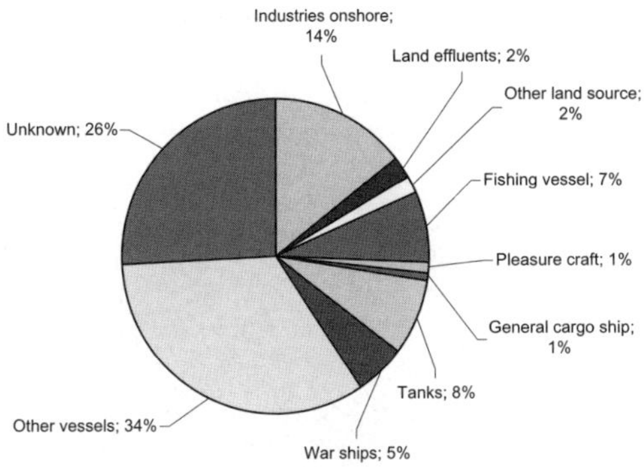

Figure 10. Source of pollution incidents.

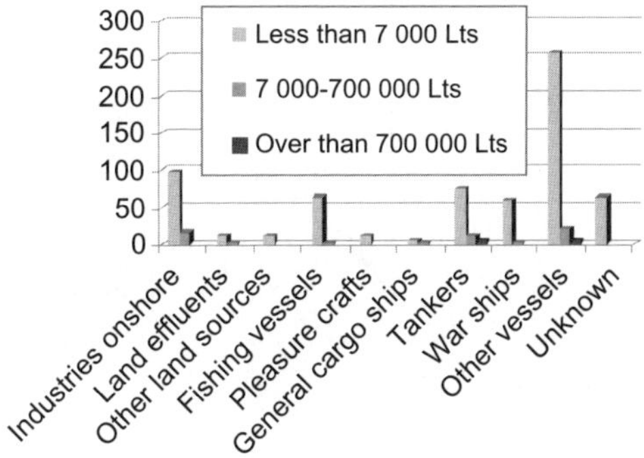

Figure 11. Source of incidents per volume.

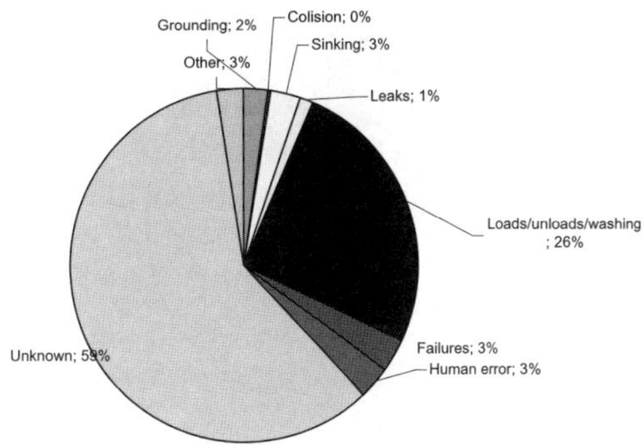

Figure 12. Causes of the incidents.

have led to the spill. Therefore, the predominant is the unknown cause (60%), followed by the operations of loading and unloading (13%) and sewage or washing causes (12%). These two factors together, which are referred as operational causes, are responsible for a quarter of the spills recorded and correspond mostly to small ones. The other causes have less significance but may lead to spills with greater environmental impact.

4.5 *Consequences*

With regard to the consequences known, 14% of the spills were negatively affected ports/docks/marinas; in 13% of the cases the spots have faded away without another type of visible consequences. Of the rest, it is stressed in cases where there were human casualties (deaths).

4.6 *Undertaken actions*

Once a pollution incident occurs, it is necessary to carry out actions that will help to solve the problem. From the analysis of this factor, it appears that in the vast majority of the spills no action was taken either by lack of resources, or because it was understood they were not needed. When actions occurred on 20% of the situations they were to carry out operations of cleaning the affected areas. Many of the actions involved the use of absorbent material—especially in ports—and the collection of the product that reached the shore. The use of dispersants also occurred in some cases despite the fact that in Portugal, according to the law in force, the use of these chemicals is not allowed without the Ministry of Environment permission.

In cases where there is sampling, the OPRS assess the need to send the samples to be analyzed by the competent laboratories: the Portuguese Environment Agency (APA), the Institute for Fisheries and the Hydrographical Institute from the Navy. This procedure aims at finding the polluter, although this is a very difficult process. Usually there are only a very few possibilities to prosecute the polluter. Moreover, in certain procedural frameworks, it is sometimes difficult to ensure a solid evidentiary procedure to achieve that purpose.

4.7 *Oil spills by volume*

Figures 14 to 18 show statistics of oil spills, by volume and by cause for spilss of different size.

Once again, it is possible to see in this analysis the high percentage of ignorance concerning the amount of oil spilled into the sea. Although only 4%—which can still be seen as corresponding to a large number of major spills-, representing the spills with more than 7 tons, and only 1% of the spills with more than 700 tones, the truth is that the small spills should not be neglected.

In scientific surveys made by the Institute for Fisheries there was found to roughly 750 meters water deph under the lanes off Algarve coast some crustaceans impregnated with hydrocarbons, which can only happen by long term accumulation of oil. The same result can be found in harbours and estuaries where the small spills abound and where the

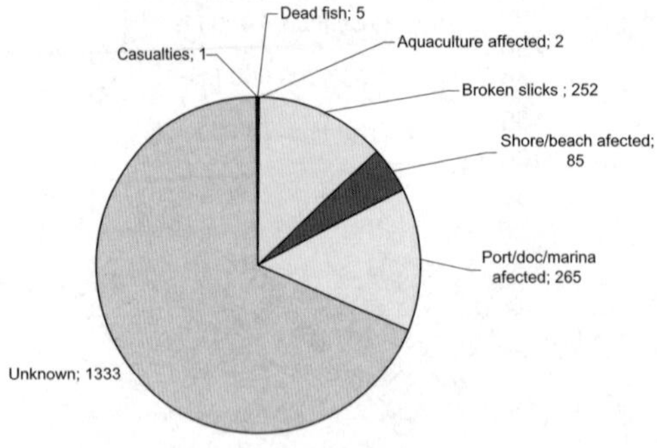

Figure 13. Consequences of the incidents.

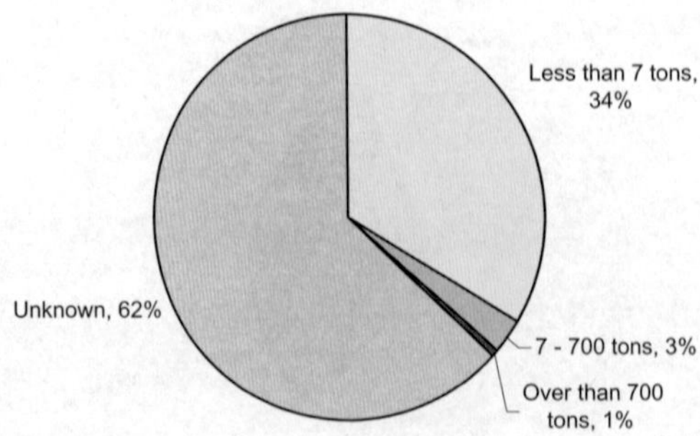

Figure 14. Percentage of incidents by volume.

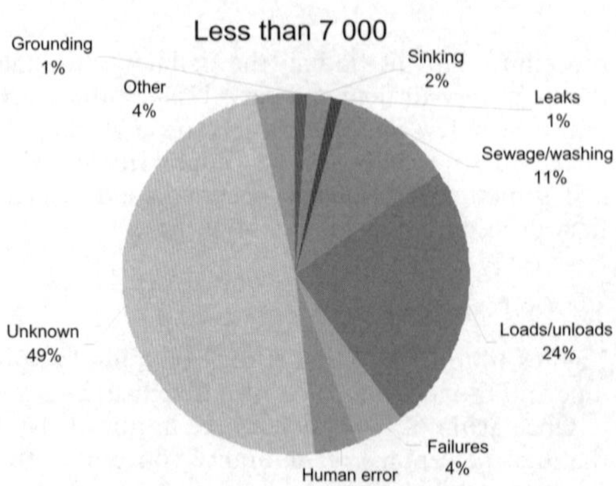

Figure 15. Characterization of small-scale incidents by cause.

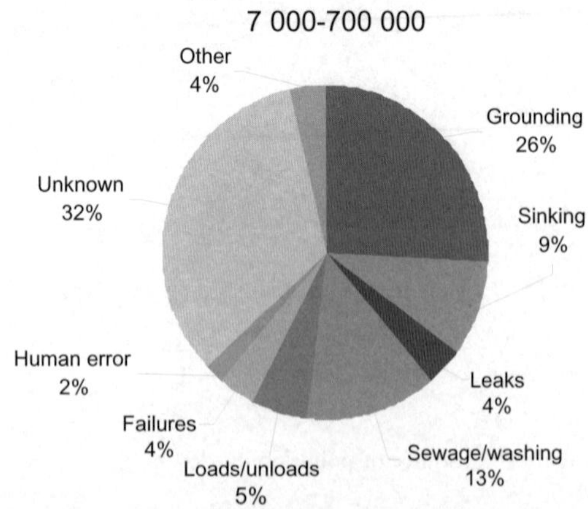

Figure 16. Characterization of medium-scale incidents by cause.

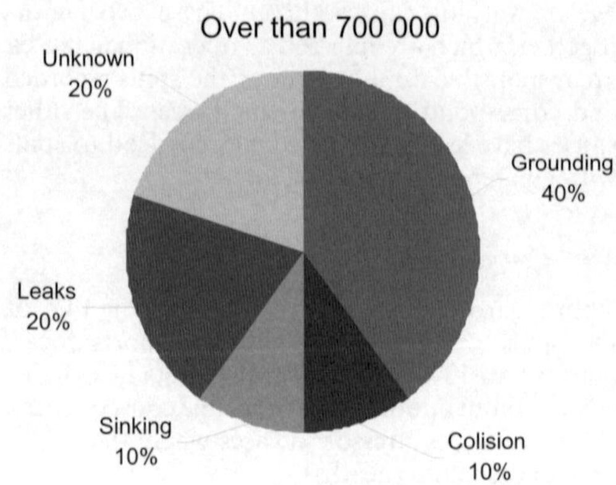

Figure 17. Characterization of large scale incidents by cause.

accumulation of waste pollution is easier. In particular, the number of 738 spills registered within Lisbon local maritime authority jurisdiction is a significant one.

Moreover, large spills are always involved in serious accidents such as grounding (40% of cases), collisions and sinking with 10% each and leaks with 20% of records.

4.8 *Other relevant results*

Since 2001 until 2007 fines were applied of approximately 2.5 million Euros (MADG, 2008). The competent authorities to do that are the captain of the port or the Advisory Board of National Maritime Authority. The Portuguese legal base is Decree nº. 235/2000, of 26 September, which established the procedures and the responsibility in this context.

Those penalties were applied in the vast majority to the infrastructure on land (inside port areas) or involve small spills originated in fishing vessels or in pleasure crafts. In particular, and despite the efforts that arise in trying to punish these increasingly offenses to the marine environment, only 15% of the incidents recorded were fined, which shows the difficulty to identify and punish the polluters.

It is relevant to note that some of the above difficulties are also experienced by other authorities in EU Member-States that have expressed that in international events on this subject.

But if there are situations where it is really difficult to apply the legal existing rules, there are others that may be referred as a good examples taking in consideration the results achieved. In the case experienced by Portuguese maritime authorities in the 9th of December 2005 when the containers ship *CP Valor* run aground at the northwest coast of the island of Faial, in Azores archipelago, the ship's company took over the responsibility for what happened and has assumed all the costs raised from the problem. Even the costs involved within the three years restoration plan imposed by the captain of the port.

5 CONCLUSIONS AND DISCUSSION

There is no doubt that the whole coastline and the maritime areas under Portuguese sovereignty or jurisdiction are highly vulnerable to pollution incidents which can cause serious damage to the interests of the country. The increasing volume of traffic associated with other aspects related to failures of the ship equipment or human errors are very concerned issues on this. The about 2,000 incidents recorded, which must correspond to a small percentage of spills actually exist, are a good indicator of this reality. It is believed that the lack of systematic surveillance that lasted for years has also contributed to facilitating illegal discharges in the Portuguese maritime areas.

It should be noted the large number of small operational spills inside port areas denoting sometimes much inattention on the part of masters or commanders of the vessels, port facilities and even port administrations. It is confirmed that the largest number of incidents is located in the approaches to major ports such as Lisbon, Sines, Setúbal and Leixões, and many occur along the main routes for shipping off the coast. With the change approved by IMO to the traffic separation schemes (TSS) of Roca and S. Vincent that entered into force in July 2005—those TSS are now further away from the shoreline—it is expected that the authorities have more time to be prepared to better respond if an pollution incident occur in these places.

Regarding the pattern of the source of incidents, there were problems with the unspecified ships, tankers, the fishing fleet and industrial facilities located on the coast in port areas. However, the major concerns come from tankers and land industries.

Regarding the causes of the incidents, and excluding the analysis of unknown cause, Portugal follows the international pattern, with the loading and unloading (operational spills) and sewage or washing operations that contribute most to the picture presented.

As demonstrated in many countries, the small spills resulted from the load/unload operations and from deliberate or negligent acts, while the medium and large ones have a very different pattern linked to big accidents like grounding, collisions, structural failure (leaks and cracks) which often results in the sinking of the ship.

In the aim of polluter's punishment, it appears that there is a long way ahead until we are effective. Two possible ways in which we must invest in this respect are the traffic monitoring and the collection of the evidences. Important, yet, is to highlight the need for more reliable statistical data, since the database has many flaws that should be enhanced. Specifically, a great effort should be made to get as much information as possible from each incident in order to fill in the fields in a proper way.

Finally, and taking in account the intrinsic link that this issue presents with the problem of identifying places of refuge for ships in distress, it is particularly important that the country could has a tool that enables the identification of these places and support the political decision within this context. This tool must be built with the most accurate and reliable information in order to fit well its purpose.

REFERENCES

Clark, R.B. (2001), *Marine Pollution*, Oxford University Press, 5ª Ed., 2001.

Gouveia, J.V. and Guedes Soares, C. (2004), "The problem of identifying the refuge áreas in the Portuguese Coast" (in Portuguese), *in As Actividades Marítimas e a Engenharia, C. Guedes Soares e V. Gonçalves de Brito*, Lisboa, 2004, pp. 417–429.

ITOPF (2008), "Statistics" (www.itopf.com).

Probabilistic approach to damage stability

Advanced Ship Design for Pollution Prevention – Guedes Soares & Parunov (eds)
© 2010 Taylor & Francis Group, London, ISBN 978-0-415-58477-7

Probabilistic approach to damage stability

T.A. Santos & C. Guedes Soares
Centre for Marine Technology and Engineering (CENTEC), Technical University of Lisbon, Instituto Superior Técnico, Lisboa, Portugal

ABSTRACT: This paper reviews briefly the development of the traditional method for ship subdivision and of the deterministic damage stability method. The main shortcomings of these methods, which caused the development of the probabilistic damage stability method, are identified. The theoretical foundations for the probabilistic method are explained, starting with reference to the meaning of the attained subdivision index. The methodologies for calculating the probabilities of flooding and survival are described, as well as the method used to derive the formula for the required subdivision index. The method employed to derive the formulae for calculating the probability of flooding taking in consideration the transverse, longitudinal and horizontal watertight subdivision, starting from damage statistics, is explained. Similarly the method employed to obtain formulae for the probability of survival, based on model test results, is described. The relationship of damaged survivability with sea state is also explained. Special consideration is given to the so-called "harmonized" regulations, pointing out the innovative aspects in relation to previous probabilistic regulations. Finally, reference is made to the conclusions of some case studies carried out to assess the impact of the new regulations in ship design.

1 INTRODUCTION

At the end of the XIX century classification societies established empirical rules for positioning bulkheads in merchant ships. As a result, fore and aft peak tanks appeared and machinery spaces became segregated from cargo spaces. Interest in watertight subdivision design was constant in the early XX century, spurred by accidents such as that of the *Titanic*, leading to the development of such concepts as the margin line and floodable length. Buoyancy in damage condition was the main concern at this time. In the first International Conference on the Safety of Life at Sea (SOLAS), held in 1914, a design method for ship subdivision (factorial system) was adopted but this design method never actually came into force.

It was only in the 1948 SOLAS conference that an explicit requirement on damage stability was introduced, namely that the metacentric height should be positive in damage condition. Following the *Andrea Doria* accident, at the 1960 SOLAS conference, a minimum damage condition metacentric height of 0.05 m became mandatory. From 1974, SOLAS also included an implicit requirement for a righting arm in damage condition of at least 0.03 m. In 1990, following the Herald of Free Enterprise (1987) accident, a complete damage stability criterion was introduced in SOLAS as Regulation 8 of Chapter II-1 and it became known as SOLAS 90.

These methods for positioning subdivision and design criteria were all based on evaluating the consequences of certain deterministic damage. It was Wendel (1960, 1968) who first presented the concept of a probabilistic method for designing ship subdivision. This method recognized that ship survival after damage is affected by numerous uncertainties related to damage extent, damage location, loading condition, permeabilities and wave and wind conditions.

The probabilistic method was further developed by Comstock and Robertson (1961) and Volkov (1963), leading to its adoption by the International Maritime Organization (IMO) in 1973. It was applicable to the subdivision of passenger ships, through resolution A.265(VIII). It was not mandatory, simply equivalent to the deterministic requirements in SOLAS.

In 1990 the Maritime Safety Committe (MSC) of IMO adopted resolution MSC.19(58), introducing a mandatory probabilistic regulation for new dry cargo ships with more than 100 m in length. In 1996, resolution MSC.47(66) extended the same method to ships between 80 m and 100 m in length. As two different sets of regulations based on the probabilistic method now existed, with the ones applicable to passenger ships not mandatory, IMO then decided to harmonize the regulations, which were finally approved in May 2005 at MSC80, MSC.194(80). In December 2006, at MSC82, minor changes were introduced in the harmonised regulations, mainly dealing with safe return to port, MSC.216(82). Harmonised regulations take the form of a new SOLAS Chapter II-1. SLF49 approved the Interim Explanatory Notes for Harmonized SOLAS Chapter II-1, Circular MSC.1/1226.

The current situation regarding design criteria for passenger ships is the use of floodable length calculations and the SOLAS90 damage stability criterion. Alternatively, for passenger ships, Resolution A.265(VIII) may also be used. For passenger ferries in the European waters the Stockholm Agreement should also be applied. For dry cargo ships, SOLAS B-1 Probabilistic Regulation (Regulation 25) should be used.

In the future, the harmonized probabilistic regulations (known as SOLAS 2009) are to be applied to passenger and cargo ships with keel laid after 1 January 2009. The ship types covered by new regulations include pure passenger ships, RoRo passenger ships, cruise vessels, dry cargo ships (L > 80 m), container ships, RoRo cargo ships, car carriers and bulkcarriers with reduced freeboard and deck cargo.

A number of special ship types are not affected, with existing damage stability rules still standing, including offshore supply vessels (IMO Resolution MSC.235(82)), special purpose ships (IMO Resolution A.534(13)), special trade passenger ships "Pilgrim Trade" (1971), high speed crafts (HSC Code 2000), tankers (MARPOL 73/78), ships carrying dangerous chemicals in bulk (IBC Code) and ships carrying liquefied gases in bulk (ICG Code).

Figure 1. Examples of extensive damage on ship's sides.

2 DETERMINISTIC SUBDIVISION AND DAMAGE STABILITY METHODS

A number of different methods are currently available for designing ship subdivision: floodable length method, SOLAS90 damage stability criterion, Stockholm Agreement, stability criteria for specific ship types.

The floodable length method consists of ensuring that the ship has some capacity to resist flooding. Floodable length is defined as the maximum longitudinal extent centered in any given point of the ship which can be flooded without causing the immersion of the margin line and maintaining positive metacentric height. The floodable length method traditionally started by calculating a floodable length curve. Permissible length was then calculated using the factor of subdivision. The position of transverse bulkheads was then evaluated against the permissible length curve to check the admissibility of the design.

Following the loss of the *Herald of Free Enterprise* (1987), IMO agreed on a damage stability criterion (SOLAS 90). For any specific ship, the number of consecutive (or adjacent) compartments to be flooded depends on the subdivision factor. Sufficient intact stability shall be provided in all service conditions so that the ship will withstand the flooding of any one main compartment. If the factor of subdivision is 0.5 or less, but more than 0.33, the ship should withstand flooding of two adjacent main compartments. If the factor of subdivision is less or equal than 0.33 the ship should survive flooding of three main compartments.

The damage stability criterion SOLAS 90 requires that the ship in any possible damage condition meets the following parameters:

- Range of stability of at least 15° from the equilibrium angle.
- Minimum reserve of stability of 0.015 m.rad.
- Minimum metacentric height in damage condition of 0.05 m.
- Minimum righting arm of 0.10 m, including the effects of passengers crowded on one side, launching of lifeboats and wind pressure.
- Maximum heeling angle before equalization should not exceed 15°.

- Righting arm greater than 0.05 m and range of stability greater than 7° in the intermediate stages of flooding.

In 1995, in the wake of the *Estonia* capsize, the criterion SOLAS90 above became mandatory for existing ships. IMO also opened the possibility for certain groups of countries to sign regional agreements on damage stability standards and as a result several northern European countries set up the so-called Stockholm Agreement. This regional agreement applies to passenger ro-ro ships and specifies that ships must comply SOLAS90 criterion with a certain amount of water on deck. For details on the impact of the Stockholm Agreement, refer to Vassalos and Papanikolaou (2002).

The height of accumulated sea water in the ro-ro deck is given as function of freeboard, as depicted in Fig. 2. Between 0.3 m and 2.0 m freeboard the height of water, h_w, is to be found by linear interpolation. For ships with geographically defined areas of operation, it is possible to reduce the height of water. Between 1.5 m and 4.0 m significant wave height, H_s, the height of water is to be determined by linear interpolation. The significant wave height in well defined areas of the sea is defined in appropriate maps, given in the Stockholm Agreement.

The water on deck between 1.5 m and 4.0 m significant wave height is calculated by plotting the line for actual H_s of the operational area of the ship. Interpolation along that line for the residual freeboard corresponding to damage condition under consideration gives the actual height of water to be considered.

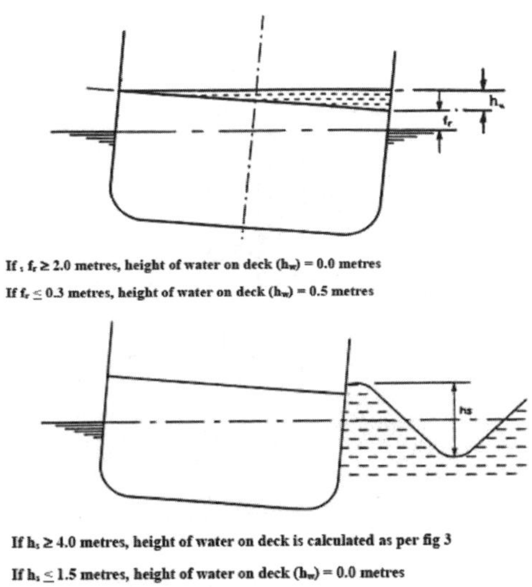

Figure 2. Guidelines for water on deck calculation (1).

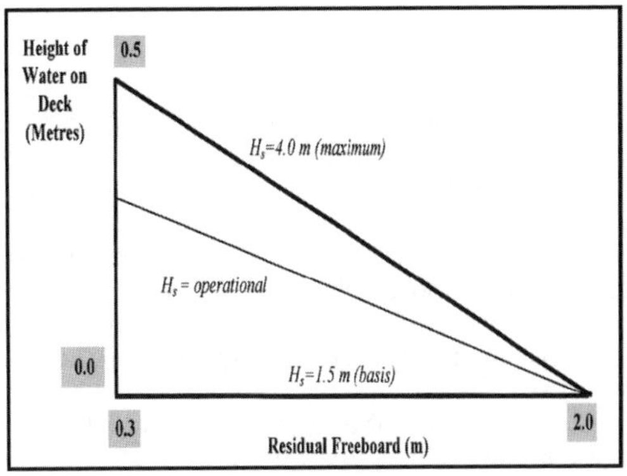

Figure 3. Guidelines for water on deck calculation (2).

3 PROBABILISTIC SUBDIVISION AND DAMAGE STABILITY METHODS

3.1 *Attained and required subdivision indices*

Assessing the probability of survival of a ship in damage condition includes the following probabilities:

- Probability of flooding each single compartment and each group of two or more adjacent compartments,
- Probability that buoyancy and stability after flooding of a compartment (or group of adjacent compartments) is sufficient to prevent sinking, capsizing or dangerous heeling due to loss of stability or heeling moment.

The effects of ship flooding depend on numerous uncertainties: compartment or group of compartments flooded, draught and intact stability of the ship, permeability of flooded compartments, sea state when accident occurs, steady heel angle and transient asymmetric flooding. These factors are complex and interdependent and their effect depends on specific cases. The main conclusion is that the exact evaluation of the probability of survival of a ship is very difficult. However, a logical approach can be applied, resulting in an estimate of ship's probability of survival to flooding hazard: the probabilistic method.

Probability theory indicates that the probability of survival can be calculated as a summation over all possible damage cases. The probability of survival to flooding is proportional to the Attained Subdivision Index, A:

$$A = \sum_{i \in I} p_i s_i$$

In the new harmonised regulations the attained index is given by the following expression:

$A = A(p,v,r,s)$

where:

p is the probability that only the compartment or the group of compartments under consideration may be flooded, disregarding any horizontal subdivision.

v is the probability that the damage will not exceed a given height above the waterline.

r is the reduction factor, which represents the probability that inboard spaces will not be flooded.

s is the probability of survival after flooding the compartment or the group of compartments under consideration.

The probabilistic method then requires that the attained subdivision index, A, should then be larger than a Required Subdivision Index, R:

$A > R$

3.2 *Methodology for calculating the probabilistic of flooding*

The calculation of the probability of flooding involves first developing a database for distribution functions of damage location, damage extent, penetration and vertical extent. Regression techniques are then applied to obtain analytical formulae for these distributions, which are combined into formulae for approximating the probability of flooding between transverse bulkheads and correction factors for longitudinal and horizontal subdivision.

When applied to a specific ship, given its bulkhead locations, these formulae produce an approximation to the probability of flooding in case of complex subdivision geometries involving transverse, longitudinal and horizontal subdivision.

The simplest case is to consider only the location and length of damage in the longitudinal direction.

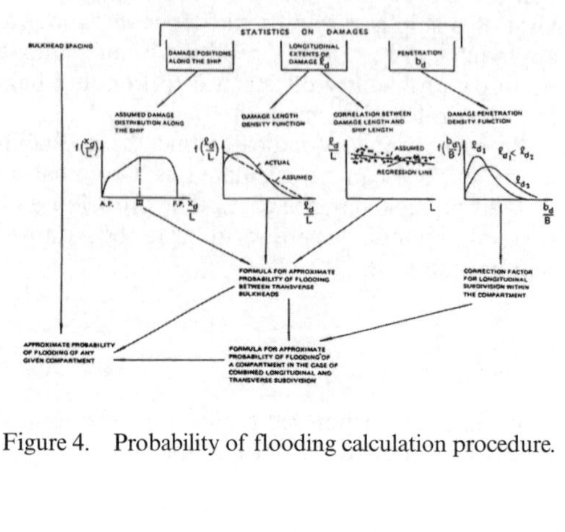

Figure 4. Probability of flooding calculation procedure.

Damages can then be represented by points within a triangle with equal base and height (L_s).

For two compartments damages the probability of flooding is given by:

$p_i = p_{12} - p_1 - p_2$

For three compartments damages the probability of flooding is given by:

$p_i = p_{123} - p_{12} - p_{23} + p_2$

Damage location x and damage length y are random variables and their distribution density functions $f(x,y)$ can be derived from statistics and plotted as shown in Figure 5. The total volume between x-y plane and $f(x,y)$ equals one, corresponding to assuming that there exists a flooding. In a similar way, the volume above a triangle and under the distribution density functions represents the probability that a specific compartment is flooded.

When there is double bottom, the probability of flooding should be split in probability of flooding the double bottom only, probability of flooding the compartment above the double bottom only and probability of flooding both spaces. Each case will present different probabilities of survival, s_i. In MSC.19(58) the approach was to assume that the most unfavourable vertical extent of damage occurs with total probability p.

In case the ship has also a horizontal subdivision above the waterline, the vertical extent of damage may be limited to the depth of that subdivision. In fact, the vertical extent of damage is also an

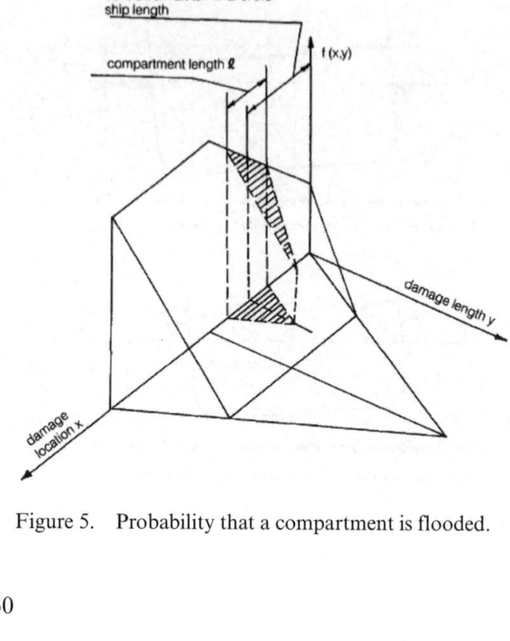

Figure 5. Probability that a compartment is flooded.

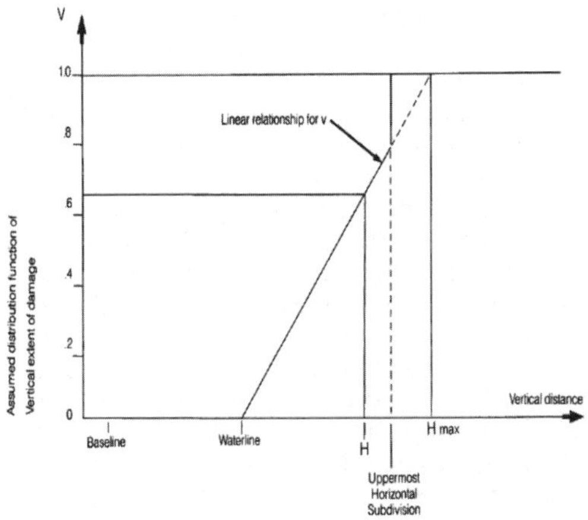

Figure 6. Assumed distribution of vertical extent of damage.

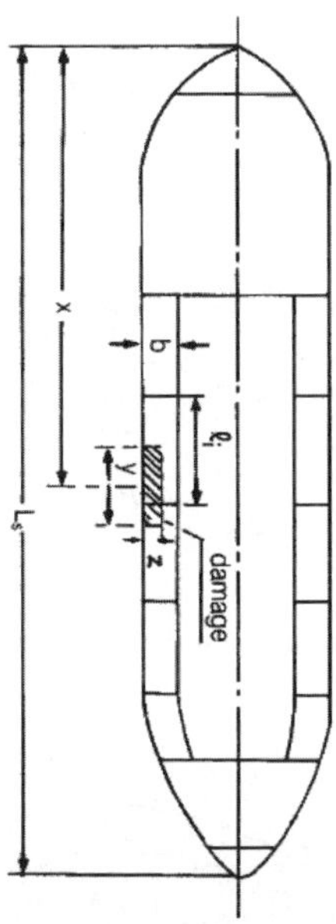

Figure 7. Parameters for calculating probability of flooding when ship has longitudinal subdivision.

uncertainty and the probability of not damaging that horizontal subdivision is represented by factor v_i. This factor represents the assumed distribution function of the vertical extent of damage and varies from zero for subdivision at the level of waterline linearly upwards to a given level above the waterline.

Ships may also have longitudinal subdivision. In this case, the uncertainty related to damage penetration has also to be considered. The probability that only a side compartment is opened is expressed as $p_i r$. The probability that an inner compartment is opened in addition to the side compartment can be expressed as $p_i(1-r)$.

All damages which open a side compartment correspond to the points within a smaller prism with height b, as shown in Figure 8. This factor b is the distance between the longitudinal bulkhead and the ship's side. Damage location, longitudinal extent and penetration are then the uncertainties and their distribution density functions $f(x,y,z)$ may be derived from damage statistics. These functions $f(x,y,z)$ may be assumed to vary from point to point within the volume. The integration of $f(x,y,z)$ over the whole volume then gives one, which is the total probability that there is a damage in all conditions. The integration of $f(x,y,z)$ over a partial volume (representing flooding of certain spaces) represents probability that these spaces are flooded.

At this stage, four uncertainties are under consideration: longitudinal location, damage length, damage vertical extent and damage penetration. Collision casualty reports containing damage data and voyage reports with loading and stability conditions provide statistical data which is then used to derive probability distributions for the variables which characterise damage. Formulae for p, r, v in the probabilistic regulations are then based on these probability distributions.

Data for the updated damage database developed to support the harmonization of probabilistic regulations include original IMO data cards compiled for resolution A.265, a small number of recent IMO damage cards and databases of Lloyd's Register of Shipping, Germanischer Lloyd's, Det Norske Veritas, among others. The total number of reports is 2946, including 930 collisions (ship-ship collisions and collisions with floating objects

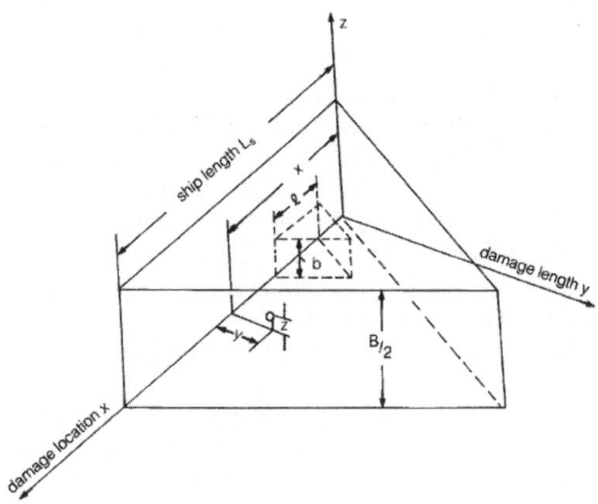

Figure 8. Probability that a compartment is flooded in case of longitudinal subdivision.

or berths). A comprehensive analysis of the distribution density functions for non-dimensional damage location, damage length, penetration and vertical extent of damage is presented in Lutzen (2002), including comparison with previous statistical data and derivations of formulae adopted in the harmonized rules.

3.3 *Methodology for calculating the probability of survival*

The probability of survival of the ship in a certain damage condition depends of uncertainties such as the loading condition and the sea state at the time of the accident. A number of characteristics of the vessel are also obviously important such as the beam, the freeboard and the details of the subdivision. Uncertainties in the loading condition may be taken into consideration by developing density distribution functions using cargo manifests and estimating the probability of the ship being in a light condition, an intermediate condition or a service condition. Seastate distributions are also available both for specific areas of the seas and for the sea as a whole, see IMO-Assembly (1973). Model experiments have been used for some time now in the development of formulae relating the survival of damaged ships with its main characteristics or with stability parameters. These formulae are then combined with seastate statistics to obtain formulae for the probability of survival in a seaway, which is then weighted using the probabilities associated with each loading condition. The final result is an approximation to the probability of survival for the specific damage condition under consideration.

Model testing has traditionally been the most common approach to evaluate damaged ship survivability in waves. Middleton and Numata (1970) and Bird and Browne (1973) report two of the earliest model experiments with damaged models in waves. Maritime accidents generally trigger more experimental work. HARDER EU project involved extensive model testing for different ship types including cruise liners, ro-pax, cargo ro-ro, containerships and bulk carriers. Model experiments allowed the observation of three mechanisms of capsize in waves. These relate to high freeboard ships, low freeboard ro-ro ships and low freeboard conventional ships.

The probability of survival is generally obtained in four steps.

The first step consisted in generalizing the results of all model tests and presented in a format which shows the relationship between survival significant wave height (h1/3) and stability parameters. As a second step, the cumulative distribution function is constructed by combining the model experiment results with the probability of certain wave heights at the time of the casualty. IMO statistics on seastate at moment of collision are generally used.

Finally, approximate formulae representing the probability of survival are obtained.

After the loss of the *Herald of Free Enterprise* (1989) and of the *Estonia* (1994) a number of model test with damaged RoRo models were conducted allowing the development of the Static Equivalent Method (SEM) to calculate the critical seastate of a damaged ship, see Vassalos *et al.* (1997). This method resulted from observations of model tests and numeric simulations of damaged RoRo ships. The primary observation of the model test was that the capsize mechanism almost always appeared to be quasi-static in nature. It was dependent upon the volume of water pumped up onto the vehicle deck due to wave action.

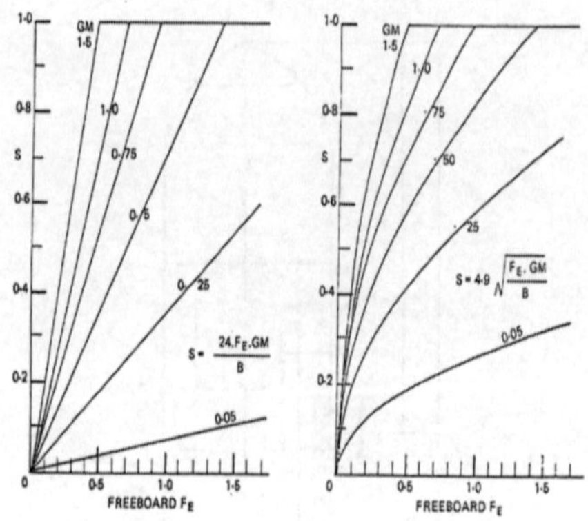

Figure 9. Formulae for probability of survival.

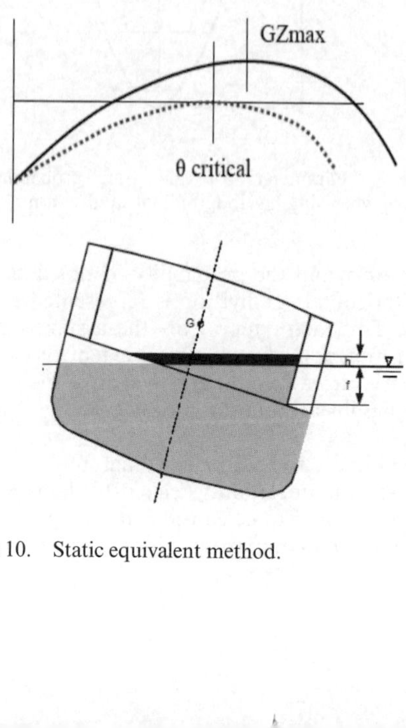

Figure 10. Static equivalent method.

The SEM calculation method starts with the definition of the intact and damage condition, followed by the calculation of the static stability curve allowing the identification of GZ_{max} and θ_{max}. Water is then added on the car deck until the ship heels to θ_{max} where GZ should be zero and h is to be calculated for that heel angle.

The significant wave height, H_s is then obtained using the following formula (from model tests):

Combining this H_s with the seastate distribution a probability of survival can be calculated.

During HARDER project, a new formula for the SEM was developed.

Results of model tests with ro-ro ships concerning the critical significant wave height were plotted against h and f and the following three dimensional regression formula was obtained:

$$H_s = 6.7h - 0.8f - 0.9$$

Previous probabilistic regulations used formulae dependent on other standard stability parameters such as GM and damage freeboard. Traditional deterministic damage stability criteria employ the use of properties of the GZ lever curves, such as GZ_{max}, range of stability or righting energy.

The fit of critical sea state to such traditional stability characteristics was also tested within HARDER. For the three tested stability characteristics, typical spreading was found to be about a 1 m band either side of the mean regression values. The fitting, for Ro-Ro ships, is clearly better using the SEM.

During HARDER, the possibility to extend SEM to non ro-ro ships was also tested but the correlation was found to be not quite as good as for ro-ro ships. After some discussion, the prevailing opinion at SLF and MSC was that the SEM was too complicated to be applied in practice and that formulae based on SEM were not significantly more accurate than more traditional formulae. The following formula based on the traditional stability parameters was finally adopted:

$$S_{final,i} = K.\left[\frac{GZ_{max}}{0.12}\frac{Range}{16}\right]^{1/4}$$

where GZ_{max} not to exceed 0.12 m, Range not to exceed 16° and K is a factor dependent on the equilibrium angle.

3.4 *Methodology for formulating the required subdivision index*

Developing a formula for the required subdivision index R involves creating a database containing different ship types, size and construction and layout arrangements. It should cover a statistically significant sample of the world fleet. The constructed database includes ships marginally meeting the existing SOLAS regulations.

Derivation of a suitable formula involves analyzing the damage stability characteristics, particularly the attained subdivision indices using existing and new regulations of the ships in database. The possibility of formulating unified formulae for the Required Subdivision Index for both cargo and passenger ships were also explored. Definition of database finally resulted in a sample of 40 passenger ships and 92 dry cargo ships.

The derivation started from the assumption that current damage stability regulations for newbuildings correspond to a satisfactory level of safety. Harmonization of relevant rules on the basis of the probabilistic damage stability concept should

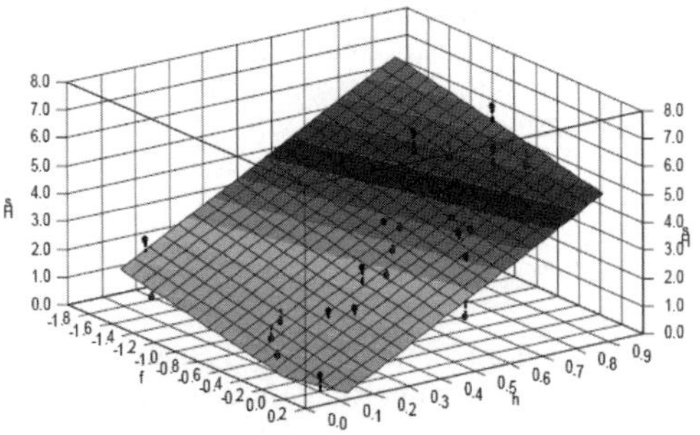

Figure 11. Regression formulae and model test results.

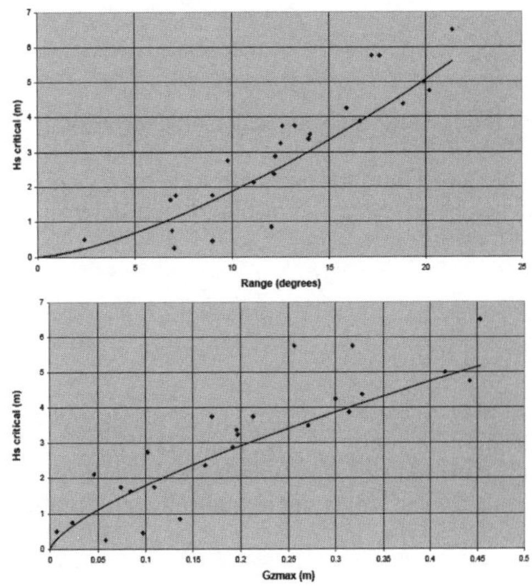

Figure 12. Correlation of the critical significant wave height with the GZ_{max} and range of stability.

aim at keeping these levels of safety on average unchanged. Thus, the calculation of the new Required Index will be based in the "equivalence of safety" principle:

$$\frac{A_{new}}{R_{new}} \approx \frac{A_{existing}}{R_{existing}}$$

The procedure for calculating the new required indices for cargo ships built under probabilistic rules (SOLAS B-1) is:

$$\frac{A_{new}}{R_{new}} \approx \frac{A_{existing}}{R_{existing}} \Rightarrow R_{new} \approx A_{new} \cdot \frac{R_{existing}}{A_{existing}} \approx A_{new} \cdot \frac{R_{B-1}}{A_{B-1}}$$

where R_{B-1} and A_{B-1} are calculated according to SOLAS B-1.

The procedure for calculating the new required indices for passenger ships built under deterministic rules is:

$$\frac{A_{new}}{R_{new}} \approx \frac{A_{existing}}{R_{existing}} \Rightarrow R_{new} \approx A_{new} \cdot \frac{R_{existing}}{A_{existing}}$$

assuming that R/A according to existing rules as 1.0 and taking into account limit GM curve.

The procedure for calculating the new required indices for passenger ships built under resolution A.265:

$$\frac{A_{new}}{R_{new}} \approx \frac{A_{existing}}{R_{existing}} \Rightarrow R_{new} \approx A_{new} \cdot \frac{R_{existing}}{A_{existing}} \approx A_{new} \cdot \frac{R_{A.265}}{A_{A.265}}$$

where $R_{A.265}$ and $A_{A.265}$ are calculated according to A.265.

The formula for R should include the following main parameters:

- Number of persons on board, N, considering also life saving equipment.
- Size of ship, Ls, indicative of value of ship and its cargo.

The required subdivision index is given by empirical formulae which depend mainly on the number of persons onboard and on the ship's length:

$$R = f(L_s, N)$$

For passenger ships, the required subdivision index was found to be in the range 0.68–0.83. For cargo ships, the required subdivision index was found to be in the range 0.47–0.74.

In the case of cargo ships greater than 100 m in length (L_s) the required index is given as:

$$R = 1 - \frac{128}{L_s + 152}$$

In the case of passenger ships the required index is given as:

$$R = 1 - \frac{5000}{L_s + 2.5N + 15225}$$

where:

- N = N1 + 2N2
- N1 = number of persons for whom lifeboats are provided
- N2 = number of persons (including officers and crew) the ship is permitted to carry in excess of N1.

For dry cargo ships it was observed that lower draughts lead to higher attained indices, namely with light service draught having typically 150% of the global attained index. Subdivision draught has the lowest attained index, on average 66% of global attained index.

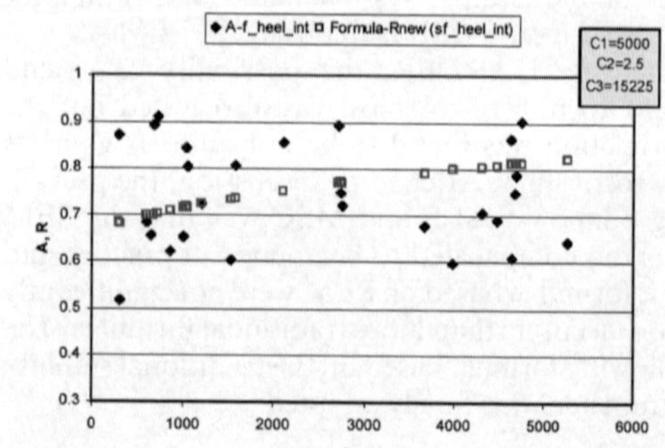

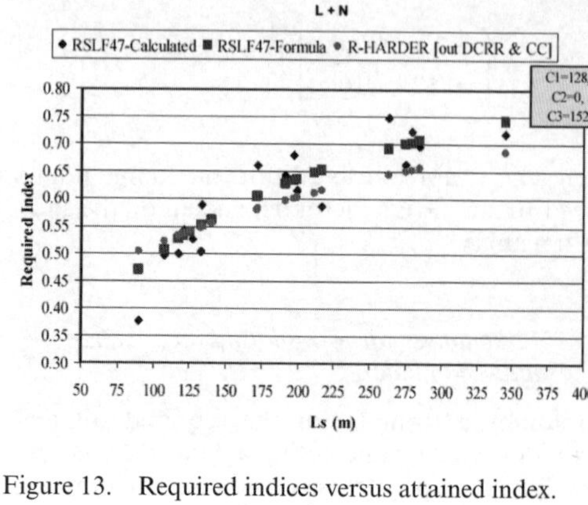

Figure 13. Required indices versus attained index.

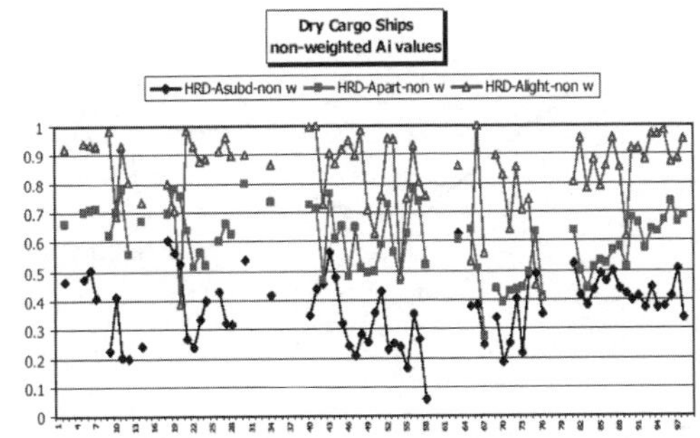

Figure 14. Attained indices for the three draughts (cargo ships).

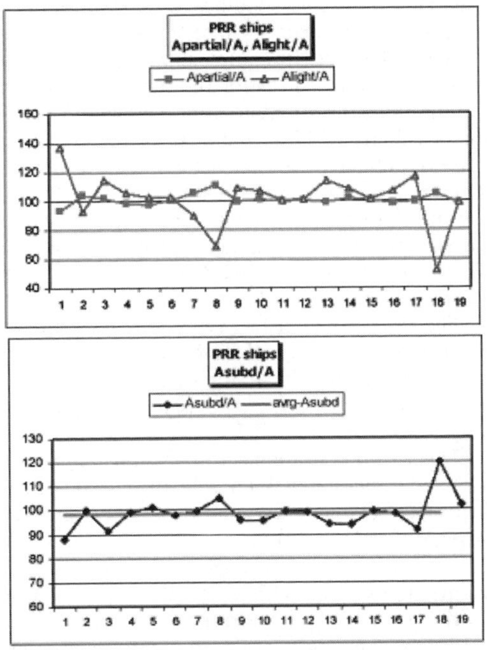

Figure 15. Attained indices for the three draughts (passenger ships).

For passenger ships all draughts dispose of comparable attained index values, thus differing from the behavior of cargo ships. The subdivision draught contribution is in the average about 99% of the global attained index, both for PCLS and PRR ships.

The light service draught and the partial draught dispose of indices in the average equal to about 101% of the global attained index, for both ship subtypes. Taking in consideration these observations, minimum values of the attained subdivision index at specific draughts are set in the harmonized rules. The partial indices A_s, A_p and A_l are to be not less than 0.9R for passenger ships and 0.5R for cargo ships.

4 HARMONIZED PROBABILISTIC SUBDIVISION AND DAMAGE STABILITY REGULATIONS

4.1 *Methodology for formulating the required subdivision index*

A number of innovations have been included in the new probabilistic harmonized regulations. Regarding the required subdivision index, it is worth mentioning that in A.265(VIII), which was applicable to passenger ships, R depended on length of ship and number of passengers. In MSC.19(58), applicable to cargo ships R depended only on length of ship. The new harmonized regulations (MSC.194(80)) which are applicable to both types of ships contain separate formulae for the required index since it was found that no single formula would be suitable for both ship types.

Draughts for which the attained subdivision index was to be evaluated in Res. A.265(VIII) were three intermediate draughts between the subdivision draught and the light draught. The weights to be applied were 0.45, 0.33 and 0.22. Probabilistic cargo ship rules (MSC.19(58)) considered only two draughts: the subdivision draught and a draught at 60% between light and subdivision with equal weights. The new harmonized probabilistic method (MSC.194(80)) considers three draughts: full load, 60% between light and full load and light condition. However, the weight of the light condition has now been reduced to 0.2.

The longitudinal damage extent in the harmonized regulations is now characterized by an increased maximum damage length, from 48 m to 60 m. Regarding the transverse subdivision in harmonized regulations, damages of B/2 are now considered (allowing penetration of longitudinal centre line bulkheads).

In the first probabilistic regulations, resolution A.265(VIII), horizontal subdivision was not considered (no probability *v*). The new harmonised regulations now consider the effects of horizontal subdivision, but the maximum vertical extent of damage above a given waterline has changed from 7 m to 12.5 m.

The formula for probability of survival in A.265(VIII) depended on freeboard and metacentric height in damage condition. In more recent probabilistic regulations, both in the resolution MSC.19(58) and in the harmonized regulations, the probability of survival depends on righting arm and range of stability. An important innovation in

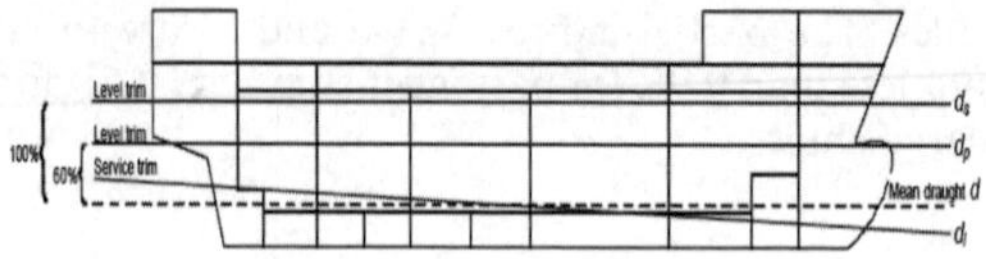

Figure 16. Draughts and trim assumptions for harmonized regulations.

the harmonized regulations is that different factors which influence the probability of survival are now also considered, including the survival capacity in the intermediate stages of flooding and the effect of the different heeling moments which can act on the ship (same as existed in old deterministic regulation 8).

In resolution A.265(VIII) the cargo space permeabilites were given by a formula dependent on draught, aiming at reflecting that ships in light condition have higher permeabilites. In resolution MSC.19(58) cargo spaces permeability was set to 0.70 and in the harmonized regulations cargo space permeability is set to a value dependent on the draught of the ship, but which is always equal or larger than 0.70. For ro-ro spaces it is assumed to be higher: always equal or larger than 0.90.

Trim effects were considered in A.265(VIII). In each initial draught condition the ship was taken to be at the most unfavourable intact service trim anticipated at that draught. However, trim effects were neglected in MSC.19(58) and level trim was to be used for both draughts. In MSC.194(80) level trim is used for the deepest subdivision draught and the partial subdivision draught whilst service trim shall be used for the light service draught.

In the new harmonized regulations, if trim variations are significant, additional calculations of A are to be submitted.

In cases where the operational trim range exceeds ± 0.5% of L_s, additional calculations are to be carried out.

First the original GM limit line should be calculated with the trims mentioned previously. Then, additional sets of GM limit lines should be constructed on the basis of the full range of trims ensuring that intervals of 1% Ls are not exceeded. The whole range including intermediate trims should be covered by the damage stability calculations. These sets of GM limit lines are then to be combined to give one envelope limiting GM curve, and the effective trim range of the curve should be clearly stated.

When checking for compliance when trim is significant the GM values for the three loading conditions could, as a first attempt, be taken from the intact stability GM limit curve. If the required index R is not obtained, the GM values may be increased. This implies that the intact loading conditions from

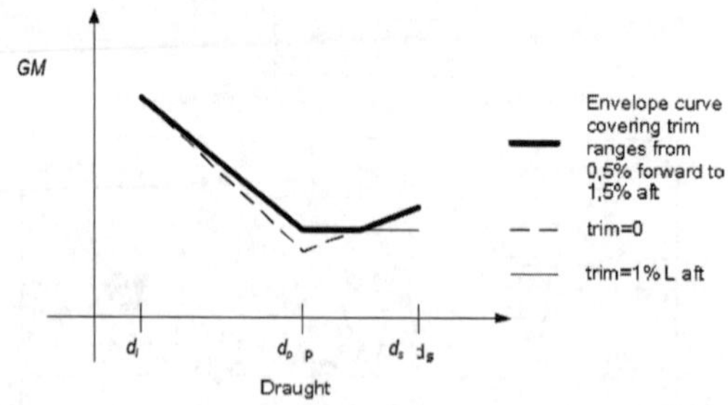

Figure 17. Derivation of GM limiting curves.

the intact stability book must now meet the GM limit curve from the damage stability calculations. This GM limit curve should be derived by linear interpolation between the three GM's.

When checking for compliance when trim is significant the GM values for the three loading conditions could, as a first attempt, be taken from the intact stability GM limit curve. If the required index R is not obtained, the GM values may be increased. This implies that the intact loading conditions from the intact stability book must now meet the GM limit curve from the damage stability calculations. This GM limit curve should be derived by linear interpolation between the three GM's.

The new probabilistic regulations have already been revised through IMO-MSC (2006) (resolution MSC.216(82)) including a new Regulation 8-1 *System capabilities after a flooding casualty on passenger ships.*

This regulation applies to passenger ships constructed only on or after 1 July 2010. A passenger ship shall be designed so that the systems specified in regulation II-2/21.4 remain operational when the ship is subject to flooding of any single watertight compartment. In IMO-MSC(2007) (resolution MSC.1/Circ.1214) the performance standards are set for systems and services to remain operational for safe return to port after a casualty and systems and services to remain operational on passenger ships for orderly evacuation and abandonment after a casualty. These requirements are expected to have impact in the compliance of new designs with rules.

4.2 *Application to assessment of passenger Ro-Ro ship*

The Maritime and Coast Guard Agency (2007) has promoted a study aiming at comparing the equivalence of the safety levels offered by new probabilistic regulations with the existing SOLAS regulations but also comparing new regulations with the Stockholm Agreement for ro-pax ships. It was also expected to anticipate as far as possible what effects the new regulations may have on the design of a range of ship types.

Five basis ships, shown in Figure 18, were selected for evaluation: panamax cruise ship, ro-ro passenger cruise ferry, car carriers, container feeder ship, small coaster.

Regarding the panamax cruise ship, damage stability was found to be limiting for upper part of draught range both using SOLAS90 and MSC194(80).

The existing ship complied with SOLAS90 but failed to comply with the new probabilistic regulations. The causes of non-compliance were the static heel angles caused by large tank asymmetries in the double-bottom. The ship was then redesigned using freedom allowed by MSC194(80), with all loading conditions found to be below the limiting KG curve for MSC.194(80). It was then found that the new design does not comply with SOLAS90. The cause for the non-compliance with SOLAS90 is the increased compartment length and the absence of margin line in the new probabilistic regulations. If margin line requirement is not taken into account during the SOLAS90 calculations, the new design is much nearer to compliance with old rules.

Assessment of the large ro-pax ship shows that differences between limiting KG curves according with damage stability (SOLAS90), for different aft or forward trims, are small. Intact stability KG limits are not very dependent on trim, but present some variations in the upper draught range.

The existing ship complies with the new regulations due to high freeboard, low number of passengers and large watertight barriers on the port and starboard of the car deck. The ship was then redesigned with a smaller number of main transverse watertight bulkheads and side casings on car deck enclosed with watertight doors at each end replaced with narrower side casings without watertight doors. The redesigned ship complies with MSC.194(80) but with less margin. The new ship has identical limiting KG as the existing vessel when using SOLAS90.

For the large car-carrier, damage stability is dominant in the upper half of draught range and the existing ship complies with SOLAS Reg. 25. It was not possible to calculate a limit KG curve for the MSC194(80) and thus the ship failed to pass the new regulations. The reason was the increased permeability of cargo spaces, which raised the required subdivision index and attained index calculated for 3 draughts instead of 2. The ship was then redesigned to be watertight to two more decks (deck 7 instead of 5). The new ship then complies with the harmonised probabilistic regulations with slightly smaller limit KG. The redesigned ship has equal KG limit curve to the existing vessel according with SOLAS90.

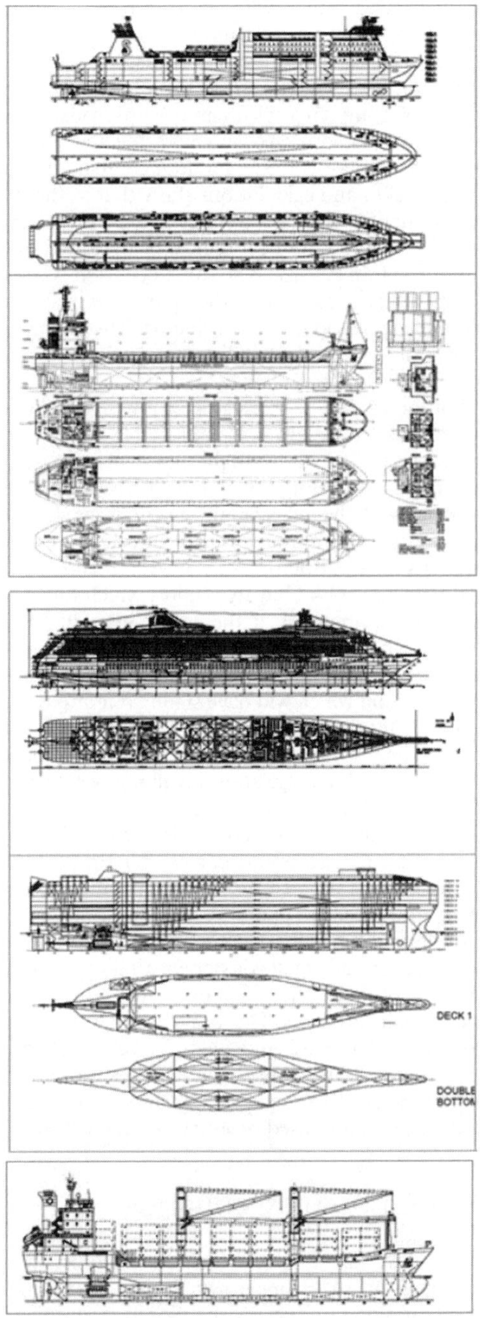

Figure 18. Test ships for study on impact of harmonised probabilistic regulations.

Assessment of the container feeder shows that the damage stability is dominant in the lower half of the draught range. Limiting KG curves for existing and new regulations are quite similar and the existing ship complies with cargo ship probabilistic regulations. It was then found that the existing ship also complied with the new probabilistic regulations. This came as a surprise, since it had been thought that ships complying with cargo ship probabilistic regulations would not comply with the new regulations. This container feeder has four cargo holds and a double skin and calculations show that at the full load draught d_s, damage to either of the two forward container holds or to the foremost of the after holds would result in capsize. The same occurred when a 2-compartment damage to the after container hold and the engine room was considered. These conclusions indicate new probabilistic regulations to be a poor survivability standard.

For the small general cargo ship it was found that the existing cargo ship probabilistic regulations are dominant for all the draught range. The existing ship, fitted with a double skin but with a single hold, complied with the old rules and calculations showed that the limiting *KG* according with new and old rules is similar. A design variant was then considered with single hull but two holds, which failed to meet the requirements of the new regulations. This seems to indicate that new ships of this type may be forced to fit double skins even if they have more than one hold which will have significant impact on functionality and cargo capacity. In a similar way to the container feeder, it is alarming to see how many of the full load damage cases involving penetration of the hold(s) still result in capsize.

In general, it was concluded that for the Panamax cruise ship and car carrier the existing SOLAS90 and the new MSC.194(80) rules do not result in equivalent operational envelopes for the basis ship and new design. For large ro-pax, container feeder and small coaster existing regulations and new probabilistic regulations result in similar operational envelopes. This indicates that the operational envelopes are equivalent for vessels already using a probabilistic approach (cargo ships) or for ship types (Ro-Ro) used in the formulation studies of the new harmonised probabilistic regulations.

5 TIME DOMAIN SIMULATION

In complex ships, such as passenger ships, there are numerous situations in which the intermediate stages of flooding have to be closely scrutinized. Typical situations include the progressive flooding when pipes, ducts or tunnels are within the assumed extent of damage and allow the flooding of other ship spaces.

In the new probabilistic rules damage cases of instantaneous flooding in unrestricted spaces do not require intermediate stage flooding calculations. However, damage cases of progressive flooding require intermediate stages of flooding calculations. These must reflect the sequence of filling as well as the filling level phases. Calculations should then be performed in stages, each stage comprising at least two intermediate filling phases in addition to the final phase.

In this task, numerical models of flooding in the time domain can be quite useful. In fact, different computer codes have been developed capable of simulating in the time domain progressive flooding of ships.

Santos, et al. (2002) and Santos and Guedes Soares (2007) presented one such code. This code is capable of simulating the flooding of any ship type and arrangement and can predict the intermediate stages of flooding. Openings between compartments or in the ship's hull are modelled as "missing" panels. The numerical model is capable of simulating transient phenomena such as:

- Progressive flooding (assessed now using method in MSC.245(83).
- Cross-flooding.
- Down-flooding.

The equations describing the ship motions are the following:

$$\sum_{j=2}^{6}(M_{ij}+A_{ij})\ddot{X}_j(t)+B_{ij}\dot{X}_j(t)+F_j(t)=F_i^E(t)$$
$$+F_i^{AC}(t)$$

for i = 2,…,6, where:

Mij represents the mass matrix,
Aij and *Bij* represent the radiation coefficients,
Fi represents nonlinear hydrostatic forces,
FiE represents the wave excitation forces (in the future),
FiAC represents the floodwater forces.

The forces and moments acting in the ship as a result of floodwater are calculated under the following assumptions:

- Water flows in a quasi-static way inside each compartment.
- Waterplane in each flooded compartment is flat and horizontal in each instant.

The forces and moments caused on the ship by the floodwater can be calculated using:

$$F_3^{AC} = \sum_{j=1}^{Ncomp} M_{wj}\rho_s g$$

$$M_4^{AC} = \sum_{j=1}^{NComp} M_{wj}\rho_s g(y_G - y_{Mwj})$$

$$M_5^{AC} = \sum_{j=1}^{NComp} M_{wj}\rho_s g(x_G - x_{Mwj})$$

where *Mwj* represents the volume of water in the *j* flooded compartment and *Ncomp* represents the number of flooded compartments. Floodwater is calculated using an hydraulic model.

The theoretical model can be applied in the time domain simulation of the flooding of ships. One application can be its use in damage control operations and in training for such operations. This allows evaluating, for example the flooding of the bow of a warship. In this case, progressive flooding of five bow compartments is evaluated. It may be seen in Figure 19 that compartment 13 is flooded extremely rapidly. Compartments 3 and 4 start flooding after 3s and 7s. Compartment 9 starts flooding after 45s.

Varela, Santos and Guedes Soares (2003) have shown that virtual reality models in connection with time domain simulation models allow objects to be controlled inside a damaged ship, permitting the limitation of flooding (Fig. 20). It is possible to open and close at any given time:

- Damage openings.
- Watertight doors.
- Watertight hatches.

This allows the training in damage control situations.

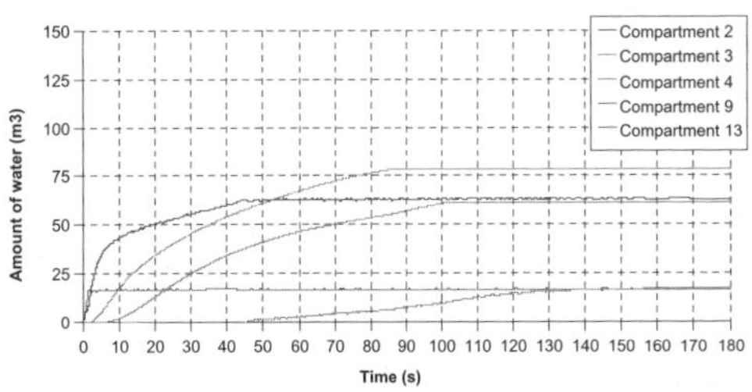

Figure 19. Progressive flooding of 5 compartments in bow of warship.

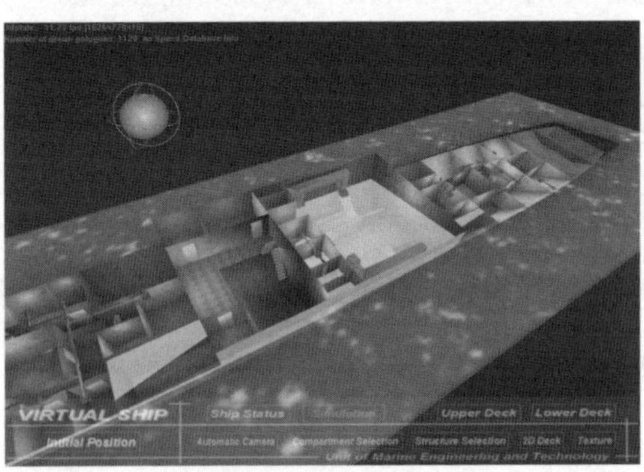

Figure 20. Flooding of a machinery space.

6 OPTIMIZATION OF SHIP SUBDIVISION

The new probabilistic methods for calculating damage stability lend themselves very easily to the optimization of ship subdivision. However, the ship designer, when optimizing a ship's subdivision has to consider multiple, often conflicting, objectives:

- Maximum attained subdivision index.
- Maximum cargo capacity of the main deck.
- Maximum cargo capacity of the lower deck.
- Minimum hull structural weight.

The optimization problem considered in studies carried out by Santos and Guedes Soares (2002, 2005b) was to maximize the attained index in order to increase safety without compromising economics. In practice, the optimization problem would be the minimization of the attained index under a constraint for A>R.

Figure 21 shows the methodology adopted for ship subdivision optimization. The optimization is carried out using FRONTIER v2.5.0 software. The damage stability calculations (attained subdivision index) are carried out using the GHS v8.00 software.

The hull and subdivision numerical models are created using a specially dedicated component of the damage stability software (Partmaker). Different commercial software interacts using a custom made program which:

- Controls the creation of the hull and subdivision numerical models.
- Calculates the cargo capacities.
- Calculates the loading conditions.
- Controls the calculation of the attained

Figure 22 shows the implementation using Frontier of a passenger ro-ro ship subdivision

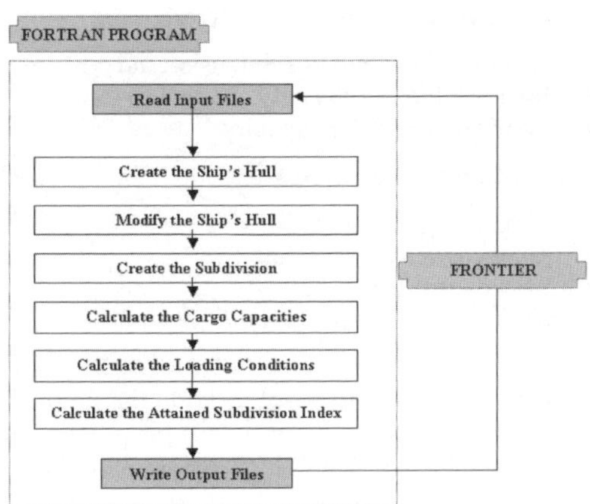

Figure 21. Methodology for ship subdivision optimization.

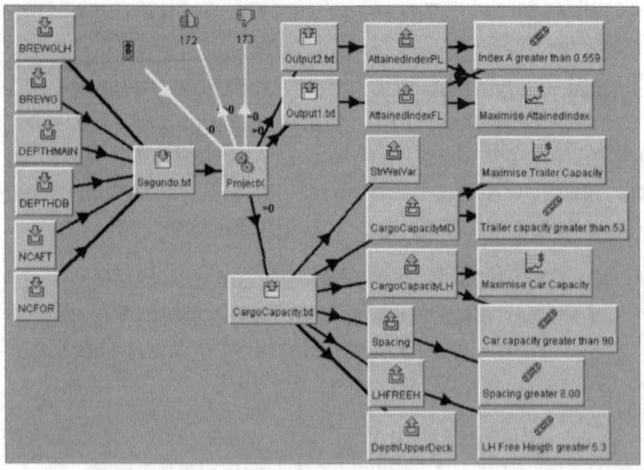

Figure 22. Implementation of ship subdivision optimization.

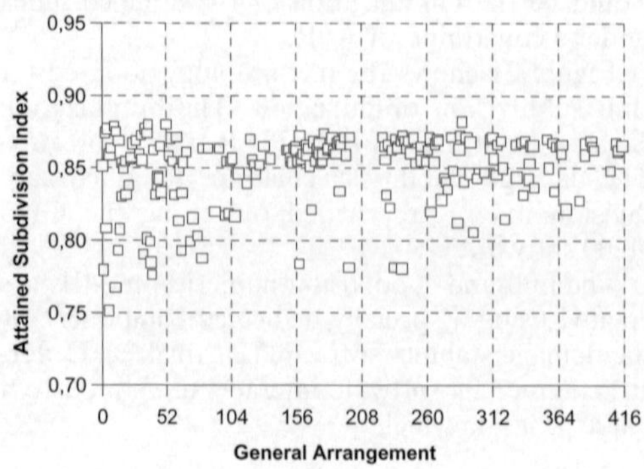

Figure 23. Attained subdivision index.

optimization problem. Design variables considered are the breadth of upper and lower side casings, depths to the main and upper deck and number of compartments forward and aft of machinery spaces.

Objective functions considered are:

- Maximize the attained subdivision index.
- Maximize the car capacity.
- Maximize the trailer capacity.

Constraints considered are:

- Index A larger than 0.559.
- Trailer capacity larger than 53.
- Car capacity larger than 90.
- Spacing of bulkheads larger than 8 m.
- Lower hold free height greater than 5.3 m.

Figure 23 shows the evolution of the attained subdivision index as the optimization progresses. It may be seen that A index increases as more ship designs are evaluated.

This study allowed the conclusion that the attained subdivision index has a modest increase of only 0.02. Cargo capacities of the ro-ro decks remain unchanged. The number of compartments remains unchanged. The height of the double-bottom decreases and the depths to the main and upper deck increases. The breadth of wing compartments in main deck increases slightly and the breadth of wing compartments in the lower hold decreases slightly.

7 SHIP SURVIVABILITY IN DAMAGE CONDITION

Calculating the global survivability of the ship involves uncertainties: loading condition, longitudinal location of damage, extent of damage, sea state. Santos and Guedes Soares (2002) have shown how Monte Carlo simulation can be used to calculate ship survivability in damage condition.

Ship's survivability is evaluated in terms of the critical sea state, using the Static Equivalent Method (SEM), well suited for Ro-Ro ships. This method is used when the main vehicle deck is flooded for the current damage condition. The critical sea state (significant wave height) is given by either of the two formulae:

$$H_s = \left(\frac{h}{0.085}\right)^{1/1.3} \qquad \mathrm{H_s = 6.7h - 0.8f - 0.9}$$

For damage cases not involving flooding of the main deck, a formula has been used depending on the traditional damage condition stability parameters, *GZmax* and range of stability.

The overall probability of survival of a ship depends on the probabilities of different damage conditions. These can be derived from the probability distributions used to derive the formulae in the new harmonized probabilistic regulations.

Figure 24 shows that the evaluation of ship global survivability involves the following steps:

1. Generate loading condition according with probability distributions.
2. Generate damage condition according with probability distributions.
3. Generate sea state.
4. Apply SEM or other method to evaluate critical sea state.
5. Check if critical sea state above or below actual sea state.
6. If sea state above critical level, loss will occur.

The first three steps allow the definition of the accident scenario while the last three include the actual evaluation of the ship's survivability in these specific scenarios.

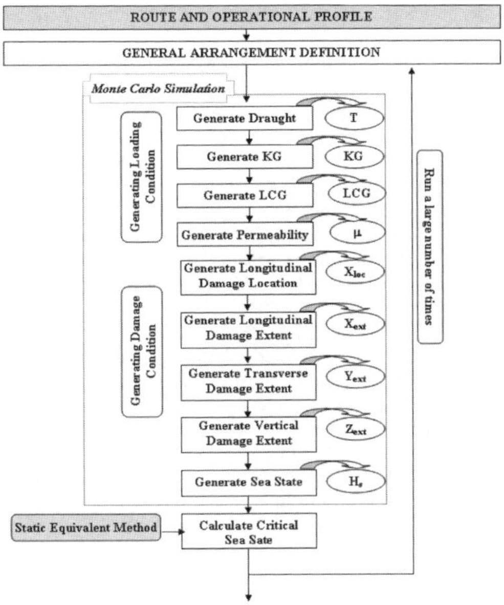

Figure 24. Methodology for ship survivability evaluation.

This methodology has been used to calculate passenger ro-ro ship and fast ferry survivability by Santos and Guedes Soares (2005a, 2005b).

Two different sea state distributions have been considered in these studies: the IMCO distribution and the North Sea December-February distribution. For the IMCO sea state distribution no capsizes are recorded. For the December-February sea state distribution, 13 capsizes are recorded for the low position of the centre of gravity. With the centre of gravity higher, the number of capsizes is much higher: 68.

These values correspond to between 0.01 and 0.07 probabilities of capsize. However, this relates to extreme conditions in the North Sea, meaning that these can be the probabilities of loosing this particular passenger ro-ro ship in case it sustains a damage involving flooding in the North Sea in winter.

8 CONCLUSIONS

The probabilistic method has now completely replaced the deterministic methods in the design of ship's subdivision. This method is based on sound theoretical principles which allow taking into account the numerous uncertainties that arise regarding ship survivability in damage condition. The different probabilistic regulations have now been harmonized into a single probabilistic set of regulations. Formulations have been updated with the latest statistical results and tank test results. The probabilistic method is now mandatory for cargo and passenger ships above 80 m. The new probabilistic regulations have proved to be difficult to implement in reliable computer programs and its consequences in the design of different ships types remain uncertain.

Probabilistic methods such as the Monte Carlo simulation can also be of use in damage ship stability since it allows taking in consideration various uncertainties to calculate the overall ship survivability. Damage statistics already used to support harmonized regulations development can be used in this context.

The probabilistic index of subdivision also lends itself very easily to ship subdivision optimization at the design stage. Moderate increases in the attained index may be expected by changing the location of bulkheads, number of compartments and depths to main decks.

Finally, the new harmonized regulations demand in the calculation of the probability of survival a careful consideration of the intermediate stages of flooding, which should be studied using time domain simulation instead of empirical approaches.

ACKNOWLEDGMENTS

This paper presents part of the teaching material prepared and used in the scope of the course "Probabilistic Approach to Damage Stability" within the project "Advanced Ship Design for Pollution Prevention (ASDEPP)" that was financed by the European Commission through the Tempus contract JEP-40037-2005.

REFERENCES

Comstock, J.P. & Robertson, J.B. (1961), "Survival of Collision Damage Versus the 1960 Convention on Safety of Life at Sea", Transactions of SNAME.

IMO (1995), "Regional Agreements on Specific Stability Requirements for Ro-Ro Passenger Ships", Resolution 14 of SOLAS Diplomatic Conference, International Maritime Organization, London, UK.

IMO-Assembly (1991), "Explanatory Notes to the SOLAS Regulations on Subdivision and Damage Stability of Cargo Ships of 100 Metres in Length and Over", A.684(17), International Maritime Organization, London, UK.

IMO-MSC (1990), "Adoption of Amendments to the International Convention for the Safety of Life at Sea", MSC.19(58), International Maritime Organization, London, UK.

IMO-MSC (2005), "Adoption of amendments to the International Convention for the Safety of Life at

Sea, 1974, as Amended", MSC.194(80), International Maritime Organisation, London, UK.
IMO-MSC (2006), "Adoption of amendments to the International Convention for the Safety of Life at Sea, 1974, as Amended", MSC.216(82), International Maritime Organisation, London, UK.
IMO-MSC (2007), "Interim explanatory Notes to the Solas Chapter II-1 Subdivision and Damage Stability Regulations", MSC.1/Circ.1226, London, UK.
IMO-SLF (2002a), "Investigations and Proposed Formulations for the Factors 'p', 'r' and 'v': the Probability of Damage to a Particular Compartment or Compartments", SLF45/3/5.
IMO-SLF (2002b), "Investigations and Proposed Formulations for the Factor 's': the Probability of Flooding After Damage", SLF45/3/3.
IMO-SLF (2002c), "Selection of Sample of Ships for the Evaluation of Required Subdivision Index and Procedure for Calculating R", SLF45/3/4.
IMO-SLF (2004a), "On the Effect of Draught on the Global Attained Index—Dry Cargo and Passenger Ships", SLF47/3/2.
IMO-SLF (2004b), "Sample Ship Recalculation Results", SLF47/3/3.
Lutzen, M. (2002), "Damage Distributions", HARDER EU project report 2-22-D-2001-01-4.
Maritime and Coast Guard Agency (2007), "Assessment of the Impact of the New Harmonized Probabilistic Damage Stability Regulations (SOLAS2009) on the Subdivision of New Passenger and Dry Cargo Ships", Final report of RP552.
Santos, T.A. & Guedes Soares, C. (2002), "Probabilistic Survivability Assessment of Damaged Passenger Ro-Ro Ships Using Monte-Carlo Simulation", *International Shipbuilding Progress*, Volume 49, N°4, pp. 275–300.
Santos, T.A. & Guedes Soares, C. (2005a), "Monte Carlo Simulation of Damaged Ship Survivability", *Journal of Engineering for the Maritime Environment*, Number M1, March, pp. 25–35 (11).
Santos, T.A. & Guedes Soares, C. (2005b), "Multi-objective Optimization of Fast Ferry Watertight Subdivision", in *Maritime Transportation and Exploitation of Ocean and Coastal Resources*, C. Guedes Soares, Y. Garbatov, N. Fonseca (Eds.), Taylor & Francis Group, London, UK, Vol. 1, pp. 893–900.
Santos, T.A. & Guedes Soares, C. (2007), "Time Domain Simulation of Ship Global Loads due to Progressive Flooding", in *Advancements in Marine Structures*, Guedes Soares, C., Das, P.K. (eds), Taylor & Francis/Balkema, pp. 79–88.
Santos, T.A., Winkle, I.E. & Guedes Soares, C. (2002), "Time Domain Modeling of the Transient Asymmetric Flooding of RoRo Ships", *Ocean Engineering*, 29, pp. 667–688.
Varela, J.M., Santos, T. & Guedes Soares, C. (2003), "Simulation of Fluid Dissemination in a Virtual Reality Environment Onboard the Ship", *Proceedings of COMPIT2003*, Hamburg, Germany, May.
Vassalos, D., Turan, O. & Pawlowski, M. (1997), "Dynamic Stability Assessment of Damaged Passenger/Ro-Ro Ships and Proposal of Rational Survival Criteria", *Marine Technology*, vol. 34, n° 4, October, pp. 241–266.
Vassalos, D. & Papanikolaou, A. (2002), "Stockholm Agreement—Past, Present and Future (Part1 & Part2)", *Marine Technology*, Vol. 39, N°3/4, July/October, pp. 137–158/199–210.
Volkov, B.N. (1963), "Determination of Probability of Ship Survival in Case of Damage", *Sudostrjenie*, N.° 5, pp. 4–8.
Wendel, K. (1968), "Subdivision of Ships", *Proceedings of the 1968 Diamond Jubilee Int. Meeting*—75th Anniversary, SNAME, New York, paper N.° 12.

Advanced Ship Design for Pollution Prevention – Guedes Soares & Parunov (eds)
© 2010 Taylor & Francis Group, London, ISBN 978-0-415-58477-7

Aircraft carrier stability in damaged condition

V. Slapničar
Faculty of Mechanical Engineering and Naval Architecture, University of Zagreb, Zagreb, Croatia

B. Ukas
Saipem S.p.A., Milano, Italy

ABSTRACT: Building today's aircraft carriers is the culmination of science and the art of shipbuilding for maritime nations with this type of vessel in its naval fleet. The wartime role of aircraft carriers during the 20th century evolved rapidly, so in less than 50 years the aircraft carrier has changed from an auxiliary warship of limited combat value, into the most important and valuable genre of warship. As aircraft carriers grew in importance, so did the need to establish and improve the stability criteria, with special attention to determining the stability criteria after sustaining damage below the water line. Stability of aircraft carriers in damaged condition is vitally important from the standpoint of the retention of both operational and combat capabilities. The fundamental purpose of this paper is to verify the capability of an aircraft carrier after sustaining damage with regard to maintaining sufficient stability to be still operational and combat capable i.e. to withstand prevailing wind and seas in damaged condition. The results of these calculations were presented in the form of regression equations with an R^2 value of between 0.9 and 1, giving the possibility to use them for the fast and efficient prediction of the expected equilibrium position of the damaged aircraft carrier with a flooded compartment, or group of adjacent compartments.

1 INTRODUCTION

Aircraft carrier is the most important warship with high-speed and a large action radius. Hull is identical to the shape of any other type of large warship. It has a large and spacious deck with variety of devices and equipment for takeoffs and landing various types of aircrafts. Interior is organized in way that it can accommodate all aircrafts, equipment, weapons, fuel, and about 2,000 crew members. Aircraft carrier strike force is the main part of every major Navy's, especially those who have pretensions to control the waterways around the world's oceans. Aircraft carrier designs features results directly from the experience gained during the war operations, particularly the damage control ability which can be described with the following features:

- Watertight bulkheads must have the full integrity with as much as possible watertightness above the highest water line that the ship can reach after the damage occurs by hostile action. At the same time the watertight bulkheads must limit and prevent the spreading of fire, since from the war experience the fire is the second cause of loss.
- Longitudinal subdivision, with a rare set of transverse watertight bulkheads proved to be an extremely dangerous because it caused the occurrence of very large transverse angles after flooding, which has regularly provoked the abandonment of the vessel, although the danger of sinking due to loss of reserve buoyancy did not exist.
- Engine and power plants on aircraft carrier must be well protected and standardized, and marine electrical network capacity have to be significantly higher in order to enable the smooth supply of electricity of all equipment (e.g. radar, sonar).
- Uptakes and openings for ventilation must be designed as mutually separated independent systems. Systems must have simple and fast watertight closings to avoid that their openings became critical points.
- Experience gained during the war period revealed the need for serious consideration of the shock impact to hull caused by underwater no-contact explosion. Therefore, it is necessary to devote great attention to protection and elastic foundation for all important equipment, devices and machines.

2 RULES AND CRITERIA FOR AIRCRAFT CARRIER STABILITY IN DAMAGED CONDITION

Although the largest and most important ship in the fleet, an aircraft carrier is like the all other vessels exposed to the same external influences and risks that significantly affect her stability and buoyancy. Hazards to which the aircraft carrier is exposed are those that result with hull penetration, flooding the inner volume (grounding, explosion of torpedoes, mines, grains cannon, rocket projectiles), but also significant external influences such as an action of wind and waves.

After the flooding, according to Uršić (1991) we can observe penetrated water or as added weight or as lost displacement. Regardless of the observation method it is evident that the penetrated water affects the longitudinal and transverse stability that will increase ships draft, trim and heel.

In the case when one, two or more watertight compartments are flooded the ship will float on a new water line which must be less than the border line. Heel angle as a result of an unsymmetrical flooding must be less than the angles where new water line reaches points of progressive flooding i.e. when water line reaches openings which can not be water tightly closed. Also, heel angle must be not so large to avoid uncontrollable moving of various masses and to keep operability of machines and equipment as much as possible.

Besides all mentioned here the metacentric height (GM) and stability for large angles should be sufficient to prevent ship capsizing due to effects of winds and waves. All this implies that the stability of the aircraft carrier is one of the most important design requirements that the designers should dealt with.

As a result of testing the effects of explosives on structural models of war ships and war reports the specific models of damage have been developed which describe probability of survival of a warship. All of these results and the knowledge are the reason for leaving the concept of damage length and accepting the new concept of the allocating the watertight compartments (aircraft carrier has about 300 watertight compartments) as the basis for the determining capability of sufficient stability in damaged condition.

Parallel to the empirically established requirements most navy countries today slowly began to apply the most important parts of the IMO and SOLAS conventions regarding the safety and stability of damaged ships.

2.1 *About the stability of damaged ships*

According to Sarchin & Goldberg (1960) aircraft carrier belongs to the warship category that have side protection system which consists of several (usually five) longitudinal bulkheads. These, together with a number of transverse bulkheads form watertight compartments which are in the vertical direction bounded with strength deck (usually the hangar deck). This side-protective system extends to about 2/3 of the ship length and its main purpose is to minimize or to limit the flooded volume after damage.

As it will be explained later on this protection system increases the internal subdivision factor of aircraft carrier and secures operational characteristics when inner hull is flooded. Damage stability calculation has the purpose of determining minimum amount of intact stability that ship must have to survive damage. The new waterline should principally satisfy two basic considerations:

- Ship immersion should not exceed the border line with aim to prevent further flooding.
- Heel should not exceed certain values in order to keep the operating and seaworthiness characteristics.

When determining this new position i.e. the maximum allowable angle of heel after flooding (Russo & Robertson 1950) we distinguish symmetrical and unsymmetrical flooding of inner volume of the vessel.

2.1.1 *Symmetrical flooding*

During the symmetric flooding, according to Uršić (1991) the basic problem with which the aircraft carrier is facing with, is primary change of metacentric height (GM). Particularly fatal for stability is reduction of the metacentric height which is as much dangerous as the centre of gravity of penetrated water is higher. It follows that for the stability aspects it will be particularly adversely the case of penetration above watertight compartment i.e. watertight deck.

This is particularly dangerous in the case of fire extinguishing, especially in the area of hangar deck where large quantities of water are thrown into. It is necessary to ensure the efficient drainage of this large quantity of inserted water because due to these large and highly placed free surfaces of water ship can easily capsize.

2.1.2 *Unsymmetrical flooding*

Unsymmetrical flooding occurs when side longitudinal bulkheads are not damaged so the flooding occurs only at one side of the ship. Unsymmetrical penetration of water is particularly dangerous for ships with side-protective system (Sarchin & Goldberg 1960) since it cause the appearance of strong heeling moment and consequently a sudden and uncontrolled angle of heel with significant reduction of stability heeling levers (GZ).

This unsymmetrical flooding is extremely dangerous due to large angle of heel at which the side-protective system is in very poor position to resist further torpedo attack since the ship is exposed to damage below the side-protective system on the high side and above it on the low side. If this angle is not possible to remove rapidly it might lead to the abandonment of the ship much earlier than there is any danger of foundering or capsizing.

For this reason the aircraft carrier must be able to quickly and effectively reduce angle of heel and if possible completely neutralize it. This is usually done in such way that the ship is equipped with the system of cross-connected pipelines which are independent of the ballast system, and which connects side tanks, tanks in double bottom and flooding spaces within the side-protective system.

2.2 *Rules applicable to stability of aircraft carrier in a damaged condition*

Safety of aircraft carrier, and particularly the possibility that retains substantially for his operational and combat features after the damage, primarily depends on stability and the amount of reserve displacement.

Since internationally accepted rules for the stability of war ships do not exist (Sarchin & Goldberg 1960) like for civilian ships (IMO and SOLAS regulations) the navies had to determine, independently their own rules for stability in damaged condition. Rules considering damage standards are generally dealing with two basic requirements, i.e. design and operation.

Design requirements are based on an evaluation of the vessels ability to withstand major amount of damage and to continue to be sufficiently seaworthy and operational. Design requirements are established at the beginning of the process of designing and they significantly influence the selection of ships main dimensions, form, layout and number of watertight compartments.

Operational requirements are based on the minimum degree of offensive and defensive capability for continuing operations. These considerations include the capability of rapid list and heel corrections to the allowable value.

It is important to note that the rules for the stability of aircraft carrier in a damaged condition by no means guarantee that the vessel will not founder but its strict application can significantly increase the survival after being damaged.

When considering the damage stability rules for the aircraft carrier it is important to know standard loading conditions at which vessel can sail before damage occurs.

According to Sarchin & Goldberg (1960) loading conditions vary in the range from light conditions over the maximum loads in which the ship is departing from the port to the optimal combat load at sea. Standard loading conditions are as follows: basic condition, deep condition, light condition, light seagoing condition, light harbour condition and optimal combat condition. It is considered that the aircraft carrier is capable to survive several successive torpedo attacks.

Development of criteria regarding the size of damage that ship should survive it considered the range from flooding one, two or three adjacent compartments, based on a subdivision factor, while today the standard for aircraft carriers is based on a percentage length of a total length of the hull that can be damaged.

Therefore, the aircraft carrier should satisfy the following requirements for the extent of damage:

- For the vessels of waterline length greater than 92 m vessel must be able to retain significant amount of stability (defined as an area under the GZ curve) after damage occurs anywhere along its length extending 15% of the waterline length or 21 m whichever is greater the (MOD 2000, Brown & Chalmers 1989). This extent of damage usually affects 3–4 watertight compartments which is very similar to the request of the IMO (1974) Chapter II-1: Regulation 8.1.4., which applies to passenger ships.
- In the case of transverse propagation of damage, after the explosion (Keil 1961), it is allowed that the extent of damage comprises about 20% of the interior of the ship, and therefore aircraft carrier has a side-protective system.
- Vertical distribution of damage is not defined by standard but the according Keil (1961), it can be concluded that it should not exceed damage control deck, which is usually determined by the first deck above the constructive water line.

Based on these requirements and data from the Keil (1961) the extent of damage caused by a contact explosion creates an opening length of 9–15 meters depending on the power of explosives, structural design and the steel quality. It can be concluded that the aircraft carrier will be lost after receiving several consecutive explosive damages. With these requirements and data ship designer can determine the length and arrangement of the watertight compartments.

2.3 *Effect of wind and waves action on the ship in a damaged condition*

Standards of sufficient stability for the vessel in a damaged condition when exposed to the forces of wind and waves are based on comparison of righting lever curve and wind heeling lever curve.

Wind heeling lever curve moment caused by wind and waves for the ship in damaged condition (Sarchin & Goldberg 1960) is calculated by the following formula:

$$\text{Wind heel lever} = \frac{0.004 \times V_{Wind}^2 \times A \times l \times \cos^2\theta}{2240 \times \Delta} \quad (1)$$

where V_{Wind} = normal wind velocity which damaged ship should withstand in knots; A = projected sail area in square feets; l = lever arm from half draft to centroid of sail area in meters; Δ = displacement in tonnes; and θ = angle of heel in degrees.

Wind speed that damaged ship must sustain without risk of capsizing is (MOD 2000):

$$V_{Wind} = 22.5 + 0.15 \times \sqrt{\Delta} \tag{2}$$

where V_{Wind} = wind velocity in knots; and Δ = displacement (deep) in tonnes.

2.4 *Damage stability criteria*

Requirements for damage stability criteria are as follows (MOD 2000):

- Angle of list and loll must be less than 20°.
- The amount of GZ at the intersection of righting lever curve and wind heeling curve must be less than 60% of the maximum GZ (point C).
- Area A_1 to be greater than the value given from:

$$A_{min.} = 0.164 \times \Delta^{-0.265} \text{ in mrad} \tag{3}$$

- where Δ = displacement (deep) in tonnes.
- Area A_1 must be greater than the area $1.4 \times A_2$.
- Longitudinal trim must be less than that required to cause downloading.
- Metacentric height must be greater than 0 and is calculated according to St. Denis (1966):

$$\overline{GM} \cong \left(\frac{100}{80000 + \Delta} \right) \times B^2 \tag{4}$$

where B = ship breadth in meters; and Δ = displacement in tonnes.

Figure 1 shows the requirements for damage stability criteria.

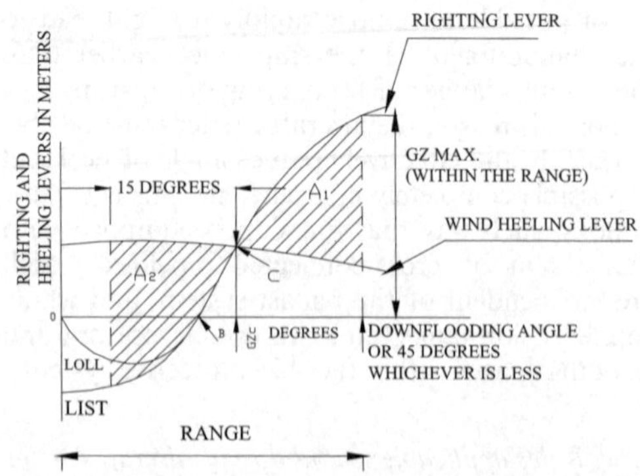

Figure 1. Damage stability criteria of aircraft carrier (MOD 2000).

2.5 *Operational criteria for aircraft carrier*

Sarchin & Goldberg (1960) give a basic criteria that aircraft carrier has to fulfil in the event of flooding one or more compartments. These operational criteria mostly determine the angle of heel where the vessel can be fully or partly operative:

- All installed equipment, regardless of the purpose must be so designed and installed to be able to be fully effective at a constant heel (transverse or longitudinal) of 15°.
- All machines (main and auxiliary) must be fully operational at a constant heel (transverse or longitudinal) up to 15°.
- To ensure survival of the vessel and to increase toughness, equipment and machines have to be able to function some time under the heel of 20° ÷ 25°.
- Upper heel limit of 20° is allowed for withdrawing the damaged vessel from the zone of combat actions.
- The amount of heel angle of 5° is limit for conducting a full scale air activity (landing/take-off) while the heel angle of 8° stands for limited air activity.
- Exceptionally in the case of helicopter operations the heel angle to 12° may be allowed.
- Heel angle (transverse and longitudinal) up to 5° is considered to be safe. It is necessary that the aircraft carrier has such an arrangement of ballast tanks which allows rapid and effective correction of the inclination from 20° to 5° or less.
- Any angle greater than 20° is critical and in this case the order will be placed to abandon the vessel.

3 SELECTED AIRCRAFT CARRIER MODEL FOR CALCULATION

At the beginning of consideration of this issue, it was necessary to choose the kind and type of aircraft carrier for which the damage stability calculation will be carrying out. US aircraft carrier Yorktown (Essex class) from the period of the Second World War was chosen due to reasons of availability of data and literature since the necessary information about the newer generation of aircraft carriers is generally deficient and unavailable. It should be emphasized that the procedure applied in this paper can be applied to newer generation of aircraft carriers as well.

When the 1930th The U.S. Navy ordered the construction of a new type of aircraft carrier it was the first purpose built aircraft carrier class in history. Previously, the navies were mainly engaged with reconstruction of existing battleships or battle cruisers to aircraft carriers. This purposely built aircraft carrier class was given name Essex according to the first ship of the series of four completely identical units. The second ship of that class was the aircraft carrier Yorktown (CV-5).

Aircraft carrier Yorktown (CV-5) was armed with 8 guns of 127 mm for close defence, 16 guns of 28 mm anti-aircraft defence and 30 guns of 20 mm. She had anti-aircraft defend system, carried 71 aircrafts and the crew consisted of 2217 members. Figure 2 shows the general plan of the aircraft carrier York town (CV-5). On the picture it can be seen the interior layout with tanks, accommodation for crew, aircraft hangar, as well as the position and mutual distance of transverse watertight bulkheads.

3.1 *Analysis of selected aircraft carrier*

Selected aircraft carrier is one of the first modern aircraft carriers designed to carry a large number of different aircrafts. Analysis of the selected carrier shows the typical layout for this type of ship. The aim of this analysis is to briefly describe all vital parts of the aircraft carrier that are interesting for the damage stability consideration.

3.1.1 *Subdivision*

The main purpose of aircraft carrier watertight subdivision is to improve combat efficiency and reduce the risk of loss when the outer shell is damaged and interior been flooded. Watertight bulkheads contribute to a better use of space dividing the various vessel activities while in the same time they reinforce the structure. Watertight subdivision increases the ability of aircraft carriers to overcome the effects of underwater damages and thus to control the amount of flooding. Also, enables control of the vertical position of gravity of flooded water, quantity of water, moment of inertia of the free surface and spreading the fire and toxic gases.

Table 1. Main particulars of aircraft carrier Yorktown CV-5.

Length over all	L_{oa} = 247.00 m
Length between perpendiculars	L_{pp} = 220.40 m
Length on WL	L_{KVL} = 231.70 m
Length of flight deck	L_1 = 245.00 m
Breadth of hull	B = 29.96 m
Breadth of flight deck	B_1 = 33.38 m
Draught	T = 7.00 m
Height of hull	H = 20.85 m
Total height of vessel	H_{UK} = 34.75 m
Height of aircraft hangar	H_{hangar} = 5.48 m
Displacement	Δ = 30375.36 t
Block coefficient	C_B = 0.610
Speed	V_{max} = 32.5 kn
Action radius	N = 12000 nm at 15.0 kn
Main engine	4 × Parsons steam turbins
Main engine power	P = 89500 kW
Longitudinal centre of gravity	LCG = 106.162 m
Vertical centre of gravity	VCG = 12.656 m
Longitudinal centre of buoyancy	LCB = 106.162 m
Vertical centre of buoyancy	VCB = 3.879 m

Structure analysis shows that the selected aircraft carrier has 17 transverse and 4 longitudinal watertight bulkheads. Longitudinal watertight bulkheads are extending the entire length of the vessel and the transverse watertight bulkheads are extending the entire width from bottom up to hangar deck. In this way the carrier is divided into a number of watertight compartments which have the task to limit the flooding after damage. Strength of bulkheads is calculated to withstand the pressure load for deeper draft in damaged condition.

According to the MOD (2000) it is allowed to have a watertight bulkhead openings closed with watertight doors only above damage control deck while below this deck they must be compact. All horizontal communication between the watertight compartments is managed above the damage control deck while the vertical communication is provided for each compartment separately.

We have to keep in mind that the aircraft carrier combat effectiveness grows proportionally with the number of underwater hits that she can survive and resist but this is not the only measure of combat efficiency. Efforts that carrier can withstand most physically possible attacks can negatively affect her offensive striking power. Combat effectiveness is not satisfactory if it fails after the only one underwater hit. So, the all watertight subdivision does virtually no purpose. As a minimum allowable resistance standard it is considered when the vessel can withstand at least one successful underwater hit which crashes one watertight bulkhead that causes flooding of two adjacent sections.

3.1.2 *Side-protective system*

Beside the densely set of watertight subdivision basic characteristic of aircraft carrier is the side-protective system which is located inside the hull on each side. Side-protective system consists of several (usually 4) watertight layers of which 2 outer are mostly kept empty and freely to flooding and 2, which serve as the internal tanks for various liquids (e.g. heavy fuel, kerosene, oil, water). This side-protective system extends to the 2/3 of the ship length and protects from the underwater damage vital parts of vessel (e.g. engine, power

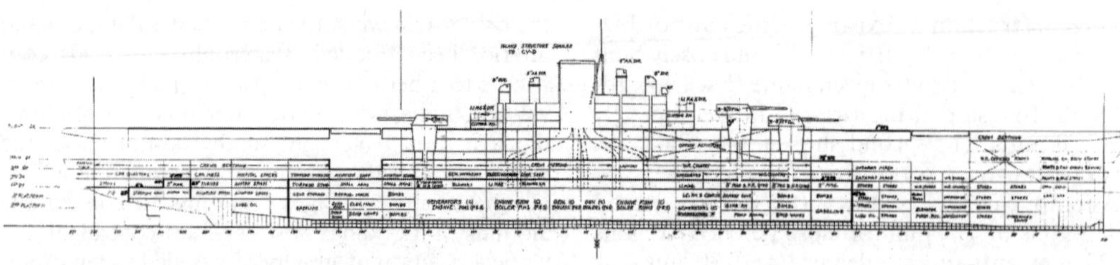

Figure 2. General plan of aircraft carrier Yorktown (CV-5).

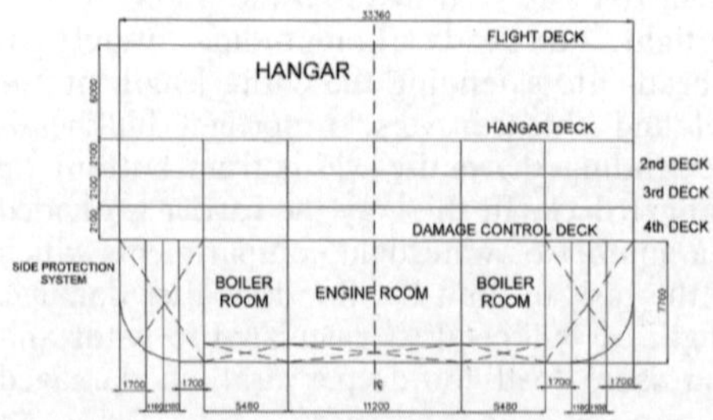

Figure 3. Cross section of aircraft carrier, Yorktown (CV-5).

station, ammunition warehouses, fuel, supplies). Figure 3 shows the cross section of aircraft carrier while the Figure 2 shows the distribution of the length of the side-protective system in longitudinal direction.

3.2 *The equilibrium position of the aircraft carrier after flooding*

In this subsection the damage simulation of the selected aircraft carrier model is performed and a new equilibrium position after flooding of inner volume is determined. It is assumed that the explosion of torpedo or underwater mine in all examined cases does not cause the damage beyond the deck 4 i.e. damage control deck that vertically limits the amount of flooding.

Such damage causes sudden unsymmetrical flooding and the carrier emerges to the new water line so the total displacement is equal to the total weight of the aircraft carrier. But on this new water line direction of displacement and weight forces are not on the same line vertically placed on water line so the heeling moment appears and consequently vessel has certain angle of heel.

Damage stability calculation will be carried on taking into account the maximum allowed amount of wind that carrier must withstand. Area between GZ curve and the wind heeling moment curve indicated as $A1$ is calculated to a total heel of 30°. Larger angle of heel will probable led to progressive flooding (Figure 1).

4 DAMAGE STABILITY CALCULATION OF AIRCRAFT CARRIER

Damage stability calculation of aircraft carrier with determining the new equilibrium is done using the program GHS (2005) (Figure 4). Simulation is performed for the damage of one, two, three and four adjacent compartments starting from stern to bow using the following commands (Figure 5): WIND, WIND HMMT, TYPE (compartment name) FLOODED, SOLVE, RAH and RAH/AREA.

The mechanism of damage in all cases of flooding is the same and it assumes that the explosion destroys completely side-protective system in adjacent compartments, damages transverse and longitudinal watertight bulkheads which leads to flooding of boiler room and the side ballast tanks in double bottom (Figure 6). It must be emphasized that this damage combination is valid for those compartments which have side-protective system while in the case of bow and stern compartments where there is no side-protective system, explosion penetrates longitudinal watertight bulkhead which leads to symmetrical flooding.

The volume of damage assumes that the torpedo or mine explosion happens at the moment of contact with hull. For such definition of flooded compartments the new equilibrium position is determined taking into account the maximum amount of allowable wind that blows from the left side of the ship. Table 2 presents calculation results for the equilibrium for one compartment flooding depending on the longitudinal location of damage x_C.

4.1 *One compartment flooding*

Simulation of one compartment flooding is done starting from stern to bow (Figure 5a). The basic assumption is that the aircraft carrier is hit by one torpedo which affects the considered compartment

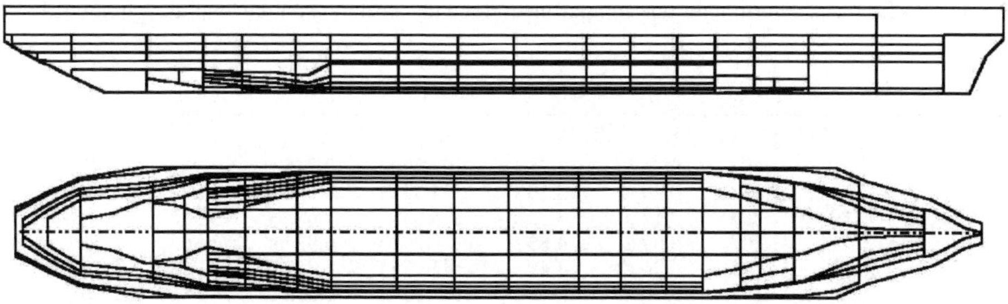

Figure 4. Geomteric model in GHS of Yorktown (CV-5).

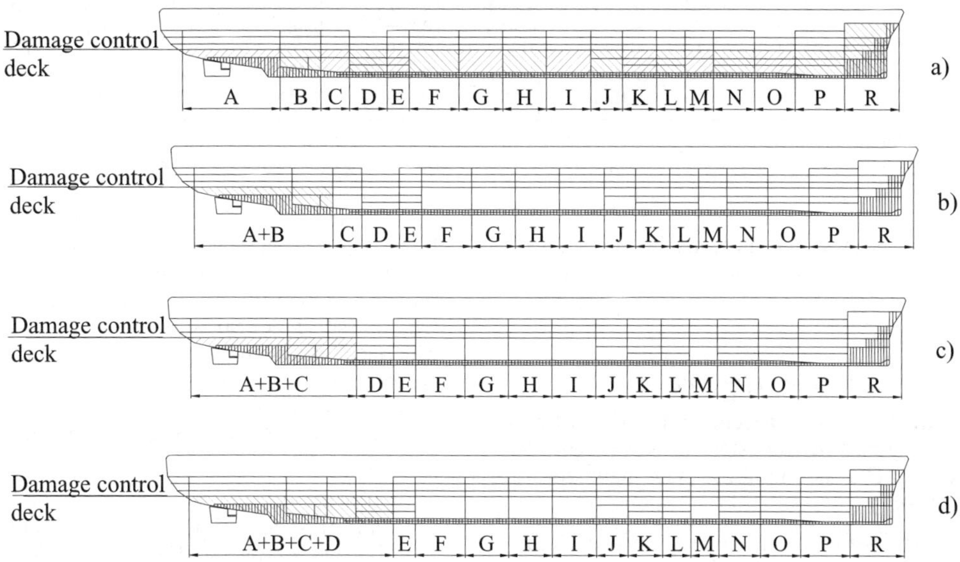

Figure 5. Longitudinal flooding sheme (one (a), two (b), three (c) and four (d) compartments).

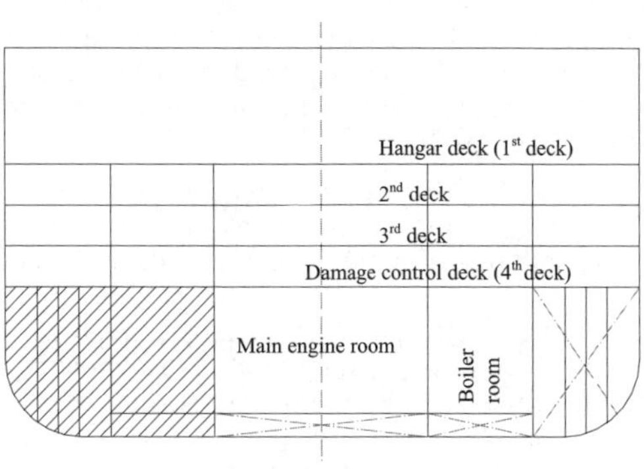

Figure 6. Transverse flooding scheme of compartments.

in the middle while surrounding transverse bulkheads are not damaged. Example of the one compartment flooding calculation is shown in the Table 2.

4.2 *Two compartment flooding*

Simulation of two adjacent compartment flooding is done starting from stern to bow (Figure 5b).

Following two assumptions were made for the calculation:

- Aircraft carrier is hit by one torpedo near the transverse watertight bulkhead which leads to its collapse and flooding of two adjacent compartments.

Table 2. Results for the equilibrium for one compartment flooding depending on the longitudinal flooded location.

Flooded compartment	x_C [m]	T [m]	ψ [°]	φ [°]	LCB [m]	TCB [m]	VCB [m]	GM [m]	A_l [m-rad.]
1	2	3	4	5	6	7	8	9	10
A	2.90	7.603	0.26 a	1.28 s	103.799 f	0.245 s	3.966	2.418	0.2883
B	26.47	7.471	0.20 a	3.10 s	104.325 f	0.591 s	3.971	2.488	0.3795
C	37.80	7.283	0.12 a	3.52 s	105.055 f	0.673 s	3.952	2.441	0.3594
D	48.78	7.207	0.09 a	4.83 s	105.297 f	0.924 s	3.966	2.416	0.3233
E	58.67	7.145	0.06 a	4.48 s	105.575 f	0.858 s	3.953	2.397	0.3370
F	70.48	7.224	0.10 a	8.19 s	105.155 f	1.552 s	4.085	2.605	0.2854
G	85.97	7.123	0.04 a	7.36 s	105.693 f	1.396 s	4.051	2.540	0.1946
H	100.30	7.045	0.00	7.00 s	106.078 f	1.330 s	4.037	2.510	0.2010
I	115.31	6.940	0.06 f	8.56 s	106.539 f	1.618 s	4.100	2.603	0.2803
J	127.59	6.946	0.05 f	5.22 s	106.556 f	0.997 s	3.973	2.408	0.3230
K	138.51	6.897	0.08 f	5.58 s	106.796 f	1.065 s	3.985	2.418	0.2187
L	149.43	6.889	0.08 f	4.86 s	106.829 f	0.929 s	3.963	2.391	0.3203
M	158.30	6.842	0.13 f	4.29 s	107.220 f	0.816 s	3.969	2.411	0.3375
N	169.89	6.710	0.23 f	4.44 s	108.065 f	0.840 s	4.003	2.488	0.3475
O	183.54	6.642	0.29 f	1.32 s	108.597 f	0.249 s	3.982	2.355	0.2916
P	198.21	6.890	0.08 f	1.33 s	106.862 f	0.256 s	3.905	2.305	0.2834
R	214.11	6.960	0.03 f	1.38 s	106.399 f	0.265 s	3.890	2.281	0.2757

Two torpedoes independently hit two adjacent compartments at their centre which caused their flooding.

4.3 *Three compartment flooding*

Simulation of three adjacent compartment flooding is done starting from stern to bow (Figure 5c). Following two assumptions were made for the calculation:

- Aircraft carrier is hit with two torpedoes, first one which hits the hull girder near the transverse watertight bulkhead which leads to its collapse and flooding of two adjacent compartments. Second one torpedo hits adjacent compartment to these two already flooded compartments in the centre making a group of three completely flooded compartments.
- Three torpedoes independently hit three adjacent compartments at their centres which caused flooding of three adjacent compartments.

Four compartment flooding.

Simulation of four adjacent compartment flooding is done starting from stern to bow (Figure 5d). Following three assumptions were made for the calculation:

- Aircraft carrier is hit with two torpedoes, where each hits one non adjacent transverse watertight bulkhead. This leads to flooding of two pairs of adjacent compartments making a group of four adjacent completely flooded compartments.
- Four torpedoes independently hit four adjacent compartments at their centres which caused flooding of four adjacent compartments.
- One torpedo hits near the watertight bulkhead which collapses and causes flooding of two adjacent compartments and the other two torpedoes hit other two adjacent compartments at their centres which caused totally flooding of four adjacent.

Graphical presentation for all groups of flooding for heel and trim depending on longitudinal location of damage is shown on Figure 7 and Figure 8. The results of these calculations are presented in the form of regression equations with an R^2 value of between 0.9 and 1, giving the possibility to use them for the fast and efficient prediction of the expected equilibrium position of the damaged aircraft carrier. Equations are listed below where $\varphi 1$ (Equations 5–8) and $\psi 1$ (Equations 9–12) indicates the angle of heel and trim after damage respectively while the index refers to the number of adjacent flooded compartments.

$$\varphi 1 = 5*10^{-8}x_C^4 - 2*10^{-5}x_C^3 + 0.0019x_C^2 + 0.0248x_C + 1.1669,\ R^2 = 0.8797 \quad (5)$$

$$\varphi 2 = 10^{-7}x_C^4 - 4*10^{-5}x_C^3 + 0.0041x_C^2 - 0.0038x_C + 2.3237,\ R^2 = 0.9638 \quad (6)$$

$$\varphi 3 = 10^{-7}x_C^4 - 5*10^{-5}x_C^3 + 0.0046x_C^2 + 0.012x_C + 3.3837,\ R^2 = 0.9808 \quad (7)$$

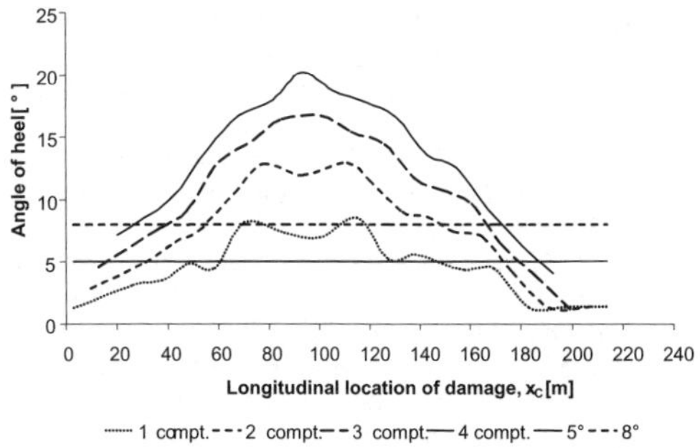

Figure 7. Angle of heel diagrams for the equilibrium position.

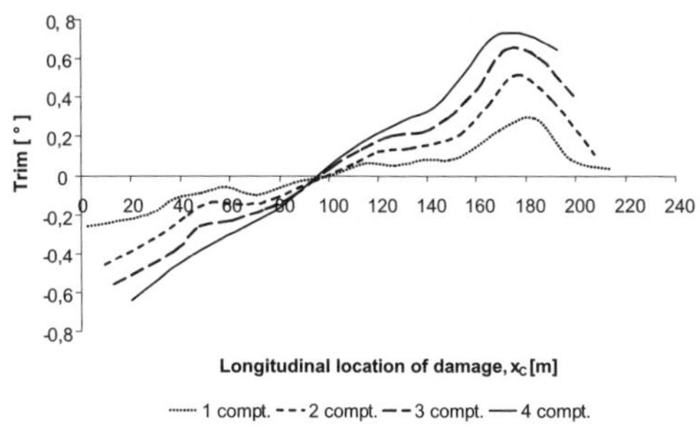

Figure 8. Trim diagrams for the equilibrium position.

$$\varphi 4 = 10^{-7}x_C^4 - 6*10^{-5}x_C^3 + 0.0061x_C^2 - 0.0263x_C + 5.2763,\ R_2 = 0.988 \quad (8)$$

$$\psi 1 = -3*10^{-9}x_C^4 + 10^{-6}x_C^3 - 0.0001x_C^2 + 0.0087x_C - 0.3017,\ R2 = 0.914 \quad (9)$$

$$\psi 2 = -7*10^{-9}x_C^4 + 3*10^{-6}x_C^3 - 0.0004x_C^2 + 0.0222\ x_C - 0.6655,\ R^2 = 0.9588 \quad (10)$$

$$\psi 3 = -6*10^{-9}x_C^4 + 3*10^{-6}x_C^3 - 0.0003x_C^2 + 0.0213\ x_C - 0.8046,\ R^2 = 0.9775 \quad (11)$$

$$\psi 4 = -6*10^{-9}x_C^4 + 2*10^{-6}x_C^3 - 0.0003x_C^2 + 0.0224\ x_C - 1.0013,\ R^2 = 0.9923 \quad (12)$$

5 CONCLUSION

In this paper damage stability calculation was performed for the selected aircraft carrier using the program GHS (2005) for the new equilibrium position after the flooding was caused by underwater damage.

At the very beginning the assumption was taken into account that the aircraft carrier as warship has side-protective system capable to keep operating and combat features after received at least one straight underwater hit to any part of hull that causes flooding of one or two adjacent sections.

Required damage simulation was conducted for four different flooded groups (groups with one, two, three and four adjacent flooded compartments), (Figure 5). The obtained results are presented graphically for the heel and trim in equilibrium position (Figures 7 and 8).

Obtained results shows that the aircraft carrier has the highest angle of heel after the flooding of the central compartments while flooding the compartments at the bow or stern gets mostly a high trim. These results are logical because the sections in the bow and stern are symmetrically flooded.

Also, it can be noticed that in the case of flooding only one compartment the results generally correspond with the results stated in Manning (1967), (2°–3° angle of heel after one compartment flooding) except the results for the case of flooding the compartments F, G, H and I (midship compartments), which are somewhat higher. These higher results do not represent problem since the angle of heel is about 8° which still allows limited air operations.

It is necessary to note that the obtained angles of heel are higher because of the allowable maximum wind that is taken into account for calculation. Only when two adjacent compartments are flooded angle of heel becomes significant but in no case exceeding an angle of 15°, which is limit for the smooth and safe functioning of lifesaving equipment. Flooding three and four adjacent compartments in the central part of the aircraft carrier the angle of heel becomes more serious and generally significantly restricts the seaworthiness of vessel with the loss of operational and combat capabilities. With such flooding combination the careful and rapid ballasting is needed to correct the angles of heel and trim to the allowable values.

Although the values of angles of heel in the case of three or four flooded compartments are significant review of the area between the GZ curve and the curve of wind heeling moment shows that the capsizing is not the imminent risk since the values are considerably higher than the minimum allowed. Therefore it can be concluded that this aircraft carrier will be in stable equilibrium even when the angle of heel is greater than 20°.

With this fact, placing the order for leaving the vessel it can be postponed and the crew can concentrate on the vessels rescue operations. Also, the obtained results confirm the assumption stated in Manning (1967) that the aircraft carrier can withstand more than one direct underwater hit and under the equilibrium angle of heel it could still continue to sail but with limited combat activities.

As the aircraft carrier is warship it is assumed that the main cause of underwater damage will be primarily due to hostile activity. Having this in mind for the presented results (Figures 7 and 8) the regression analyses was performed and we got corresponding equations. These equations can be used for fast and efficient prediction of the expected equilibrium position for the damaged aircraft carrier depending on flooded compartment or group of adjacent compartments.

REFERENCES

Brown, D.K. & Chalmers, D.W. 1989. The Management of Safety of Warships in the UK, *RINA Transactions*: 29–46.

General Hydrostatics—User's Manual, 2005.

IMO resolution A.265 (VIII), Regulations on Subdivision and Stability of Passenger Ships as Equivalent to Part B of Chapter II of International convention for the Safety of Life at Sea, 1969., London, 1974.

Keil, A.H. 1961. The Response of Ships to Underwater Explosions. *SNAME Transactions, Vol. 69*: 366–410.

Manning, G.C. 1967. *Theory and technique of ship design*. Tehnička knjiga, Zagreb, in Croatian, 1967.

Russo, V.L. & Robertson, J.B.1950. Standards for Stability of Ships in Damaged Condition, *SNAME Transactions, Vol. 58*: 478–566.

Sarchin, T.H. & Goldberg L.L. 1960. Stability and Buoyancy Criteria for U.S. Surface Ships, *SNAME Transactions, Vol. 70*: 418–458.

St. Denis, M. 1996. The Strike Aircraft Carrier: Considerations in the Selection of Her Size and Principal Design Characteristics, *SNAME Transactions, Vol. 104*: 260–304.

Stability Standards for Surface Ships—Part 1, Conventional Ships, UK Ministry of Defence, Defence Standard 02-109 (NES 109), Issue 1, 2000.

Uršić, J. 1991. *Ship stability*. University of Zagreb: FAMENA, textbook, in Croatian.

Advanced Ship Design for Pollution Prevention – Guedes Soares & Parunov (eds)
 ISBN 978-0-415-58477-7

The effect of L/B and B/T variation on subdivision indexes

I. Munić & V. Slapničar
Faculty of Mechanical Engineering and Naval Architecture, University of Zagreb, Zagreb, Croatia

V. Stupalo
Brodosplit Shipyard Ltd., Split, Croatia

ABSTRACT: This paper examines the addresses the Ro-Ro ship regarding the effect of variation of bulkheads' positions on the attained subdivision index. In addition *L*/*B* and *B*/*T* ratios were also varied to obtain the relation between the subdivision index and the varied ratios. The initial ship (car-truck carrier) was analysed first and the subdivision zones were established first along with the calculation of the attained and required subdivision indexes using the probabilistic method through GHS application. In order to improve the attained index and get a safer ship variations of positions of transverse bulkheads were investigated to get better indexes of the adjacent subdivision zones. Diagrams that show potential solutions for making the initial ship safer were presented.

1 INTRODUCTION

The attained subdivision index depends on number of parameters with many of them restricting each other. To improve the design tools, understanding the correlation between those parameters is crucial.

For the attained subdivision index calculation the method described in the IMO Resolution MSC 19(58) was used. Various configurations of Ro-Ro ships were examined and an extensive database (Slapničar 1998) was prepared by thorough literature survey. The database contains principal data for 200 ships. Total number of ships for which it was able to identify type of midship section, from clear and available general plans, is about 100 and a typical car-truck carrier was chosen for the analysis.

To check standard of subdivision the attained subdivision index, *A*, is calculated in accordance with the IMO rules. It should not be lower than required subdivision index, *R*.

According to IMO the required subdivision index solely depends on the subdivision length of a ship.

$$R = \left(0.002 + 0.0009 \cdot L_s\right)^{\frac{1}{3}} \tag{1}$$

The attained subdivision index, *A*, is calculated for two subdivision load lines. The first one is the deepest subdivision load line that corresponds to the summer draught assigned to the ship. The second one is partial load line that corresponds to light ship draught plus 60% of the difference between the light ship draught and deepest subdivision load line.

For the economical reasons ship should be operating on draughts close to summer draught and that is why the rules use only two draughts. The attained subdivision index, *A*, is calculated as the mean value for two mentioned draughts.

$$A = \frac{1}{2}\left(A_f + A_p\right) \tag{2}$$

$$A_f = \sum_i p_i s_f \tag{3}$$

$$A_p = \sum_i p_i s_p \tag{4}$$

where A_f = index for the deepest subdivision load line; A_p = index for the partial load line; p_i = probability of flooding; s_f, s_p = probability of survival.

For the damage stability calculation software application GeneralHydroStatics (GHS) was used.

The rules concerning intact and damaged stability are inconsistent in this case. While all classification societies require intact stability to be calculated for various realistic loading conditions and real (trimmed and heeled) waterlines the IMO rules require calculation at one realistic waterline (summer draught) and the partial waterline at the draught previously described which is not directly related to any loading condition of the ship. Using real loading conditions would lead to more complex damaged stability calculation but it would also give the ships crew better data of the damaged ship behaviour in situations they might find themselves in.

2 INITIAL DESIGN ANALYSIS

Initial design for the analysis is a typical car-truck carrier (Fig. 1) built for long intercontinental routes. Main particulars are presented in Table 1.

Vehicles are transported on eleven decks as follows (Fig. 2):

- Decks 1 (double bottom top), 2, 3, and 10—cars only.
- Decks 4, 6 and 8—truck decks (where deck 6 is main and subdivision deck).
- Decks 5, 7, and 9—hoistable decks (cars only).
- Deck 11—the upper (weather) deck (cars only).

Six transverse watertight bulkheads extend from the bottom to the main deck. These bulkheads are equipped with hydraulic operated watertight doors which enable vehicle manipulation while ship is in harbour. Double-bottom top and the main deck are the only watertight vehicle decks while the others are treated as non-watertight.

Onboard accommodation for 25 crew members and 12 passengers is provided. The ship is propelled by one reversible low speed diesel directly connected to the propeller over the shaft. One bow thruster is driven by an electrical motor. There are also three auxiliary engines connected to the generators. Ballast system consists of three pairs of ballast tanks spreading through the whole double bottom, one fore ballast tank and three aft ballast tanks.

Table 1. Main particulars of initial ship.

Subdivision length, L_S	171.43 m
Length over all, L_{OA}	176.00 m
Length, L_{PP}	165.00 m
Breadth moulded, B	31.10 m
Main deck height, H	14.46 m
Upper deck height, H_1	28.00 m
Design draught, T	8.77 m
Displacement on T	24,825 t
Deadweight on T	12,594 t
Maximal car capacity	4632 m

3 CASE STUDY

Bulkheads divide the ship in six floodable zones (Fig. 3). No zone is homogenous and they consist of various sub-zones with different permeability.

Critical points are the points of progressive flooding i.e. flooding of these points would probably cause the loss of the ship. Every weathertight closing is considered to be a critical point while all of the watertight closings are discarded as such. They are closely related to the accounts for probability of survival, *s*, which equals zero if the lowest edge of the opening goes underwater in the final phase of the flooding.

Critical points (Table 2) were defined as follows:

First we calculate the attained index for the case of damage of each zone exclusively and then for the combination of two or more consecutively damaged zones (Table 3).

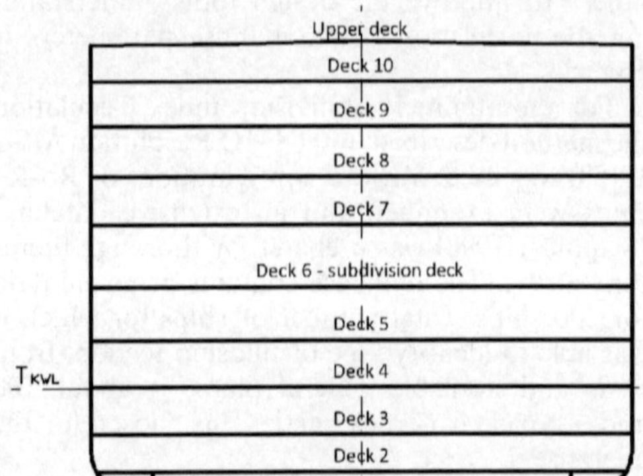

Figure 2. Initial design—cross section.

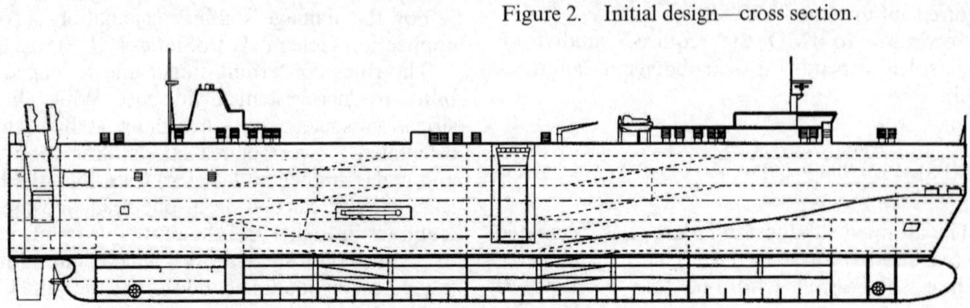

Figure 1. Initial design—side view.

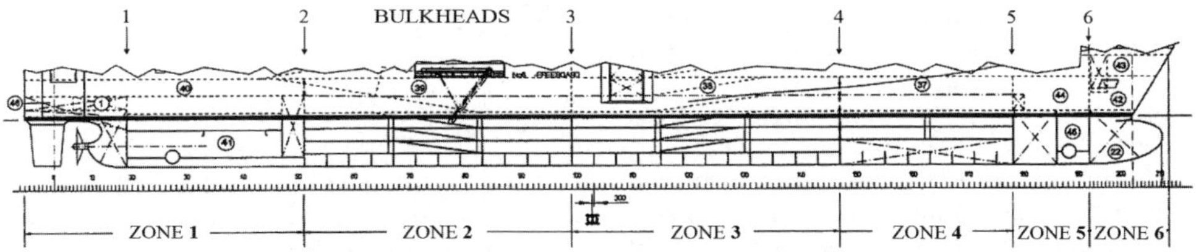

Figure 3. Tank and compartment subdivision in floodable zones.

Table 2. Critical points.

	Distance from AP m	Distance from CL m	Distance from BL m
CP 1	–4.635	15.550	14.460
CP 2	134.650	15.550	14.460
CP 3	139.750	14.300	14.460
CP 4	148.250	11.650	14.460
CP 5	158.450	7.900	14.460
CP 6	9.900	15.100 s	15.260
CP 7	11.100	15.100 s	15.260
CP 8	39.250	15.130 s	15.260
CP 9	114.680	14.950	15.260
CP 10	118.920	14.950	15.260
CP 11	144.400	12.060 s	15.260
CP 12	157.650	0.600	14.910
CP 13	157.650	1.400	14.910
CP 14	158.750	10.150 s	18.710
CP 15	18.200	15.550 s	21.910
CP 16	25.000	15.550 s	21.910
CP 17	33.500	15.550 s	21.910

Coordinate system origin is at the AP and the BL; CP = critical point; s = symmetrical pair of critical points

Table 4. Initial design—attained indexes by zones.

Zones	p_i	s_i	$p_i \times s_i$	A
6	0.021	1.000	0.021	0.021
5	0.021	1.000	0.021	0.042
4	0.090	0.886	0.080	0.122
3	0.185	0.762	0.039	0.263
2	0.151	0.259	0.141	0.302
1	0.115	1.000	0.115	0.417
			1-zone damage:	0.417
6+5	0.033	1.000	0.033	0.450
5+4	0.056	0.598	0.034	0.484
4+3	0.091	0.000	0.000	0.484
3+2	0.092	0.000	0.000	0.484
2+1	0.063	0.000	0.000	0.484
			2-zone damage:	0.067
6+5+4	0.027	0.094	0.003	0.486
5+4+3	0.005	0.000	0.000	0.486
4+3+2	0.000	0.000	0.000	0.486
3+2+1	0.000	0.000	0.000	0.486
			3-zone damage:	0.003
6+5+4+3	0.000	0.000	0.000	0.486
			4-zone damage:	0.000
			Attained index in this condition:	0.486
			Required index:	0.542

Table 3. Initial design—attained subdivision indexes.

No. of zones damaged	100% draught	60% draught
1	0.417	0.508
2	0.067	0.089
3	0.003	0.027
4	0.000	0.000
	$A_f = 0.486$	$A_p = 0.624$

This averages:

$$A = \frac{1}{2}\left(A_f + A_p\right) = 0.555 \quad (5)$$

The attained subdivision index is greater at 60% draught than at full draught which is understandable since the freeboard and the positions of the critical points are higher above the waterline at lower draughts. From the contribution to the overall subdivision indexes it is clear that the probability of ship's survival is very low when more than one zone is flooded. The attained subdivision index for the full load line, A_f, is lower than required index, R. The distribution of contribution to the overall index by zones (Table 4) shows us that the zones 3, 5 and 6 contribute the least.

4 VARIATION OF THE BULKHEADS' LONGITUDINAL POSITIONS

Small values of the attained indexes for zones 5 and 6 is hardly improvable since it is caused by their position (fore peak). Position of the collision bulkhead (bulkhead 6) is very strictly defined by the rules of classification societies. Bulkheads 4 and 5 are tank bulkheads and variation of their longitudinal position would change the volume of the tank. Thus, zone 3 should be varied to attain higher contribution to the subdivision index.

Moving the bulkhead 3 forward was performed to make the zone smaller. Total capacity and complexity of the loading/unloading should have remained the same as on initial design. Since the ships' car carrying capacity was of the primary interest the only logical solution was to move the bulkhead one average car lengths forward i.e. 4 metres.

The modification was conducted in GHS and the results are presented in Table 5 and Table 6.

Table 5. Attained indexes after longitudinal position variation.

No. of zones damaged	100% draught	60% draught
1	0.392	0.478
2	0.073	0.085
3	0.018	0.027
4	0.000	0.000
	$A_f = 0.483$	$A_p = 0.590$

Table 6. Attained indexes by zones after longitudinal position variation.

Zones	p_i	s_i	$p_i \times s_i$	A
6	0.021	1.000	0.021	0.021
5	0.021	1.000	0.021	0.042
4	0.068	0.998	0.067	0.109
3	0.212	0.606	0.129	0.238
2	0.151	0.259	0.039	0.277
1	0.115	1.000	0.115	0.392
			1-zone damage:	0.392
6+5	0.033	1.000	0.033	0.426
5+4	0.052	0.778	0.040	0.466
4+3	0.086	0.000	0.000	0.466
3+2	0.092	0.000	0.000	0.466
2+1	0.063	0.000	0.000	0.466
			2-zone damage:	0.073
6+5+4	0.027	0.652	0.018	0.483
5+4+3	0.009	0.000	0.000	0.483
3+2+1	0.000	0.000	0.000	0.483
			3-zone damage:	0.018
6+5+4+3	0.001	0.000	0.000	0.483
			4-zone damage:	0.000
			Attained index in this condition:	0.483
			Required index:	0.542

This averages:

$$A = \frac{1}{2}\left(A_f + A_p\right) = 0.537 \tag{6}$$

The final attained index is lower than required which is mainly due to the zone 2 getting larger and participating significantly less to the overall attained index (larger amount of flooding water means greater loss of stability). Contribution of the new reduced zone 3 to the overall attained index, A, was larger than in initial design, as expected.

5 VARIATION OF PRINCIPAL DIMENSIONS

Variation of the L/B, B/T ratios was performed with the constant values of displacement and block coefficient. Chosen ratios were varied from –8% to +8% in steps of 2%. This variation span gives us a clear information of the trends on required and attained indexes with a relatively small change of the initial design principal dimensions.

New values of L, B and T were calculated:

$$L = \sqrt[3]{\frac{V(L/B)^2(B/T)}{C_B}} \tag{7}$$

$$B = \frac{L}{L/B} \tag{8}$$

$$T = \frac{B}{B/T} \tag{9}$$

where V = initial design's displacement (23,120 m^3); C_B = initial design's block coefficient (0.514).

Scaling of the initial design (Table 7) was performed in GHS with decks remaining at their initial positions.

With the variation of the L/B ratio the required subdivision index changes accordingly due to the changes of subdivision length (L_S). The attained subdivision index changes mainly due to the changes of draught and breadth of the ship. Designs I_1 to I_4 are narrower, thus lesser stability and designs J_1 to J_4 have larger draught which leads to lesser freeboard.

The calculation shows that the only ratio that satisfies the requirements is the one of the initial design (Fig. 4). Value of the attained subdivision index of the design J_1 (–2% variation of L/B) is close to the required value and with some small corrections of that design (e.g. critical points minor repositioning) it might also pass the requirement.

Table 7. Variation of principal dimensions.

Design	Ratio	L m	B m	T m
Initial	$L/B = 5.305$ $B/T = 3.548$	165.00	31.10	8.766
I_1	L/B+2% = 5.411	167.20	30.89	8.708
I_2	L/B+4% = 5.517	169.37	30.70	8.650
I_3	L/B+6% = 5.624	171.54	30.50	8.596
I_4	L/B+8% = 5.730	173.70	30.31	8.544
J_1	L/B–2% = 5.199	162.80	31.31	8.825
J_2	L/B–4% = 5.093	160.57	31.53	8.885
J_3	L/B–6% = 4.987	158.33	31.75	8.948
J_4	L/B–8% = 4.881	156.08	31.98	9.013
K_1	B/T+2% = 3.619	166.10	31.31	8.651
K_2	B/T+4% = 3.690	167.18	31.50	8.539
K_3	B/T+6% = 3.761	168.24	31.71	8.431
K_4	B/T+8% = 3.832	169.29	31.91	8.329
L_1	B/T–2% = 3.477	163.90	30.89	8.885
L_2	B/T–4% = 3.406	162.77	30.70	9.007
L_3	B/T–6% = 3.335	161.63	30.46	9.135
L_4	B/T–8% = 3.264	160.48	30.25	9.267

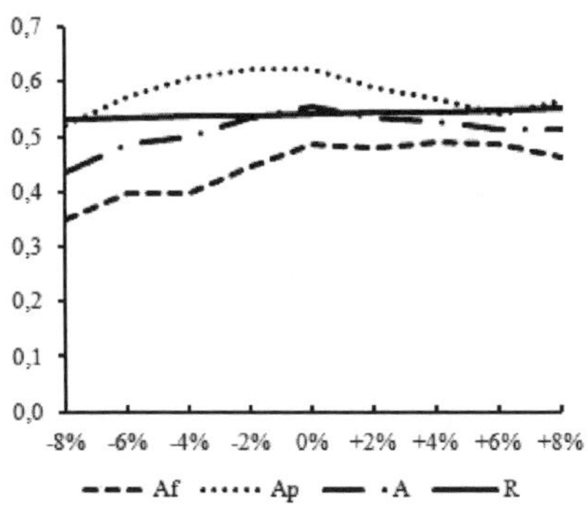

Figure 4. Effect of L/B variation on subdivision indexes.

But still the initial design has a larger margin between the attained and the required value.

Variation of the B/T ratio has a similar trend on required subdivision index i.e. the index increases with the increase of the ratio (L_S as well). As it can be seen on Figure 5, small changes in B/T significantly change the attained index so the only favourable designs are the ones with the increased B/T. It is due to the increase of breadth which leads to better stability (metacentric heights) and decrease of draught which leads to larger freeboard, i.e. larger margin for flooding.

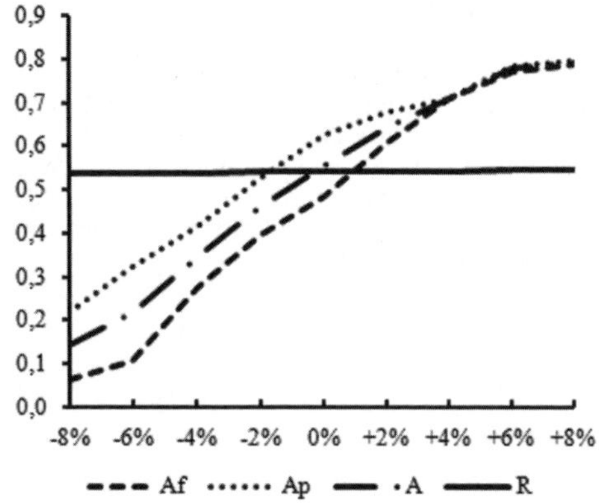

Figure 5. Effect of B/T variation on subdivision indexes.

6 CONCLUSIONS

This paper along is continuation of the previous work on subdivision and safety (Slapničar 1998, Stupalo 2005).

Longitudinal repositioning of the bulkheads is almost insignificant in this investigation since it was restrained only to moving bulkhead 3. It caused a decrease of the attained subdivision index and implied that better values of the index should be achieved through adequate positioning of the critical points and not through moving the bulkhead. The best longitudinal subdivision of the ships mid part is with equal zones. At the fore part floodable zones should be smaller.

Varying of the L/B ratio gives us a slightly convex curve of attained subdivision index which, at its maximum, could give a value that fulfils the requirement.

Curve of the attained index increases with an increase of a B/T ratio so when in need for a larger values of attained indexes this might be the way to obtain them.

While the ships design goes in the direction of fulfilment of the requirements the optimum design is the initial design as a well designed ship since she doesn't have large margins between A and R but still satisfies the requirement.

Larger margins between indexes would mean that the other aspects of a design (besides the damaged stability) could probably be improved while the value of the attained index would still stay above the required. On the other hand, in the situation where every improvement or redefinition of damaged stability rules also applies to the existing ships, too small margin between those indexes might lead to the costly modification of the ship.

REFERENCES

General Hydrostatics. 1993. Users Manual.

Jensen, J. et al. 1995. Probabilistic damage stability calculations in preliminary ship design. *PRADS 95*.

Resolution MSC 19(58) on the adoption of amendments to the 1974 SOLAS Convention regarding subdivision and damage stability of dry cargo ships. 1990. London: IMO.

Slapničar, V. 1998. *Relation between subdivision and safety of Ro-Ro Ships*. Master thesis. University of Zagreb.

Stupalo, V. 2005. *Subdivision analysis of Ro-Ro Ships*. Diploma thesis. University of Zagreb.

Vibration & measurements

Advanced Ship Design for Pollution Prevention – Guedes Soares & Parunov (eds)

Evaluation of propeller-hull interaction in early stage of ship design

R. Grubišić
Faculty of Mechanical Engineering and Naval Architecture, University of Zagreb, Zagreb, Croatia

ABSTRACT: The ship design evaluation from the vibration excitation aspect is presented. This evaluation should be carried out in the early stage of the ship design. It includes the assessment of non-uniform full scale effective wake in the propeller plane, of hull surface pressure and integrated vertical force and of propeller shaft excitation forces and moments. Appropriate evaluation criteria are applied.

1 INTRODUCTION

When a new ship design is developed, then it is a common practice to carry out its evaluation from the vibration aspect. This is usually done by evaluating the natural frequencies and response of the ship structure, and more rarely by analyzing the propeller-hull interaction and evaluating the excitation parameters resulting from this phenomenon.

Namely, the ship and propeller designers have an increasing need to know not only that a given hull form and propeller combination will meet the propulsive performance requirements, but also that it will operate within some acceptable limits of the propeller excitation. The research results in this field indicate that the main causes of the hull aft-end excitation are the non-uniformity of the inflow to the propeller and the heavy propeller blade loading.

From the vibration-excitation aspect, following features of the propeller-hull interaction should be included in the evaluation process: the non-uniform flow to the propeller plane, the span-wise blade loading distribution, the fluctuations of the propeller shaft forces and of the hull surface pressure.

This evaluation should be performed in the early design stage, immediately after carrying-out the routine model tests, using the test results as a starting point. Following model test results are used: nominal wake in the propeller plane, propeller open-water diagrams and self-propulsion data. The evaluation procedure starts by carrying out the model nominal wake adaptation into an effective full scale wake profile. Then, using the adapted wake data as well as the hull and propeller geometry the propeller loading and the propeller excitation features are assessed. Finally, the evaluation of the calculation results by means of appropriate criteria is done. This enables both to the ship and propeller designer to validate the initial design or any of its modifications without performing additional expensive model tests.

2 VELOCITY FIELD IN THE PROPELLER PLANE

Most calculation methods for the propeller performance, propeller excitation forces, etc. use the nominal wake distribution in the propeller plane and satisfy the boundary conditions by adding the induced velocity. But, the flow field with the presence of the operating propeller cannot be understood as a superposition of the propeller induced velocity field and the nominal velocity in the absence of the propeller. Therefore, instead of the nominal wake, the effective wake distribution should be employed.

To obtain from the measured model nominal wake the full-scale effective wake, two steps should be carried out:

- assessment of full-scale nominal wake
- conversion of nominal wake into effective wake.

2.1 *Full-scale nominal wake*

The assessment of the full-scale nominal wake is based on the so-called scale effect correction (Hoekstra 1975). Namely, when the Reynold's similarity law is not achieved, the boundary layer of a ship is relatively thinner than the model layer, but the velocity profiles are approximately similar. Therefore, the applied scale correction method yields the contraction of the model boundary layer.

The contraction procedure comprises:

- the total contraction factor
- the three-fold contraction
- the tangential velocity correction.

The **total contraction factor** reads

$$C = (C_{FS} + \Delta C_F)/C_{FM} \tag{1}$$

where C_F is the frictional resistance coefficient (*M*—model, *S*—ship), using the ITTC'57-line:

$$C_F = 0.075/(\log_{10} \mathrm{Re} - 2)^2 \tag{1.1}$$

and $\Delta\ C_F$ the roughness allowance from the 1978 ITTC Performance Prediction Method:

$$\Delta\ C_F = [105\ (k_S/L_{WL})^{1/3} - 0.64] \cdot 10^{-3} \tag{1.2}$$

$k_S = 150 \cdot 10^{-6}$ m.

The total contraction factor *C* may be divided into **three components of contraction**

$$C = (i + j + k)\ C \tag{2}$$

where *i, j* and *k* represent the rates of concentric, horizontal and vertical contraction respectively, see Figure 1.

Each contraction direction is connected with certain parts of the harmonic content of the circumferential velocity distribution of the measured nominal wake, (Hoekstra et al 1975).

The practical computation of the presented coordinate contraction means, that a velocity vector should be shifted in the propeller plane from its old position (r, θ) to a new position (r_{new}, θ_{new}), and on the old position (r, θ) replaced by a new velocity vector v_{new}. This procedure requires the interpolation of velocity vectors twice in the radial and circumferential direction respectively. The described three-fold contraction may equally be used for contracting the axial velocity field and the transverse velocity field as well.

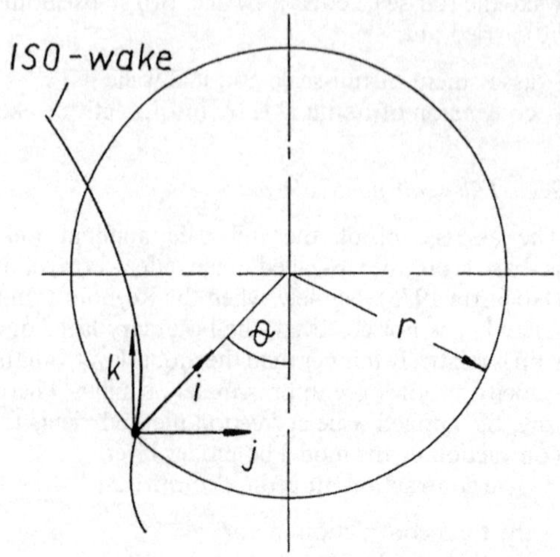

Figure 1. Model boundary layer contraction components.

The contribution of the tangential velocity component to the non-uniformity of the inflow velocity profiles is taken into account by correcting the axial velocity

$$v_x(r,\theta) = v_{x0}\ (r,\theta) + v_t(r,\theta)\ \mathrm{tg}\beta \tag{3}$$

where v_{x0} is the nominal axial velocity, v_t the tangential velocity and β mean pitch angle ($\beta = P_{\text{mean}}/2R\pi$, P_{mean} – mean pitch, R – propeller diameter).

2.2 *Full-scale effective wake*

In theory, the effective velocity field is obtained by subtracting the induced velocities from the velocity with the propeller in operation (total velocity). But, in practice the effective wake can be determined only approximately due to the fact that the total cannot be assessed in an exact way. Namely, the total velocity is only measurable by a complex laser technique, and no reliable numerical methods are on disposal for its exact calculation.

Here, for a reliable prediction of the effective wake, a method based on thrust/torque identity is applied (Hoekstra 1975). The effective velocity v_e is a fictitious quantity of the wake at which the propeller, working behind the hull and advancing at speed v_s, produces the same thrust (or torque) as in the open-water conditions, when advancing at advance speed v_a. By satisfying the advance coefficients identity $v_e/nD = v_a/nD$ and with the same number of revolutions *n*, this yields in $v_e = v_a$.

In practice, five successive steps are involved:

i. Define thrust T_S, torque Q_S, revolutions n_S and ship speed v_S for the working conditions from the self-propulsion tests
ii. Calculate thrust and torque coefficients K_T and K_Q respectively
iii. Interpolate K_T and K_Q in existing open-water diagrams to obtain advance coefficients J_T and J_Q
iv. Calculate mean effective wake

$$v_e = v_a = \frac{1}{2}\left(v_{a_T} + v_{a_Q}\right) = \frac{1}{2} n_s D\left(J_T + J_Q\right)$$
$$w_e = 1 - \frac{v_e}{v_s} \tag{4}$$

v. Calculate effective wake distribution in propeller plane

$$w_e(r,\theta) = w_x(r,\theta)\frac{w_e}{w_x} \tag{5}$$

where $w_x(r,\theta)$ is the full-scale nominal axial Taylor wake fraction and w_x is its mean value over the propeller plane.

3 HULL SURFACE PRESSURE EXCITATION

Due to a permanent interaction between the oncoming flow and the propeller induced velocities, the propeller is surrounded by a non-stationary and non-homogeneous velocity field that, according to the Bernoulli equation, produces pressure fluctuations even of higher order. In addition, the cavitation development on the propeller blade amplifies the pressure amplitudes.

The action of the propeller induced pressure on the after-hull surface generates a vertical excitation force that is usually assessed by the integration of the pressure amplitudes over an arbitrary surface region.

3.1 *Hull surface pressure*

An exact analytical solution of the propeller induced pressure means solving a complicated hydrodynamic problem. So, instead of employing complicated hydrodynamic theories, in the early design stage there is a tendency in developing alternative methods, usually based on statistical processing of systematic model test and full-scale measurement results, yielding in handy diagrams or regression formulae.

Here, two such methods are presented:

- DnV method (Skaar & Raestad 1983)
- BSRA method (PHIVE 1978–1982)

3.1.1 *DnV method*

This method applies the regression formulae concept (Skaar & Raestad 1983). The pressure amplitude for non-cavitating conditions reads

$$p_0 = 12.45\rho n^2 D^2 Z^{-1.53}(t_h/D)^{1.33} \cdot R/d \qquad (6)$$

The single blade frequency pressure amplitude for cavitating conditions is as follows

$$p_c = 0.098\rho n^2 D^2 (J_E - J_M) f_2 K_S \sigma^{-0.5} (d/R)^{-\chi} \qquad (7)$$

and the double blade frequency pressure amplitude has the form

$$p_{2c} = 0.024\rho n^2 D^2 (J_E - J_M)^{0.85} K_S \sigma^{-0.5} (d/R)^{-\chi 1} \qquad (8)$$

where: ρ = fluid density; n = propeller revolutions; D = propeller diameter; J_E, J_M = advance coefficients at effective and maximum wake, respectively; d = distance from blade tip to hull point; f_2 = blade geometry parameter; σ = cavitation number; χ, χ_1 = regression coefficients; K_S = free surface correction.

K_S is given as a 3rd order polynomial of the argument h/h_a (h = depth of the investigated hull surface point, h_a = depth of the propeller shaft),

$$K_s = 9.34\frac{h}{h_a} - 30.143\left(\frac{h}{h_a}\right)^2 + 33.19\left(\frac{h}{h_a}\right)^3$$

3.1.2 *BSRA method*

According to this method (PHIVE 1978–1982), the total, normalized blade frequency pressure amplitude K_{pc} can be composed of a non-cavitating part K_{p0} and an amplification factor P_R due to the effect cavitation

$$K_{pc} = K_{p0} \cdot P_R \qquad (9)$$

Non-cavitating part K_{p0} depends mainly on the vertical clearance a_z/D, as it can be seen in Figure 2a.

Amplification factor P_R is determined by cavitation number σ_{nz} and wake non-uniformity parameter w_Δ, see Figure 2b.

For calculating the normalized double blade frequency pressure amplitude, the following formula is used:

$$K_{p2c} = K_{pc}(2P_R - 2)/(3P_R - 1) \qquad (10)$$

To obtain pressure value p from pressure coefficient K_p, the following common relation is valid

$$p = \frac{1}{2}\rho n^2 D K_p \qquad (10.1)$$

3.2 *Vertical hull surface force*

The vertical excitation force, generated by the hull surface pressure, is obtained by integrating the pressure over a relevant area of the after-hull immersed surface. The phase shift of the pressure amplitudes in each point of the calculation area is taken into account. In Figure 3 the model of a calculation area in accordance with the ship after body form is presented. Single and double blade frequency forces are usually assessed.

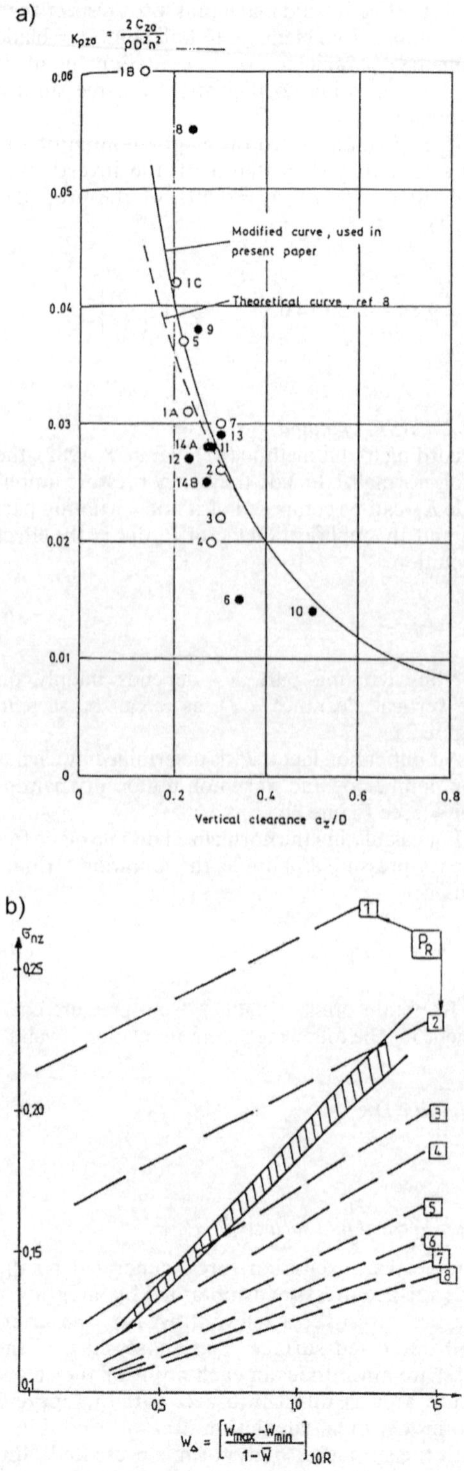

Figure 2. BSRA method, PHIVE 1978–1982.

4 PROPELLER SHAFT EXCITATION FORCES

The phenomenon of the propeller shaft forces (often called bearing forces) is closely connected with the lift development along the propeller blade. Their fluctuations, called propeller shaft excitation, see Figure 4, depend on the time-varying inflow velocity field in the propeller plane.

In reality, this inflow is non-stationary, what means that in each point of the propeller plane the intensity and direction of the inflow velocity changes with time. To handle this phenomenon in an exact way, the "non-stationary" blade theories are required. Due to the fact that these theories are very complicated and computer-time consuming, in practice very often the application of simpler "quasi-stationary" methods is preferable, which deal with the time averaged velocity in each point of the propeller plane. Anyway, these time-averaged velocities differ going from point to point within the propeller plane. So, during the propeller rotation, each blade section experiences the periodically varying inflow velocity profiles in the circumferential direction. Their harmonic components cause the above-mentioned fluctuations of the propeller shaft forces.

In the presented evaluation procedure two concepts of such a quasi-stationary approach are made use of:

– the blade loading concept
– the open-water concept.

4.1 *Blade loading method*

This method is based on the loading distribution along the propeller blade (Schwanecke & Laudan 1972). The loading distribution depends on the harmonic components of the effective inflow velocity to the propeller blade. Therefore, it can be presented as a series of its harmonic components

$$dL(r,t) = \sum_{m} dL_m(r) \cdot \cos\left(m\omega t - \varphi_m(r)\right) \tag{11}$$

Amplitude dL_m and phase shift $\varphi_m(r)$ have to undergo various corrections due to the non-stationarity and three-dimensionality of the real flow.

After defining the harmonic components of sectional loading dL, the determination of the propeller shaft excitation forces is carried out in three steps:

i. blade section level
ii. blade level
iii. propeller level.

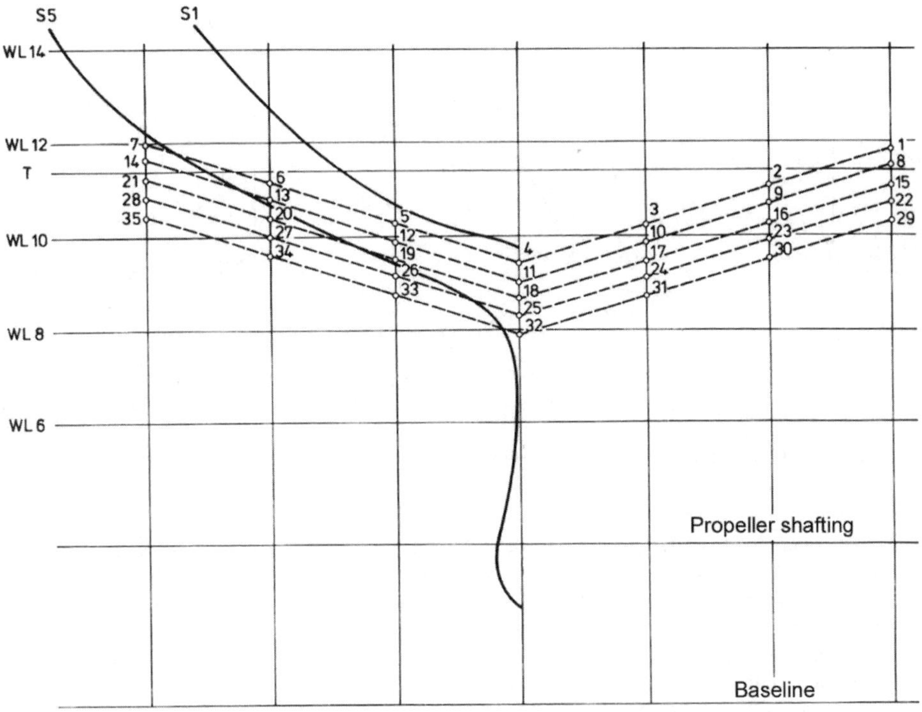

Figure 3. Hull surface pressure integration area.

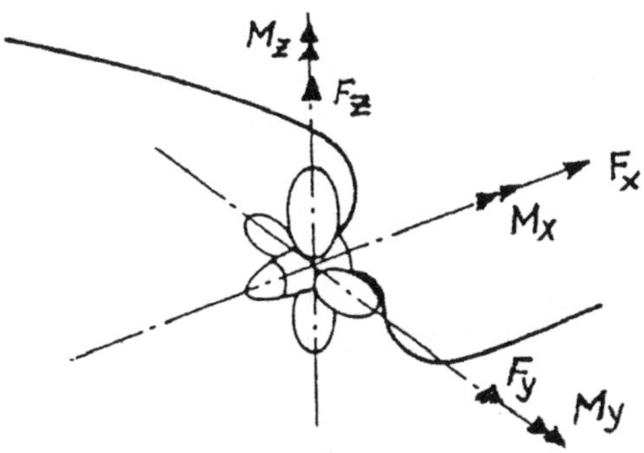

Figure 4. Propeller shaft excitation forces.

On the **blade section level**, by decomposing dL according to Figure 5a the axial and circular force, df_x and df_θ respectively, are obtained. The axial force produces bending moment dm_θ about the load line, and the circular force produces torque dm_x about rotation axis (x) according to Figure 5b. Circular force df_θ can further be decomposed into horizontal and vertical transverse forces df_y and df_z respectively, see Figure 5c, and bending moment dm_θ into horizontal and vertical bending moments dm_y and dm_z respectively, see Figure 5b.

They can be divided into longitudinal excitation (df_x, dm_x) and transverse excitation (df_y, df_z, dm_y, dm_z).

On the **blade level** the sectional forces are spanwise (along the blade) integrated, yielding in the single blade excitation: f_x, m_x, f_y, f_z, m_y, and m_z.

On the **propeller level** the vector summation of the single blade excitation is performed taking the phase angle between blades into account. In such a way the harmonic components of the propeller shaft forces and moments are obtained, which are called the propeller shaft excitation, see Figure 4. Usually, the blade rate frequency and double rate frequency excitation is evaluated, but the steady-state (bearing) shaft forces are not available.

4.2 *Open-water data method*

According to this method (Schwanecke & Laudan 1972), the basic assumption is, that advance coefficient J varies during the propeller rotation due to the circumferential variation of the inflow velocity

$$J(\theta) = v_x(\theta)/nD \tag{12}$$

where $v_x(\theta)$ is the radially averaged velocity $v_x(r, \theta)$ at each blade angular position θ.

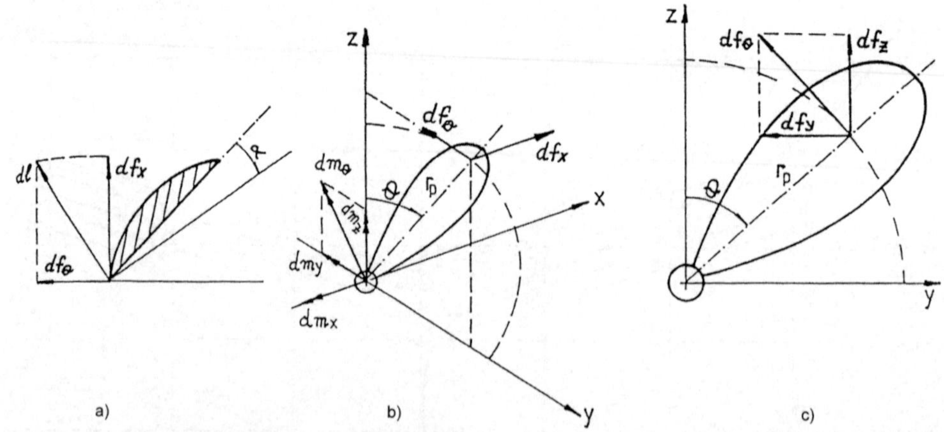

Figure 5. Decomposition of the sectional blade loading.

On the **single blade level** coefficient $J(\theta)$ is interpolated in the propeller open-water diagrams $K_T(J)$ and $K_Q(J)$ obtaining the angular distribution of relevant thrust and torque coefficients $K_T(\theta)$ and $K_Q(\theta)$ respectively. Using these coefficients, the thrust and torque circumferential distribution $t(\theta)$ and $q(\theta)$ respectively for a single blade is defined. This is the basis for the assessment of the circumferential distribution of the remaining forces and moments $f_h(\theta)$, $f_v(\theta)$, $m_h(\theta)$ and $m_v(\theta)$ using their geometrical relationships like in Figure 5.

On the **propeller level**, the vector summation of the single blade forces is performed, resulting in the circumferential distribution of the shaft forces for the entire propeller.

The propeller shaft forces are periodic functions of propeller angular position θ with period $T = 2\pi / Z$ (Z = number of blades). Therefore, they may be represented by their harmonic components. The zero component of each force is its steady-state part, usually known as the bearing forces. The 1st and 2nd component are the fluctuating parts with the blade and double blade frequency respectively, and they are the actual propeller shaft excitation.

5 EVALUATION CRITERIA

Each of the described phenomena is covered with adequate evaluation criteria, what makes three groups of criteria.

5.1 *Wake non-uniformity criteria*

Here, two criteria are available: the $w_\Delta - \sigma$ criterion and the local wake gradient criterion (PHIVE 1972).

The $w_\Delta - \sigma$ criterion is presented in Figure 6a, where the wake non-uniformity is expressed by w_Δ and the cavitation inception by σ using the following relationship

$$w_\Delta = \frac{w_{\max} - w_{\min}}{1 - \bar{w}}, \sigma = \frac{10.4 + T_A - z_p + h_{wv}}{0.052\,(0.7\pi n D)^2}$$

where $w_{\max}$, $w_{\min}$ and $\bar{w}$ are the maximum, minimum and average wake respectively at 1.0 R, T_A-draught aft, z_p-height of propeller shaft, h_w-stern wave height, n-propeller revolutions.

The local wake gradient criterion is given by

$$|(dw/d\theta)/(1 - w)| < 1.0 \tag{13}$$

and should be applied at each radius between 0.7 R and 1.15 R in the angular interval θ_B, see Figure 6b.

5.2 *Hull surface pressure criteria*

For this purpose five criteria are on disposal such as

i. Huse criterion (Huse & Guoqiang 1982)

$$\Delta p_{\max} \leq 8.3 \text{ kPa} \tag{14}$$

ii. Lindgren-Johnsson criterion (Johnsson 1979)

$$2\Delta p_{\max} \leq 6250(a_x/a_z) = 10^3 D^2 \tag{15}$$

iii. Holden criterion (Holden 1980)

$$\Delta p_{\max} \leq (a_x / a_z)\, 60400/((f_c\,(1 + B/2))/(T_A / H)^{1.5}) \tag{16}$$

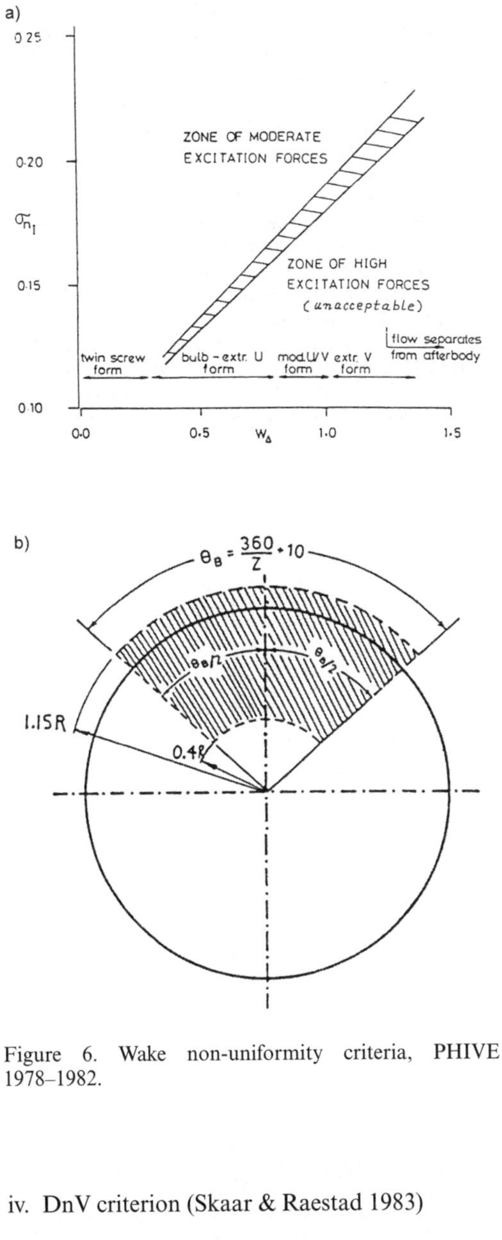

Figure 6. Wake non-uniformity criteria, PHIVE 1978–1982.

iv. DnV criterion (Skaar & Raestad 1983)

$$\Delta p_{max} \leq 543000/((f_c \cdot B_0/2)/(T_A/H)^{2.47}) \quad (17)$$

v. Björheden criterion (Leenars & Forbes 1979)

$$\Delta p_{max} \leq (0.84\nabla_A/(10^3 \cdot D^2)) \quad (18)$$

In the given pressure expressions a_x and a_z are the horizontal and vertical propeller clearance respectively, D – propeller diameter, T_A – draught aft, B – ship breadth, H – ship height and ∇_A – ship displacement volume. All given criteria correspond to the pressure in the hull point vertically above the blade tip in the upright position.

5.3 *Propeller shaft forces criteria*

The evaluation is carried out by means of the Schwanecke-Weitendorf criteria (Schwanecke et al 1972) such as

i. Axial excitation

$$K_x = (2F_x + 2M_x/0.3R)/T \leq 0.2 \quad (19)$$

ii. Transverse excitation

$$K_{yz} = (2F_y + 2F_z + (2M_y + 2M_z)/0.3R)/T \leq 0.25 \quad (20)$$

iii. Torsional excitation

$$K_\theta = 2M_x/Q \leq 0.1 \quad (21)$$

In criteria (19) to (21) the amplitudes of the 1st order shaft excitation forces have to be introduced.

6 CONCLUDING REMARKS

Both the hull surface and the propeller shaft excitation, although very often underestimated, may be a very annoying vibration cause, particularly in the local structure of the ship after part and in the propeller line shafting. Therefore, its analysis represents an important step in the overall ship design evaluation.

Due to the fact that the wake non-uniformity has a major influence on the excitation parameters, the conversion of the nominal wake, which is usually on disposal from model tests, into the effective wake, which approaches us to the velocity profile behind a propeller in operation, has shown to be important.

The concept to introduce alternative methods based on statistical processing of systematic model test and full-scale measurements, instead of complicated hydrodynamic theories, to evaluate the hull surface pressure has shown to be satisfactory in the early stage of the ship design. For comparative reasons it is recommended to use at least two such methods.

When using the quasi-stationary and two-dimensional inflow velocity approach for evaluating the propeller shaft excitation, it has shown to be beneficial to introduce corrections due to the non-stationary and three-dimensional real flow.

Many practical calculations and comparisons with full-scale measurements, (Parunov et. al. 2004), (Grubišić & Senjanović 2004) and (Toda et al. 1988) have confirmed the validity of the presented propeller excitation evaluation.

REFERENCES

Grubišić, R. & Senjanović, I. 2004. Analysis of propeller shaft excitation forces and moments for 65200 dwt oil tanker, Yards 441–444. Technical report, Faculty of Mechanical Engineering and Naval Architecture, University of Zagreb.

Hoekstra, M. 1975. Predictions of full-scale wake characteristics based on model wake survey. *International Shipbuilding Progress* 22.

Holden, K. O., Fagerjord, O. & Frostad, R. 1980. Early Design-Stage Approach to Reducing Hull Surface Forces Due to Propeller Cavitation. SNAME Annual Meeting.

Huse, E. & Guoqiang, W. 1982. Cavitation induced Excitation forces on the hull. NSFI Report, Marine Technology Centre, Trondheim.

Johnsson, C.A. 1979. Some experiences from Vibration Excitation tests in the SSPA Large Cavitation Tunnel. RINA, London.

Leenars, C.E. & Forbes, P. E. 1979. An Approach to Vibration Problems at the Design Stage. Symposium on propeller induced ship vibration. RINA, London.

Parunov, J., Rudan, S., Ljuština, A. & Senjanović, I. 2004. Skeg vibration analysis, oil tanker 65200 dwt, Yards 441–444. Technical report, Faculty of Mechanical Engineering and Naval Architecture, University of Zagreb.

PHIVE 1978–1982. Propeller Hull Interactive Vibration Excitation. Report 120.14 on common project BSRA/NMI and LR.

Schwanecke, H. & Laudan, J. 1972. Ergebnisse der instationären Propellertheorie. Jahrbuch der Schiffbautechnischen Gesellschaft, Band 66.

Skaar, K.T. & Raestad, A.E. 1983. Propeller clearances of importance to fuel economy and ship vibration. PRADS 83, Tokyo/Seoul.

Toda, J., Stern, F., Tanaka, I. & Patel, V.C. 1988. Mean-flow Measurements in the Boundary Layer and Wake of a Series 60 CV = 0.6 Model Ship with and without Propeller. IIHR Report No. 326, Iowa City.

Advanced Ship Design for Pollution Prevention – Guedes Soares & Parunov (eds)

Investigation of propulsion system torsional vibration resonance of a catamaran vessel in service

A. Šestan, B. Ljubenkov & N. Vladimir
Faculty of Mechanical Engineering and Naval Architecture, University of Zagreb, Zagreb, Croatia

ABSTRACT: During the service torsional vibration resonance occurred in the propulsion system of a catamaran vessel. After series of measurements and testing some shortcomings of the ship and propulsion system, which could influence vibration response, have been detected. The ship-owner decided to improve the propulsion system and to eliminate detected imperfections step by step. The paper deals with measurements of shaft line torsional vibrations of a catamaran vessel in service. Some typical inducements of shafting torsional vibrations are indicated, providing full detail description of considered problem. Along with a description of the problem, measurement aim, plan, methodology and used instrumentation are described in details. Special attention is paid to measurement conduction and three measurement programs have been established. The results of the measurement are listed and properly commented, and circumstances which lead to torsional vibration resonance have been detected.

1 INTRODUCTION

Significant torsional vibrations often occur in the dynamical chain of ship propulsion system. Comparing to other kinds of shafting vibrations (lateral, axial), torsional vibrations are more difficult to notice, but their effects can lead to severe damage of ship propulsion system. Shafting torsional vibrations are the consequence of a number of phenomena and processes but one can say that typical inducements of such vibrations are pressure variations in cylinders of main engine, inertial forces of a crank mechanism and seawater-propeller dynamic interaction. Although, lateral and torsional vibrations often occur independently, some things can cause that lateral vibration produces a periodic torque and vice versa, and this phenomenon is called cross coupling (Bently, 2003).

It is generally accepted fact that chosen dimensions of shafting system have significant influence on its torsional behavior (Long, 1980, Hakkinen, 1987), or coupling of torsional, and lateral vibration. However, not only shafting system dimensions have to be taken into account when assessing and analyzing vibration response. The response is influenced by vibration characteristics of each part of the system (main engine, shafting, gearbox, clutches, couplings, bearings, propeller, working of the governor etc.).

In the ship design and ship building process, naval architects and marine engineers use available design tools in order to avoid resonant behavior of the ship and ship components. Despite this fact and bearing in mind unpredictable environment of ship service, wide range of loading conditions, changes in ship speed and main and auxiliary engines working modes, etc., resonant behavior of ship components can occur in service. Elimination of vibrations is rather complex task, and beside numerical calculation procedures which are on disposal, measurements and testing in service have to be done to detect reasons and circumstances which lead to appearance of the resonance. If vibration problem failed to be resolved by some minor intervention, complete propulsion system redesign or mounting of a vibrations damper has to be done.

This paper deals with a problem of propulsion system torsional vibrations of a catamaran vessel. Ship crew noticed significant vibration motions at some operating modes associated with changes in engine noise, temperature increase of flexible torsional coupling as well as with movement of angle bar for power change on the controller. Measurements and testing have been made to investigate and to eliminate the problem. The problem is described in details in the following chapter. A description of the problem is followed by a full detail measurement description. Further on, measurement and testing results are listed and commented. Finally, circumstances which lead to significant shaft line vibrations are detected.

2 DESCRIPTION OF THE PROBLEM

As mentioned above, during the exploitation of a catamaran vessel significant torsional vibrations have been noticed. At the same time the temperature of flexible torsional coupling, placed on junction shaft between main engine and gearbox, is increased. Significant vibrations occurred in two operating modes i.e. when engine works at 700 revolutions per minute or less, and at 1100 RPM, but only in the case when revolutions per minute are decreased, for example from 1500 RPM or more, to about 1100 RPM.

The considered ship is two-screw catamaran vessel, Figure 1, equipped by two independent propulsion systems, Figure 2, i.e. left and right propulsion system. Main engines are the following: MAN, D 2842 LE 412, with nominal power of 588 kW at 1800 RPM, and have 12 cylinders placed in "V" shape. Main particulars of the vessel are the following:

Length overall	$L_{oa} = 49.20$ m
Length between perpendiculars	$L_{pp} = 41.00$ m
Breadth	$B = 15.30$ m
Breadth of the each hull	$B_h = 5.00$ m
Depth	$H = 3.60$ m
Displacement, full load	$\Delta = 650$ t
Ship speed	$v = 11.5$ kn.

A series of measurements and testing have been conducted in order to investigate possible inducements of the resonance of torsional vibrations in ship propulsion system, and following facts are detected: shaft line is not aligned properly and bearing sleeve is not deteriorated uniformly; bearings have to large clearances which cause significant lateral vibrations and coupled lateral and torsional vibrations; elastic basement of main engines is not made properly; bearing in mind the type of bearings and type of shaft line setting, shaft line seems to be too stiff; the Thordon-bearing cooling and lubrication seawater system is not appropriate; stern structure of the ship is not as stiff as it should be; main engine governor operation is perturbed, and the governor has inappropriate regulation characteristics.

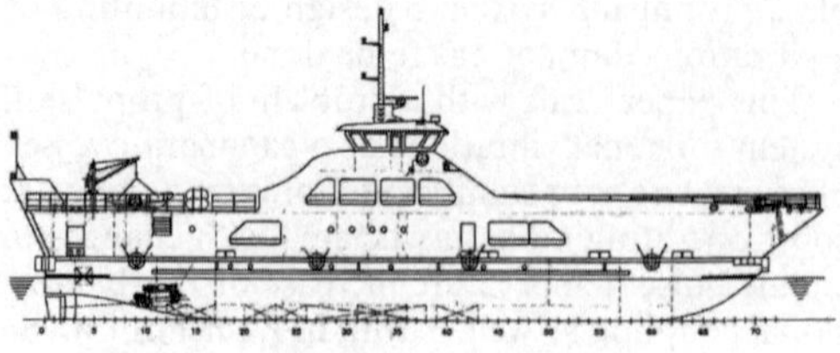

Figure 1. Catamaran vessel plan.

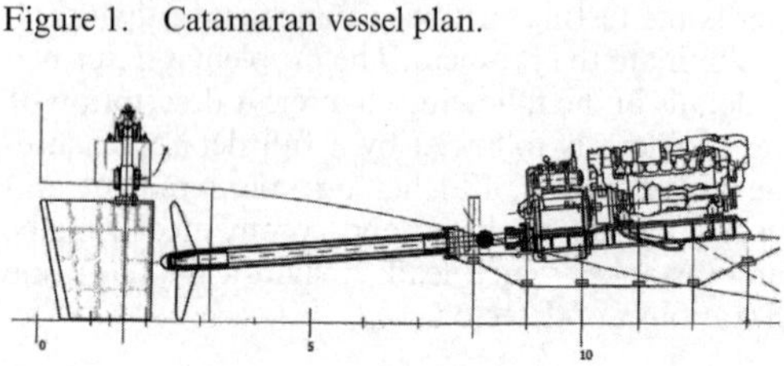

Figure 2. Ship propulsion system.

Based on the situation defined, the ship-owner decided to eliminate these shortcomings step by step in order to improve the ship propulsion system features and to avoid possible resonance effects. The reconstruction of the ship aft structure and shaft line as well as reconstruction of main engine elastic basement were done. Also, the elastic springs of main engine controller are changed with more stiff ones, and regulation characteristics of the controller are improved. Although some of above listed shortcomings are eliminated, torsional vibration resonance occurs, but in different RPM range now. Significant torsional vibrations now occur at 1500 RPM in stationary navigation mode about 2 hours after starting the engines, i.e. after start of navigation. Vibrations are associated with changes in the engine noise, temperature increase of flexible torsional coupling as well as with movement of angle bar for power change on the controller.

The trouble is that occurrence of resonance is still in the working range of main engines, and bearing in mind the above mentioned phenomena, measurements in the service are necessary to investigate problem completely.

3 MEASUREMENT AIM, PLAN, METHODOLOGY AND INSTRUMENTATION

3.1 *Aim of the measurement*

The aim of the measurement and testing is to confirm of the relation between occurrence of torsional vibrations in ship propulsion system and increased cyclic movement of regulation bar of high pressure fuel pump on the main engines. Also, time instant of vibration occurrence has to be determined correctly, as well as the amplitude of vibrations and the amplitude of regulation bar movement when propeller shaft is disconnected.

3.2 *Measurement plan*

The measurements are done on the propulsion system of the vessel in the conditions which correspond to exploitation ones during the service. Testing location was Velebit channel seaway at the Adriatic Sea, Figure 3.

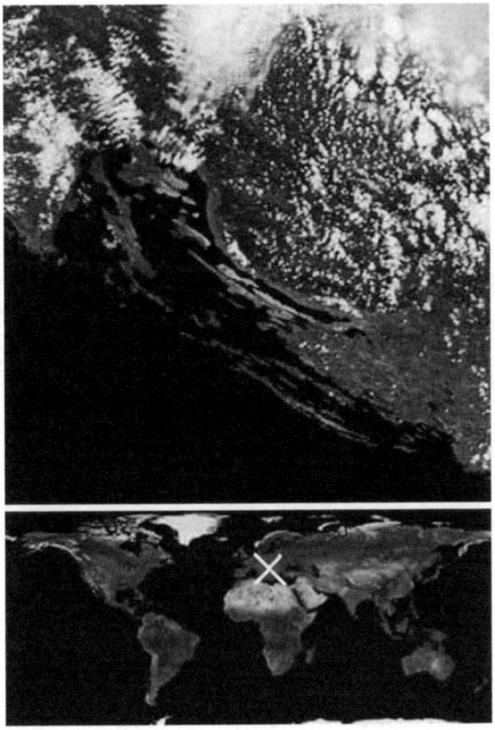

Figure 3. Area of trial trip.

Measurement spots are regulation bar of the high pressure fuel pump, propulsion shaft between gearbox and stuffing box on the right propulsion system, gearbox, main engine, and flexible torsional coupling, Figures 4 and 5.

Movements of the regulation bar of the left propulsion system are measured in the whole working range of the main left engine. On the right propulsion system amplitudes and frequencies of regulation bar vibrations are measured in the whole working range. Further on, on the right propulsion shaft torsional vibrations as well as torque are measured. During the examination, oil temperatures in the engine, controller box and gearbox are controlled, and at the same time oil pressure in the gearbox have been monitored.

Three measurement programs have been established by the authors, i.e. program A, B and C, and measurement was conducted by Brodarski Institute, Zagreb. Program A is initial testing program when the ship is in sea port. Main engines are working until working temperature is achieved, and range of revolutions per minute is being changed from 700 RPM to 1700 RPM with increment of 100 RPM. After each increment, stabilization is necessary, which takes one minute. Propeller shafts are disconnected. Auxiliary engines are working

Figure 4. Location of torsiometer on propeller shaft.

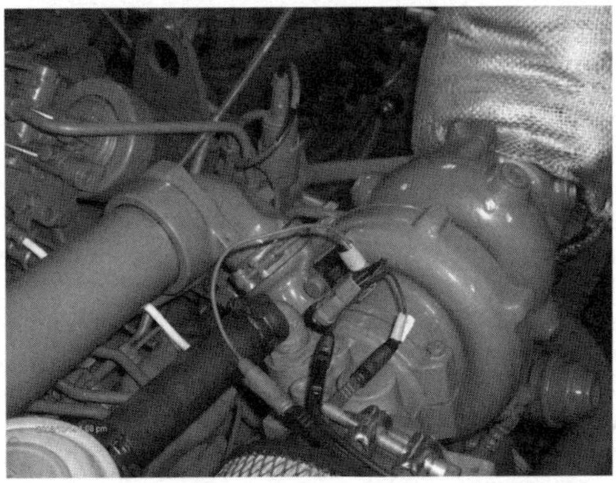

Figure 5. Sensors on the main engine.

intermittently. Program B is usual operating mode of the ship, when ship navigates according to the ferry route. Program C follows after Program B is finished and it is done for free operating mode i.e. out of the ferry route.

3.3 *Measurement methodology and instrumentation*

Amplitudes and frequencies of vibrations of the regulation bar for fuel dosage are recorded by interpreting voltage signal on the movement sensor (RWG 2-417-224-002) which is made up into high pressure Bosch pump box. Voltage signal from the sensor, located on left main engine, is transmitted to the amplifier and after amplification to digital universal voltmeter (Unimer DMS), without possibility of saving and data printing. Monitoring of the signal serves only for the indication of the increased movement of the regulation bar, and the results are omitted in this paper. Voltage signal from the right engine and from the right shaft are

Table 1. Instrumentation for measurement of regulation bar vibrations, (Owner: Brodarski institute, Zagreb).

No.	Instrumentation	Label
1	Torsiometer	RF: B3 T002/05
2	Accelerometer	B&K4371
3	Accelerometer calibrator	B&K4291
4	Charge preamplifier	B&K2635
5	Tape recorder with vibration unit	B&K7005
6	Tape recorder with FM unit	B&K7005
7	Oscilloscope	SONY-TEKTRONIX 314
8	FFT Analyzer	B&K2515
9	Measuring computer	FUJITSU/SIEMENS with NI A/D PCI-MIO-16XE10

monitored on the display and saved in the measurement system memory at the same time. Measurement of the oil temperature and pressure in the engines and gearbox systems is done by using identical instruments which the ship crew uses during the regularly measurement and control of those parameters. The instrumentation for measurement of vibrations of regulation bar is given in Table 1. Relative lateral vibrations in X, Y and Z (longitudinal, transverse and vertical) directions as well as additional dynamic torsional stresses (rotational) are measured.

4 RESULTS AND COMMENTS

The results of the measurements and testing confirm appearance of torsional vibration, accompanied with flexible coupling temperature increase, in both propulsion systems in identical navigation conditions.

It is deduced that significant vibration motions occur in the main engine revolution range from 1440 RPM to 1560 RPM. Vibrations are not generated by hastily changes in revolutions. Moreover, significant vibration motions occur when ship navigates stationary at critical revolutions. With regard to conditions of line navigation, vibrations appear about 10 minutes after stationary navigation. It should also be noticed that vibrations occur at least two hours after first starting of the engines. The first starting of engines means that they where turned off longer than 2 hours.

Motion of regulation bar for fuel dosage at 1500 RPM, when measuring according program A (ship in sea port, propeller shaft disconnected), about 1 hour after starting the engines, is shown in Figure 6. Approximate readings of the motion are $H_{max} = 4.02$ mm and $H_{min} = 3.22$ mm. Mean value of the movement is $H_{mean} = 3.66$ mm, and difference between maximum and minimum is $\Delta H = 0.8$ mm.

Figure 7 shows highest levels in vibration motion of regulation bar in the range 700–1500 RPM in program A of the measurement.

The Figure 8 shows vibration motions of the regulation bar for fuel dosage in measurement program B, about two hours after starting the engines.

Mean value of vibration motions equals $H_{mean} = 7.36$ mm. Significant motion of regulation bar is not present yet, but comparing to program A, where $\Delta H = 0.8$ mm, Figure 6, slightly increase

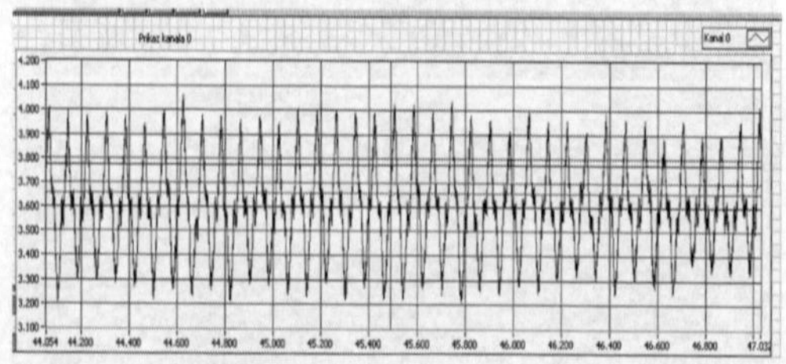

Figure 6. Regulation bar motion (mm) at 1500 RPM, propeller shaft disconnected (Program A).

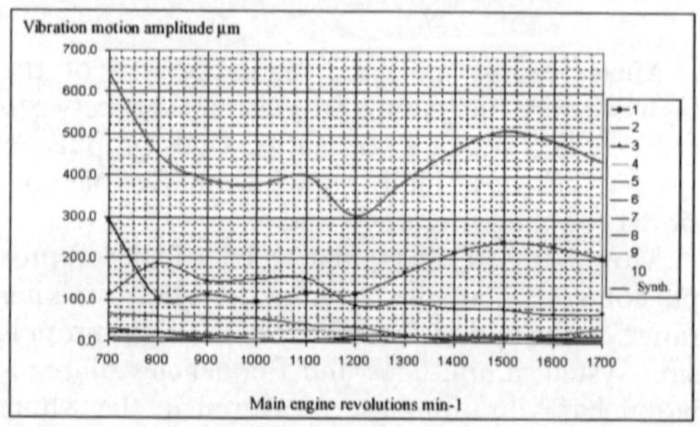

Figure 7. Vibration motion of regulation bar $\Delta H = H_{max} - H_{min}$ mm, (Program A).

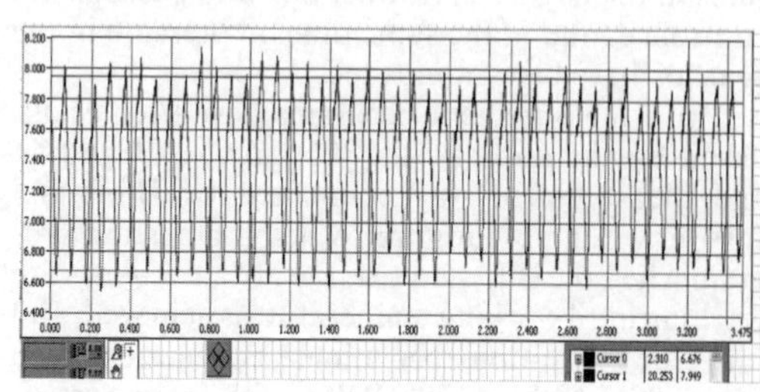

Figure 8. Time variation of regulation bar motion (mm) at 1500 RPM about two hours after starting the engines, (Program B).

is evident, since here $\Delta H = 7.95 - 6.68 = 1.27$ mm. Larger increase in vibration motions of regulation bar, and development of motion instability are shown in Figures 9 and 10. It is evident that increased vibrations do not occur instantaneously. Moreover, 30 to 40 seconds is needed for initial instability development, and it is finalized by development of full resonance of bar motion and torque in ship propulsion system.

In the zoomed view of regulation bar movement, Figure 11, one can see that maximum and minimum value of movement is 8.85 mm and 6.35 mm, respectively. The mean value equals 7.60 mm, and difference between maximum and minimum value yields $\Delta H = 2.5$ mm (directly before resonance appearance).

Joint presentation of vibration motions of regulation bar and torque of propulsion system throughout resonance development period are shown in Figures 12 and 13. It is evident that appearance of significant vibration of fuel dosage regulation bar precedes the torque resonance.

Figures 14 shows increase in regulation bar motion to approximately $\Delta H = 8.0$ mm, and at the same time stresses derived by torque are negative (about -30 MPa), Figure 15, which confirms resonant behavior existence.

At the same time, hastily increase of temperature of silicone core of flexible coupling is recorded. The rate of the temperature increase is approximately 1°C per second. During the resonance

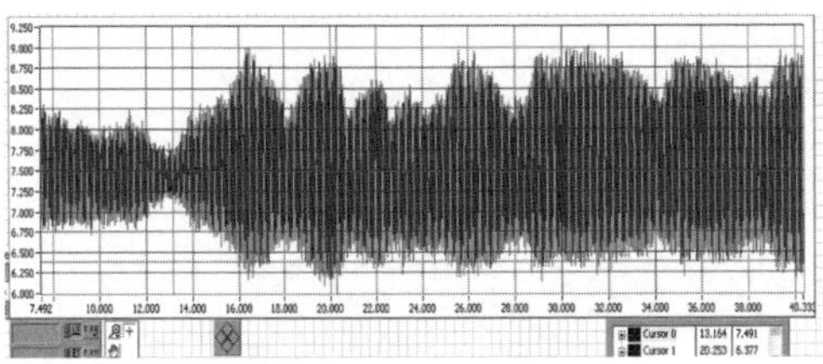

Figure 9. Significant vibration motions of regulation bar (mm), about 4 hours after starting the engines (Program B).

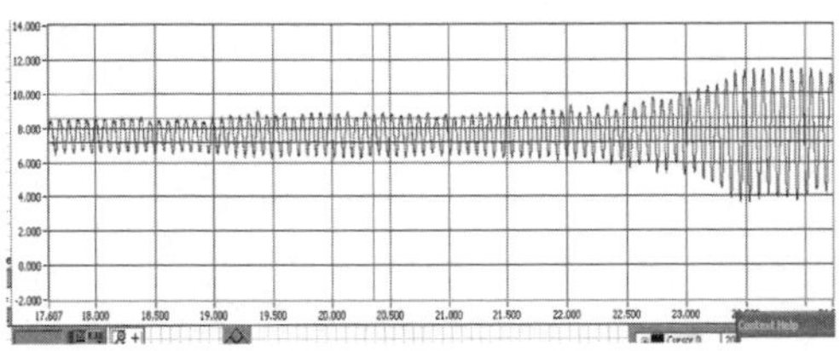

Figure 10. Development of regulation bar vibration instability (mm), about 4 hours after starting the engines (Program B).

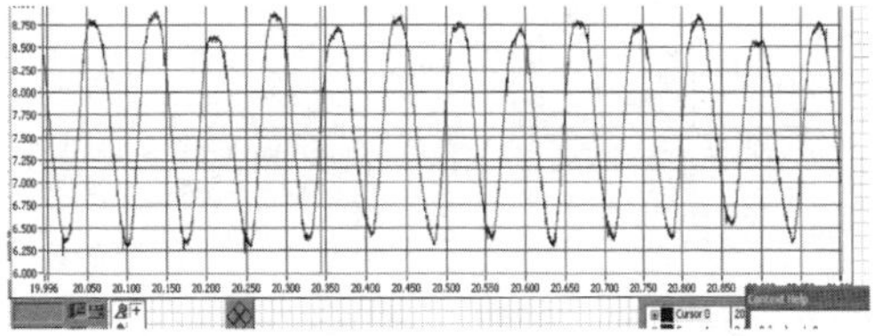

Figure 11. Zoomed view of regulation bar vibration instability (mm) development, about 4 hours after engines start (Program B).

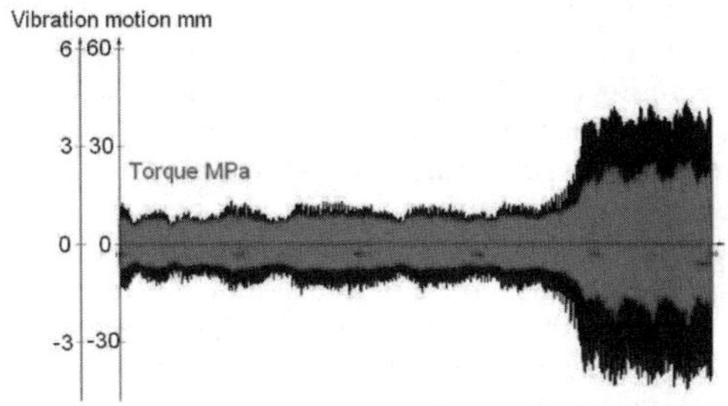

Figure 12. Common presentation of regulation bar motion and torque throughout resonance development.

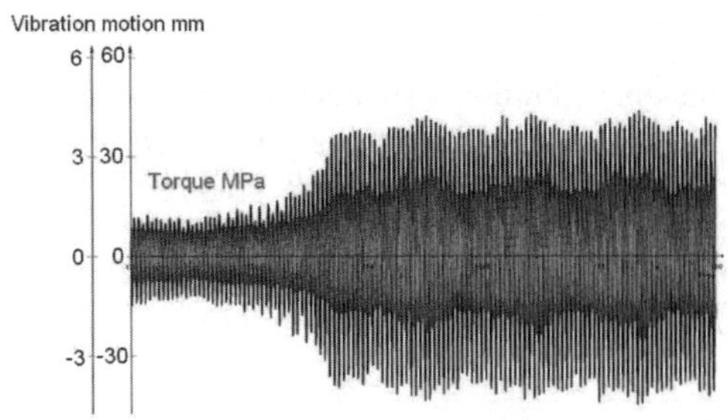

Figure 13. Common presentation of regulation bar motion and torque throughout resonance development, zoomed view.

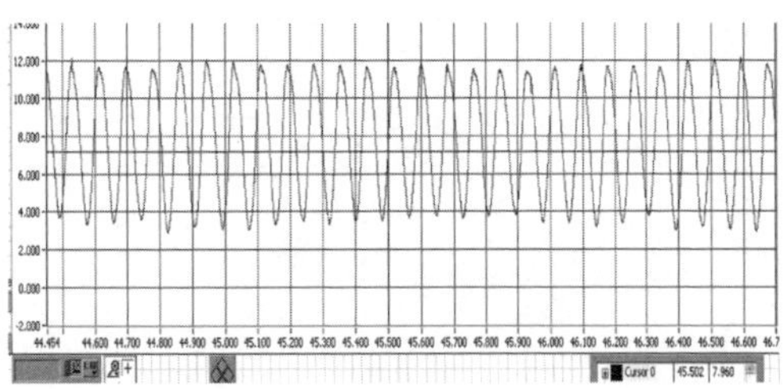

Figure 14. Motion of regulation bar (mm) throughout resonant period at 1500 RPM, (Program C).

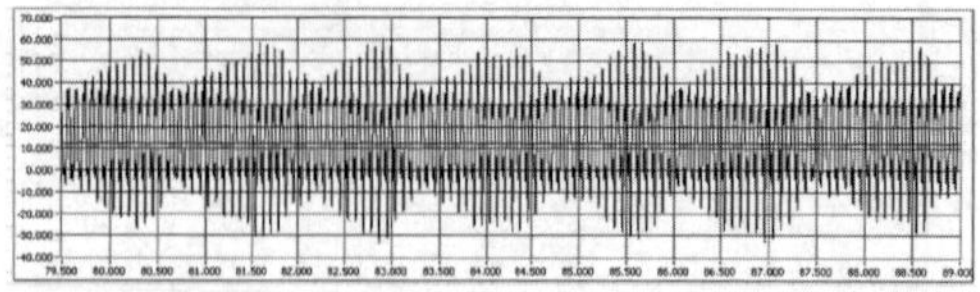

Figure 15. Time variation of torque at 1500 RPM, (Program C).

period (about 30 seconds) temperature of silicone core increased from 28°C to 58°C with tendency of further nonlinear increase.

During the whole testing process, oil temperatures were in the prescribed range for particular operating mode. Same remark is applied for values of oil pressure in gearboxes of ship propulsion system. The synchronization of revolutions of main engines and auxiliary engines has been made in order to register the influence of auxiliary engines on vibration response of the system. Influence of auxiliary engines working on the main engine fuel dosage regulation bar and torsional vibration resonance is not recorded.

The testing showed that when all measuring conditions are achieved, resonance will always occur, i.e. recurrence of the resonance is proved.

5 CONCLUSIONS

Torsional vibration resonance issues of a catamaran vessel have been investigated. The problem is described in details and measurement technique is discussed. Based on the obtained results, one can conclude that torsional vibration resonance in the ship propulsion system occurs under following circumstances: propulsion system has to be working at least two hours, revolutions of main engines are in the range from 1440 RPM to 1560 RPM, vibration occurs spontaneously when ship navigates at critical revolutions after 5 to 10 minutes. Occurrence of significant vibrating motion of the main engine regulation bar for fuel dosage (about 14 times more than out of resonance) leads to torque instability, and final result of this process is propulsion system torsional vibration resonance. Future investigations should be concentrated on causes of increased vibration motions of regulation bar in the "critical revolution range".

REFERENCES

Bently, ED., Hatch, CT. 2003. *Fundamentals of Rotating Machinery Diagnostics*. ASME Press, New York.

Hakkinen, P. 1987. Are Torsional Vibration Problems All in the Past? *The Motor Ship 9th International Marine Propulsion Conference*, London, 43–52.

Kicinski, J. 2003. Coupled Forms of Vibrations in Rotating Machinery. *2nd International Symposium on Stability Control of Rotating Machinery (ISCORMA-2)*, Gdansk.

Long, CL. 1980. Propellers, Shafting, and Shafting System Vibration Analysis. in: Harrington R.L. (ed.), *Marine Engineering*, SNAME, New York.

Platon, MV. 1995. *Vibration Analysis of a Swath-Type Ship*. Master Thesis, Massachusetts Institute of Technology.

Advanced Ship Design for Pollution Prevention – Guedes Soares & Parunov (eds)

Active vibration control using PI regulated electro-dynamic actuator—bond graph approach

I. Tomac, I. Pehnec & Ž. Lozina
Faculty of Electrical Engineering, University of Split, Mechanical Engineering and Naval Architecture, Split, Croatia

ABSTRACT: Active control vibration on the laboratory test model is described using a single channel feedback control. Experimental results of control system are presented. Numerical model is established using bond graph method. Using bond-graph method numerical simulation of such a model can be obtained. Model parameters used in numerical simulation are measured on the static model. Those values are used to make numerical simulation using ideal PI controller on bond-graphs model. Efficiency of both systems, numerical and experimental, is presented.

1 INTRODUCTION

There are three primary types of vibration absorbers: passive, semi-active and active. The earliest class of vibration absorbers is passive absorbers, which do not require any additional source of power to work. Next class is semi-active actuators that behave essentially as passive elements. Their active control property came form the fact that their passive mechanical properties can be adjusted, e.g. use of electrorhological fluids, shape memory alloys, etc. The latest class of vibration absorbers is the active absorbers, which are based on modern control theory, and one simple type will be presented in this paper. Main objective of active mechanical vibration control is to reduce vibrations of mechanical system modifying systems structural response. Important parts of any such system are: vibration detector, electronic controller and actuator. Bond-graphs are used to describe mechanical and electrical models. Using software, Bond-Sim (Damić & Montgomery, 2003), numerical simulation of described model can be obtained.

Theory and experiments are applied on a mechanical model of a 2DOF vibration system. Digital PI controller is used for control. The paper shows experimental and numerical results of active control vibration on impact exaction and mass unbalance. Experimental data are obtained measuring time response of stimulus and response signals with proper sensors. Numerical data were gained with direct measurement of mass and stiffness of described model.

2 PRESENTATION OF STUDIED MODEL

Firstly mathematical set up of the studied physical model is made. The same model is represented with bond graphs. Then numerically and experimentally measured FRF (frequency response function) is presented.

2.1 *Mathematical model*

In following text mathematical representation of 2DOF laboratory test model is displayed. Also, numerical values of mass and stiffness matrices that were obtained with direct measurement are shown here. The same will be used in numerical simulation.

Corresponding differential equation is:

$$\mathbf{M}\ddot{\mathbf{x}} + (\mathbf{D}\dot{\mathbf{x}}) + \mathbf{K}\mathbf{x} = 0 \tag{1}$$

Two beams are represented as lumped masses. Numeric's used for mass are:

$$\mathbf{M} = \begin{bmatrix} 4.8402 & 0 \\ 0 & 2.8288 \end{bmatrix} \text{kg}$$

Stiffness matrix is:

$$\mathbf{K} = \begin{bmatrix} 35.1318 & -13.5843 \\ -13.5843 & 18.7371 \end{bmatrix} \cdot 10^3 \ \frac{\text{N}}{\text{m}}$$

Figure 1. Test model equipped with electro magnetic actuator.

Type of system damping is unknown to us but as we are dealing with real system damping exists. So for this system damping is assumed viscous—proportional.

$$\mathbf{D} = \alpha \cdot \mathbf{K} + \beta \cdot \mathbf{M} \tag{2}$$

where coefficients are assumed: $\alpha = 0.05$ and $\beta = 0.0007$

Damping matrix:

$$\mathbf{D} = \begin{bmatrix} 24.8343 & -9.509 \\ -9.509 & 13.2574 \end{bmatrix} \tag{3}$$

Transfer function of system is:

$$\mathbf{G}(s) = \frac{1}{\mathbf{M}s^2 + \mathbf{D}s + \mathbf{K}} \tag{4}$$

Frequency response function of mathematical model using these values follows

2.2 *Bond graph representation of 2DOF mechanical system*

Main goal of bond graph method is to describe any physical system in two main terms: effort and flow variable. Product of effort and flow gives power. That way different physical systems can be combined into a single model. Short description of the method follows.

The horizontal line (bond) is a connection between subsystems that exchange energy. Positive direction of energy flow is noted with a half arrow. Normal arrow is used to describe signals. Energy can flow into three different containers: flow storage *C*, effort storage *I*, resistive element *R*. It can be transformed using transforming elements: transformer *TF* or gyrator *GY*. Energy can be channeled thought two different junctions: serial junction *s* and parallel junction *p*. Serial junction property is to keep the same flow variable to all connected elements while sum of efforts equal zero. For the parallel junction is vice versa.

Model from Figure 2 is modeled using bond-graphs; it is displayed on next figure. Similar diagram using program BONDSIM response from bond graph model is obtained on impulse exaction for free model and for actively controlled.

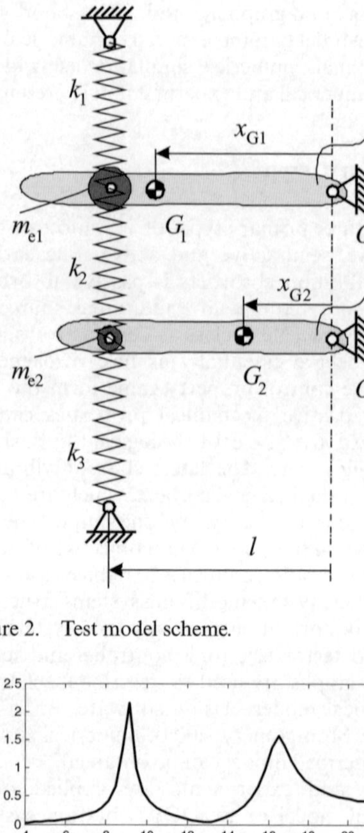

Figure 2. Test model scheme.

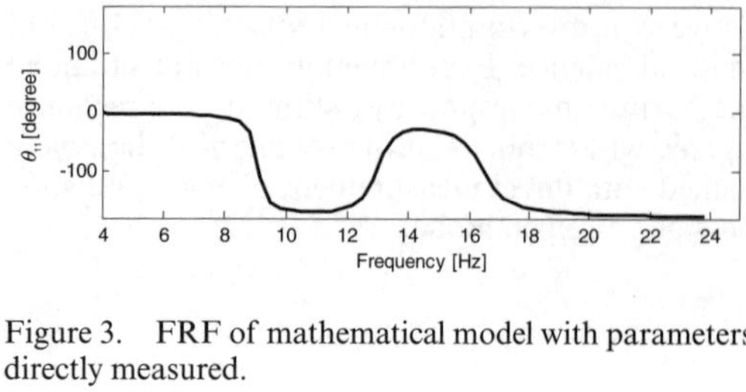

Figure 3. FRF of mathematical model with parameters directly measured.

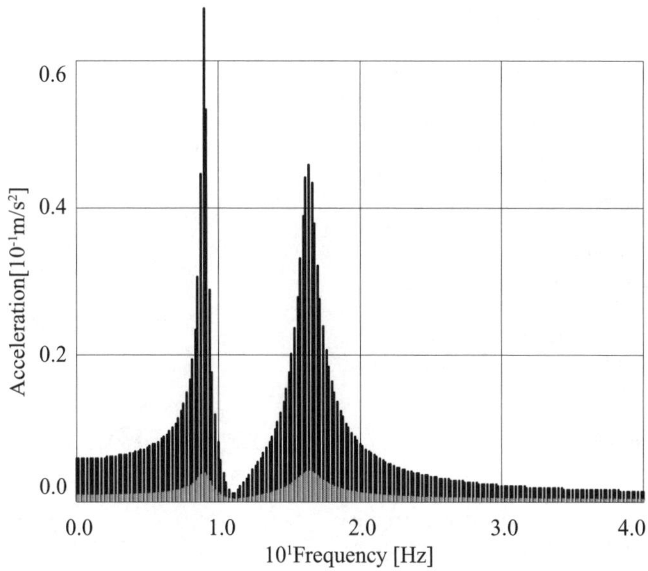

Figure 4. FRF of bond-graph model using BondSim. Black response is for free system, and grey is for controlled.

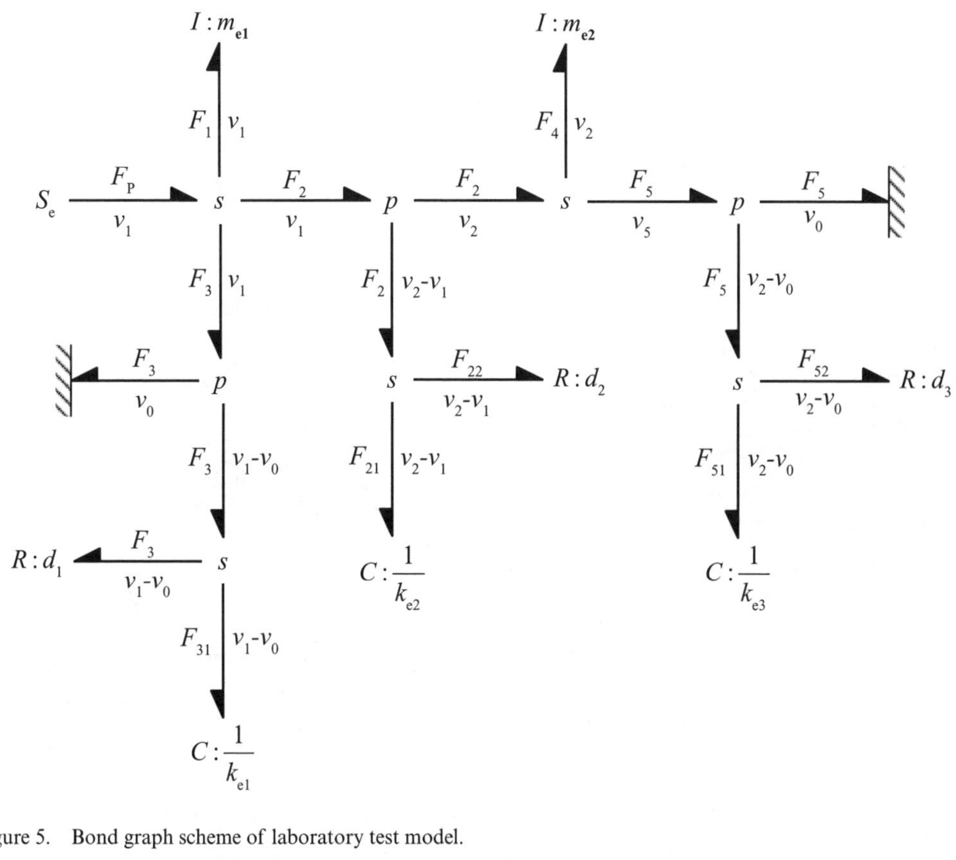

Figure 5. Bond graph scheme of laboratory test model.

3 ACTIVE VIBRATION CONTROL

For vibration control, single channel feedback control system is used. System's behavior is monitored with a single control sensor and a single secondary actuator, for which only a single channel feedback controller is required. Typically the control sensor measures the total response of a mechanical system, which is then fed via controller to the secondary actuator. Mechanical block diagram of such a feedback control system is shown on Figure 6.

Linear theory (Fuller, Elliott & Nelson, 1996.) can be used to gain overall response of the system including feedback control. Transfer function of such a system in s space is:

$$W(s) = G(s)[F_p(s) + F_s(s)] \tag{5}$$

where, $W(s)$ = Laplace transform of system response; $G(s)$ = transfer function of the original mechanical system; $F_p(s) + F_s(s)$ = Laplace transform of the net exaction.

Corresponding electrical diagram in which the differencing operation between primary and secondary inputs is shown to obtain net exaction to the mechanical system (Figure 7).

Laplace transform of the secondary exaction, expressed in term of controller transfer function is

$$F_s(s) = H(s) \cdot W(s) \tag{6}$$

Combined with (5), transfer function of the mechanical system with feedback control is obtained:

$$\frac{W(s)}{F_p(s)} = \frac{G(s)}{1 + G(s)H(s)} \tag{7}$$

Experimental results will be presented in terms of FRF (amplitude and phase) for actively controlled system, compared with uncontrolled (original) system. Prior to showing results, performance of the control system will be tested.

3.1 *Performance of active control device*

Performance is tested measuring FRF that is obtained from white noise signal reproduced to a shaker that will be used for vibration control. On the following figure measuring system is represented. It is consisted of two sensors (force and acceleration), electro-dynamic shaker with signal amplifier, A/D measuring PC card, one D/A card which is used for white noise generation. Signal is processed with software LabVIEW.

FRF results are represented in form of amplitude and phase.

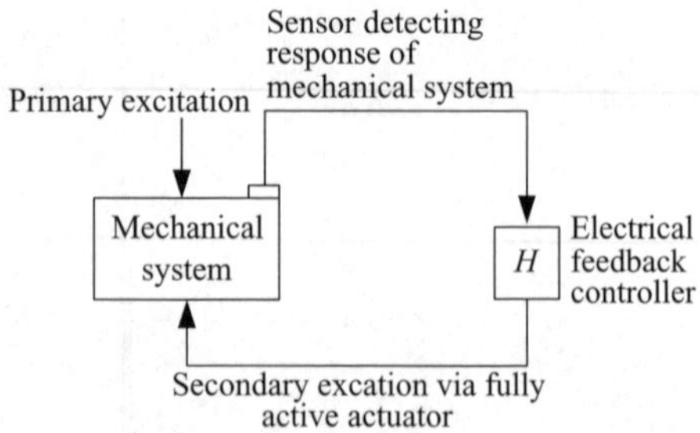

Figure 6. Components of a feedback control system.

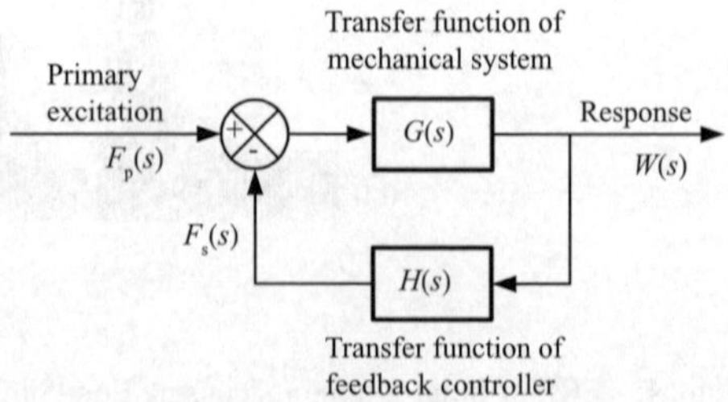

Figure 7. Equivalent electrical block diagram of a feedback control system.

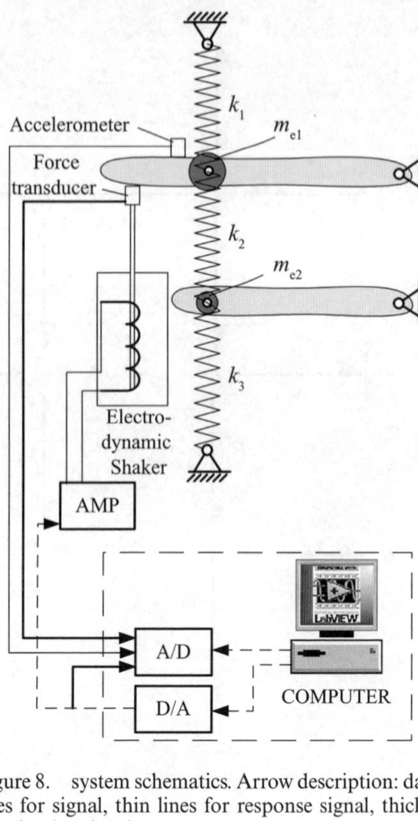

Figure 8. system schematics. Arrow description: dashed lines for signal, thin lines for response signal, thick line for stimulus signal.

3.2 *Experimental results of active system control*

Equipment used for active control is:

- Homemade electro-dynamic shaker with amplifier
- Capacitive accelerometer with analog integrator implemented for velocity signal
- Digital PI controller programmed in LabWIEV (Figure 10) using NI PXI-6259 multifunction DAQ card.

PI controller constants K_P and K_I are tuned using Ziegler–Nichols method.

Active control system attached to the laboratory test model modeled using bond graphs is shown on the Figure 11.

Controlled source effort presents electro-dynamic shaker—secondary exaction F_e. Primary exaction F_p is independent source effort. To test efficiency of the control system in a frequency range, for harmonic exaction (Figure 12). Test results are obtained using impact test to the test model with PI control enabled and disabled. Test results are:

Diagrams on the Figure 12 show a response for the beam where control system is attached to. Decrease of amplitudes owing to active control, especially in resonance areas, can be observed. Interesting diagrams are obtained for the second beam which is free.

Figure 13 shows: When control system is off two resonances are present from 2DOF system. When control system is on, then a second beam behaves as a 1DOF system with kinematic excitation.

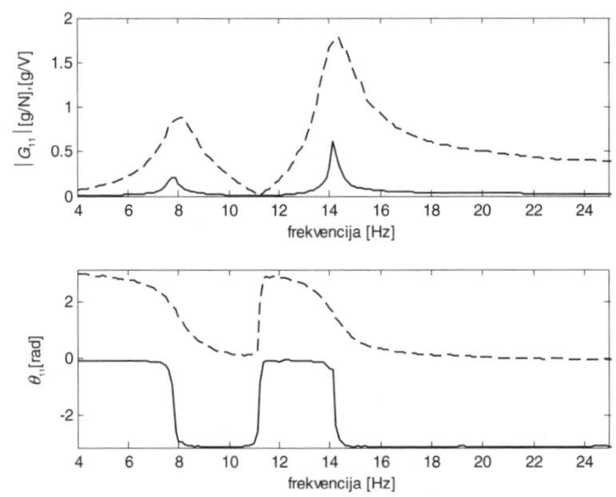

Figure 9. Measuring results. Dashed line has voltage output signal from D/A card as a stimulus and blue has stimulus signal from force transducer.

3.3 *Simulated results of active control vibration*

Using software BONDSIM simulation of the system with active control vibration has been made.

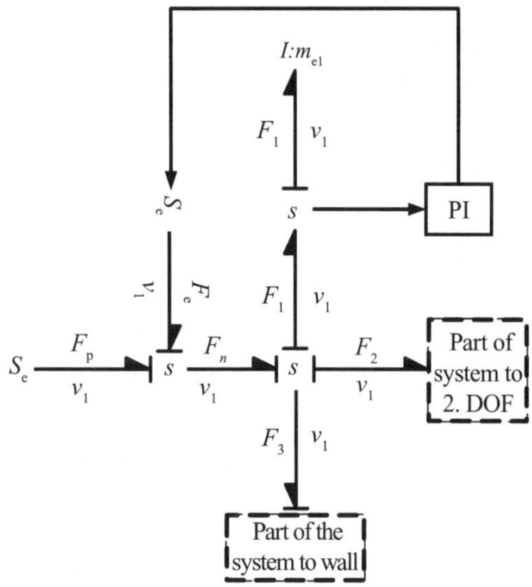

Figure 11. Ideal active control system attached to test the model—bond graph representation.

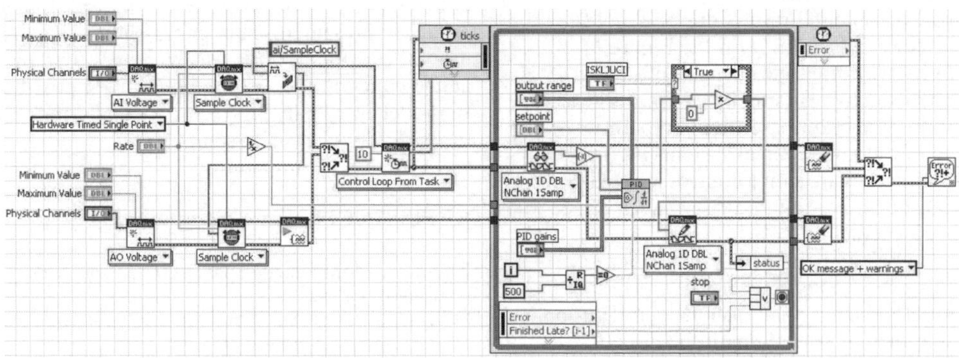

Figure 10. Block diagram of LabVIEW PI controlled.

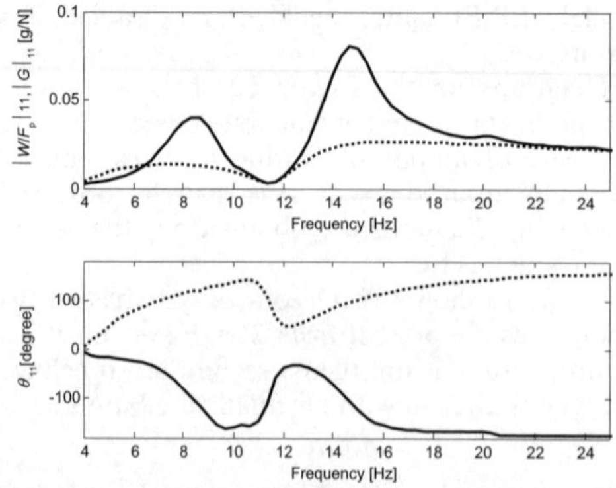

Figure 12. FRF in form of amplitude and phase showing system response to an impact test with (dashed line) and without (full line) active controll. First DOF.

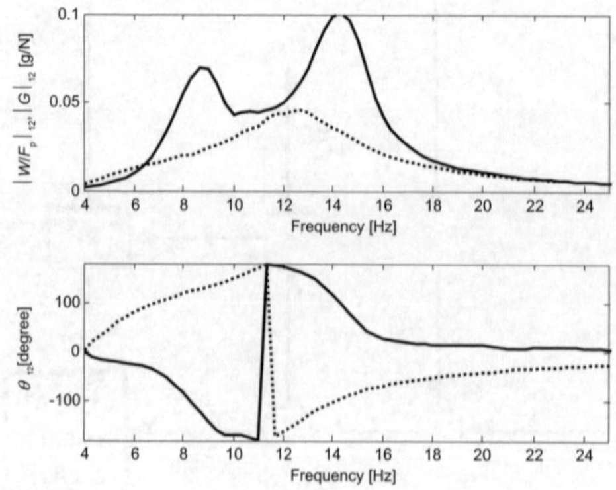

Figure 13. Second DOF, FRF with (dashed line), without (solid line) active control.

It is important to say that simulation results are gained from data that were used for mathematical model. Figure 14 shows displacement measured in time for harmonic exaction induced with mass unbalance at 120 rad/s.

Also experimental results are presented for the same excitation frequency on the Figure 15.

We observe amplitude decrease of 80% at 19 Hz due to active control in simulation what is in accordance with prediction based on comparative FRFs for undamped and damped system (Figure 4).

The efficiency of active control at 19 Hz on the physical model is around 30% what is in accordance with prediction based on comparative FRFs for undamped and damped system (Figure 12).

Difference among numerical and experimental results indicates that system identification of the experimental system has to be done to extract more accurate system parameters for use in numerical model.

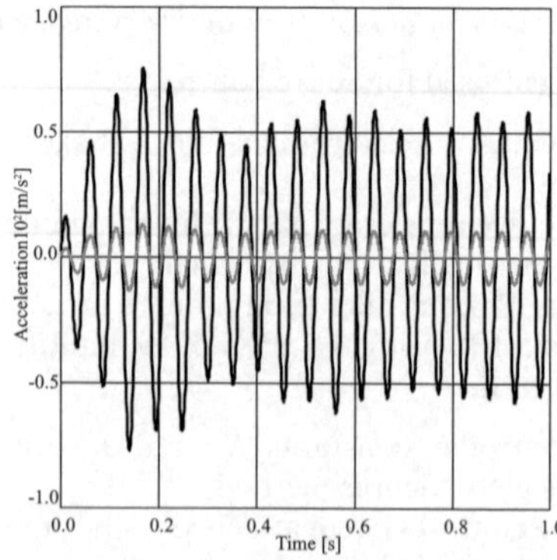

Figure 14. Simulated results of mass unbalance for frequecy od 120 rad/s. Black line stands for free system and gray stands for controlled system.

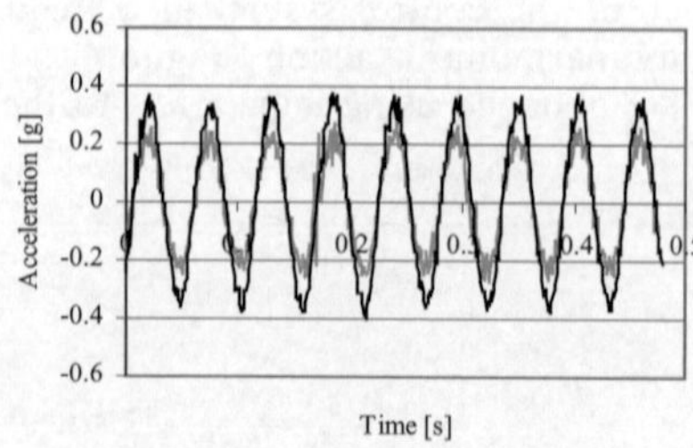

Figure 15. Experimental results of mass unbalance for frequency od 120 rad/s. Black line stands for free system and gray stands for controlled system.

4 CONCLUSION

Mathematical model is established using directly measured parameters. Using the same parameters bond graph model with PI control is established. Numerical results are presented. Experimental measurements on the real test model are conducted with active control system attached to it. Measurements for experimental model are presented and efficiency of both numerical and experimental model is presented. This paper shows that bond graphs can be used to easily model system for active control. This method gives us a possibility to model every part of system in details, making it more closely to a real system.

REFERENCES

Damić V., Montgomery J. 2003. *Mechatronics by Bond Graphs*, Berlin.

Fuller C.R., Elliott S.J., Nelson P.A. 1996. *Active Control of Vibration*, London.

Ljung L., Glad T. 1994. *Modeling of Dynamic System*, Englewood Cliffs, New Jersy.

Advanced Ship Design for Pollution Prevention – Guedes Soares & Parunov (eds)
 ISBN 978-0-415-58477-7

Measurements for reliability of ship production and operation interaction

B. Ljubenkov, T. Zaplatić & K. Žiha
Faculty of Mechanical Engineering and Naval Architecture, University of Zagreb, Zagreb, Croatia

F. Bosančić
Brodosplit Shipyard, Split, Croatia

ABSTRACT: The ship propulsion system alignment is important activity for quality and reliable operation. The alignment procedure is under classification society rules, which determines guidelines for design, calculation and installation of the propulsion system elements. Part of the classification society guidelines is related to measurement of the hull deflections, which could be caused by ship loading condition, temperature variation affecting the propulsion shafting system and launching. The introduction to this paper considers the system engineering approach to the ship propulsion system alignment problem and its meaning to the ship production and operational reliability. Paper presents use of digital photogrammetry for measurement of the structure deflection in the machinery space and displacements of the main engine after launching. Theory and measurement equipment of digital photogrammetry are presented. The automated photogrammetric system TRITOP, presented in the paper, is used for 3D measurements, results processing and analysis. Measurement procedure and results are illustrated by an example. Conclusion contains review of the measuring method applicability, advantages of the digital photogrammetry and suggestions for the propulsion alignment development.

1 INTRODUCTION

Ships are parts of transportation system and themselves are systems comprised of a number of dependent or interacting subsystems. The ship structure is a subsystem of a ship that provides possibilities for implementation of other subsystems. Of particular interest for overall ship safety and general pollution prevention is the placement of the main engine into the ship's hull. However, two important systems mutually intensively interact through subsystem for propulsion over the intermediate shaft with flanges, bearings, coupling, propeller shaft and with the external environment through the screw propeller. Therefore, not only their particular qualities but also the quality of installation and assembly of all the elements, i.e. their joining into a whole system, contribute to the obtaining of a high quality and reliable ship operation. Such conditions require system engineering approach.

2 THE OPERATING ENVIRONMENT ARCHITECTURE

A significant aspect of system engineering (SE) is to fully understand and analyze the operating environment (Wesson, 2005). That implies recognition of system mission, objectives and application, decomposition of operating environment relative to the mission into manageable tasks and development of a solution that ensures mission success within constraints of cost, schedule, technology, support and risk factors. The challenge is that complex systems involve large number of disciplines in problem solving and require a strategy to achieve convergence and consensus about the operating environment. The approaches to complex systems impose moral and ethical obligations to organization and society and well being of the user and the public. Overlooking of important aspects of operational environment may impact human life and properties that may have severe consequences and penalties.

All natural and human-made systems function as individual systems of interest (SOI) within a hierarchy of system of systems. Each higher level abstraction has a scope of authority and operational bounds (Wesson, 2005). The systems of interests in our case are the ship hull structural system, the shafting system and the propulsion system.

The operating environment can be considered as consisting of two high level domains:

1. Higher order systems
2. Physical environment

The higher order system in this consideration is the shipyard characterized by its:

1. Organizational purpose or mission (building ships)
2. Objectives (building efficiently and profitably)
3. Organizational structure (as it is)
4. Rules, policies, procedures of operations (rules, regulations and conventions of classification societies; laws, conventions and standards of national and international authorities)
5. Operating constraints (space, workmanship, material, manpower, financing, market conditions, experience, knowledge)
6. Accountability and objective evidence of value added tasks performed (technical adequacy, quality control, cost/benefit analysis)
7. Delivery of systems, products and services (inspections, sea trial, guaranty period)

Human-made systems interact with external systems within the physical environment. The physical environment consists of three levels of analytical abstractions:

1. Human-made systems (business, educational, financial, transportation, governmental)
2. Induced environmental elements (intrusions, disruptions, perturbations, discontinuities)
3. Natural environment system elements

Natural environment comprises three classes of environmental elements:

1. Local environment (current location of the shipyard, contractors, subcontractors, banks etc)
2. Global environment (oceans, world economy)
3. Cosmosheric environment (not relevant in this consideration)

The purpose of any system is to accomplish the intended user's missions and performances as bounded by the operating environment. Reliability is defined as the probability of successful completing a bounded mission with specified objectives within a specific operating duration for a prescribed set of operating environment conditions without a failure. A failure is determined by criteria that establish the criticality of system components to achieving the mission. Or, what degree of performance degradation is allowable and still meet performance standards. The success of any system is ultimately determined by its ability to cope with threats of vulnerability and survivability within prescribed operating environment (Wesson, 2005).

3 HULL, MAIN ENGINE AND PROPULSION SYSTEM ELEMENTS

An increase in the main engine power and in the dimensions of propulsion system elements has an influence on the problem of shaft alignment. On the other hand, the ship structure optimization results in a reduced ship plate thickness, in changed geometric features of stiffeners and in a narrower ship which is more likely to be subject to deflections, which in turn influence the elastic line of propulsion shafting and the wear of bearings (ABS Guidance, 2006) in an interactive manner.

The procedure of propulsion system alignment is to be carried out in accordance with classification society rules, which give instructions for the design and revision of the calculation model. After that, the process of assembly and alignment of propulsion system elements, monitoring activities and requirements for measuring the accuracy of the process parameters are defined. A whole chapter deals with the ship hull deformations, which can be caused by different conditions of ship loading or by changes in the temperature in the machinery room and in bearings during the ship assembly and in ship service.

In the design stage, all elements of the propulsion system are modeled according to the real situation in the design. Calculations to be used are defined, and calculations results are key parameters for the control of the propulsion system alignment procedure. These parameters include the distribution of loads and reaction in bearings, the shape of the intermediate shaft deflection and of the crankshaft. The assembly of propulsion system elements follows after the modeling and calculation of the alignment procedure. It is important to point out that a necessary condition for an accurate alignment procedure is that the welding of structures in the machinery room has been completed in order to avoid the effect of additional welding on the alignment procedure accuracy. During assembly, parameters affecting the accuracy of the assembly of propulsion system elements are monitored.

Hull deformations have a dominant influence on the propulsion system elements when a ship is in service. Changes in the elastic line shape of the intermediate shafts occur, resulting in the irregular position of the intermediate shafts in bearing and thus causing their premature wear. It is not easy to calculate hull deformations; yet, it is necessary to predict them in the alignment calculations in order to prolong the life of bearings. Classification societies conduct extensive research into ship hull deformations occurring with different ship types and sizes. Research results form a database for predicting the size and the shape of ship hull deformations. Researches have shown that tankers and bulk carriers are most likely to be subject to hull deformations due to their short and stiff propulsion systems which are usually directly connected to the main engine. Container ships also suffer hull deformations because of their long and flexible propulsion systems and a narrow form.

Different measurement methods have been used in the research into hull deformations, and one of the research goals has been to define a reliable measurement method. One of the methods used in determining hull deformations is the measurement of vertical positions of intermediate shaft bearings. These values are first measured during the ship construction in the dock and are taken as initial values. After that, the vertical position of bearings is measured after the ship launching, in the ballast condition and in the fully loaded condition. From these values one subtracts the initial values, and the result gives the values of hull structure deformations.

The hypothesis proposed in this paper is that hull structure deformations can be reliably measured by using the photogrammetric method. By putting a large number of measuring points on the ship structure and on the propulsion system elements, and by calculating their position in the space for different ship conditions, extensive data are amassed. Using these data one can more easily make a conclusion about the direction and the value of deformations of ship hull structure.

4 PHOTOGRAMMETRY

The optical measurement techniques are an important part of the experimental mechanics measurement techniques (Gomerčić, 2000). A lens is used in the measurement procedure, and an optical effect is recorded by a camera. The optical measurement technique characteristics are high accuracy and no contact with the object to be measured; therefore, it does not affect the object.

4.1 *Photogrammetry theory*

Photogrammetry is an optical measurement technique where 3D coordinates of observed object points are determined from photographs taken with a camera (Gomercic, 1999). The recorded distinguished measuring points are used for reconstruction of optical rays.

The mathematical model for reconstruction of a light rays uses two sets of coordinates, i.e.:

- OBJECT (X, Y, Z)—coordinates of an object points,
- IMAGE (x, y)—position of an observed point projection in a photograph, as shown in Figure 1.

Equations of image coordinates are:

$$x = f(c, Xo, Yo, Zo, \omega, \psi, \kappa, X, Y, Z, x_o, \Delta x)$$

$$y = f(c, Xo, Yo, Zo, \omega, \psi, \kappa, X, Y, Z, y_o, \Delta y),$$

where:

c—coefficient of the camera, value close to focal length

Xo, Yo, Zo—coordinates of the lens origin in auxiliary coordinate system

ω, ψ **and** κ—turning angles of the camera coordinate system

x_o, y_o—image coordinates

X, Y, Z—object coordinates

Δx, Δy—deviation from the central projection. Due to lens imperfection, position of the point, which is not on a central projection axis, has to be corrected. In corrections, parameters like radial distortion, disparate scale of the photograph, no perpendicularity and deformity of the photograph are considered.

Reconstruction of an optical straight-line consists in establishing a functional relation between the object and image coordinates. This model is sufficient for considering measurement in the plane with two unknown values of the object coordinates X and Y, (Z = 0), and they could be calculated using two equations of image coordinates (x, y).

A 3D measurement includes three unknown values of object coordinates X, Y, Z. Two equations of image coordinates are not enough for calculation of three coordinates. Additional equations of image coordinates are necessary to solve the problem. The object point has to be recorded from another position, which gives additional two equations. System of the equations is predefined.

The optical measurement techniques for calculation of point position in space are based on stereoscopic effect. A 3D position of a point is determined by triangulation, as shown in Figure 2.

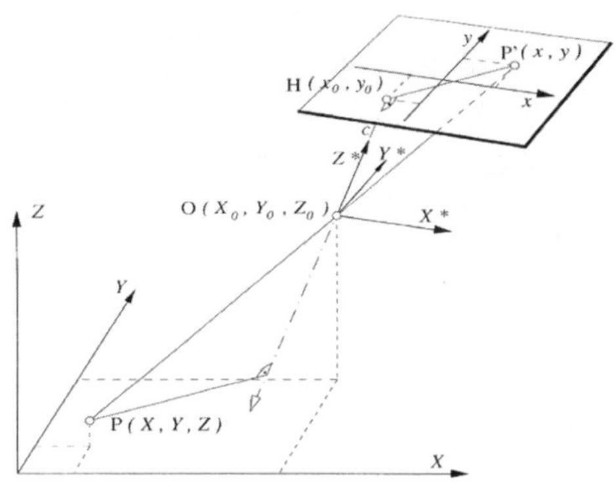

Figure 1. Relation between object and photograph point.

The position of a point is determined from an intersection of straight lines determined by a point on object P and its projections on photographs. It is necessary to set up additional equations in order to record the point from another camera position and make redefining of the system, i.e. a larger number of equations than unknowns possible. The system of nonlinear equations redefined in this way is solved iteratively by an error minimization method, and the outputs of this analysis are 3D coordinates of measuring points and other parameters of the mathematical model.

4.2 *Photogrammetry system and measurement procedure*

For preparation of this paper, the TRITOP-V5.3 system (GOMmbH, Germany) was used consisting of a digital camera, notebook, scale bars and reference markers, as shown in Figure 3 (Ljubenkov, 2006).

A condition for quality measurement and exact and reliable results is use of a high-resolution digital camera, with good mechanical stability of the lens. A notebook is used to save and process the records, and calculate the measurement results. It is desirable to use a notebook with stronger processor and larger memory space. The records processing and measurement results calculations are performed by TRITOP-V5.3 program package.

During the measurement procedure, the scale bars are placed on the measured object in order to define the measuring project scale and exact geometrical relations between the measuring points.

An important condition for reliable measurement using the photogrammetric system is availability of measuring points with sharp contrast. So, white points on black background or black points on white background are used. The measuring points are divided into coded and uncoded. The TRITOP software recognizes the coded points

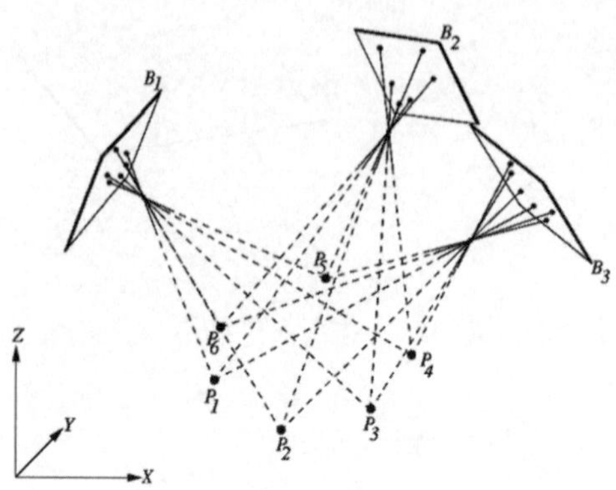

Figure 2. Triangulation principle.

Figure 3. Photogrammetric system.

Figure 4. Coded measuring point.

by the defined bar code, as shown in Figure 4, and they are used to determine position of a camera in space. Uncoded points are placed in characteristic points on the object the position of which is to be determined. The TRITOP software automatically determines position of an uncoded point, and the measuring point is attached an identification number which later facilitates the analysis of the results and reconstruction of characteristic lines, edges, and bores. The uncoded points are usually equal in size to the coded points.

Inspection and monitoring of the TRITOP photogrammetric system is performed by VDI/VDE guideline, which introduces a standardized way of determining the length measuring deviation as stated in the ISO 10360-2. The measuring uncertainty for TRITOP photogrammetric system is:

- 0,015 mm for size of the object—$1{,}0 \times 0{,}5 \times 0{,}5$ m^3
- 0,2 mm for size of the object—$10 \times 5 \times 5$ m^3

The measurement (Zaplatic, 2008) using the photogrammetric system is performed in six steps:

- Measurement task and recording strategy
- Placing measuring points and scale bars
- Preparation of camera
- Recording the object
- Record processing and
- Result analysis and presentation.

Measurement task and recording strategy
In the first step, the object is selected and measurement task and recording strategy are determined.

The measurement task defines what to control on the measuring object. The recording strategy includes defining of the measuring point position on the object and it is closely related to the measurement task. Size and number of measuring points is determined depending on the object accessibility, i.e. free space around it. Distance between the person taking photographs and the object determines the size of the measured volume.

Placing measuring points and scale bars
The scale bars, coded and uncoded points are properly arranged on all the surfaces and edges on the measured object according to the recording strategy. An important condition that needs to be fulfilled during the measuring points placing is that minimum 5 coded measuring points caught on each photograph.

Preparation of camera
This step includes determining and adjustment of camera parameters according to the recording conditions. The camera diaphragm opening and focal length of the lens are defined. The camera parameters are manually set and need to be fixed and kept unchanged during the entire recording process. The photogrammetric system uses wide-angle lens. Its characteristics are small distance between focus and lens, wide angle of view and in-depth sharpness.

Recording the object
The measuring object is recorded according to demands of the photogrammetric technique. Photographs have to cover all measuring point positions. Number of photographs to be recorded depends on size and shape of the measured object. Recording of voluminous objects larger in size is characteristic for shipbuilding. It asks for a larger number of photographs and recording with overlaps, since it is not possible to encompass the entire object from one position. Each measured volume is to be recorded from different angles and recording elevations, and the overlapped photographs are than merged by the program package into a single measuring project.

Record processing
The processing and calculating modules of the TRITOP software process the photographs and calculate the object coordinates. Result is a cloud of measuring points or a text file with data on measuring points.

Result analysis and presentation
After the photographs have been processed and object coordinates calculated, lines, curves and planes relevant for presentation and analysis of the results are reconstructed from the measuring points cloud. The TRITOP software constructs measuring elements by minimum square deviation method where lines, planes or surfaces are set approximately through measuring points. Measuring elements are used for distance or position determination. The results are presented in tables and graphs.

5 MACHINERY ROOM STRUCTURE DEFORMATION MEASUREMENT USING THE PHOTOGRAMMETRIC METHOD

The measurement of structure deformation using the photogrammetric method was carried out in the machinery room space below the third platform in the vicinity of propulsion system elements. The space, shown in Figure 5, is limited by the following: the machinery room transverse bulkhead, two longitudinal bulkheads, side plating, web frames and the sternpost bulkhead. The machinery room length is 23.2 m, its width at the transverse bulkhead is 16.8 m, and its height to the third platform is approximately 8 m.

Magnetic coded measuring points of 18 mm in diameter and self-adhesive uncoded measuring points of 20 mm in diameter were used. More than 200 measuring points were placed. In addition, calibrated scale bars, shown in Figure 8, which are used for the scale definition, are placed in the machinery room.

After photographs had been taken (Figure 9), they were transferred to a computer and processed.

In order to define the hull structure deformations after launching, two measurements need to be carried out. The first measurement is carried out before launching, and the second after launching. Measurement results are put into a common coordinate system. The results from the first measurement are taken as referential data from which the results of the second measurement vary.

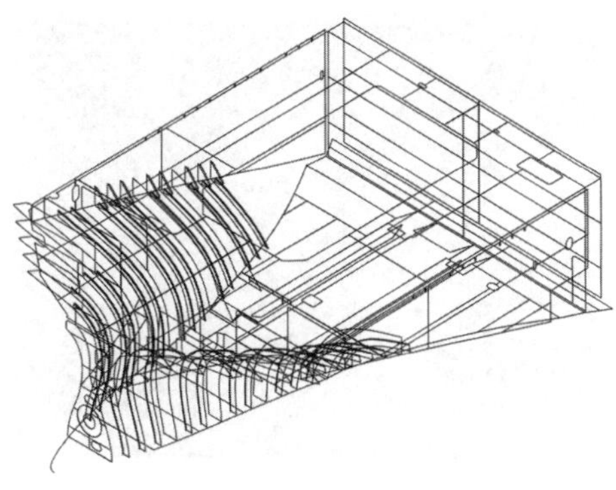

Figure 5. A CAD model of a machinery room.

Both measurements were carried out in the same way. In the machinery room space, coded and uncoded measuring points are put on the main engine and on the longitudinal and transverse bulkheads at three different heights, as well as on web frames in the stern, as shown in Figures 6 and 7.

The measurement result is a cloud of measuring points, shown in Figure 10, which represents coordinates of all measuring points in the machinery room space and on the main engine. For the purpose of conducting the measurement of the machinery room, more than 600 digital photographs with a 16 megapixel resolution were recorded.

5.1 *Analysis of machinery room structure deformations after launching*

After the measurements of the machinery room structure had been conducted before and after launching, measurement results were analyzed. The measuring points of the first and the second measurement are corresponded in a common coordinate system by means of chosen points on the machinery room structure. The TRIPTOP Def program package can then identify the deflection of the marked measuring points with regard both to the direction and the value. The correspondence of measurement results was established by means of 18 measuring points. The measuring points that were chosen were those at the joint of the transverse and longitudinal bulkheads as it is considered that they could show the least deflection, together with the points on the web frames FR31 and FR24.

Figure 6. Measuring points on the machinery room structure.

Figure 7. Measuring points on the main engine.

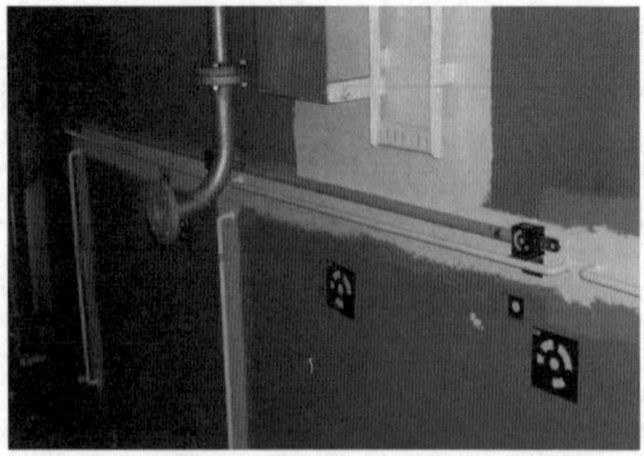

Figure 8. A scale bar.

Figure 9. Taking photographs of the object.

Since there is a large number of measuring points and the volume to be measured is large, a comprehensive and detailed analysis may be required. Examples presented in this paper indicate the scope of the result analysis in the case when the photogrammetric method is applied. Measurement results could be presented by graphs and tables. The measurement results in the given examples are presented isometrically, where deflections are shown by a direction vector for each measuring point. To enable an easier control of the measurement results, a deflection column is defined in the program, and the points, which show the greatest deflection, are in interval from 2.5 mm to 3 mm.

Figure 11 presents the measurement results of the hull structure deformations and the main engine deflection after launching.

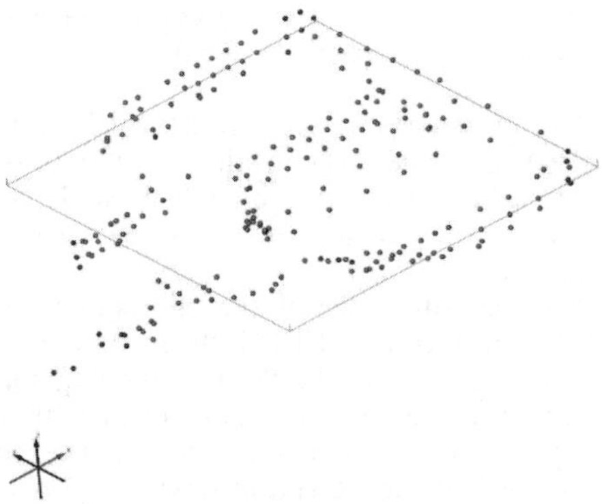

Figure 10. A cloud of measuring points on the main engine and the hull structure.

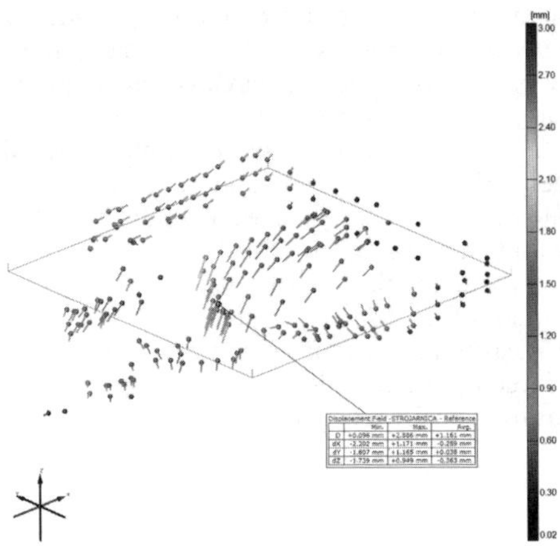

Figure 11. Deflection vectors of the structure measuring points and of the main engine after launching.

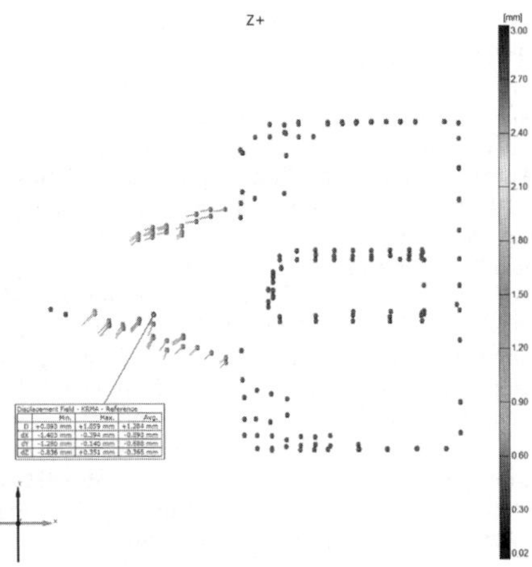

Figure 12. Deflections in the stern area of the ship—top view.

As far as the structure is concerned, the most outstanding deformations occur in the stern area of the ship where the hull form is narrower and where there is no double plating. There, deformations of measuring points of up to 1.5 mm were measured which are illustrated by Figures 12 and 13 (top and front view).

The largest deflections, shown in Figure 14, refer to the deflections of the main engine after launching. At the moment of launching, the main engine is not in its final position since the propulsion shafting alignment has not been completed. It is fixed by bolts and is still free to a large degree, which has been confirmed by measurements. During launching, the main engine is lowered by 1 to 2 mm with respect to its original position. Smaller deflection values were measured on the front part of the main engine, and higher on the rear part.

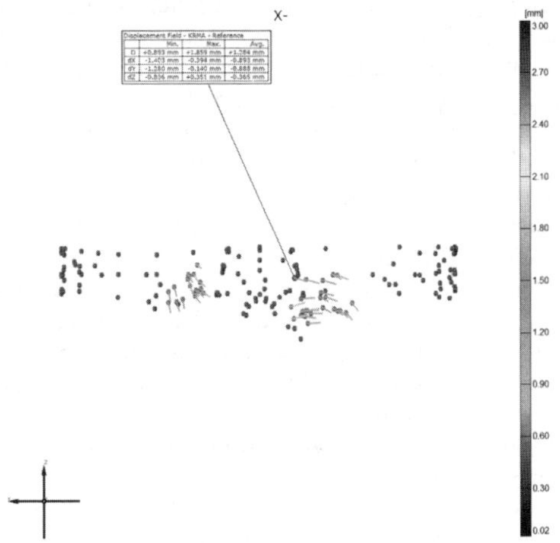

Figure 13. Deflections in the stern area of the ship—front view.

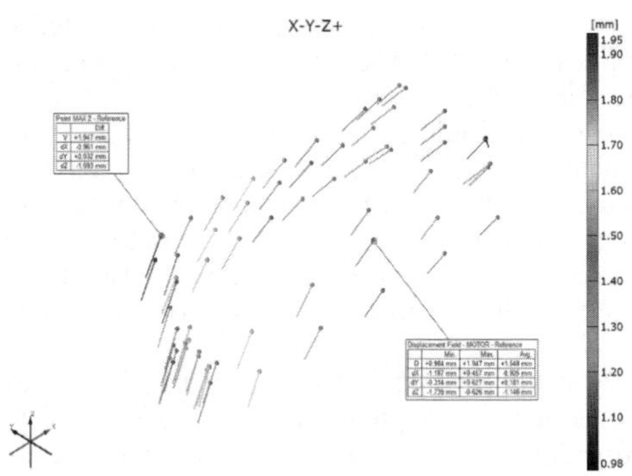

Figure 14. Results of the main engine deflection analysis.

6 CONCLUSIONS

The paper presents the application of the photogrammetric method to the measurement of hull structure deformations. The research goal of defining a reliable measuring method has been achieved. Also, the hypothesis proposed in this work has been confirmed since the measurement of the machinery room structure deformation after launching has been successfully carried out by using the photogrammetric method.

The system engineering consideration identified the shipyard responsibility as a higher order system in fulfilling the systems of interests that in our case are the ship hull structural system, the shafting system and the propulsion system as well as their interactions. Systems have an inherent failure rate at delivery due to latent defects from workmanship processes and methods. Therefore the aim of the measurements presented in the paper with respect to the systemic approach is to confirm the reliability of a bounded mission in changing the shipyard's technology of shaft alignment with specified objectives to shorten the building process for a prescribed set of operating environment conditions without a failure. As the system evolves, interaction between the system and its operating environment and internal elements must be compatible and interoperable. In the final consequence the measurements have to assure compliance with governing rules, conventions and standards as well as to verify the quality of transition to more efficient workmanship practice in order to assure reliable ship operations. Moreover, the reliable ship operations have an important role in pollution prevention.

Positions of 202 measuring points in the machinery room and on the main engine have been measured. The measuring uncertainty of 0.2 mm was obtained, and the measured values of the structure deformation and the main engine deflection amount to 2 mm.

Advantages of the photogrammetric method application over the commonly applied procedures are numerous.

First, it should be pointed out that this method enables obtaining a large number of measuring points, which in turn enables a better quality analysis of the machinery room structure deformations. A large number of measuring points also enables extending the control task and defining additional measurements if/when necessary.

Another distinct advantage of the photogrammetric method application is a capability of the program package to present measurement results visually. The program package enables the control of the processing of photographs and the calculation of measurement results, and it even makes it possible to intervene into the procedures if necessary. Photographs make it easier to cope with the measurement results because photographs can identify at any instant which part of the measured object is considered. Photographs can be filed as documentation, which can be used for subsequent verifications of some measurement results.

As a final advantage of the photogrammetric method to be mentioned here is the use of computers, which develop constantly. The ongoing development of computers is accompanied with the development of digital cameras, program packages and abilities of photograph processing. Also, calculations, analyses and measurement result presentations are being improved. All this contributes to faster and easier measurements.

Further development of the photogrammetric system strives towards faster and simpler measurement of large objects by the implementation of new possibilities, such as wireless transmission of photographs to the notebook.

Also, development of photographs processing enables that lines drawn randomly on an object could be recognized by software on the basis of the line and background contrast.

Fast and simple construction of elements from a cloud of measuring points is important for the measurement results analysis. This might be achieved by specially designed adapters that would be automatically recognized by the software.

REFERENCES

Gomercic, M. 1999. *Contribution to Automatic Processing of Optical Effect in Experimental Stress Analysis*. PhD thesis. Faculty of Mechanical Engineering and Naval Architecture, Zagreb, Croatia.

Gomercic, M. & Jecic, S. 2000. A New Self-Calibrating Optical Method For 3d-Shape Measurement. *17th Symposium "Danubia-Adria" on Experimental Methods in Solid Mechanics,* Prague.

Guidance Notes on Propulsion Shafting Alignment. 2006. American Bureau of Shipping, Houston, USA.

ISO 10360-2 2001. *Coordinate measuring machines used for measuring linear dimensions.* Volumetric length measuring error E.

Ljubenkov, B. 2006. *Modified Photogrammetry Method in Shipbuilding*. PhD thesis. Faculty of Mechanical Engineering and Naval Architecture, Zagreb, Croatia.

Wasson, C.S. 2005. *System Analysis, Design and Development,* Wiley-Interscience Publication.

Zaplatic, T. Gomercic, M. Ljubenkov, B. & Bakic, A. 2008. Dimensions and Shape Control of Sub-assembled Sections using Digital Photogrammetry. *Schiffbauforschung (1)*:68–83; Rostock, Germany.

Ship design

Advanced Ship Design for Pollution Prevention – Guedes Soares & Parunov (eds)
© 2010 Taylor & Francis Group, London, ISBN 978-0-415-58477-7

AFRAMAX tanker Baltic Ice class—power determination doubts

D. Bezinović
Zagreb, Croatia

ABSTRACT: The goal of the paper is to present parameters that have influence on basic design regarding power determination of Aframax size tanker with Baltic Ice 1A class, particularly taking into account the Swedish—Finnish Ice Class Rules. According to rules determination of the engine output required of ships for navigation in ice should be obtained according to rule formula or by other method which in our case was model testing in ice. Differences of results using formula and model testing are significant so the investigation of correlation factors is necessary and the formula should be updated to enable the preliminary power prediction more accurate.

1 INTRODUCTION

In last decade interest for ice class vessels increased. In oil tanker fleet particularly, an aframax tanker as type was frequently required.

Reason for this is increasing of oil export from Russian ports in Baltic.

Overview of Terminal and Port development in Gulf of Finland shown in Table 1.

In this paper the design of the aframax tanker size with ICE 1A Baltic class that is efficient in open water and able to pass through channel ice in Baltic, particularly in Primorsk port area is considered.

1.1 *Why aframax size?*

Strait Skagerak connects Atlantic ocean with Baltic. Limited draught of vessel in transit is abt. 15.15 m. This value satisfies draught of Aframax size vessels that found is very acceptable.

1.2 *Design target*

Vessel is designed and optimized for open sailing mode. She has hull with bulb, slow speed main engine with fixed pitch propeller.

Design target is to satisfy ice class rules [FSICR] with the same propulsion components needed for open sailing mode, because more than 90% of life will sail out of ice.

Table 1. Export of oil rate in million tones.

Year	Gulf of Finland	Port Primorsk
2000	40	0
2005	125	60
2010	160	75

1.3 *Design requirements*

Design requirements are as follows:

- Size of vessel—Aframax tanker.
- Operational area—Baltic, Primorsk port.
- Ice Class—Baltic 1A acc. to Finnish Swedish Ice class rules 01 October 2002.
- Draught limit—15.15 m.
- Hull design—with bulb, optimized for open water sailing mode.
- Special ice capability—just acc. to requested ice class.

Yard has developed several forms of an aframax size tankers with high block coefficient and optimised hull for open water sailing. The question that could be placed is what will be the effect of ice class rules on power determination?

One of aframax forms was chosen and checked for ice capability started.

2 FINNISH SWEDISH ICE CLASS RULES

For determination of ice class vessel it is important to ensure:

- Hull of higher strength than in open water.
- Equipment for low temperature.
- Power of main engine and propulsion system for required ice class.

Hull and equipment are defined by the rules and it is possible to determine their exact characteristics.

Power of main engine and propulsion system for required ice class is another problem. Exact data for this size of vessel do not exist. Formulas presented in rules were confirmed by experience

Table 2. Thickness of brash ice in channel for different ice class notation.

Ice class notation acc. to FSICR	Thickness of brash ice in channel
IA Super	1.0 m and a 0.1 m thick consolidated layer of ice
IA	1.0 m
IB	0.8 m
IC	0.6 m

with vessels of smaller size and for aframax size are not fully confirmed yet.

What is the quality of power prediction using rules formulae?

Tankers pose many problems in addition to other vessel types, such as the blunt bow form that crushes ice rather than breaks it. On one side the size advantage of large tankers is disadvantage regarding to breadth of icebreakers, being around 18–25 m. Vessels with larger beam than this result in need for two icebreakers.

The Finnish Maritime Administration (FMA) and the Swedish Maritime Administration (SMA) have developed the Finnish—Swedish ice class rules in co-operation with classification societies. The development of the rules started as early as in the 1930's and has been amended several times:

- Board of Navigation Ice Class Rules, 1971.
- Board of Navigation Ice Class Rules, 1985.
- Finnish-Swedish Ice Class Rules (FSICR), 2002.
- Finnish-Swedish Ice Class Rules (FSICR), 2008.

FSICR 2002 rules are still valid for power determination. The Finnish Swedish Ice Class rules have a minimum engine power requirement developed for the Baltic. This is based on the principle of an integrated system of traffic management to enable ships to follow icebreakers with minimum speed of 5 knots in the brash ice channel.

The Finnish and Swedish administrations provide icebreaker assistance to ships passage to ports in these two countries in the winter seasons. Icebreakers only assist ships that meet the Finnish-Swedish ice class rules.

Determined Baltic ice class notation is the following: **IA** Super, **IA**, **IB** and **IC.**

The design requirement for ice classes is minimum speed of 5 knots in the following brash ice channels shown in Table 2.

3 DESIGN ALTERNATIVES

Following design requirement for 1A Ice class, first question is how frequently vessel is attended to sail in ice area. If the she will sail all the winter time on short frequent voyages in Baltic, than better capabilities in ice is required, using special hull form and type of propulsion.

The following propulsion systems are used in ice-going ships:

- Diesel-electric propulsion system is very common in icebreakers but not in merchant vessels. It provides very efficient propulsion and maneuvering characteristics at slow speed, but due to the high cost, it is very seldom used in merchant vessels.
- Medium speed engine, gearbox and controllable pitch (CP) propellers is the most common propulsion system used in merchant vessels having an ice class. It provides reasonable propulsion characteristics at slow speed as well as reasonable manoeuvring characteristics.
- Direct driven slow speed engine with a fixed pitch (FP) propeller gives poor propeller thrust at a low ship speed. It is recommended that a controllable pitch (CP) propeller would be installed in ships having a direct driven slow speed diesel engine propulsion system.

In particular case Buyer's requirement is for the vessel with priority to be efficient in open water and passing in channel ice just according to the rules. That means that project will have basically the best open water hull form.

3.1 *Project main characteristics*

Chosen vessel was constructed before. Form of 111000 dwt was optimised in "MARIN"—the Netherlands, for open water and her wave resistance was well adapted for particular speed. Maximum draught is 14.90 m that satisfies Baltic passage.

Length overall (Loa) 246.00 m
Length between perpendiculars (Lbp) 236.00 m
Breadth moulded (B) 42.00 m
Depth moulded (D) 21.00 m
Draught scantling (Ts) 14.90 m

MAN-B&W type 7S60 MC-C which develops 16660 kW at 105 rpm.

Trial speed will be 16.0 knots at 90% MCR. Stern could accommodate propeller CPP or FPP with 7.9 m diameter.

4 DETERMINATION OF POWER IN ICE

4.1 *Basis*

Vessel's speed in open water sailing mode is obtained by the thrust delivered from the propeller.

Power delivered to the propeller depends on main engine possibility, i.e. pressure in cylinders and rpm.

Two stroke engines lose power and torque moment by reducing of rpm. This is the reason why in combination of FPP and two stroke slow speed engine, delivered power at small rpm is smaller than MCR.

On the other hand, combination of two stroke engine and CPP allows that power delivered to propeller is close to MCR, because of the possibility of running with full rpm.

Passing through the ice, vessel's speed drops, propeller works in heavier conditions, and depending of pitch and rpm the thrust can be obtained.

In Figure 1 the main parameters are presented. Curve of delivered thrust via FPP and CPP at the same constant engine power are drown as well as curve of vessel's resistance in open water. Difference between delivered thrust and vessel's resistance is the remaining thrust, which can be used for resistance in ice. So, the ice resistance should be defined.

4.2 *Determination of power procedure*

Determination of needed shaft power could be obtained according to "Finnish-Swedish Ice Class Rules of 01 October 2002":

- Using proposed formula by application of the FS (Finnish-Swedish) Rules, based on a vessel moving in Brash Ice,
- Or by other method, in our case by model testing in ice in towing tank according to FM standard (Finish Maritime).

Formula is valid up to vessel's width of 40.0 m—proposed Aframax has 42.0 m width. That was the reason to use model test as method for power definition and in the same time chance to compare results with formula (this 42 m width will not have influence on formula results).

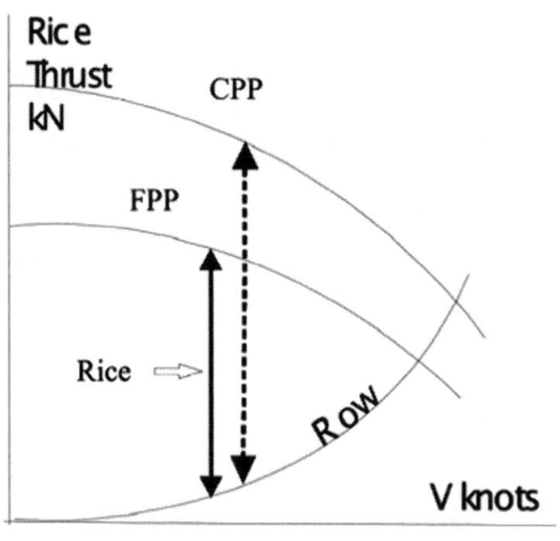

Figure 1. Ice thrust capability.

4.3 *Definition of engine output formula*

The engine output Ps (kW) is the maximum output the propulsion machinery can continuously deliver to the propeller(s).

$$Ps = Ke\,(RCH/1000)^{3/2}/Dp \tag{1}$$

RCH = The resistance in Newton of the ship in a channel with brash ice and a consolidated layer (N); Dp = Propeller diameter (m); Ke = Propulsion type factor;

$$RCH = C_1 + C_2 + C_3(HF + HM)^2 \times (B + C\psi HF) + C_4 LparHF^2 + C_5(LT/B^2)^3 Awf/L \tag{2}$$

C_1 and C_2 take into account a consolidated upper layer of the brash ice and are to be taken as zero for ice classes IA, IB and IC. $C_3 = 845\ kg/(m^2\ s^2)$; $C_4 = 42\ kg/(m^2\ s^2)$; $C_5 = 825\ kg/s^2$; $C\mu\ C\psi$, = function of bow shape/angles; HF = thickness of the brash ice layer displaced by the bow (m); HM = thickness of the brash ice in channel (m); B = maximum breadth of the ship (m); L = length of the ship between the perpend. (m); T = actual ice class draughts of the ship (m); Aw = area of the waterline of the bow (m^2); $Lpar$ = length of the parallel midship body (m), (Ref. DNV Pt5 Ch.1 Sec. 3 J100 Engine output).

Result of calculation:

- RCH = Resistance = 1800 kN
- Ps = 22000 kW using FP Propeller
- Ps = 19700 kW using CP Propeller

According to formula, existing main engine MAN-B&W type 7S60 MC-C which develops MCR of 16660 kW will not satisfy the rules, and particular vessel couldn't be offered to the Buyer.

4.4 *Model testing—other methods of determining Ke or RCH*

For an individual ship, in lieu of the *Ke* or *RCH* values defined in 4.3.1, the use of *Ke* or *RCH* values based on more exact calculations or values based on model tests may be approved. The design requirement for ice class comprises minimum speed of 5 knots in the brash ice channel of ice height HM = 1.0 m and a 0.1 m thick consolidated layer of ice.

Model testing of ships in open water is an established science. Ice mode tests are performed basically for two purposes:

- Determination of the ice resistance of hull in specific ice conditions (ice thickness, ice strength, speed).
- Determination of the power needed to propel the vessel in those specific conditions.

When a full scale phenomenon is reduced to a smaller scale, there must be a certain similarity between the actual and simulated event:

- The model ice must have the same characteristics as sea ice, although on reduced scale. The model ice materials can be divided in model ice made from water (different chemicals may be dissolved in the water) or in model ice made from synthetic materials.
- The kinetic friction between ship hull and ice has a major effect on the performance of an ice-transiting vessel. From comparative full scale tests it is known that corrosion of hull plating can cause up to a100% increase in ice resistance. It is thus clear that proper modelling of friction in model tests is crucial for correct performance prediction on full scale.

The correct model friction coefficient has been determined by comparing the full and model-scale friction dependent components of ice resistance.

The same values for the model-scale panels are when the model has the ITTC friction coefficient in the range of 0.04 to 0.06. In full scale this friction coefficient corresponds to the friction coefficient of a new ship with an Inertia 160 coating in ice with a snow cover about 10 cm in thickness.

4.5 *Towing tests—ice resistance tests*

Model tests of subject project were performed as self-propulsion tests.

In this test type, the model is equipped with propulsion. During the tests the propeller thrust, the shaft torque, and propeller revolutions are measured.

The benefit with this type of tests is that the breaking phenomenon is real. In self-propulsion tests, the model moves like a vessel in reality; it can move with a constant speed; it also accelerates and decelerates and the model might even stop.

4.5.1 *Resistance in ice—measurements*

Target is to define resistance for ice 1A class rules in 1 m ice thickness in brush ice. According to FM standard that means:

- Width of channel in ice is 2 × B vessel's model.
- Ice thickness in model scale corresponds to 1.36 m in full scale.
- Model speed corresponds to 5 knots in full scale.

The ice resistance is the most important parameter for definition of vessel's ice capability. Method using model testing can give the answer better than calculation, but still with same questions about correlation from model to full size vessel.

The ice resistance can be determined by three types of model tests:

- Towed resistance test.
- Towed propulsion test.
- Towed self propulsion test.

When towed resistance in ice is used, direct total resistance in ice is measured. By subtracting of open water resistance the net ice resistance is calculated—Figure 2.

$$Ri = FI - Riw \tag{3}$$

where: Ri = Ice resistance; FI = Total resistance; Riw = Ice free water resistance.

When towed self propulsion test in ice is used, the following parameters can be measured:

- Propeller thrust
- Propeller revolution
- Shaft torque

Differences in these parameters between sailing in open water and in ice are evident. After coming in brash ice channel, higher thrust and shaft torque is needed to keep the same speed. This difference shows influence of the ice resistance.

The forces acting on the model in the towed self propulsion tests are shown in Figure 3.

From the scheme presented in Figure 3 we obtain:

$$Fi + Tni = Ri + Riw + CW \tag{4}$$

where: Tni = the net propeller thrust in ice; CW = counterweight.

As the following is applicable to open water:

$$Tn = (1 - t) \cdot T \tag{5}$$

where: Tn = the net propeller thrust in open water.

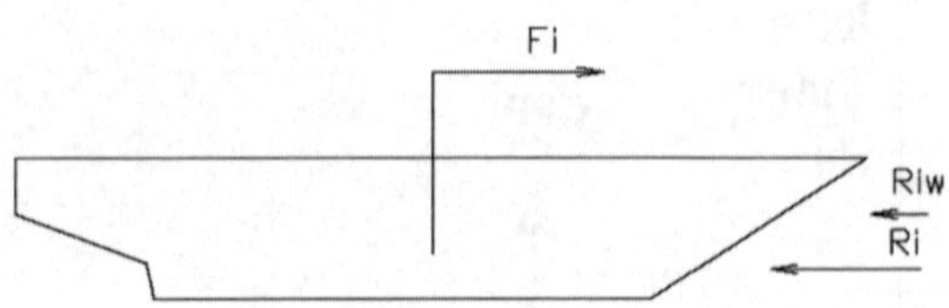

Figure 2. Scheme of forces during towed model test.

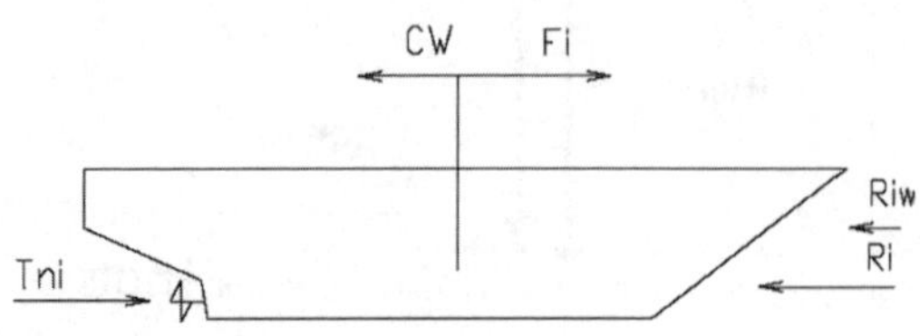

Figure 3. Towed self propulsion test.

Then *Tni* can be determined as follows:

$$Tni = Tn - dTn \qquad (6)$$

where *dTn* is the decrease of net thrust due to ice.

Based on the equations presented above we obtain:

$$Ri + dTn = Tn + Fi - Riw - CW = Fi - Fw \qquad (7)$$

where: *Fw* = towing force in ice free water. *Ri* + *dTn* is usually the main result of towed propulsion tests.

4.5.1.1 Self propulsion tests

The methods of analysis are the same as those presented in towed resistance test. There is of course no towing force ($Fi = 0$) or counterweight ($CW = 0$).

4.6 *Results of testing*

For this size of vessel, for ice class 1A, the thickness of brash ice should be 1.36 m in full scale. The thickness of ice is 1.0 m in first test and 2.0 m in second. In both thicknesses the thrust is measured for different speed. By interpolation, the exact thrust is found for speed of 5.0 knots (2.56 m/s).

Results of measurements are presented by resistance lines in Figure 4 (Rice (kN) versus Vs(m/s)), where also thrust curves for FP and CPP for different power are calculated.

The results of resistance are also shown in diagram. Measurements are provided for two different ice thicknesses (1.0 m and 2.0 m) and interpolation for corresponding 1.36 m is made.

Line Hi = 1.0—Measured Rice for different speed in brush ice corresponding to 1.0 m ice height,

Line Hi = 2.0 m—Measured Rice for different speed in brush ice corresponding to 2.0 m ice height,

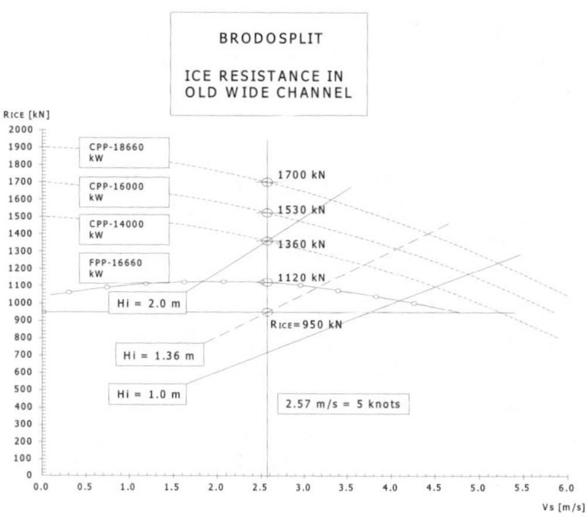

Figure 4. Resistance and propellers thrust curve.

Line Hi = 1.36 m—Interpolated Rice for different speed in brush ice corresponding to 1.36 m ice height, what satisfy conditions for 1A ice class.

At this line for speed of 2.57 m/s (5.0 knots) the Rice of 950 kN is found. This is minimum thrust that should be obtained by the propeller.

It is clearly seen, that net thrust in ice obtained from all powers/propellers is higher than Rice.

5 POWER DETERMINATION

Having needed thrust, the power of main engine could be determined. Common work of main engine and propeller should be considered. The possibilities with fixed and controllable pitch propeller will be investigated.

5.1 *Selection of slow speed diesel engines with fixed pitch propeller*

In general, a well optimized slow diesel and fixed pitch propeller combination is not very well suited for operation in ice conditions. The diesel is normally chosen so that the propeller optimizing point is quite close to the maximum rating point of the specific engine layout diagram. This has the consequence that when the speed of the vessel drops, due to additional ice resistance, the propeller curve becomes "extra heavy" and starts to reduce engine speed and power along the torque/engine speed limit line. When the vessel is completely stopped by ice and the propeller runs in so called "bollard pull" code a normally optimised propeller can reach about 60% of nominal engine speed and absorb less than 40% of nominal power which reduces the propeller thrust by almost the same percentage.

Based on model test results and general diesel engine data, the propeller and diesel engine power curves can be drawn as presented in Figure 5 using the K_Q versus J values and the wake fraction.

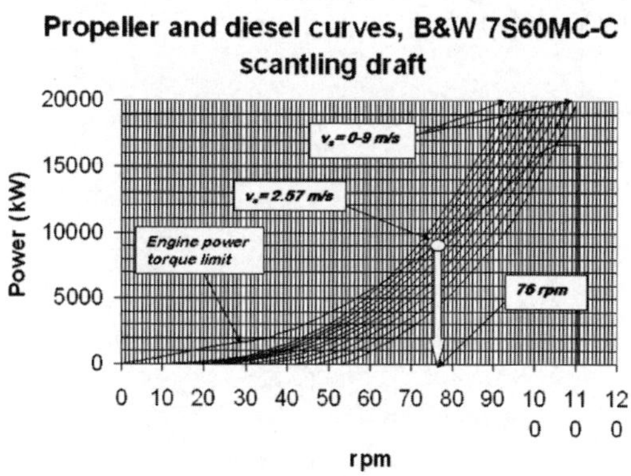

Figure 5. Propeller and diesel engine power curve.

In Figure 5 there are 10 propeller curves representing the power versus rpm relationship at constant vessel speed, 0 to 9 m/s vessel speed. The propeller curve at 0 m/s vessel speed is the so called "bollard pull" curve.

The diesel engine power versus rpm curve is also plotted into Figure 5.

The interception points between the diesel curve and each vessel speed curve is determined.

Now, each resulting rpm-value represents the maximum rpm, which the diesel engine is able to deliver at that particular vessel speed. For us, the intersection at 2.57 m/s vessel's speed is interesting point.

Figure 6 shows the corresponding propeller thrust curves. There are 10 propeller thrust curves for constant rpm, from 10 up to 100 rpm. The curves are based on K_T versus J values and the thrust deduction and wake fractions. The thrust deduction fraction is treated the same way as the wake fraction.

The vessel resistance in open water is also plotted into Figure 6. Using this diagram for each propeller rpm at defined vessel's speed produced thrust is defined.

The difference between produced thrust and open water resistance present capable thrust for ice resistance.

This capable thrust (net resistance) for full range of vessel's speed is presented in Figure 7, and can be compared by ice resistance curve from model testing in diagram in Figure 4.

For example, the point of intersection between the engine limit curve and the curve which represents power for vessel's speed of 2.54 m/s, define rpm and power which the engine can produce for requested Ice conditions (Figure 5).

Even the engine MCR (Maximum Continuous Rating) is 16,660 kW, using defined FPP (Fixed Pitch Propeller), only 9000 kW is possible to obtain.

Further, by drawing the same point of 76 rpm at vessel's speed of 2.54 m/s in thrust/resistance diagram (Figure 6), capable net thrust of 1120 kN for ice is found.

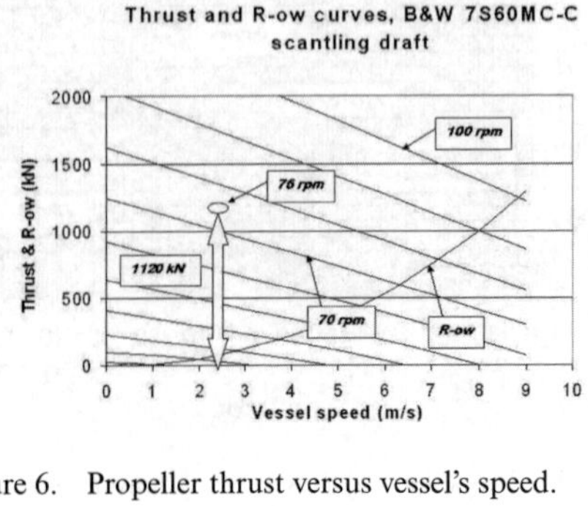

Figure 6. Propeller thrust versus vessel's speed.

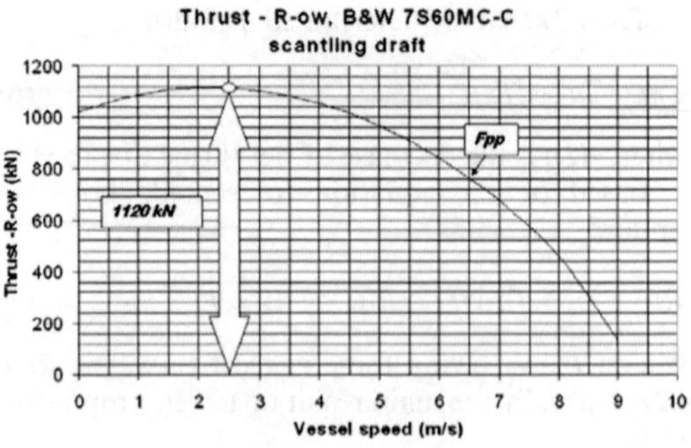

Figure 7. Net thrust in ice (Thrust – Row).

From Figure 4 it is presented that this value is greater than Ice resistance.

So, conclusion is that for Ice 1A combination of subject main engine and FPP will satisfy the requested ice capabilities.

5.2 *Selection of slow speed diesel engines with controllable pitch propeller*

Determination of net thrust with CPP was done in the similar way. The advantage is, that propeller can work at full rpm all the time, and delivered thrust at small vessel's speed is significantly higher than using FPP.

The propeller thrust curves for the constant power at different vessel's speed are presented in Figure 5 (for 18,600/16,000/14,000 kW). Again the difference between propeller's curve and resistance in open water is calculated, and these net ice thrusts are presented in Figure 4.

From the data presented, we can see that obtained thrust is significantly higher than the needed resistance.

By extrapolation, needed 9000 kW of power is found (as with FPP), but using CPP this is in the same time power of the engine, because there is no limit curve for reduced rpm.

6 COMPARISON OF ENGINE POWER/ SHAFT POWER FOR DIFFERENT SAILING MODE

In Figure 8 the results are compared. Using formula the calculated resistance *Rh* is twice than measured by testing in basin.

Using CPP calculated and measured power follows this law. Advantage of combination engine—CPP is that shaft power can be close to MCR of the engine.

Using FPP with proposed pitch, abt. 60% of MCR can be delivered to the propeller because of balance

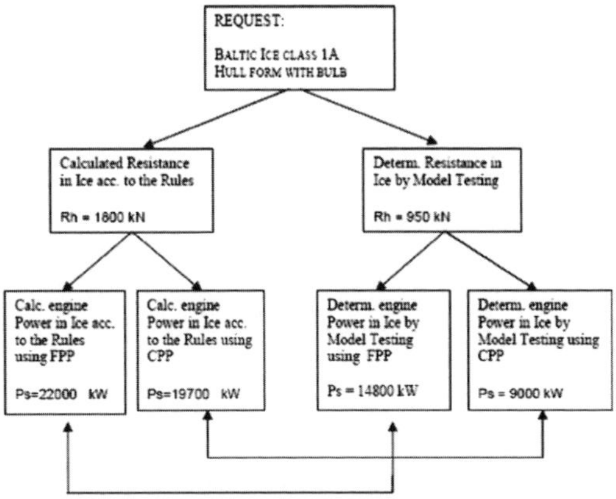

Figure 8. Result of calculated and tested ice resistance and power.

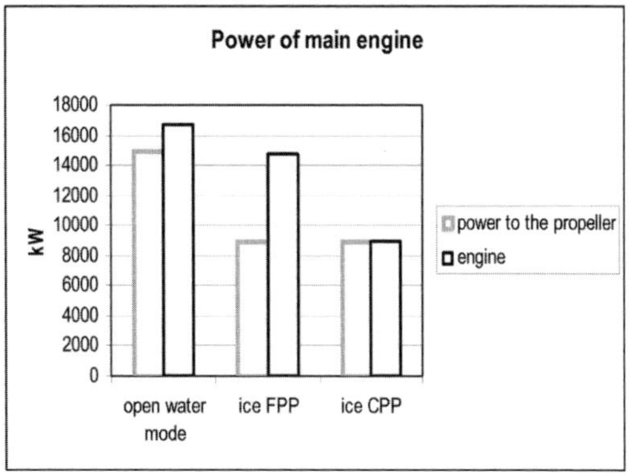

Figure 9. Power of main engine for different sailing mode.

point—engine power limit curve/rpm. Higher power can be delivered by chose of smaller pitch, but then this propeller will be to light in open water.

Finally, according to formula, the existing project will not satisfy request for engine power.

On the other hand, according to model testing project fully satisfy the requirements for navigation in brash channel in Baltic for Ice class 1A.

Project has the following engine parameters (Figure 9):

Main engine MAN-B&W type 7S60 MC-C
MCR 16,660 kW at 105 rpm
SCR 14,994 kW, 90% of MCR

Needed power, Figure 6 (acc. to model test results):

- *Open water mode*

Power for open water is 90% of MCR. This sailing mode ask full MCR of present engine. There is no significant difference between FPP and CPP, because both propellers should have the same design pitch for contracted power.

- *Ice sailing mode (Figure 9)*

Required power which to be delivered to propeller for Ice Class 1A 9,000 kW.

Required engine power using FPP 14,800 kW.

Depends of design point of propeller this power could be higher or smaller. In our case design point is 100% rpm at 90% MCR.

Required engine power using CPP 9,000 kW.

Theoretically, the engine power could be the same as needed shaft power. For ice sailing mode for this tanker this engine power could satisfy the requirements and present the best result.

Status of main engine:

Needed engine power of CSR = 16,600 kW is the higher power which should be used to cover requirement for speed in open water mode.

Power for ice 1A according to model testing is lesser than for open water mode, and project of subject Aframax tanker could be used for requested purpose.

7 CONCLUSIONS

Compared to the model testing the values obtained from FSICR formula are quite conservative for ice propulsion power determination for Baltic 1A Ice class tanker. Difference in results are significant so the investigation and harmonisation of correlation factors for large tankers is necessary. This will enable more accurate and reliable calculation of preliminary power prediction which is essential in the design phase. Model testing is always the right option regardless of the quality of formulas. The final selection of design will depend on question how frequently vessel is expected to sail in ice area i.e. if the better capabilities in ice are required as design priority.

REFERENCES

DNV, 2002. Rules for Classification of Ship.

Finnish-Swedish Ice Class Rules 2002 (FSICR).

MARC Report A-329, 2004. Model Test in Ice for Brodosplit Tanker of 111000 dwt.

Soinen, H. & Nortala-Hoikkanen, A. 1990. Development of Ice Model Test into a Reliable Tool for Icebreaking Ship Design, *ICETECH Symposium, Calgary.*

Wilkman, G. & Mattson, T. 1991. Ice model tests as a tool to study and improve the performance of ships in ice, *POAC '91, St. John's, Canada.*

Advanced Ship Design for Pollution Prevention – Guedes Soares & Parunov (eds)

Conceptual changes in the CRS classification and Croatian statutory technical rules

N. Vulić & A. Vidjak
Croatian Register of Shipping, Split, Croatia

ABSTRACT: Designers, manufacturers and operators in maritime business shall always be aware of the types of components they are dealing with. Failure of the primary components of type-A (statutory components) must not happen. Failure of the primary components of type-B (class components) shall be avoided. Statutory components shall satisfy the requirements stated in the IMO conventions (SOLAS, MARPOL, Load Lines, ColReg, etc.). Class components shall comply with the IACS Unified Requirements and Procedural Requirements. The requirements for both types of components are incorporated and expressed in the Technical Rules. The aim of this paper is to present the newly established concept of the Technical Rules prepared by the Croatian Register of Shipping (CRS) The basic difference to the previous concept is that the class and the statutory technical rules are strictly separated. CRS presently develops classification rules for the CRS classed maritime units, as well as the statutory rules for the Republic of Croatia. The new concept is in force since 1st July 2009.

1 INTRODUCTION

1.1 *Classification societies, their mission and the Technical Rules*

The mission of classification societies is to contribute to the development and implementation of technical standards for the protection of life, property and the environment (IACS 2004).

Classification societies establish and apply technical requirements for the design, construction and survey of maritime-related facilities, principally ships and offshore structures. These requirements are published in form of Technical Rules, known as classification rules. Classification societies participate in the on-going development of technical standards (IACS 2004).

Classification societies may also act as Recognized Organizations for Flag State Administrations, verifying the same vessel's compliance with international and/or national statutory regulations (IACS 2004). This type of supervision is based upon the Flag State Administration national statutory rules.

1.2 *Classification societies and maritime components*

The basic activities of classification societies are related to the supervision of the primary components of maritime units: ships, boats and yachts, floating objects, offshore and other types of units on the basis of their Technical Rules. There exist two types of primary maritime components.

Primary maritime components of the type-A shall comply with the requirements for the safety of maritime units, including prevention of environment pollution. Design, manufacture, assembly, testing and operation of these components are based upon the so-called statutory requirements. These are the requirements that the flag states impose on the maritime units flying their flag and are commonly based upon the conventions of the International Maritime Organization (IMO), well-known SOLAS, MARPOL, Load Lines, ColReg and similar conventions.

Primary maritime components of the type-B are to comply with the functional requirements for the maritime units, often called the class requirements. Classification societies are organizations that prescribe these requirements for the components of maritime units under their class supervision. This is done in their Technical Rules. These rules are to be based upon the Unified requirements and Procedural requirements of the International Association of Classification Societies (IACS), providing thus the highest standards and the state-of-the-art in maritime business.

Secondary maritime components are the components that need not to comply either with the statutory or class requirements. For secondary components of e.g. newly built ships it is only important what has been written in the Technical

Description contracted between the shipyard and the future ship owners.

1.3 *Croatian Register of Shipping, its history and the present position*

Croatian Register of Shipping (CRS) is the Croatian classification society, with the Head office situated in Split, Croatia.

CRS is a heritor of ship classification activities at the eastern Adriatic coast. The Austrian Veritas was founded in this area, already in 1858, as the third classification society in the world. In 1918 the Austrian Veritas changed its name into the Adriatic Veritas and was acting as such till year 1921. CRS was founded in 1949, was acting till 1992 as JR (Yugoslav Register of Shipping).

CRS was an associate of the International Association of Classification Societies (IACS) since 1973. This status was kept till 2005, when IACS generally discontinued associate status. (CRS 2009)

CRS is presently engaged in the process of its recognition by the European Maritime Safety Agency (EMSA) to become the recognized European classification society. One of the important jobs that had to be done within this process was the restructuring of the CRS Technical Rules. This job was successfully completed on 1st July 2009, bringing out the completely new concept of the Technical Rules. (Vidović 2006)

This new concept strictly distinguishes the statutory and class requirements. This would be the essential advantage of the new concept.

2 QUALITY SCHEME AND PROCESS APPROACH

Internationally recognized classification societies shall develop, implement and maintain their quality management system. This system shall be certified in accordance with an internationally recognized scheme, such as IACS QSCS (Quality Systems Certification Scheme), based upon the presently valid edition of ISO 9001 international standard. To better understand the role of the Technical Rules within the classification societies the processes taking place within the CRS will be listed hereafter. These processes are:

- Development, formulation and updating of Technical Rules and other publications:
- Assessment and approval of technical documentation (i.e. plan approval);
- Type approval of products;
- Approval of manufacturers, testing institutions and service suppliers;
- Survey during manufacture and testing of materials and components;
- Supervision during construction of newly built maritime units;
- Supervision during conversion of existing maritime units;
- Initial, periodical and occasional surveys of ships in service;
- Reporting and verification of reports;
- Tonnage measurement;
- Research and development.

By performing classification and statutory services CRS is verifying that the ship, material or component surveyed is in compliance with the requirements of the Rules for intended service and navigation area.

Basically, the Technical Rules are to be developed in accordance with the international requirements (IACS Unified Requirements and Procedural Requirements, or IMO conventions). Technical documentation is to be approved with or without remarks in accordance with the Technical Rules requirements. Supervision during construction of newbuildings or conversions is to be based upon the approved technical documentation. The ship is provided with classification and statutory documents upon the successfully completed process of supervision. So, the proper development of Technical Rules is essential for the CRS or any other classification society, because all other activities rely on the rules. Quality schemes, such as IACS QSCS, state the development of the Technical Rules in their chapter Design of services.

3 CRS CLASSIFICATION RULES

In general, classification rules are developed to contribute to the structural strength and integrity of essential parts of the ship's hull and its appendages. They also cover the reliability and the function of the propulsion and steering systems, power generation and those other features and auxiliary systems which have been built into the ship in order to maintain essential services on board for the purpose of safe operation of the ship. (IACS 2004)

Rules for the Classification of Ships, developed by the CRS (in the text of the Rules addressed as the *Register*), prescribe the requirements for the classification of ships on the basis of the adopted international standards. CRS classification rules prescribe the basis for design, construction and maintenance of ships with respect to (see Figure 1): (Vidović 2009)

1. Structural strength, and where necessary the watertight integrity of all essential parts of hull and its appendages;

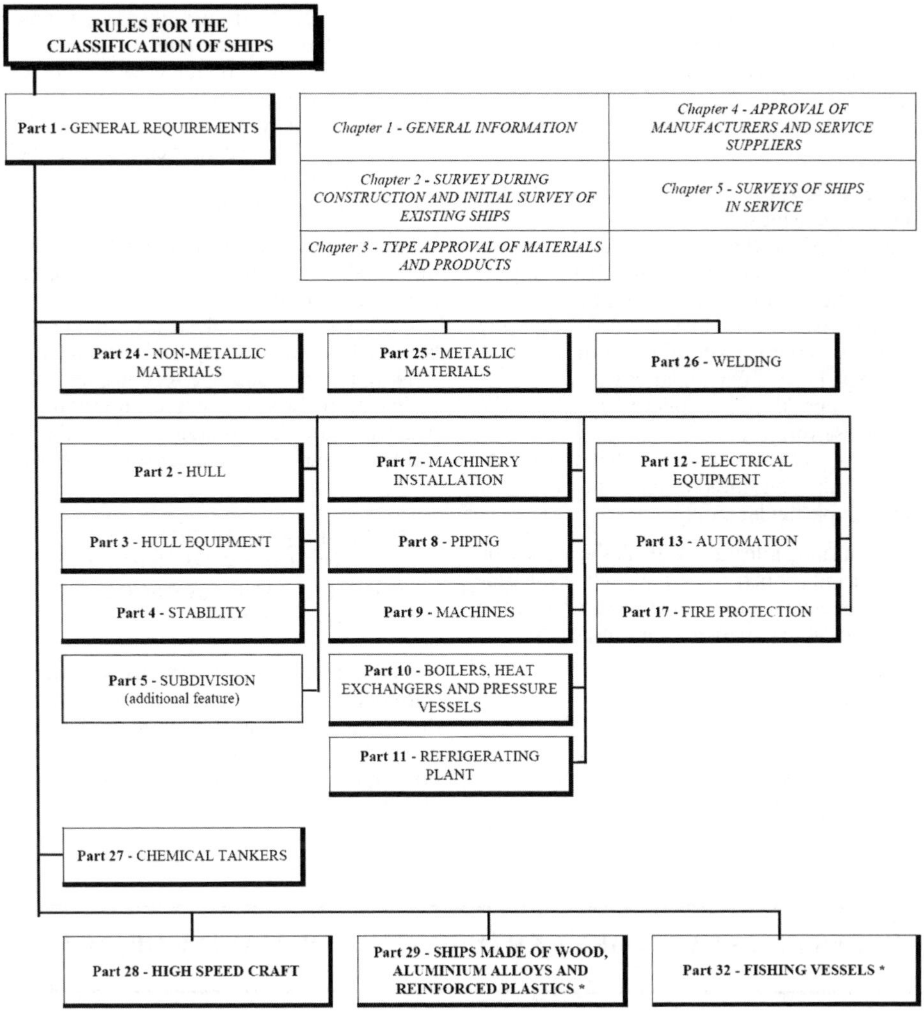

Figure 1. CRS Rules for the classification of ships graphic layout (CRS-Class 2009).

2. Safety and reliability of the propulsion and steering system, and those features and auxiliary systems necessary for establishing and maintaining basic conditions on board;
3. Ship stability;
4. Subdivision (additional class notation);
5. Fire protection;
6. Refrigerating equipment.

It is to be noted that the CRS classification rules are *not* applicable to: (Vidović 2009)

- mobile offshore drilling units;
- liquefied gas carriers;
- certain types of oil tankers, i.e. those that are subject to the requirements of Condition Assessment Scheme (CAS) as stated in the IMO resolution MEPC.94(46).

CRS classification rules are continuously updated and regularly published in English.

4 CROATIAN STATUTORY RULES

Every maritime administration may prescribe its own technical requirements for maritime units. In case of Croatia, these requirements are prescribed in the form of the Rules for the Statutory Certification of Sea-Going Ships (Croatian statutory rules).

Croatian statutory rules are a part of the Rules for the statutory certification of maritime units (objects). This relationship is presented in Figure 2.

Regardless of the ship navigational area the Croatian statutory rules are implemented to the:

1. New ships;
2. Existing cargo ships, in case they are converted into passenger ships;
3. Existing ships, if they are subject to repairs, replacements, conversions of significant importance, change of purpose, navigational area or number of passengers, in the scope which the recognized organization finds rational and appropriate in each particular case;
4. Existing ships, if it is explicitly stated in the rules themselves;
5. Floating objects, in the scope that the recognized organization finds rational and appropriate in each particular case.

Rules are dealing integrally with all the aspects of safety of life and property at sea, as well as the protection of sea environment. They prescribe all the relevant technical requirements with respect to:

1. Stability;
2. Subdivision;
3. Freeboard;
4. Fire protection;
5. Life-saving appliances;
6. Radio equipment;
7. Navigational aids and signaling means;
8. Cargo handling equipment;
9. Protection at work and crew accommodation;
10. Carriage of passengers;
11. Carriage of cargo;
12. Prevention of maritime environment pollution;
13. Safety management;
14. Security protection of ships.

The scheme presenting the complete system of the Croatian statutory rules for sea-going ships is presented in Figure 3.

CRS classification rules are continuously updated by the CRS and regularly published in Croatian.

Note that the body performing the supervision on behalf of the Croatian maritime administration (i.e. Croatian flag) is addressed in the Croatian statutory rules as the Recognized Organization.

5 BASIC PRESUMPTIONS FOR A CLASSIFICATION SOCIETY WHEN ACTING AS RECOGNIZED ORGANIZATION

In addition to the requirements prescribed by the statutory rules, sea-going ships shall comply with the standards of a Recognized organization for design, building and maintenance of hull, machinery installation, electrical and control equipment (including automation).

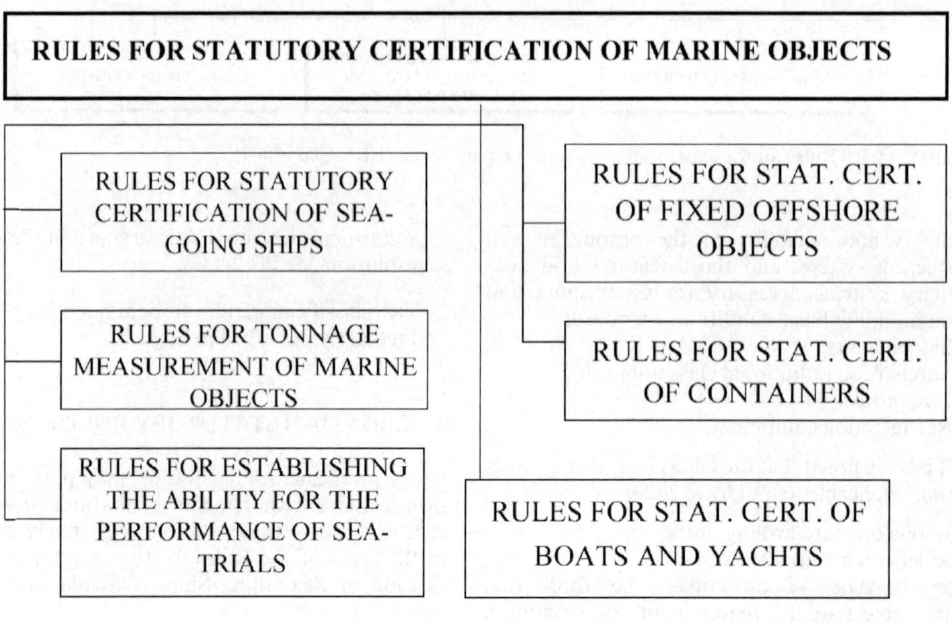

Figure 2. Croatian statutory Technical Rules graphic layout (as translated from Croatian) (CRS Statutory 2009).

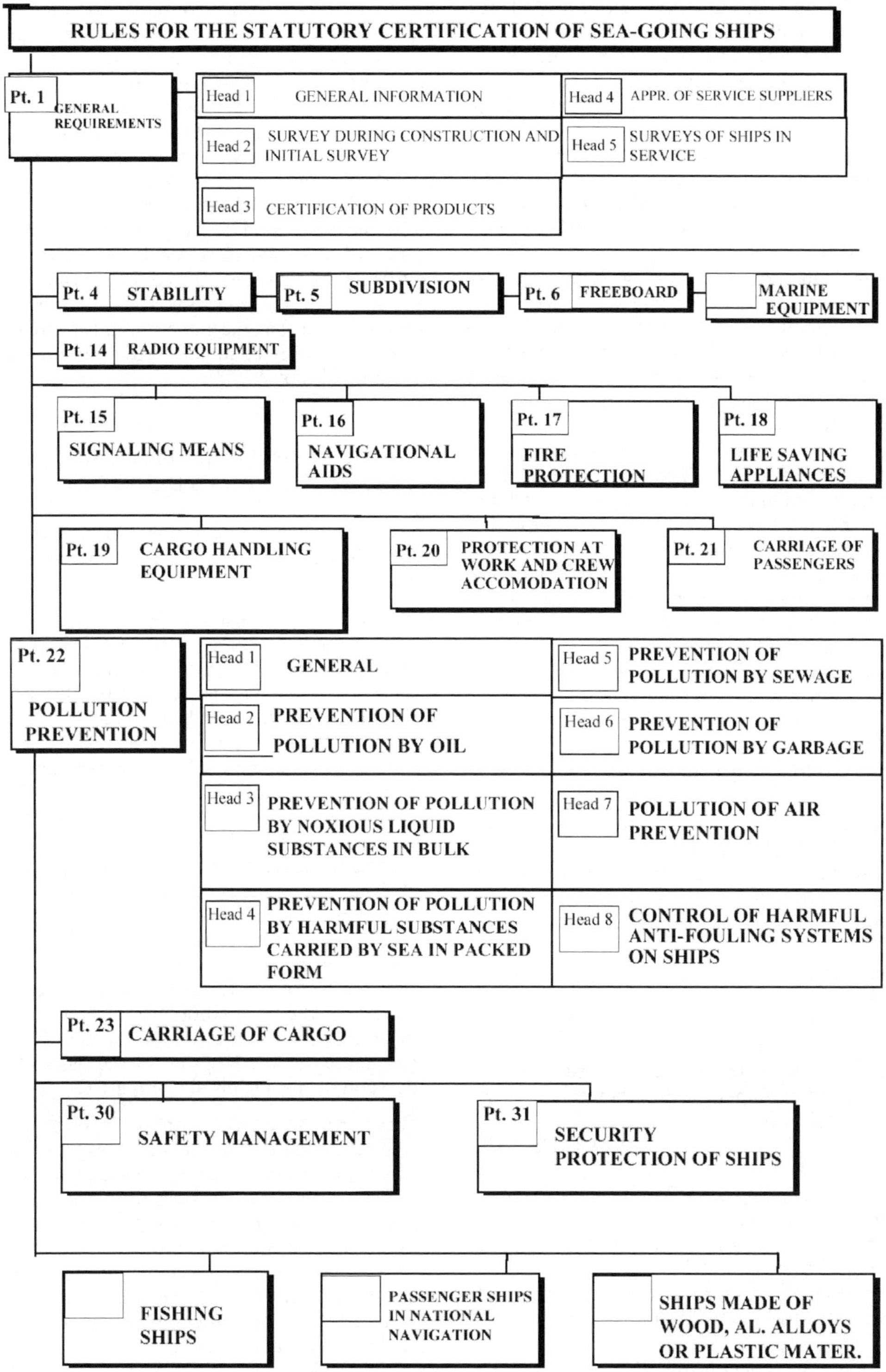

Figure 3. Croatian statutory rules for sea-going ships (as translated from Croatian) (CRS Statutory 2009).

Classification societies, certified in accordance with internationally recognized quality schemes may gain authorization of maritime administrations (Flag State Administrations), to perform statutory surveys and other activities on behalf of the flag state. In this case the classification society gains the status of the Recognized organization. The organizational requirements and the levels of competence for such Recognized organizations have been defined yet long ago; see (IMO 1993) and/or (IMO 1995).

When authorized by the Flag State Administration concerned, the CRS will act on its behalf within the limits of such authorization. In this respect, the CRS will take into account the relevant national requirements, survey the ship and issue or contribute to the issue of the corresponding certificates. (CRS-Class 2009)

The above surveys do not fall within the scope of the classification of ships, even though their scope may overlap in part and may be carried out concurrently with surveys for assignment or maintenance of class. (CRS-Class 2009)

In statutory matters, when authorized by the Flag State Administration concerned and acting on its behalf, the CRS applies available IACS Unified Interpretations, unless Flag State Administration has provided another interpretation, another requirements or decides otherwise. (CRS-Class 2009)

For ships, the arrangement and equipment of which are required to comply with the requirements of the following regulations:

1. International Convention for the Safety of Life at Sea, 1974 (SOLAS 74),
2. International Convention for the Prevention of Pollution from Ships, 1973, as modified by the Protocol of 1978 relating thereto (MARPOL 73/78),
3. International Convention on Load Lines, 1966, (ILLC 66),
4. International Convention on Tonnage Measurement of Ships (TMC 69),
5. International Code for the Construction and Equipment of Ships Carrying Dangerous Chemicals in Bulk (IBC Code),
6. International Safety Management Code (ISM Code),
7. International Ship and Port Facility Security Code (ISPS Code),
8. Applicable conventions of the International Labour Organization (ILO),
9. International Convention on the Control of Harmful Anti-Fouling Systems on Ships,

and applicable amendments thereto, the CRS requires that the applicable statutory certificates are to be issued either by the CRS, Flag State Administration, or by some other recognized organization so authorized by the Flag State Administration. (CRS-Class 2009)

Compliance with the class related requirements does not relieve the ship-owner (company), or any other interested party, from compliance with any statutory requirement demanded by the Flag State Administration. (CRS-Class 2009)

On the other hand, for ships that are not in possession of the valid class certificate, or the valid certificate of the Recognized organization, the ship-owner or the company shall, at their own expense, prove that the subject ship (or any other maritime unit) is compliant with the equivalent standards. (Vidović 2009)

6 BASIC STEPS IN THE PROCESS OF THE TECHNICAL RULES DEVELOPMENT

The operating procedure of preparation and adoption of the CRS Technical Rules consists of the following steps:

- The authors (technical specialists in charge for each part of the Rules) prepare the so-called beta version (amended) of the next edition of the present Rules, or the entirely new parts of the Rules;
- Groups for internal evaluation review the beta version. The authors rework the text in accordance with the remarks of these groups and thus the beta version becomes alpha version;
- Technical Committees (external), consisting of representatives of all interested parties (administration, manufacturers of products, shipping and shipbuilding industry, professional institutions, scientific institutions and CRS representatives) consider the alpha version. Once their remarks are complied with, the alpha version becomes the proposal of the Technical Rules marked with the expected year of edition (e.g. 2009);
- Technical Rules are adopted by the General Committee of the Register on the basis of the decision of the Technical Committee.

By the presented mechanism the involvement of all the interested parties has been ensured. There exist several Technical Committees, each for a specific field covered by the Rules (e.g. hull, machinery, electricity, automation, materials and welding, etc.) and this ensures the competence of the persons considering the proposed text of the Rules.

If not explicitly stated otherwise, the new Rules, as well as the amendments to existing Rules, shall enter into force, after they have been adopted by the General Committee of the Register, on the date indicated in the Rules themselves.

In general, the applicable Rules for the assignment of class to a new ship are those being in force at the date of contract for construction. This also applies to existing ships when undergoing major conversions, or to the altered part of the ship in the case of partial alterations. (CRS-Class 2009)

The delayed entry into force after the adoption by the CRS General Committee is necessary, so that all the interested parties (including the CRS itself) shall have enough time to prepare for the forthcoming implementation of the Technical Rules.

Requirements of the Rules related to class assignment, maintenance and withdrawal of class for existing ships, are applicable from the date of their entry into force. (CRS-Class 2009)

7 CONCLUSIONS

The paper presents the essentials of the newly established concept of the CRS classification rules and the Croatian statutory rules, within the process in which the CRS will become the European classification society. This is taking place in another wider process within which the Republic of Croatia will join the European union.

CRS mission in the field of classification and statutory certification is to promote the highest internationally adopted standards in the safety of life and property at sea and inland waterways, as well as in the protection of the sea and inland waterways environment. The basic functions of CRS are classification and statutory certification of ships on behalf of national maritime administrations.

Previously, CRS acted both as the classification society (associate to IACS), as well as the exclusively recognized organization by the Republic of Croatia for the statutory supervision of maritime units.

In the new concept the processes of classification and statutory supervision are separated. Classification is the matter that the classification society deals with. Statutory supervision is the matter of a Recognized Organization. Consequently, the Technical Rules prepared by the CRS are now strictly separated into the CRS classification and the Croatian statutory rules.

CRS classification rules comprise the functional requirements applied to the ship hull, machinery, electrical appliances and automation. Class can be understood as the function related with the purpose or type of the ship. These rules are independently formulated by the CRS as the classification society. They do not interfere with or refer to any of the statutory rules. CRS classification rules are prepared in English and they may be translated into Croatian if necessary.

Croatian rules for the statutory supervision of the ships comprise the requirements related with the safety and environment protection for the ships under the Croatian flag. Croatian ministry in charge for maritime affairs adopts these rules, by the decision of the Minister. They are prepared in Croatian language. These rules also describe the matters in charge of the Recognized organizations, to which the Republic of Croatia gives authority for the statutory certification. From the practical point of view, when the statutory rules refer to the classification rules, the latter are to be stated as the rules of the Recognized organization. For the time being, the appropriate item from the CRS classification rules is still stated in parentheses after the linked reference in the Croatian statutory rules.

This concept is applied not only to the ships, but also to other maritime units subject to CRS classification and/or Croatian technical supervision.

The main advantage of the newly established concept is that it recognizes the fact that the international regulations ensuring highest standards for classification and statutory supervision are established and developed independently from each other. They are based on different requirements: either on the classification rules (relying on IACS Unified Requirements/Procedural Requirements), or the statutory rules (relying on IMO conventions).

Strictly distinguishing the statutory and class requirements, the new concept easily points out what the designers, manufacturers and operators of maritime unit should always be aware of: the level of their responsibility in case of a failure, as well as the consequences of such a failure. In case of failure of statutory component the consequences can impose responsibility in accordance with the penalty law. On the other hand, the consequences of the failure of a class component can impose only contractual responsibility. All the persons involved in any of the described activities related with the statutory or class maritime components shall always have bear this in their mind.

REFERENCES

CRS 2009. *Directory*. Split: Croatian Register of Shipping.

CRS-Class 2009. *Rules for the Classification of Ships*. Split: Croatian Register of Shipping.

CRS-Statutory 2009. Rules for the Statutory Supervision of Sea-Going Ships (in Croatian), Split: Croatian Register of Shipping.

IACS 2004. *Classification societies-what they do and do not do*. London: International Association of the Classification Societies.

IMO 1993. IMO Resolution A.739(18) Guidelines for the Authorization of Organizations Acting on Behalf of the Administration. London: International Maritime Organization.

IMO 1995. *IMO Resolution A.789(19) Specifications on the Survey and Certification Functions of Recognized Organizations Acting on Behalf of the Administration.* London: International Maritime Organization.

Vidović, I. 2006. Technical conditions. *Presentation given in the scope of the Republic of Croatia negotiating team.* Bruxelles: The negotiating team for the accession of the Republic of Croatia to the European Union.

Vidović, I. 2009. New Rules for Classification and Statutory Certification. *Presentation at the Meeting of the Surveyors and Auditors of the Croatian Register of Shipping.* Petrčane: Croatian Register of Shipping.

Advanced Ship Design for Pollution Prevention – Guedes Soares & Parunov (eds)

Ship geometry modelling using radial basis functions interpolation methods

D. Ban, B. Blagojević & J. Barle
Faculty of Electrical Engineering, University of Split, Mechanical Engineering and Naval Architecture, Split, Croatia

ABSTRACT: The radial basis functions (RBFs) networks are the basis of recently developed and more widely used meshless methods. Among many useful properties they have, RBFs are recognized as the best approximators and therefore possibly suitable for describing complex ships' geometry. Moreover, they are the solution of scattered data interpolation problem and can achieve high accuracy. Various types of radial basis functions and their parameters will be investigated and their applicability in modelling ship geometry studied in this paper.

1 INTRODUCTION

The work of Bochner and Schoenberg Williamson from the middle of 20th century, see Bochner (1933), Schoenberg (1938a, 1938b), on positive definite functions is continued in the last three decades by several authors like Micchelli (1986), Schaback (1995), Wendland (1995), and Wu (1995), when new calculating meshless techniques are developed. Among them, radial basis functions (RBF) are recognized as the solution of scattered data interpolation problem, therefore applicable for mathematical representation of 2D and 3D objects.

Being recognized as the best approximators, Poggio and Girosi (1990), RBFs are more widely used in modelling of 3D objects with complex geometry using local interpolation and approximation techniques. The usage of explicit global interpolation methods is difficult due to bijection and breaks in form problems, together with singularity of inversion matrix, that usually occurs there. The complex geometries like ships' hull form usually cannot be described properly (accurately) using direct, exact, explicit representations without decomposition to bijective parts, called manifolds.

Other representation methods are used for ship geometry representation, nowadays, with parametric methods mostly used in shipbuilding modelling computer programs. The parametric representations based on Bezier, Basis (B-spline) or Non-Uniform Rational Basis splines (NURB spline) solve above mentioned representation problems, like bijection and breaks description, with advantage in recursive characteristics, but having major disadvantages in the fact that they cannot be used for direct calculation of intrinsic properties nor intersection problems. Moreover, maybe the largest disadvantage of NURBs and B-spline is the fact that offset points cannot be in any way connected with control points used for curve/surface generation in used procedures based on basis spline. In that way, the NURBs and B-spline representations are suitable for modelling from scratch only, with in general approximate nature.

The applicability of different types of RBFs to ship geometry representation using global RBF interpolation procedures is the subject of this paper. The characteristics of different radial basis functions and their parameters will be observed, checking their accuracy and properties for 2D representation of ship's test frame without and with camber. The quality of RBF representation will be thus tested for the description of form discontinuity as one of the major advantages of NURB representation.

2 RBF NETWORKS DEFINITION

The RBF networks can be by analogy with neural networks, defined as direct, feed-forward single-layered neural networks with possibly infinite input and output data sets, as shown on Figure 1.

Their input and output variables are connected with weighted sum of radial basis functions translated around the points called centres, whose number depends on mathematical procedure chosen for object representation.

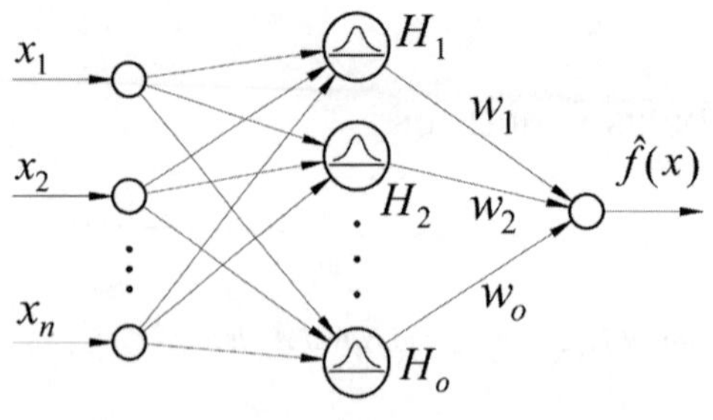

Figure 1. Single-layer feed-forward RBF neural network.

Mathematically, RBF network as linear combination of certain basis functions defined with the statement:

$$\hat{f}(x) = \sum_{i=1}^{O} w_i B_i = \sum_{i=1}^{O} w_i \Phi_i(x) = \sum_{i=1}^{O} w_i \varphi(\|x, t_i\|) \quad (1)$$

where: $x_j, j = 1,\ldots, N$; $x \in IR^s$ is input data set, B_i are basis functions, Φ_i are radial basis functions, t_i are the development centers of RBF with i = 1,…,O, where O is the number of centers, w_i are RBF network weight coefficients, φ is radial basis function based on Euclidian norm between input data and centers, and $\hat{f}(x)$ is the generalized interpolation/approximation function.

For basis functions B_j to be invertible their interpolation matrix must be in Haar space, i.e. satisfy condition:

$$\det(B_j(x_i)) \neq 0 \quad (2)$$

When the number of input data set elements equals the number of centres set the interpolation network is formed, with interpolation procedure and corresponding matrix is obtained. Otherwise, the approximation network is obtained, with corresponding approximation matrix.

The interpolation procedures with high generalization accuracy are the target of this paper, and therefore the number of centres O will always equal the number of elements in input data set N and centers set equals input data set.

2.1 *Calculation procedure*

The parameters of the network to calculate are the weights of the functions w_i, equalling the number of development centres.

In the scattered data interpolation theory, the neural network's weight coefficient values for are obtained by direct inversion of the neural network interpolation (activation) matrix **H** multiplied by target vector **y**, i.e. with:

$$\mathrm{w} = \mathbf{H}^{-1} \cdot \mathrm{y} \quad (3)$$

where: y – target vector (output data set), H – neural network activation (interpolation) matrix, $N \times N$, with elements r_{ji}:

$$\mathbf{H} = \begin{bmatrix} \varphi(r_{11}) & \cdots & \varphi(r_{1j}) & \cdots & \varphi(r_{1N}) \\ \vdots & \ddots & \vdots & \cdot^{\cdot^{\cdot}} & \vdots \\ \varphi(r_{i1}) & \cdots & \varphi(r_{ij}) & \cdots & \varphi(r_{iN}) \\ \vdots & \cdot^{\cdot^{\cdot}} & \vdots & \ddots & \vdots \\ \varphi(r_{N1}) & \cdots & \varphi(r_{Nj}) & \cdots & \varphi(r_{NN}) \end{bmatrix} \quad (4)$$

where: r_{ji} is the norm $\|x_j - x_i\|$, $j, i = 1, \ldots, N$.

The main disadvantage of RBF networks is the problem with weight coefficients calculation connected with possible singularity of above interpolation matrix, (4). That matrix must be well-posed and the main criterion for it is that interpolation matrix is positive definite.

2.2 *Generalization and accuracy*

The RBF network weights w_i calculation procedure consists of two parts: *evaluation* and *generalization*. In the evaluation phase shown above, the network weight coefficients have to be calculated and corresponding accuracy on the input data set checked. After the weights are calculated, the RBF network goodness test is performed checking generalization accuracy over testing data set.

2.2.1 *Evaluation accuracy*

The usual accuracy measure used is RMSE (Root Mean Squared Error):

$$RMSE = \sqrt{\frac{\sum_{j=1}^{N} (f(x) - y_i)^2}{N}} \quad (5)$$

with N - input data set number, y_i, $i = 1,\ldots, N$ - output data set, $f(x)$ - radial basis function, and it will be used here, too.

2.2.2 *Generalization accuracy*

Usually, the evaluation accuracy does not ensure achieving overall required accuracy for geometric object to be described, so generalization accuracy has to be performed using additional data set.

$$Err_{\max} = \max_{x \in T} \left(f(x_T) - y_T \right) \quad (6)$$

where T is test data set $\{x_T, y_T\}$.

In the case of ships' hull forms, the generalization accuracy checking will be performed calculating local error on the places of interest, like bilge or transition from bilge to flat of side, with acceptable value set to 0.1 mm.

3 RADIAL BASIS FUNCTIONS AND THEIR CHARACTERISTICS

Radial basis functions are the functions based on norms (usually Euclid's L_2 norm) between input set data points and centres of development points.

$$\Phi(x) = \varphi(\|x\|) \tag{7}$$

They are usually defined with only one parameter called shape parameter, c, and their definition for the interpolation case is:

$$\varphi = \varphi(x,t;c) = \varphi(x,x_i;c) = \varphi(\|x - x_i\|_2, c),\ x \in IR^s \tag{8}$$

RBFs have some good intrinsic properties, being invariant on:

- Translation,
- Rotation, and
- Reflexion.

These properties will be used in the selection of representation coordinate system.

The main criterion for the basis function to be acceptable as RBF is that it ensures interpolation matrix invertibility. That criterion can be fulfilled if the function chosen is positive definite and radial, and that can be achieved by satisfying one of the following three criteria:

- Strictly positive definite,
- Completely monotone, and
- Multiply monotone functions.

When Fourier transform is not available, there are two alternative criteria for decision whether a function is strictly positive and radial on IR^s: complete monotone for: $(-1)^l\ \varphi^{(l)}\ (r) \geq 0,\ r > 0,\ l = 0, 1, 2, \ldots$ (but not constant), see Schoenberg (1938) (converse also holds, see Wendland (2005)) for the case of all s, and multiply monotone (non-negative, non-increasing, and convex—simply $\varphi'' \geq 0$, see Williamson (1956)) for some fixed s.

3.1 *Strictly positive definite radial basis functions*

If we choose basis functions B_i in the RBF network formulation that generate strictly positive definite interpolation matrix, Mitcchelli (1986), we will always have a well-posed interpolation problem. By the theorem from Wendland (2005), the radial basis function $\varphi(\|x\|)$ is strictly positive definite and radial on IR^s if and only if $\varphi(\|x\|)$ its s-dimensional Fourier transform is non-negative and not identically equal to zero.

The examples of strictly positive definite radial functions are stated in Fasshauer (2007):

- Gaussian functions—radial functions,
- Laguerre-Gaussians—infinitely differentiable, oscillatory functions (not strictly positive definite and radial on IR^s for all s), Andrews et al. (1999),
- Poisson functions—oscillatory function that is radial and strictly positive definite on IR^s (and all $IR^\sigma \leq s$), not defined in origin, but can be extended to be infinitely differentiable in all of IR^s, Fornberg et al. (2004),
- Matérn functions—depending on the modified Bessel function of the second kind (sometimes called modified Bessel function of the third kind, or MacDonald's function, or Sobolev splines of order v, see Schaback (1995).

Above functions are Gaussian generalizations and can be set in one category, i.e. only basis Gaussian will be investigated in this paper. The Table 1 shows the functions that are generalizations of Gaussian function.

Another group of strictly positive definite and radial basis functions that are not Gaussian based functions are as stated in Fasshauer (2007):

- Inverse Multiquadrics—infinitely differentiable, see Hardy (1971),
- Generalized Multiquadrics, see Fornberg and Wright (2004),
- Truncated power functions—the functions with compact support,

Table 1. Strictly positive definite radial functions based on Gaussian function.

Basis functions, $\Phi(x)$	Equations
Gaussian	$e^{-c\|x\|^2},\ c > 0$
Laguerre-Gaussian	$e^{-\|x\|^2} L_n^{s/2}\left(\|x\|^2\right)$, $L_n^{s/2}(t) = \sum_{i=0}^{N} \frac{(-1)^k}{k!} \binom{n+s/2}{n-k} t^k$
Poisson	$\frac{J_{s/2-1}(\|x\|)}{\|x\|^{s/2-1}},\ s \geq 2$
Matern	$\frac{K_{\beta-s/2}(\|x\|)\|x\|^{\beta-s/2}}{2^{\beta-1}\Gamma(\beta)},\ s \geq 2$

– Potentials and Whittaker radial functions, see Abramowitz and Stegun (1972).

The Table 2 shows strictly positive definite radial functions that are not Gaussian generalizations.

3.2 *Conditionally positive definite radial functions*

The other class of radial basis functions are conditionally positive definite functions of order *m*, see Micchelli (1986) and Guo et al. (1993). These are the functions that provide the natural generalization of RBF interpolation with polynomial precision, important for high accuracy required for hull geometry description.

The RBF network definition can be changed accordingly to:

$$\hat{f}(x) = \sum_{i=1}^{N} w_i \varphi(\|x, t_i\|) + \sum_{l=1}^{M} \omega_l p_l(x), \quad x \in IR^s \qquad (9)$$

where: $p_1 \ldots p_M$ form the basis for the $M = \binom{s+m-1}{m-1}$—dimensional linear space Π^s_{m-1} of polynomials of total degree less than or equal to $m-1$ in *s* variables.

There are conditionally and strictly conditionally positive definite functions. The condition for function Φ to be conditionally positive definite is that it possesses a generalized Fourier transform of order m, continuous on $IR^s/\{0\}$, i.e. if $\hat{\Phi}$ is non-negative and non-vanishing.

To ensure unique solution M additional conditions need to be added:

$$\sum_{j=1}^{N} w_j p_l(\mathbf{x}_j) = 0, \quad l = 1, \ldots, M \qquad (10)$$

with polynomial degree at most $m-1$.

Table 2. Strictly positive definite radial functions not based on Gaussian function.

Basis Functions, $\Phi(x)$	Equations
Inverse Multiquadrics	$(r^2 + c^2)^\beta$, $\beta \le 0$, $\beta \notin 2IN$, $x \in IR^s$
Generalized Multiquadrics	$(1 + r^2/c^2)^\beta$, $\beta > 0$, $\beta \notin 2IN$, $x \in IR^s$
Truncated Powers	$(1-r)^l_+$
Potentials and Whitteker	$\int_0^\infty (1-rt)^{k-1}_+ f(t)dt, k \ge \lfloor s/2 \rfloor + 2$

where $r = \|x - x_i\|_2$.

The function $\hat{\Phi}$ is strictly conditionally positive definite function of order *m* on IR^s if its quadratic form is zero only for $\mathbf{w} \equiv 0$.

If their conditional positive definiteness can be connected to complete monotone and multiple monotone functions and not generalized Fourier transform, we are obtaining the criteria for function Φ to be strictly conditionally positive definite radial function.

The examples of conditionally positive definite functions are shown in Fasshauer (2007):

- Generalized multiquadrics, Hardy (1971),
- Radial powers (without shape parameter *s*, polyharmonic splines), with no even powers,
- Thin plate splines (without shape parameter *s*, polyharmonic splines), see Duchon (1976).

The Table 3 shows the list of some conditionally positive definite radial functions.

3.3 *Compactly supported radial basis functions*

Compactly supported RBFs are strictly conditionally positive functions and radial of order $m > 0$, but not for all *s* on IR^s, see Micchelli (1986). The acceptable range of compactly supported RBFs is defined for some restricted *s* range, with condition $\lfloor s/2 \rfloor \le k + m - 2$, i.e. maximal *s* value. This restriction ensures that Φ is integrable and therefore possesses classical Fourier transform Φ which is continuous. For integrable functions, the generalized Fourier transform coincides with the classical Fourier transform.

The compactly supported functions $\varphi_{s,k}$ are all supported on [0,1] and have polynomial representation there, with minimal degree for given space dimension s and smoothness 2*k*.

These functions $\varphi_{s,k}$ are strictly positive definite and radial on IR^s and are of the form:

$$\varphi_{s,k} = \begin{cases} p_{s,k}(r), & r \in [0,1], \\ 0, & r > 1 \end{cases} \qquad (11)$$

with an univariate polynomial $\varphi_{s,k}$ of degree $\lfloor s/2 \rfloor + 3k + 1$.

Table 3. Conditionally positive definite radial functions.

Functions $\Phi(x)$	Equations
Generalized multi-quadrics	$(1 + r^2/c^2)^\beta$, $\beta > 0$, $\beta \notin 2IN$, $x \in IR^s$
Radial powers	$(-1)^\beta r^\beta$, $\beta > 0$, $\beta \notin 2IN$, $x \in IR^s$
Thin-plate spline	$(-1)^{k+1} r^{2k} \log r$, $k \in IN$, $x \in IR^s$

Table 4. Compactly supported functions.

Functions	Equations
Wendland	$\varphi_{s,0} = (1-r)_+^l$, with $l = \lfloor s/2 \rfloor + k + 1$ $\varphi_{s,1} = (1-r)_+^{l+1}[(l+1)r+1]$ $\varphi_{s,2} = (1-r)_+^{l+2}$ $\times [(l^2+4l+3)r^2+(3l+6)r+3]$
Wu	$\varphi_{s,2} = (1+r)_+^5(8+40r+48r^2+25r^3+5r^4)$ $\varphi_{s,3} = (1+r)_+^4(16+29r+20r^2+5r^3)$
Buhmann	$\Phi = 12r^4 \log r - 21r^4 + 32r^3 - 12r^2 + 1$

Table 5. Testing radial basis functions.

Functions	Equations
MQ & Inverse MQ	$(r^2+c^2)^\beta$, $\beta \notin 2IN$, $x \in IR^s$
Generalized MQ	$(1 + r^2/c^2)^\beta$, $\beta > 0$, $\beta \notin 2\,IN$, $x \in IR^s$
Thin-plate splines	$(-1)^{k+1} r^{2k} \log r$, $k \in IN$, $x \in IR^s$
Radial powers	$(-1)^\beta r^\beta$, $\beta > 0$, $\beta \notin 2IN$, $x \in IR^s$
Wendland's compactly supported RBFs	$\varphi_{3,0} = (1-r)_+^2$ $\varphi_{3,1} = (1-r)_+^4[4r+1]$ $\varphi_{3,2} = (1-r)_+^6[35r^2+18r+3]$

In order to obtain compact support the distance between points need to be divided by some value, d, that can be larger than points distance.

For example, the Wendland's CSRBFs are obtained from truncated power function $\varphi_l = (1-r)_+^l$ by dimension walk and repeatedly applied operator I, and we obtain: $\varphi_{s,k} = I^k\, \varphi_{\lfloor s/2 \rfloor + k + 1}$.

The Table 4 shows the list of some compactly supported functions, see Wendland (1995), Wu (1995), Buhmann (1998).

4 POLYNOMIAL PRECISION

In general, solving interpolation problem with extended expansion with polynomial term leads to solving a system of linear equations of the form:

$$\begin{bmatrix} \mathbf{H} & \mathbf{P} \\ \mathbf{P}^T & 0 \end{bmatrix} \begin{bmatrix} \mathbf{w} \\ \omega \end{bmatrix} = \begin{bmatrix} \mathbf{y} \\ \mathbf{0} \end{bmatrix} \tag{12}$$

where $H_{ji} = B_j(x_i)$, j, $i = 1, \ldots, N$, $P_{jl} = p_l(x_i)$, $l = 1, \ldots, M$, $\mathbf{w} = [w_1, \ldots, w_N]^T$, $\boldsymbol{\omega} = [\omega_1, \ldots, \omega_M]^T$, $\mathbf{y} = [y_1, \ldots, y_N]^T$ and $\mathbf{0}$ is a zero vector of length M.

Above system of linear equations can be solved using criterion (10), only, and therefore unique solution can be obtained by imposing this criterion.

The goal of this paper is to select RBFs with high accuracy so polynomial precision is required. The RBFs having this property are acceptable, only, and the basis functions will be selected among them.

5 GLOBAL INTERPOLATION OF SHIP GEOMETRY WITH RADIAL BASIS FUNCTIONS

The conditionally positive definite functions and also radial have polynomial precision of scattered interpolation problem that is required for high accurate hull form geometry description.

When used in global interpolation this functions have global support. Globally supported functions chosen are:

- MQs,
- Inverse MQs,
- Generalized MQs,
- Radial powers,
- Thin-plate splines.

Additionally, compactly supported functions have polynomial representation on [0,1] and are suitable for achieving high accuracy, also.

Therefore, the functions from Tables 3 and 4 are chosen for ship geometry modelling in this paper, with compact supported functions limited to Wendland's functions for $s = 3$.

Table 5 shows selected functions to be tested for accuracy in ship geometry description using RBF interpolation procedures.

6 RESULTS

The results of the RBF interpolation for above chosen representative functions in Table 5 will be shown for 2D representations of few frames of the test ship represented by single propeller general cargo ship. The chosen ship hull form has "U" type of frames, classical stem with bulb, parallel middle body and wide stern with long aft bulb and transom at the end, and is therefore suitable for testing RBF representation properties of discontinuities and the transition from rounded to flat part of the frame.

The results of RBF interpolation of the test frame for the ship geometry with and without camber are shown in the Table 6.

The overall goal of hull form representation is to represent closed hull form with camber, and have one RBF representation of whole frame, or even

Table 6. RBF Interpolation results of the ship's test frame.

	Without camber		With camber and additional point	
Function	RMSE	Err_{max}	RMSE	Err_{max}
MQ, $\beta = 1/2$, c = 0.01	$3.62 \cdot 10^{-11}$	$5.83 \cdot 10^{-4}$	$1.05 \cdot 10^{-8}$	$1.32 \cdot 10^{-2}$
MQ, $\beta = 3/2$, c = 0.01	$1.03 \cdot 10^{-7}$	$9.28 \cdot 10^{-5}$	$1.74 \cdot 10^{-1}$	–
Inverse MQ, $\beta = 1/2$, c = 1.01	$7.77 \cdot 10^{-6}$	0,138	$2.36 \cdot 10^{-9}$	37.2
Inverse MQ, $\beta = 3/2$, c = 1.01	$2.12 \cdot 10^{-4}$	$1,21 \cdot 10^{-1}$	$4.22 \cdot 10^{3}$	$8.58 \cdot 10^{3}$
Generalized MQ, $\beta = 1/2$, c = 0.01	$5.83 \cdot 10^{-11}$	$3.02 \cdot 10^{-4}$	$5.91 \cdot 10^{-9}$	$1.32 \cdot 10^{-2}$
Generalized MQ, $\beta = 3/2$, c = 0.01	$1.25 \cdot 10^{-7}$	$9.28 \cdot 10^{-5}$	$4.33 \cdot 10^{-1}$	$7.89 \cdot 10^{-1}$
Generalized MQ, $\beta = 5/2$, c = 0.001	$6.1 \cdot 10^{-5}$	$6.07 \cdot 10^{-3}$	$5.58 \cdot 10^{3}$	$9.45 \cdot 10^{3}$
Radial Power, $\beta = 1/2$	$9.41 \cdot 10^{-13}$	1.94^{-12}	$1.86 \cdot 10^{-11}$	$2.45 \cdot 10^{-11}$
Radial Power, $\beta = 3/2$	$3.45 \cdot 10^{-8}$	$6.34 \cdot 10^{-6}$	$3.08 \cdot 10^{-2}$	$5.28 \cdot 10^{-2}$
Radial Power, $\beta = 5/2$	$5.20 \cdot 10^{-5}$	$5.83 \cdot 10^{-3}$	$9.64 \cdot 10^{4}$	$1.63 \cdot 10^{4}$
Thin-plate spline, $k = 2$,	$7.70 \cdot 10^{-7}$	$1.54 \cdot 10^{-3}$ Osc.	$8.87 \cdot 10^{-5}$	$2.28 \cdot 10^{-1}$ Osc.
Thin-plate spline, $k = 4$	$2.52 \cdot 10^{-6}$	$3.41 \cdot 10^{-3}$ Osc.	5.29	8.93 Osc.
Wendland $k = 0, d = 2.5$	$1.45 \cdot 10^{-12}$	$1.96 \cdot 10^{-1}$	$2.25 \cdot 10^{-12}$	$2.26 \cdot 10^{-1}$
Wendland $k = 1, d = 2.5$	$1.45 \cdot 10^{-9}$	$1.51 \cdot 10^{-2}$	$8.84 \cdot 10^{-8}$	$1.64 \cdot 10^{-2}$
Wendland $k = 2, d = 2.5$	$6.39 \cdot 10^{-7}$	$1.17 \cdot 10^{-2}$	$7.16 \cdot 10^{-4}$	2.87
Wendland $k = 0, d = 6$	$2.84 \cdot 10^{-12}$	$2.52 \cdot 10^{-2}$	$7.43 \cdot 10^{-12}$	$2.15 \cdot 10^{-2}$
Wendland $k = 1, d = 6$	$6.99 \cdot 10^{-6}$	$5.31 \cdot 10^{-4}$	$1.72 \cdot 10^{-3}$	$3.09 \cdot 10^{-3}$
Wendland $k = 2, d = 6$	$2.36 \cdot 10^{-3}$	$3.93 \cdot 10^{-3}$	$1.94 \cdot 10^{1}$	$3.14 \cdot 10^{1}$

Note: "Osc." marks large bottom end oscillations.

whole ship hull geometry. Therefore, the results of representations are checked for frames with and without deck camber.

Except accuracy, the smoothness of representation is the additional requirement for RBF interpolation procedures to fulfil.

Additionally, the ship's hull form RBF interpolation representation of the transition between bilge and straight side need to be described without oscillations, opposite to NURB presentations of this geometric phenomenon. Also, the break of form between the deck and the side should be represented without large oscillations characteristic for exact interpolation procedures ("Gibbs phenomenon").

6.1 *The frame without camber*

The results for the frame without camber are acceptable for almost all RBF choices except thin-plate spline that oscillate near bottom end.

The Figure 2 shows the acceptable MQRBF representation of test frame with $\beta = 1.5$.

The oscillations of the description at the transition from rounded bilge to flat side is shown in the Figure 3.

The Figure 2 zoom shows slight oscillations near curves transition from rounded bilge to straight side, as shown in Figure 3, of order 10^{-4} that can be taken as acceptable value. The additional points around transition can straighten the RBF curve in the flat part, this enabling required transition.

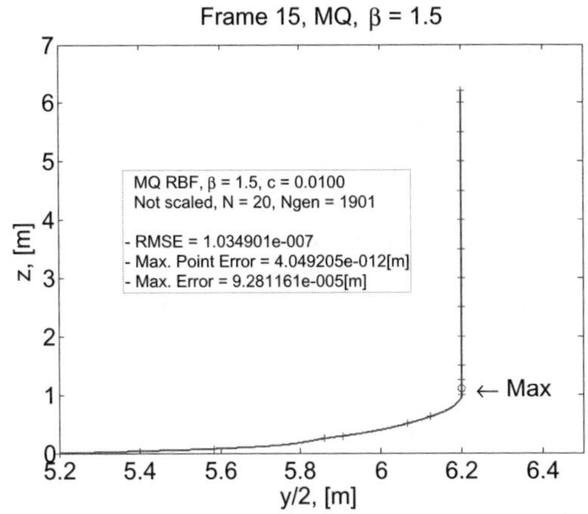

Figure 2. The results of the MQRBF generalization of the ship's test frame with $\beta = 3/2$, c = 0.02.

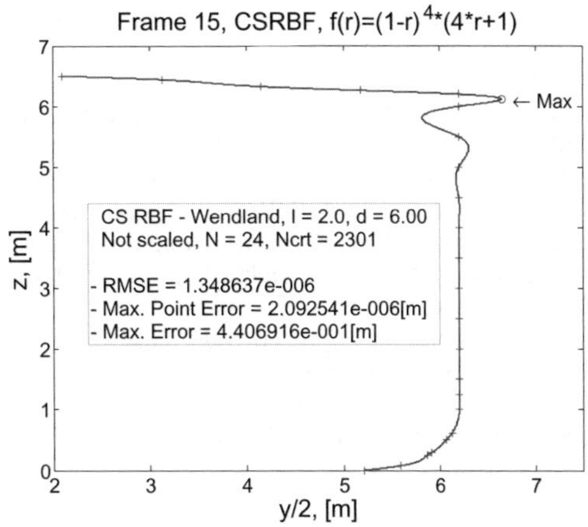

Figure 4. The results of the Wendland's CSRBF generalization of the ship's test frame with $s = 2$, $k = 1$ and $d = 6$.

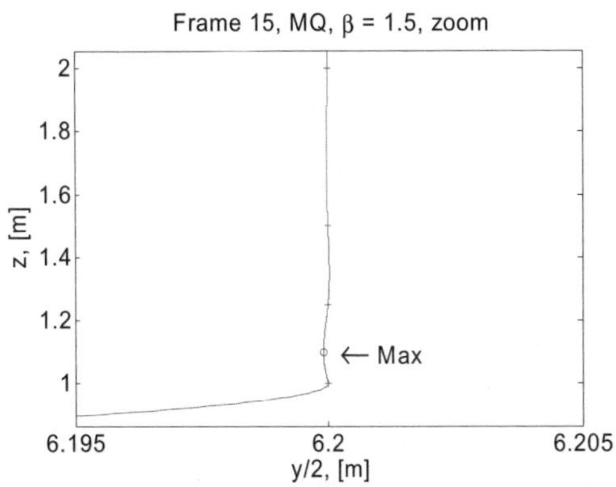

Figure 3. The zoom of MQRBF generalization of the ship's test frame with $\beta = 3/2$, c = 0.02.

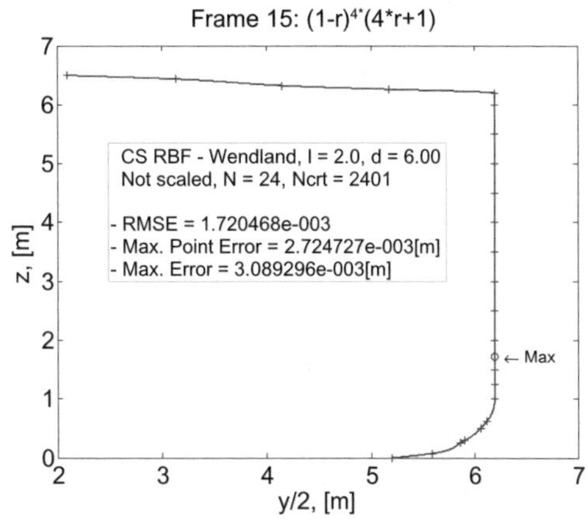

Figure 5. The results of the Wendland's CSRBF generalization of the ship's test frame with $s = 2$, $k = 1$ and $d = 6$, and added point near form break.

6.2 *The frame with camber*

For the RBF description of the frame with camber, the break of the form is the problem that cannot be solved acceptably for all chosen functions. The Figure 4 shows large oscillations that occur near form break from side to the deck, that are not acceptable.

6.3 *Oscillation stabilization using additional points*

Because of large oscillations of RBF descriptions at the upper frame end, at shear strike, a single point is added very near it to stabilize the generalization curve and the distance 10^{-4} from the top at the side.

After stabilization, the only accurate and smooth functions are Wendland's functions with $k = 1$, with slightly lower local accuracy Err_{max} than required, as shown on Figure 5.

The RBF description of conic sections depends on the number and the position of input points, but once required accuracy is obtained, scaling can be used to transform the RBF description to actual input.

6.4 *Examples of other RBF ship frames representations*

Above examples of frame description are focused to the accuracy of the description of the discontinuity and the transition from curved to flat parts of the frame. Following examples of frames description will show RBF description capabilities of multiply curved frames with knuckles. The Figure 6 shows RBF description of aft frame of the general cargo test ship with bulb, high curvature and camber.

Another example of RBF ship frame description is the description of the aft frame of Uljanik's car carrier with bulb, knuckle, camber and flat deck, as shown on Figure 7.

The examples on Figures 6 and 7 show the capability of RBF networks in very accurate global description of frames with knuckles and camber. The bijection problem in the case of exact RBF representation of flat deck is solved by simple rotation of that part around flat deck starting point on the camber.

In order to adequately describe curved and flat parts of the frame additional points are added, thus showing the need for adjustments in RBF description. The advantage of these adjustments is in the fact that the required accuracy is easy to achieve, and small number of added points is needed for high quality of description, thus showing flexibility of RBF representation.

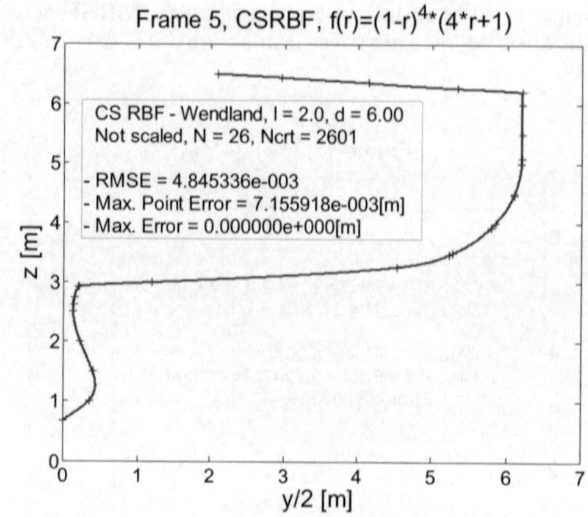

Figure 6. The results of the Wendland's CSRBF generalization with $s = 2$, $k = 1$ and $d = 6$ of multiply curved aft frame with camber and bulb.

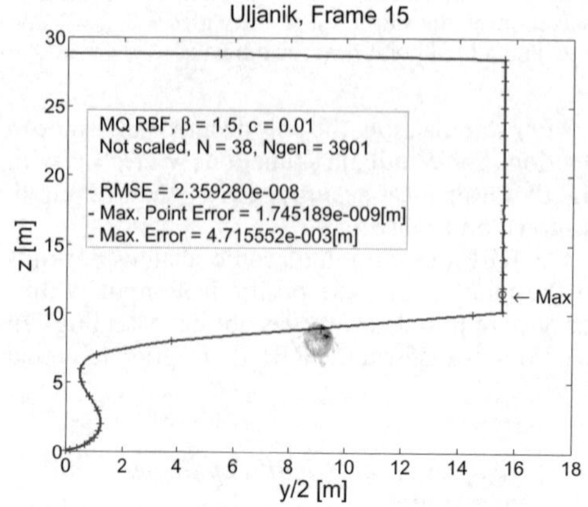

Figure 7. The results of the MQRBF generalization with $\beta = 1.5$, $c = 0.01$ of multiply curved aft frame with bulb, knuckle, camber and deck line.

7 CONCLUSION

The ships' geometry description using RBF interpolation procedures and functions are acceptable in representing hull form without knuckles when tested in 2D frame description.

Extended RBF networks with linear polynomial and with conditionally positive definite radial functions give accurate representation of the transition from rounded bilge to the flat of side comparably to NURB and B-splines, with possible adjustments using added points in order to achieve required smoothness of rounded part, and flatness of straight part of the frame.

In order to obtain good representation of knuckles it is necessary to adjust input data set with additional points near discontinuity.

The bijection problem of flat deck is solved by flat line rotation around its starting point, showing possible solution of that problem by curve development around break or inflection points.

It can be concluded overall that RBF interpolation procedures are comparable to those based on B-spline, and the effort should be done to solve corresponding 3D representations of hull form.

REFERENCES

Abramowitz, M. & Stegun, I.A. 1972. *Handbook of Mathematical Functions with Formulas, Graphs, and Mathematical Tables*. Dover (New York).

Andrews, G.E., Askey, R. & Roy, R. 1999. *Special Functions*. Cambridge Unversity Press.

Bochner, S. 1933. Monotone Funktionen, Stieltjes Integrale un Harmonische Analyise. *Math. Ann. 108*: 378–410.

Buhmann, M.D. 1998. Radial functions on compact support. *Proc. Edin. Math. Soc. II 41*: 33–46.

Duchon, J. 1976. Intepolation des fonctions de deux variables suivant le principe de la flexion des plaques minces. *Rev. Franc. Autom. Inf. Rech. Oper. Anal. Numer. 10*: 5–12.

Fasshauer, G.E. 2007. Meshfree Approximation Methods with Matlab. *Interdisciplinary Mathematical Sciencies—Vol. 6, World Scientific Publishing Co. Pte. Ltd.*

Fornberg, B. & Wright, G. 2004. Stable computation of multiquadric Interpolants for all values of shape parameter, *Comp. Math. Appl. 47*: 497–523.

Fornberg, B., Larson, E. & Wright, G. 2004. A new class of oscilatory radial basis functions. *Com. Math. App. 51*: 1209–1222.

Guo K. et al. 1993. Conditionally positive definite functions and Laplace-Stieltjes integrals. *J. Appr. Th. 74*: 249–265.

Hardy, R.L. 1971. Multiquadric equations to topography and other irregular surfaces, *J. Geophys. Res. 76*: 1905–1915.

Micchelli, C.A. 1986. Interpolation of scattered data: distance matrices and conditionally positive definite functions. *Constr. Approx. 2*: 11–22.

Poggio, T. & Girrosi, F. 1990. Networks and the best approximation property. *MIT, AI Laboratory, MIT, Cambridge, MA, A.I. Memo No. 1164, C.B.I.P. Paper No. 45.*

Schaback, R. 1995. Creating surfaces from scattered data using radial basis functions. In M. Dæhlen, T. Lysche and L. Schumaker (eds.), *Mathematical methods for curves and surfaces, Vanderbilt University Press*: 477–496.

Schoenberg, I.J. 1938a. Metric spaces and completely monotone functions. *Ann. of Math. 39*: 811–841.

Schoenberg, I.J. 1938b. Metric spaces and positive definite functions. *Trans. Amer. Math. Soc. 44*: 522–536.

Wendland, H. 1995. Piecewise polynomial, positive definite and compactly supported radial functions of minimal degree. *Adv. in Comput. Math. 4*: 389–396.

Wendland, H. 2005. *Scattered data Approximation*. Cambridge University Press (Cambridge).

Williamson, R.E. 1956. Multiple monotone functions and their Laplace transform. *Duke Math. J. 23*:189–207.

Wu, Z. 1995. Compactly supported positive definite radial functions. *Adv. in Comput. Math.*: 283–292.

Advanced Ship Design for Pollution Prevention – Guedes Soares & Parunov (eds)
© 2010 Taylor & Francis Group, London, ISBN 978-0-415-58477-7

Author index

Andrić, J. 103

Bajič, D. 65
Ban, D. 181, 307
Barle, J. 181, 307
Bezinović, D. 291
Bigot, F. 23
Blagojević, B. 307
Bosančić, F. 281
Božić, Ž. 155

Carvalho, J.P. 205
Chen, X.B. 23
Ćorak, M. 95, 113

Derbanne, Q. 23

Ehlers, S. 191

Fadljević, M. 155
Fonseca, N. 3

Garbatov, Y. 131
Gouveia, J.A.V. 217
Grubišić, R. 261
Grubišić, V. 173
Guedes Soares, C. 3, 43, 55, 79, 113, 131, 205, 217, 227

Hsu, V. 23

Jurišić, P. 149

Kitarović, S. 103
Klanac, A. 191
Kolacio, I. 55
Kujala, P. 17

Ladan, M. 181
Ljubenkov, B. 269, 281
Lozina, Ž. 71, 275

Mage, P. 149
Malenica, Š. 23, 31
Munić, I. 253

Parunov, J. 95, 113, 149, 155
Pavazza, R. 121
Pehnec, I. 275
Perera, L.P. 205
Prpić-Oršić, J. 43, 55, 65

Ribeiro e Silva, S. 43
Rudan, S. 163

Santos, T.A. 227
Sedlar, D. 71
Senjanović, I. 31
Šestan, A. 269
Sireta, F.X. 23
Slapničar, V. 243, 253
Stupalo, V. 253

Tabri, K. 191
Teixeira, A.P. 79, 113
Tomac, I. 275
Tomašević, S. 31
Tomić, M. 31
Turk, A. 43
Tušek, B. 95

Ukas, B. 243

Varsta, P. 191
Vidjak, A. 299
Vladimir, N. 31, 269
Vlak, F. 121
Vučina, D. 71
Vukasović, M. 121
Vulić, N. 173, 299

Žanić, V. 103
Zaplatić, T. 281
Žiha, K. 95, 149, 281